STRATHCLYDE UNIVERSITY LIBRARY
30125 00076692 2

KV-038-762

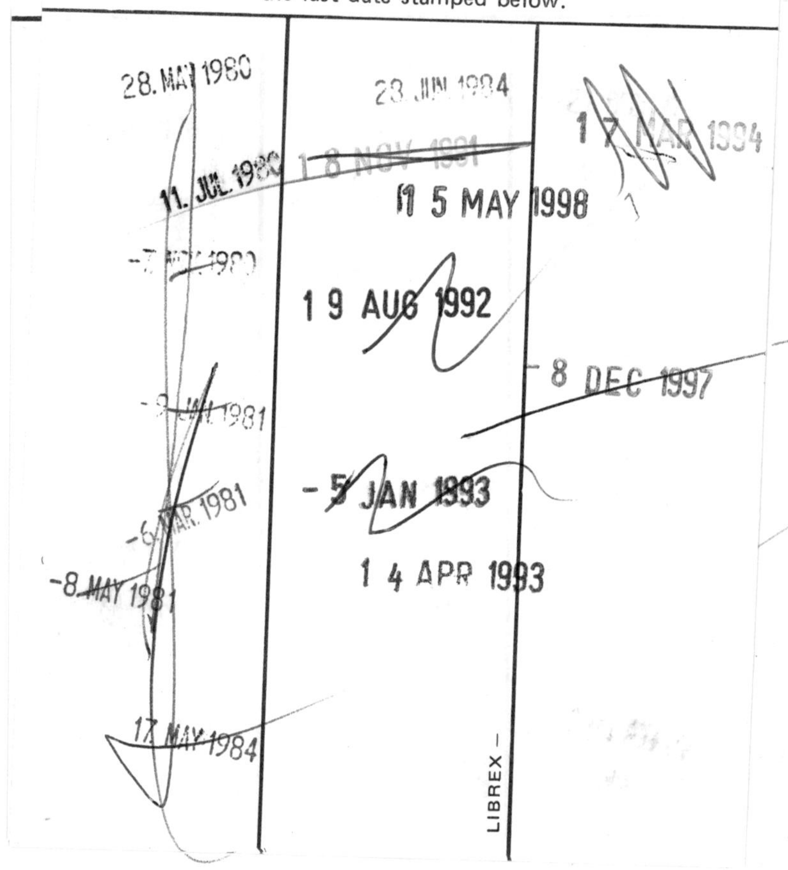
This book is to be returned on or before
the last date stamped below.
28. MAY 1980
11. JUL. 1980
-9 JAN. 1981
-6 MAR. 1981
-8 MAY 1981
17 MAY 1984
23 JUN 1984
18 NOV 1991
15 MAY 1998
19 AUG 1992
-5 JAN 1993
14 APR 1993
17 MAR 1994
-8 DEC 1997
LIBREX

POLLUTION DETECTION AND MONITORING HANDBOOK

POLLUTION DETECTION AND MONITORING HANDBOOK

Marshall Sittig

NOYES DATA CORPORATION

Park Ridge, New Jersey London, England

1974

Library of Congress Catalog Card Number: 74-75905
ISBN: 0-8155-0529-9
Printed in the United States

Published in the United States of America by
Noyes Data Corporation
Noyes Building, Park Ridge, New Jersey 07656

FOREWORD

This handbook contains methods for the detection and monitoring of pollutants in industrial effluents, notably air and water.

Such methods are prerequisites for any kind of management of environmental quality. Of equal importance are the data handling systems that must be used to collect and interpret the measurements, regardless of whether manual or automated identifying and monitoring techniques are chosen.

Generally the most important single factor in determining whether manual or automated methods of analysis and level monitoring should be used, is the required frequency interval. In the manual approach costs vary almost directly with the measurement frequency. When a continuous watch must be maintained, installation of automatic sensing is more economical, provided reliable automatic sensors are available.

For meaningful control and prevention of pollution the accepted standards and parameters for air and water quality must be known, whether they have been legislated or not. Acceptable levels for each type of impurity must be known by the industrialist or public health official, and this book, with its 1633 references makes a massive attempt to communicate such levels.

As a guideline for the industrial user, data on the toxicity of the various pollutants are included. Such data on the deleterious effects were taken from various government publications. This book, however, is not a treatise of pharmacology or industrial toxicology, only a guide to the accurate evaluation of pollutant levels in a given effluent or ambient substrate.

Whenever applicable, the name of the pollutant is followed by chemical and other information. After this come directions for sampling in air or water or both, as the case may be. Thereafter are given qualitative and quantitative analytical methods suitable for identification and measurement of the degree or level of pollution. This comprehensive information is followed by measurement techniques suitable for repeated or continuous monitoring. Preference is given to methods giving reproducible results in environmental quality control, as recommended by the EPA and other U.S. government agencies. This is followed in each case by many up-to-date references to government publications, patents and journal articles.

The arrangement is encyclopedic. The book is thus a worthy companion to the author's *Pollutant Removal Handbook* available from the same publisher.

CONTENTS

Contents

INTRODUCTION

Monitoring is a basic tool in pollution control as pointed out by Miller (1). Without monitoring there are no data upon which to base control techniques, and no means to measure progress resulting from their application. Monitoring of aqueous industrial effluents is required as a condition for a discharge permit.

Ambient air monitoring stations are required by state implementation plans for air pollution control. Also newly constructed air pollution sources must be monitored under the requirements of new national source performance standards. In-stack monitoring of opacity, SO_2 and NO_x is required for utilities, SO_2 for sulfuric acid manufacture and NO_x for nitric acid manufacturing plants. Petroleum refineries will require in-stack monitoring of opacity, H_2S and CO. Basic oxygen furnaces in steel mills will require in-stack opacity monitors.

In each section of this book considerable attention has been given, where data were available, not only to analytical methods per se but to the rationale behind the analysis. In other words, an attempt has been made to present data on the analytical method and its detection limits against some background of information on:

What levels of pollutant are normally encountered in industrial effluents?

What are the pollutant levels at which there are danger signals for man or the environment?

The coverage of detection and monitoring of individual pollutants in the volume is admittedly somewhat uneven. To do full justice to each and every pollutant would require more space than is available in this volume. Thus, more extensive coverage has been given to:

Aldehydes
Cadmium
Fluorides
Mercury
Nitrogen oxides
Odorous compounds
Oil and grease
Particulates
Pesticides
Polynuclear aromatics
Radioactive materials
Sulfur dioxide

because of their importance and because of the availability of reference material.

DEFINITION OF PROBLEMS AND TERMS

Air pollution control activities depend upon the answers to the following questions.

(1) What are the real effects of various specific pollutants on human health and the environment?

(2) Based on the answer to (1), what should be the allowable ambient air quality standards for the various pollutants?

(3) Based on the answer to (2), what should be the source emission standards for various pollutants?

This then brings one squarely to the quantitative analytical problem: How to measure ambient and source pollutant concentrations to safeguard health and the environment and how to use such measurement from continuous monitoring devices to insure proper process control and proper maintenance of human health and environmental quality?

Water pollution control activities rely heavily on the ability to monitor and survey the quality of natural and wastewaters. This includes rivers, lakes, estuaries, as well as industrial and domestic waste effluents. Presently, there are several state and federal water surveillance programs designed to (a) acquire, evaluate and disseminate information on the quality of waters for and from the varied local, state, interstate and federal agencies together with educational, commercial, industrial and individual entities, (b) determine the long term trends and variations of water quality, and (c) provide a rapid intelligence system for the preservation of the waters and the protection of the water users, including compliance with water quality standards. Water quality surveillance activities may be based on manual surveillance, aerial surveillance and automatic surveillance techniques.

Pollutant Effects on Health and the Environment

There are still many unknowns concerning the effects of various pollutants on people and things. One area of concern is that of trace metals in the air and water. The topic was discussed in a short review in *Chemical and Engineering News* (2). The experience with the sudden discovery of mercury pollution pointed up general awareness of our uncertainties of the presence and consequences of various trace metals in the environment, the article points out.

A comprehensive discussion of *Determination of Trace Metal Pollutants in Water Resources Using X-Ray Fluorescence* has been published by C.R. Cothern of the Physics Dept. of the University of Dayton (3). As a part of that report, a discussion of the health effects of trace metals was presented and it is quoted in the paragraphs which follow. This is, of course, just one example of health effects of pollutants. For more detail on the heatlh effects of a variety of pollutants, the reader is referred to such works as that by Stern (4).

The health effects of trace metals necessary to life on earth are more important to the existance of that life than their organic micronutrient counterparts, vitamins. The elements cannot be synthesized as vitamins can, but must exist in a narrow range of concentration in the environment to support life. Both deficiencies and excesses of trace elements can kill (5). The only source of trace elements is the earth's crust and seawater, and without them life would cease to exist.

A trace element can be defined as one making up less than 0.01% of the organism being considered. Because of the abundance of low atomic number (Z) elements in the earth's crust, it is not surprising that more than 99% of the structure of living things is composed of 12 of the first 20 elements in the periodic table. In spite of their miniscule presence in the human body, trace metals play an important role in human existence and can cause harmful effects. For example tiny amounts of iron are essential to the construction of hemoglobin in our blood to carry oxygen while only 50 parts per billion of mercury can cause harmful effects in humans.

The extreme biological consequences of even small changes in the concentrations of trace elements as well as the very small amounts which are required suggest that their action can only be that of vital links in the enzyme systems (6); e.g., iron in hemoglobin, copper in ascorbic acid oxidase, zinc in carbonic anhydrase. With many metals present one may substitute for another preventing proper physiological mechanisms to function. Calcium which is not a trace metal is related to the trace metals in a very important and special way. The absorption of trace metals by living things from the environment is inversely related to the concentration of environmental calcium (6). This means that the toxicity difference between hard and soft water can be very different as shown in Table 1.

TABLE 1: TRACE ELEMENTS IN HARD AND SOFT WATER

Metal	Ratio, toxic concentration in hard:soft water	Approximate toxicity in soft water (ppm)	Toxicity type (see text)
Titanium	14.6	8	II, III
Vanadium	4.2-9.2	5	I
Chromium	15	5	I, II
Iron	77	1.3	I, II
Nickel	24	4	I, II
Copper	15-500	<0.1	I, II
Zinc	3-67	2	I, II
Arsenic	--	--	II, III
Zirconium	8.2	14	??
Cadmium	5.5	<0.1	III
Tin	±4	1	III
Lead	33	2	III

Source: Report PB 213,369, 1972

There are in general three types of toxicity as shown in Table 1. The three types are: I, deficiency, certain elements are necessary for existance of life even though the amounts needed are in the ppm (parts per million) or ppb (parts per billion) ranges; II, excess, given large enough amounts of any element it is toxic; and III, disorders, caused by metals for which there are poor homeostatic mechanisms and which accumulate in man with age.

A somewhat oversimplified way (7) of looking at the relative effects of essential and non-essential trace elements is shown in Figure 1. The horizontal scale is the logarithm of ppm of the element in the enviroment and the verticale scale goes roughly from life at the bottom to death at the top. Normal represents the amounts giving rise to healthy life and anything above this is detrimental to life. As can be seen, too much of a nonessential element or too little of an essential element can be fatal. In general,if a nonessential element is present in the human environment (water, air, or food) in excess of a few ppm it can cause deleterious effects.

The body can with varying degrees of efficiency eliminate excess amounts of unnecessary elements. Most metals with the exception of cadmium and strontium are excreted by the intestines. Cadmium and strontium are eliminated by the kidney (5). The body alters the valence of a soluble metal to that required during or after absorption from the intestine. This phenomenon has been demonstrated for manganese, iron, cobalt and copper and partly

FIGURE 1: SIMPLIFIED REPRESENTATION OF THE HEALTH EFFECTS OF TRACE ELEMENTS

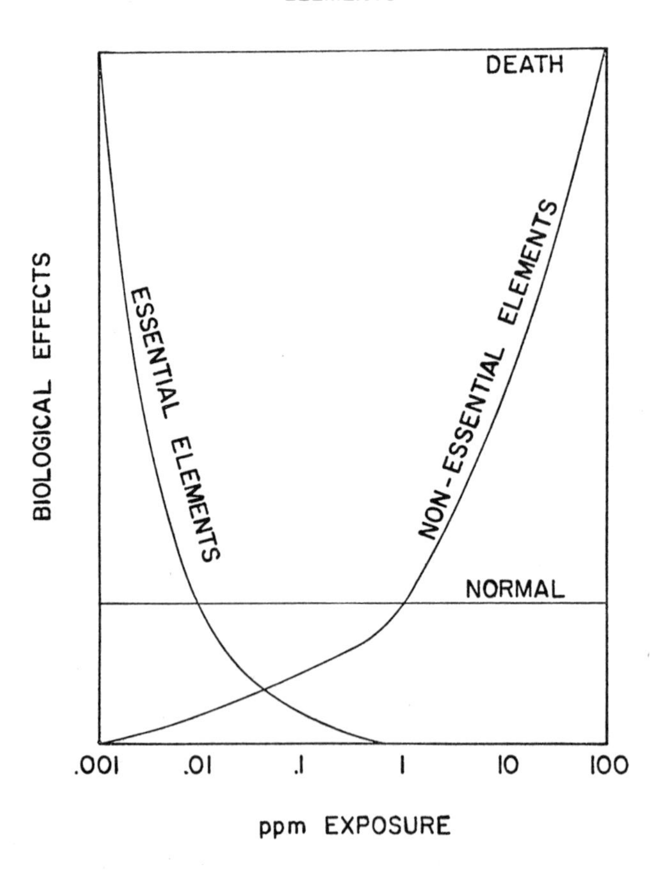

Source: Report PB 213,369, 1972

for chromium (5). This fact shows that the determination of the charge state of trace elements in water supplies, food and humans is important to measure. Interactions between trace metals complicate toxicological and health studies. For example, a zinc deficiency causes severe anemia in rats, rabbits and pigs which is indistinguishable from iron deficiency anemia, even when there is abundant iron in the tissues (6). In another example "teart" in field animals is apparently due to high levels of molybdenum in the pasture and is controlled by administering copper salts to the animal suffering from the condition (6).

Ambient Quality Standards

Table 2 shows U.S. national air quality standards for a number of pollutants as reproduced from *Instrumentation for Environmental Monitoring – Air,* (8). Table 3 shows the parameters for water quality characterization for domestic water supplies which serve as guidelines for purposes of water treatment and control. These were prepared by the Committee on Water Quality Criteria of the Federal Water Pollution Control Administration in 1968 and were quoted by Mancy (9).

TABLE 2: NATIONAL AIR QUALITY STANDARDS[a]

Pollutant	Averaging Time	Primary[b] Standards	Secondary[c] Standards	Reference[d] Method
Sulfur Dioxide	Annual Arithmetic Mean	80 $\mu g/m^3$ (0.03 ppm)	60 $\mu g/m^3$ (0.02 ppm)	
	24 hours	365 $\mu g/m^3$ (0.14 ppm)	260 $\mu g/m^3$ (0.10 ppm)	Pararosaniline Method
	3 hours	---	1300 $\mu g/m^3$ (0.5 ppm)	
Particulate Matter	Annual Geometric Mean	75 $\mu g/m^3$	60 $\mu g/m^3$	High Volume Sampling Method
	24 hours	260 $\mu g/m^3$	150 $\mu g/m^3$	
Carbon Monoxide	8 hours	10 $\mu g/m^3$ (9 ppm)	Same as Primary Standards	Non-Dispersive Infrared Spectroscopy
	1 hour	40 $\mu g/m^3$ (35 ppm)		
Photo-chemical Oxidants (corrected for NO_2 & SO_2)	1 hour	160 g/m (0.08 ppm)	Same as Primary Standard	Gas Phase Chemilumi-nescent Method
Hydrocarbons (corrected for Methane)	3 hours	160 $\mu g/m^3$ (0.24 ppm)	Same as Primary Standard	Flame Ionization Detection Using Gas Chromatography
Nitrogen Dioxide	Annual Arithmetic Mean	100 $\mu g/m^3$ (0.05 ppm)	Same as Primary Standard	Jacobs-Hochheiser Method

a. National standards other than those based on annual arithmetic means or annual geometric means are not to be exceeded more than once per year.

b. National Primary Standards: The levels of air quality necessary, with an adequate margin of safety, to protect the public health

c. National Secondary Standards: The levels of air quality necessary to protect the public welfare from any known or anticipated adverse effects of a pollutant

d. Reference method as described by the EPA. An "equivalent method" means any method of sampling and analysis which can be demonstrated to the EPA to have a "consistent relationship to the reference method."

Source: Lawrence Berkeley Laboratory, Berkeley, Calif. (May, 1972)

TABLE 3: PARAMETERS FOR WATER QUALITY CHARACTERIZATION FOR DOMESTIC WATER SUPPLIES

Quality parameter	Permissible criteria*	Desirable criteria*
1. Color (Co–Pt scale)	75 units	< 10 units
Odor	Virtually absent	Virtually absent
Taste	Virtually absent	Virtually absent
2. Turbidity	– –	Virtually absent
3. Inorganic chemicals		
pH	6.0 – 8.5	6.0 – 8.5
Alkalinity ($CaCO_3$ units)	30 – 500 mg/l	30 – 500 mg/l
Ammonia	0.5	<0.01
Arsenic	0.05	absent
Barium	1.0	– –
Boron	1.0	
Cadmium	0.01	
Chlorides	250	<25
Chromium (hexavalent)	0.05	absent
Copper	1.0	Virtually absent
Dissolved oxygen	>4.0	Air saturation
Fluorides	0.8 to 1.7 mg/l	1.0 mg/l
Iron (filtrable)	<0.3	Virtually absent
Lead	<0.05	Absent
Manganese (filtrable)	<0.05	Absent
Nitrates plus nitrites (as mg/l N)	< 10	Virtually absent
Phosphorus	10 – 50 μg/l	10 μg/l

(continued)

TABLE 3: (continued)

Quality parameter	Permissible criteria*	Desirable criteria*
Selenium	0.01	Absent
Silver	0.05	--
Sulfates	250	<50
Total dissolved solids	500	<200
Uranyl ion	5	Absent
Zinc	5	Virtually absent
4. Organic chemicals		
Carbon chloroform extract (CCE)	0.15	<0.04
Methylene blue active substances	0.5	Virtually absent
5. Pesticides:		
Aldrin	0.017	--
Chlordane	0.003	--
DDT	0.042	--
Dieldrin	0.017	--
Endrin	0.001	--
Heptachlor	0.018	--
Heptachlor expoxide	0.018	--
Lindane	0.056	--
Methoxychlor	0.035	--
Organic phosphates plus carbamates	0.1	--
Toxaphene	0.005	--
Herbicides 2,4,D plus 2,4,5–T, plus 2,4,5–TP	0.1	--
6. Radioactivity		
Gross beta	1,000 pc/l	<100 pc/l
Radium 226	3 pc/l	<1 pc/l
Strontium–90	10 pc/l	<2 pc/l

*All units are mg./l. unless otherwise specified.

Source: NBS Publication 351, 1972

A second set of criteria apply to water from the aesthetic and recreational use standpoint as follows. The general requirements are that surface waters should be capable of supporting life forms of aesthetic and recreational values. Hence, surface waters should be free from: (a) materials that may settle to form objectionable deposits or float on the surface as debris, oil and scum, (b) substances that may impart taste, odor, color or turbidity, (c) toxic substances including radionuclides physiologically harmful to man, fish or other aquatic plants or animals and (d) substances which may result in promoting the growth of undesirable aquatic life. Presently, there are no well defined water quality criteria for recreation or aesthetic purposes. A third set of criteria apply to waters from the standpoint of the environmental well-being of aquatic life, fish and wildlife as shown in Table 4.

TABLE 4: PARAMETERS FOR WATER QUALITY CHARACTERIZATION FOR AQUATIC LIFE, FISH AND WILDLIFE

1. Turbidity:
   Discharge of waste in receiving waters should not cause change in turbidity in the order of 50 Jackson units in warm–water streams, 25 Jackson units in warm–lakes, and 10 Jackson units in cold–water streams and lakes.
2. Color and Transparency:
   Optimum light requirements for photosynthesis should be at least 10 percent of incident light on the surface.
3. Settleable Matter:
   Minor deposits cf settleable matter may inhibit growth of flora and biota of water body. Such materials should not be discharged in surface waters.

(continued)

TABLE 4: (continued)

4. Floating Matter:
   All foreign floating matter should not be discharged in surface waters. A typical pollution problem is that of oil waste discharges which may result in the formation of
   a. visible objectionable color film on the surface,
   b. alter taste and odor of water,
   c. coat banks and bottoms of water course,
   d. taint aquatic biota and
   e. cause toxicity to fish and man.
5. Dissolved Matter:
   The effect of dissolved matter on aquatic biota can be due to toxicity at relatively low concentrations or due to osmotic effects at relatively high concentrations. In general, total dissolved matter should not exceed 50 millimoles (the equivalent of 1500 mg/l NaCl).
6. pH, Alkalinity and Acidity
   The pH range of 6.0 to 9.0 is considered desirable. Discharge of waste effluents should not lower the receiving water alkalinity to less than 20 mg/l.
7. Temperature:
   Heat should not be added to a receiving water in excess of the amount that will raise the temperature by 3–5 °F. In general, normal daily and seasonable temperature variations should be maintained.
8. Dissolved Oxygen:
   It is generally required to maintain a dissolved oxygen level above 4 to 5 mg/l. In cold water bodies it is recommended to maintain the dissolved oxygen above 7 mg/l.
9. Plant Nutrients:
   Organic waste effluents such as sewage, food processing, canning and industrial wastes containing nutrients, vitamins, trace elements, and growth stimulants should be carefully controlled. It is important not to disturb the naturally occuring ratio of nitrogen (nitrates and ammonia) to total phosphorus in the receiving water.
10. Toxic Matter:
    Waste effluents containing chemicals with unknown toxicity characteristics should be tested and proven to be harmless in the concentration to be found in the receiving waters. Discharging pesticides in natural waters should be avoided if possible or kept below 1/100 of the 48–hrs. TL_m values. Levels of ABS and LAS should not exceed 1.0 mg/l and 0.2 mg/l, respectively, for periods of exposures exceeding 24 hours.
    It should be noticed that the presence of two or more toxic agents in the receiving water may exert an additive effect.
11. Radionuclides:
    No radionuclides should be discharged in natural waters to produce concentrations greater than those specified by the USPHS Drinking Water Standards.

Source: NBS Publication 351, 1972

A fourth set of criteria apply to water for agricultural use as shown in Table 5.

TABLE 5: PARAMETERS FOR WATER QUALITY CHARACTERIZATION FOR AGRICULTURAL USE

1. Total Dissolved Solids or "Salinity"
   This is the most important water quality consideration since it controls the availability of water to the plant through osmotic pressure regulating mechanisms. The effect of salinity on plant growth varies from one type to another and is dependent on environmental conditions.
2. Trace Elements Tolerance for Irrigation Waters may be Summarized as Follows:

Element	Continuous water use mg/l	Short–term water use, fine texture soil mg/l
Aluminum	1.0	20.0
Arsenic	1.0	10.0
Beryllium	0.5	1.0

(continued)

TABLE 5: (continued)

Element	Continuous water use mg/l	Short–term water use, fine texture soil mg/l
Boron	0.75	2.0
Cadmium	0.005	0.05
Chromium	5.0	20.0
Cobalt	0.2	10.0
Copper	0.2	5.0
Lead	5.0	20.0
Lithium	5.0	5.0
Manganese	2.0	20.0
Molybdenum	0.005	0.05
Nickel	0.5	2.0
Selenium	0.05	0.05
Vanadium	10.0	10.0
Zinc	5.0	10.0

3. pH, Acidity and Alkalinity

   pH is not greatly significant and waters with pH values from 4.5 to 9.0 should not present problems. Highly acidic or alkaline waters can induce adverse effects on plant growth.

4. Chlorides

   Depending upon environmental conditions, crops and irrigation management practices, approximately 700 mg/l chlorides is permissible in irrigation waters.

5. Temperature

   Very high as well as very low temperatures of irrigation waters can interfere with plant growth. Temperature tolerence is highly dependent on the type of plant and other environmental conditions.

6. Pesticide

   A variety of herbicides, insecticides, fungicides and rodenticides can be present in irrigation waters at concentrations which may be detrimental to crops, livestock, wildlife and man. As far as effect on plant growth and permissible levels are concerned, these are variable and highly dependent on the type of chemical, type of plant, environmental factors and exposure time.

7. Suspended Solids

   Suspended solids in irrigation waters may deposit on soil surface and produce a crust which inhibit water infiltration and seedling emergence. In waters used for sprinkler irrigation colloids and suspended matter may form a film on leaf surface which impair photosynthesis and defer growth.

8. Radionuclides

   USPH Drinking Water Standards are usually applied to irrigation waters.

Source: NBS Publication 351, 1972

Emission Standards for Stationary and Mobile Sources

Some interesting discussions of emission problems were published as a part of the *Proceedings of the Environmental Quality Sensor Workshop,* Las Vegas, Nevada, November 30 to December 2, 1971 (10). It was stated in the proceedings of that workshop that sources of pollutants among stationary sources vary from thermal power plants and incinerators to cement plants and petroleum refineries. Air pollutants associated with such sources include particulates, sulfur oxides and nitrogen oxides, but also include fluorides, carbon monoxide, and organics.

Emissions of pollutants from stationary sources may involve flue gases or other vented gases. For some types of extended stationary sources analytical measurements would have to be made by a network of sampling sites or by long-path techniques. Such problems have not received much attention in practice. The present approach appears to be the use of the best available control technology with no continuing attempt after installation to determine compliance in terms of measurement of pollutant concentration. For flue gases or vented gases up stacks techniques of measurement do exist. In such situations it is important not only to measure the pollutants, but also the gas flow velocity, excess air, moisture, etc.

The sampling technique is of great importance and it must be suitable for the specific application. If sampling problems are handled adequately, the analytical measurement requirements can be simplified. The analytical measurements of pollutants can be accomplished by at least four approaches: (1) intermittent sampling on site with manual analysis in the laboratory, (2) continual instrumental analysis of pollutants collected through probes, (3) in-stack instrumentation, and (4) remote instrumental techniques. The second group of instrumental methods would include instruments capable of operating at emission concentration levels and the use of a proper sampling addition interface permitting use of ambient air type instrumentation. The in-stack instruments presumably integrate so that probing the stack configuration is not necessary.

Remote instrumentation cannot be utilized continually, but this class of instrument can be used for rapid surveys and checking compliance. The problem of positioning instruments or probes in stacks is eliminated. Problems associated with the correct relationship between in-stack particle concentration and particle size distribution to that emitted out of the stack into diluation air also are avoided. However, the practical development and utilization of remote optical instruments with the appropriate characteristics is difficult, scientifically and technologically.

As one quantitative expression of emission standards, performance standards for various new stationary sources are given in Table 6 (8). Standards are expressed as the maximum two hour average.

TABLE 6: PERFORMANCE STANDARDS FOR NEW STATIONARY SOURCES

Fossil-Fuel Fired Steam Generators

Particulate matter (mass loading)
0.10 lb. per million BTU

Particulate matter (opacity)
20% (40% for not more than 2 minutes)

Sulfur Dioxide
Oil Fired: 0.80 lb. per million BTU
Coal Fired: 1.4 lbs. per million BTU

Nitrogen Oxides (expressed as NO_2)
Gas Fired: 0.20 lb. per million BTU
Oil Fired: 0.30 lb. per million BTU
Coal Fired: 0.70 lb. per million BTU

Incinerators

Particulate matter (mass loading)
0.08 grain per cubic foot, corrected to 12% CO_2

Portland Cement Plants

Particulate matter (mass loading)
From kiln: 0.30 lb. per ton of solids fed to the kiln
From clinker cooler: 0.10 lb. per ton of solids fed to the kiln

Particulate matter (opacity)
10%

Nitric Acid Plants

Nitrogen oxides
3 lbs. (expressed as NO_2) per ton of acid produced
Opacity: 10%

Sulfuric Acid Plants

Sulfur dioxide
4 lbs. per ton of acid produced

Acid mist
0.15 lb. (expressed as H_2SO_4) per ton of acid produced
Opacity: 10%

Source: Lawrence Berkeley Laboratory, Berkeley, Calif. (May, 1972)

Emission standards for aqueous effluents are being issued by the Environmental Protection Agency. The following tabulation indicates industry categories covered and the Federal Register data citation for the published preliminary standards. The following tabulation covers the 31 industries involved in Phase One, Group One of this operation; some 37 additional industry classifications will be covered in the balance of Phase One and in Phase Two.

Industry	Fed. Register Publ. Date
(1) Fiberglass	August 22, 1973
(2) Beet sugar	August 22, 1973
(3) Cement	September 7, 1973
(4) Feedlots	September 7, 1973
(5) Phosphates	September 7, 1973
(6) Electroplating	October 5, 1973
(7) Inorganics	October 11, 1973
(8) Plastics and synthetics	October 11, 1973
(9) Meat	October 29, 1973
(10) Rubber	October 11, 1973
(11) Ferroalloys	October 18, 1973
(12) Flat glass	October 17, 1973
(13) Power plants	In preparation
(14) Asbestos	October 30, 1973
(15) Fruits and vegetables	November 9, 1973
(16) Nonferrous	November 30, 1973
(17) Timber	In preparation
(18) Seafood	In preparation
(19) Cane sugar	December 7, 1973
(20) Cooling water intakes	December 13, 1973
(21) Fertilizer	December 7, 1973
(22) Leather	December 17, 1973
(23) Soaps and detergents	December 26, 1973
(24) Grain milling	December 4, 1973
(25) Petroleum	December 14, 1973
(26) Organics	December 17, 1973
(27) Dairy	December 20, 1973
(28) Pulp and paper	In preparation
(29) Builders' paper	In preparation
(30) Iron and steel	In preparation
(31) Textiles	In preparation

In addition, a set of guidelines for "toxic" substances is being promulgated by the Environmental Protection Agency. The initial listing which was published in the Federal Register for December 27, 1973 specified aqueous effluent standards for nine materials as follows:

(1) Aldrin - Dieldrin
(2) Benzidine
(3) Cadmium
(4) Cyanide
(5) DDT
(6) Endrin
(7) Mercury
(8) Polychlorinated biphenyls
(9) Toxaphene

The proposed standards for toxic substances permit no discharge into a river or stream having a flow rate of 10 $ft.^3$/sec. or less or into a lake having an area of less than 500 acres. Above these thresholds, discharging would be permitted with the allowable rate determined by the resulting concentration of the toxic material in the receiving water. Now that these standards have been proposed for the above-cited limited list of materials, however, environmental action groups are suing to have organic chemical industry effluents, paper mill effluents, petroleum refining effluents and pesticide manufacturing effluents added, making them subject to far stricter controls.

Monitoring

As pointed out by D.C. Holmes of EPA (11) monitoring the environment is a key to effective management for environmental quality. It is nearly impossible to detect environmental changes, desirable or undesirable, natural or man-made, without established baselines and repeated observations. Measurements are essential for the identification of environmental needs and the establishment of program priorities, as well as for the evaluation of program effectiveness, and they provide an early warning system for environmental problems which allows corrective action to be taken.

The choice of appropriate environmental monitoring systems presents several problems: coverage of systems, both geographically and as to pollutants or parameters measured, must be adequate; sensors should be optimally located; measurement instruments have to be accurate and calibrated to ensure compatibility of data from one location to another. All of these essential features must support the need to collect and analyze environmental data on a near real-time basis.

Without all of these capabilities, one cannot accurately determine the severest problem or a cost-effective method of attacking it, nor can one evaluate the success of the efforts. Monitoring is not a substitute for action, but action without the knowledge provided by adequate monitoring is more likely to be ineffective.

As pointed out by D.S. Barth and S.D. Shearer, Jr. of EPA (12) prior to discussing appropriate sensors for use in air monitoring networks it is necessary to generally define the objectives to be achieved by various monitoring systems. Thus in the following we shall give first a broad discussion of various air monitoring objectives and then proceed to suggest which may be used for the various applications.

Air monitoring sampling devices may operate continuously or intermittently; they may collect integrated samples which may be analyzed at a laboratory or they may collect samples and analyze them simultaneously; they may give a digital reading on a dial, an analog printout on a recorder chart or a digital readout on computer tape; they may measure air concentrations at single locations or they may measure air concentrations over some specified integrated path length. Resulting monitoring data may be used for regulatory purposes or for research purposes or for some combination of both. In general, the following objectives of air monitoring systems may be defined:

(1) Regulatory Purposes – Air quality measurements are required to determine compliance with primary and secondary national ambient air quality standards. Stationary source emissions must be measured to determine compliance with emission standards set as parts of implementation plans, or national emission standards for hazardous air pollutants or standards of performance for new sources for specified categories of industry. Mobile source emissions must be measured to determine compliance with national emission standards. For all of these applications a reference measurement method must be defined when the standard is set. The reference measurement method, or an equivalent one, must be used for each regulator application.

(2) Establishment of Baselines and Trends – Baseline air quality measurements must be established for all significant air pollutants for both rural and urban locations. Trends of these baseline measurements must be followed as a function of time to determine quantitatively what changes result as the number of sources change or as air pollution controls are instituted. Inherent to this objective is the need for a systematic approach to the detection, identification and quantitation of new or newly recognized air pollutants. In many cases the development and application of new measurement methods are essential ingredients to achieve this objective.

(3) Inventory of Air Pollution Sources – Stationary and mobile man-made sources of all significant air pollutants as well as natural or man-made sources must be identified and measured on a nationwide scale. This inventory must be

periodically updated. Emission factors may be estimated from process weight and material balance considerations or preferably measured with suitable source sampling and measurement methods. For this application as for all source measurement applications it is often more difficult to obtain a representative sample than it is to analyze the sample.

(4) Exposure and Effect Data – Such data may cover a wide range of effects such as corrosion of selected materials, soiling, atmospheric turbidity, pH of precipitation, damage to growing vegetation, etc. It is essential to simultaneously measure the concentration of the air pollutant or pollutants of importance at the receptor over the time during which a specific effect is being measured on the same receptor. In this fashion it is possible to obtain data on the basis of which exposure-effect relationships may be constructed. From these relationships it is possible to determine to what concentrations air pollutants must be reduced to reduce a correlated adverse effect to an acceptable level. Clearly adequate accurate measurement methods must be available or be developed to measure both exposure and effect.

(5) Development or Validation of Predictive Models – Predictive models are necessary tools to allow calculations of required air pollution source reductions to achieve specified reductions in air quality measurements. This application requires as a minimum the simultaneous measurement as a function of space and time of sources, air quality and all meteorological parameters of importance. When important removal processes or chemical or physical transformations take place, adequate measurements must be taken to define the concentration changes which take place for the pollutants of interest.

(6) Emergency Situations – For instances where air pollution episodes or accidents resulting in release of dangerous air pollutants occur, it is necessary to monitor air pollution concentrations so that adequate protective action may be taken prior to the occurrence of substantial endangerment to human health.

All of the monitoring objectives described briefly above require the availability of field sampling devices and analytical measurement methods of known precision and accuracy. Effects of changing meteorological factors such as temperature and humidity and any interfering substances must be known. Calibration procedures for the measurement methods must be available and an adequate system for quality control must be instituted and maintained.

GENERAL TYPES OF ANALYTICAL SYSTEMS

Manual

Methods in common use for measurement of air pollutants from stationary sources of emissions have been manual rather than instrumental. Filters, impingers, evacuate flasks and condensation are the collection techniques commonly used. To express pollutant mass loadings and relate results to process variables additional measurements are required. These measurements include stack gas velocity with a pitot tube, moisture in-stack gases determined gravimetrically on condensate, excess air by Orsat analysis and carbon dioxide by the nondispersive infrared spectrometric method (13)(14)(15).

Sulfur dioxide usually is determined by collection of the flue gas in impingers containing hydrogen peroxide (16)(17)(18)(19). The sulfur dioxide is oxidized to sulfate and analyzed after reaction with barium chloroanilate colorimetrically at 530 mμ in terms of the chloroanilate ion. A second procedure often used involves determination of the sulfate by the barium perchlorate thorium titration method (16)(17). The sulfur trioxide or sulfuric acid mist after separation by filter collection is determined by the same procedures. Cations such as Al^{+3}, Fe^{+3}, Pb^{+2}, Cu^{+2}, and Zn^{+2} produce negative interference in the colorimetric procedure by precipitating choroanilate ion from solution. These interferences can be minimized by filtering the flue gas with glass wool at the probe and pretreating the solution with a cation resin. Anions causing interference are not likely to be present in flue gases. The titration with barium perchorate is interfered with by cations such as K^+, Na^+, NH_4^+ by reducing the volume of titrant needed. Cationic interferences are minimized by use of a particulate filter in the probe and by percolation of the collected solutions through a cation exchange column. Anions which coexist in the collected solutions such as nitrate, chloride and fluoride can interfere also.

These sulfur dioxide analytical procedures have practical ranges from 10 to 3,000 ppm by volume. The sensitivities are 10 ppm for 25 to 30 liter samples. The precisions are about ±3% at 1,500 ppm. For sulfur acid aerosol the range is 10 to 300 ppm with a precision at 10 ppm of ±5%. Nitrogen oxides often are collected from flue gases in an evacuated flask containing dilute sulfuric acid hydrogen peroxide absorbing solution. The nitric oxide or nitrogen dioxide are oxidized to nitric acid which is measured colorimetrically at 420 $m\mu$ as nitrophenol disulfonic acid (20). This technique has a range of from 15 to 1,500 ppm by volume with a sensitivity of 1.5 ppm. Halogens interfere in this procedure.

In common practice particulates are measured gravimetrically after removal of uncombined water (21)(22)(23). The particulate is removed from the flowing gas stream under isokinetic conditions by filtration and condensation. Impinger collection is used as part of this procedure with extract of organic particulate from the impinger solution with chloroform and ethyl ether. Acetone washing of the probe and filter holder also is included in the procedure. The total particulate weight of aqueous and organic sample components is obtained by totaling the weights of components. Total sample volume or stack velocity and other parameters are utilized in calculating particulate mass in stack gas.

Manual surveillance of wastewaters is based on the periodic collection of water samples, their transportation and storage and subsequent laboratory analysis. Sampling sites and frequency are selected to give an overall evaluation of water quality in drainage basins and effects of waste discharge.

Automated

A number of types of increasingly sophisticated analytical systems are being applied to the detection and monitoring of various pollutants in air and water.

Automated (Air): At this point it is in order to give a general brief discussion of the current state of the art of air monitoring and then present a brief review of developing areas in the remote sensing of air pollutants. Such a discussion has been presented by D.S. Barth and S.D. Shearer, Jr. (12).

Traditionally, the gathering of information regarding air pollutant concentrations in the environment has been directed toward a relatively few air contaminants known to be detrimental to human health and welfare. Such information has been used primarily by local, state, and federal agencies in their enforcement and regulatory functions. This information has been gained through monitoring activities ranging from national networks designed primarily to determine long-term nationwide and state trends and averages to small intensive short-term activities designed for specific problem solving in certain localized areas.

An article by Hochheiser, Burmann, and Morgan (24) has well described the state of the art in air pollution monitoring. In this article they state that the "methods and instruments for measuring air pollutants must be carefully selected, evaluated, and standardized." They list eleven factors which must be considered. These factors, given below, could apply equally well whether one was considering remote or traditional sensors.

(a) Specificity — Does the method respond only to the pollutant of interest in the presence of those other substances likely to be encountered in samples obtained from ambient air or pollution sources?

(b) Sensitivity and Range – Is the method sensitive enough over the pollutant concentration range of interest?
(c) Stability – Is the sample stable? Will it remain unaltered during the sampling interval and the interval between sampling and analysis?
(d) Precision and Accuracy – Are results reproducible? Do they represent true pollutant concentration in the atmosphere or source effluent from which the sample was obtained?
(e) Sample Averaging Time – Does the method meet the above-stated requirements for sample-averaging time of interest?
(f) Reliability and Feasibility – Are instrument maintenance costs, analytical time, and manpower requirements consistent with needs and resources?

For continuous automatic monitoring instruments, further requirements must be considered:

(g) Zero Drift and Calibration – Is instrument drift over an unattended operation period of at least three days slight enough to ensure reliability of data? Are calibrating and other corrections automatic?
(h) Response, Lag, Rise and Fall Time – Can the instrument function rapidly enough to record accurately changes in pollutant concentration that occur over a short time in the sample stream being monitored?
(i) Ambient Temperature and Humidity – Does the instrumental method meet all of these requirements over temperature and humidity ranges normally encountered?
(j) Maintenance Requirements – Can the instrument operate continuously over long periods with minimum down time, maintenance time and maintenance cost? Is service for the instrument available in the area of use?
(k) Data Output – Does the instrument produce data in a machine-readable format?

Tables 7 and 8 present a concise summary of the current state of the art in traditional air pollutant monitoring for the common air pollutants.

TABLE 7: COMMERCIAL CONTINUOUS MONITORING METHODS FOR COMMON AIR POLLUTANTS

Methane-CO-total hydrocarbon instruments —cost, approximately $10,000 with recorder

Manufacturer	
Beckman Instrument Co.	(all based on gas chromatographic separation and flame ionization detection)
Mine Safety Appliance Co.	
Tracor Inc.	
Union Carbide Corp.	

NO_2-NO_x instruments—cost range, $2000–5000

	DETECTION PRINCIPLE		
Manufacturer	Colorimetric	Coulometric	Electrochemical
Atlas Electric Devices Co.	■		
Beckman Instrument Co.	■	■	
Dynasciences			■
Environmetrics Inc.			■
Mast Development Inc.		■	
Monitor Labs Inc.	■		
Intertech Corp.			■
Pollution Monitors Inc.	■		
Precision Scientific	■		
Raditron Inc.	■		
Scientific Industries	■		
Technicon Corp.	■		
Wilkens-Anderson Corp.	■		

(continued)

TABLE 7: (continued)

Oxidant instruments—cost range, $1000–7000

Manufacturer	DETECTION PRINCIPLE Colorimetric	Coulometric	Chemiluminescence	Uv absorption
Atlas Electric Devices Co.		■		
Beckman Instrument Co.	■	■		
Bendix Environmental Science Div.			■	
Dasibi Corp.				■
Mast Development Co.		■		
McMillian Electronics			■	
Monitor Labs Inc.	■			
Ozone Research		■		
Raditron Inc.	■			
REM Inc.			■	
Spectrometrics of Florida Inc.				■
Technicon Corp.	■			
Triangle Instrument Co.			■	

CO instruments—cost range, $2500–5000

Manufacturer	DETECTION PRINCIPLE NDIR	Uv (mercury replacement)
Bacharach Instrument Co.		■
Beckman Instrument Co.	■	
Bendix Environmental Sci. Div.	■	
Calibrated Instrument Co.	■	
Intertech Corp.	■	
Morgan Co.	■	
Mine Safety Appliance Co.	■	

Total hydrocarbons—cost range, $2000–5000

Manufacturer	
Beckman Instrument Co.	
Bendix Environmental Science Div.	
Davis Instrument Co.	(all based on flame ionization detection)
Delphi Industries	
Mine Safety Appliance Co.	
Power Design Inc.	

SO_2 instruments—cost range, $2000–5000

Manufacturer	DETECTION PRINCIPLE Colorimetric	Coulometric	Flame photometric	Electrochemical	Conductimetric
Atlas Electric Devices Co.	■	■			
Barton Instr. Corp.		■			
Beckman Instr. Co.		■			■
Bendix Environmental Sci. Div.			■		
Calibrated Instr .Inc.					■
Combustion Equipment Associates					■
Davis Instrument Co.					■
Devco Engineering					■
Dynasciences				■	
Environmetrics Inc.				■	
Erickson Instruments				■	
Instrument Devel. Co.				■	■
Intertech Corp.				■	■
Kimoto					■
Leeds & Northrup					■
Meloy Labs			■		
Monitor Labs Inc.	■				■
Phillips Electronic Co.		■			
Pollution Monitors Inc.	■				
Precision Scientific	■				
Process Analyzers Inc.		■			
Scientific Industries					■
Technicon Corp.	■				
Tracor Inc.			■		
Wilkens-Anderson Corp.	■				

Source: EPA Proceedings, November 30 to December 2, 1971

TABLE 8: OTHER AIR POLLUTANT SAMPLING AND ANALYTICAL METHODS

Pollutant	Sampling method	Analytical method	Major improvement needed
Hg	Gas bubbler, Hg amalgamation	Flameless atomic absorption, uv absorption	Yes
Be	Hi-vol	Emission spectroscopy, atomic absorption	No
Asbestos	Membrane filter sampler	Electron microscopy	Yes
Pb	Hi-vol	Emission spectroscopy	No
Cd	Hi-vol	Atomic absorption, emission spectroscopy	No
As	Hi-vol	Colorimetric, neutron activation	No
Ni	Hi-vol	Atomic absorption, emission spectroscopy	No
Polynuclear organics	Hi-vol	Chromatography and fluorescence analysis	No
Polychlorinated biphenyl	Not developed	Not developed	Yes
F	Coated-open tubular column	Specific ion electrode	Yes
Odor	Human odor panel	Olfaction	No
H_2S	Continuous automatic analyzer and mechanized tape sampler	Flame photometric, coulometric, colorimetric	Yes
Fine particulates	Impactor	Gravimetric	Yes
V	Hi-vol	Emission spectroscopy	No
Mn	Hi-vol	Emission spectroscopy	No
Cr	Hi-vol	Emission spectroscopy	No
Se	Not developed	Not developed	Yes
Cl_2	Not developed	Not developed	Yes
HCl	Not developed	Not developed	Yes
Cu	Hi-vol	Emission spectroscopy, atomic absorption	No
Zn	Membrane filter sampler	Emission spectroscopy, atomic absorption	No
B	Not developed	Not developed	Yes
Ba	Membrane filter sampler	Emission spectroscopy	Yes
Sn	Hi-vol	Emission spectroscopy	No
P	Not developed	Not developed	Yes
Li	Not developed	Not developed	Yes
Aeroallergens	Not developed	Not developed	Yes
Reactive organics	Not developed	Chromatography	Yes
Pesticides	Not developed	Chromatography	Yes
Radioactive materials	Hi-vol	Gross beta and gamma scanning	Yes

Source: EPA Proceedings, November 30 to December 2, 1971

A number of instruments based on optical techniques have potential for use at stack gas pollutant concentration levels. Nondispersive infrared analyzers or ultraviolet analyzers have been designed for sulfur dioxide, nitrogen dioxide and carbon monoxide. Optical techniques has received attention for visibility of stack gases in plumes. Portable electrochemical transducers recently developed have potential for use in measurement of sulfur dioxide and nitrogen oxides in stack gases (25). The instruments utilizing the chemiluminescent reaction of ozone with nitric oxide could be utilized for measurement of nitric oxide in stacks (26)(27)(28)(29). Gas chromatographic instruments have received considerable evaluation for sulfur dioxide, hydrogen sulfide and organic sulfur gases in Kraft mill effluents (30)(31)(32)(33). Gas chromatographic analyzers also could be utilized to measure carbon monoxide and sulfur dioxide in various stack gases. A source sampling technique for particulate and gaseous fluorides involves use of a heated glass probe to convert hydrogen fluoride to silicon tetrafluoride (34). This type of sampling procedure has been used with a fluoride-selective ion electrode to analyze water-soluble fluorides in stack gases.

An attractive sampling approach would be to convert the stack gases by dilution and cooling to conditions approaching these in ambient air analysis. The same instruments could be used as for air quality measurements. This approach would considerably reduce the number of types of instruments in use, calibration requirements and maintenance problems. Studies are needed on the approached sampling interfaces to provide dilution and cooling without changing pollutant composition. Work is in progress on the evaluation of sulfuric oxide and nitrogen oxide analyzers for stack gases.

Instruments capable of operating within the stack itself usually have been based on optical principles. Such equipment must be built to withstand dust, heat, corrosion and vibration. Thermal gradients can cause considerable problems in optical alignments. Calibration of such instruments requires spectra under experimental conditions closely simulating stack conditions.

Monitoring capability is especially important for carbon monoxide, sulfur dioxide, hydrocarbons, ozone, and nitrogen dioxide because air quality standards for these widespread pollutants have been established throughout the United States by the Environmental Protection Agency. In addition, a number of other gases or vapors are of considerable concern or of potential concern. These substances include hydrogen sulfide, organic sulfur compounds, hydrogen fluoride, hydrochloric acid, chlorine, and nitric acid. With the possible exception of hydrogen fluoride (and water-soluble fluorides) the lack of data makes it difficult to estimate the levels or widespread prevalence of these substances.

The topic of air pollution monitoring by advanced spectroscopic techniques has been reviewed by J.A. Hodgeson, W.A. McClenny and P.L. Hanst of the Environmental Protection Agency (35). Some of the fundamentals of various types of instruments applicable to air monitoring have been discussed by L.A. Elfers (35a).

For example, the principle of conductometry, that is, electric conductance by soluble electrolytes, has had wide application as an analytical procedure. The conductance of electrolytes in solution is proportional to the number of ions present and their mobilities. In dilute sample solutions, the measured conductivity can be directly related to the concentration of ionizable substance present. Sulfur dioxide has been measured by this procedure in continuous recording instrumentation for more than 25 years. The basic concept involves absorption of sulfur dioxide in deionized water (or a very dilute reagent) to produce an acid having conductance sufficient to be detected by a conductivity cell.

Monitors employing deionized water as the reagent result in the formation of sulfurous acid (H_2SO_3). Interferences from carbon dioxide, salt aerosols, acid mists, and basic gases are common. Most sulfur dioxide analyzers use distilled or deionized water reagent modified by the addition of hydrogen peroxide and a small amount of sulfuric acid. This modified reagent forms sulfuric acid (H_2SO_4) upon reaction with sulfur dioxide. The acidic property of this modified reagent reduces the solubility of carbon dioxide gas within the reagent and minimizes this interference. However, this method is still subject to gross interferences and is not recommended. Figure 2 is a schematic diagram of a typical conductivity type monitor.

Colorimetry can be defined as a mode of analysis in which the quantity of a colored substance is determined by measuring the relative amount of light passing through a solution of that substance. The constituent may itself be colored and thus be determined directly,

FIGURE 2: SCHEMATIC DIAGRAM OF A TYPICAL SO_2 CONDUCTIVITY MONITOR

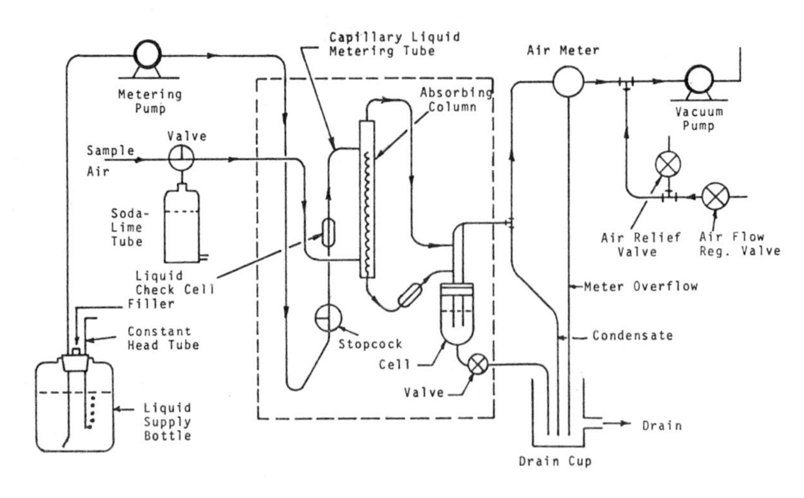

Source: Report PB 204,650, 1971

or it can be reacted with a reagent to form a colored compound and thus be determined indirectly. The physical law that underlies colorimetric analysis is commonly referred to as Beer's law. It states that the degree of light absorption by a colored solution is a function of the concentration and the length of the light path through the solution. These principles are commonly employed for the continuous measurement of three air pollutants; sulfur dioxide, nitrogen dioxide, and oxidants. Figure 3 is a schematic diagram of a typical colorimetric type monitor.

FIGURE 3: SCHEMATIC DIAGRAM OF A TYPICAL COLORIMETRIC MONITOR

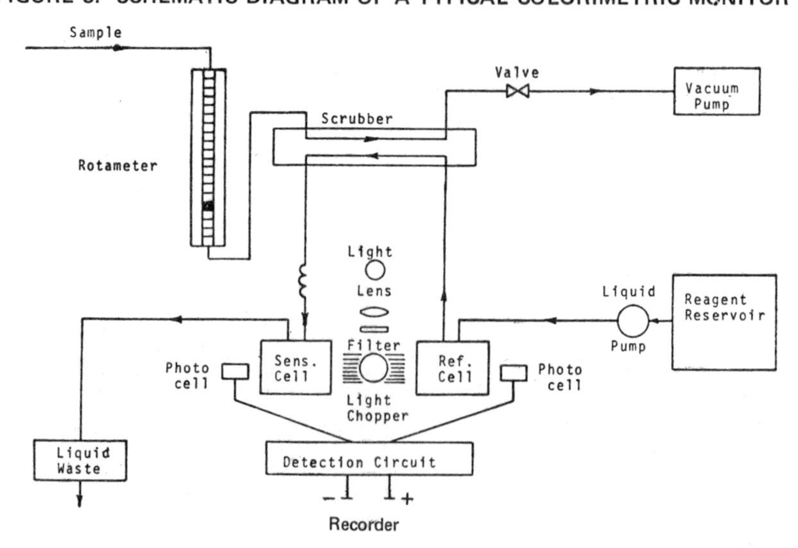

Source: Report PB 204,650, 1971

As shown, an atmospheric sample is drawn into the air-reagent flow system, first entering the absorber or scrubber where the desired constituent is reacted with the appropriate reagent. The air sample is separated from the reacted reagent and passed through the air pump and finally discharged to the atmosphere. The reacted reagent passes from the bottom of the absorber into the continuous flow colorimeter where the absorbance of the solution is measured. In most cases a dual flow colorimeter, as shown, is employed whereby the absorbance of the unreacted reagent is initially measured.

Coulometry is a mode of analysis wherein the quantity of electrons required to oxidize or reduce a desired substance is measured. This measured quantity, expressed as coulombs, is proportional to the mass of the reacted material according to Faraday's law. Coulometric titration cells for the continuous measurement of sulfur dioxide, oxidants, and nitrogen dioxide have been developed using this principle. Figure 4 is a typical flow diagram of a coulometric type monitor.

One class of cells generally employed in continuous air monitoring is designed to respond to materials which are oxidized or reduced by halogens and/or halides. Upon introduction of a reactive material, the halogen-halide equilibrium is shifted. The system is returned to equilibrium by means of a third electrode which regenerates the depleted species. The current required for this generation is measured and is directly proportional to the concentration of the depleted species, which in turn is proportional to the quantity of desired constituent.

This mode of analysis can be classified as secondary coulometry, usually employing a dynamic iodimetric or bromimetric titration. These systems are designed to respond with a sensitivity in the lower parts per billion range. Most commercial coulometric systems are made specific by the use of prefiltration devices, scrubbers, and/or chromatographic techniques that retain interfering compounds and permit passage of the desired constituents. Some loss of pollutants may occur in these devices, probably not exceeding 10%.

FIGURE 4: SCHEMATIC DIAGRAM OF A TYPICAL COULOMETRIC MONITOR

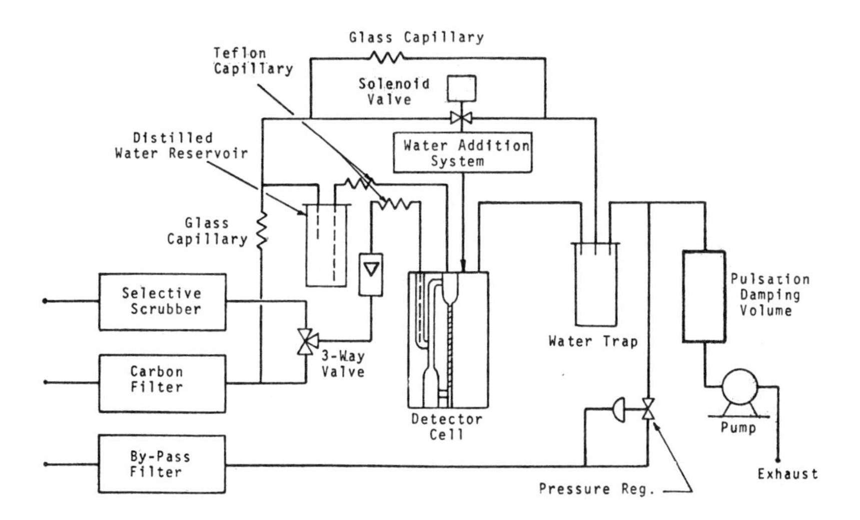

Source: EPA PB 204,650, 1971

Another type of coulometric cell, designed for oxidant analysis, employs amperometry. The oxidant reacts with an iodide solution within the cell releasing iodine which depolarizes the cathode, thus permitting current flow which is proportional to the oxidant concentration. By passing reagent and sample over the electrodes, a continuous measurement is achieved. This type of monitor is also subject to interferences. Materials which undergo oxidation will appear as negative interferences; those that undergo reduction will appear as positive interferences. Figure 5 is a diagram of a typical amperometric-type monitor.

Flame photometry is based on the measurement of the intensity of specified spectral lines resulting from quantum excitation and decay of elements by the heat of a flame. Volatile compounds are introduced into the flame by mixing them with the flammable gas or with the air supporting the flame. Nonvolatile compounds are aspirated from a solution into the flame. The specific wave length of interest can be isolated by means of narrow-band optical filters, diffraction gratings, or by means of a prism. The intensity of the specific wavelength can be measured by means of a phototube or photo-multiplier tube and associated electronics.

Recent development of flame photometric detectors having a semispecific response to volatile phosphorus and sulfur compounds has led to their use in continuous monitoring of gaseous sulfur compounds. This flame photometric detector consists of a photo-multiplier tube viewing a region above the flame through narrow-band optical filters. When sulfur compounds are introduced into the hydrogen-rich flame, they produce strong luminescence between 300 and 423 nanometers. A specificity ratio for sulfur to nonsulfur compounds

FIGURE 5: SCHEMATIC DIAGRAM OF A TYPICAL AMPEROMETRIC MONITOR

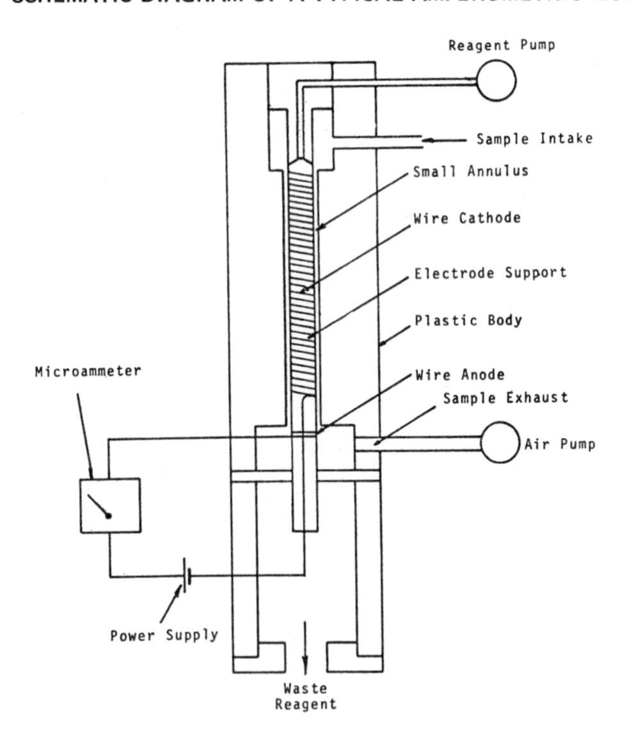

Source: Report PB 204,650, 1971

of approximately 20,000 to 1 is achieved with a narrow-band optical filter in the range between 5 and 900 parts per billion. Other sulfur compounds result in positive interferences while monitoring for SO_2, since the detector responds to all sulfur compounds. Discretion should be used in interpreting data obtained from this detector when it is located in areas where other sulfur compounds exist. For example, H_2S and CH_3SH are found in conjunction with SO_2 in the vicinity of kraft paper mills and CS_2 and H_2S in the vicinity of oil fields and refineries. Figure 6 is a schematic diagram of a typical flame photometric sulfur monitor.

Separation of parts per billion levels of SO_2, H_2S, CS_2, and CH_3SH has recently become feasible through the use of chromatographic techniques. This mode of separation followed by flame photometric detection of the separated sulfur compounds has led to the development of a new type of semicontinuous monitor for sulfur containing gases which is specific and free from interferences.

Instruments employing flame ionization detectors, originally designed for gas chromatographic detection of organic compounds, have had wide application as a continuous monitor for hydrocarbons. The sample to be analyzed is mixed with a hydrogen fuel and passed through a small jet; air supplied to the annular space around the jet supports combustion. Carbon-containing compounds carried into the flame result in the formation of ions. An electrical potential across the flame jet and an ion collector electrode produce an ion current proportional to the number of carbon atoms in the sample.

FIGURE 6: SCHEMATIC DIAGRAM OF A TYPICAL FLAME PHOTOMETRIC SULFUR MONITOR

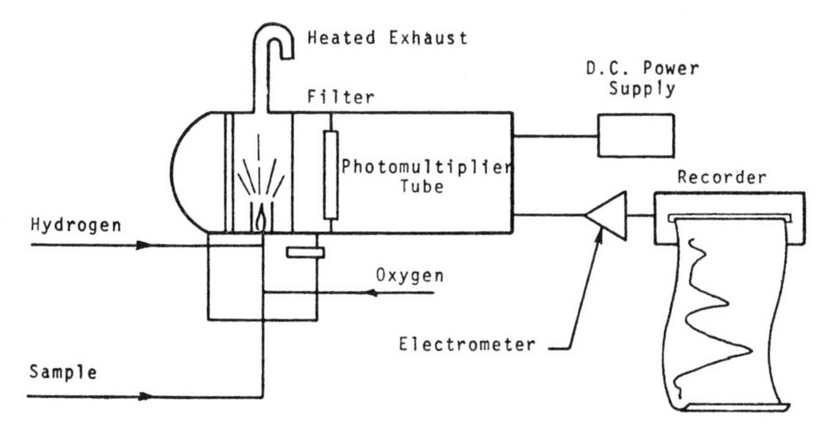

Source: Report PB 204,650, 1971

Advantages inherent in this detector are that it is free from interferences from inorganic gases, has a rapid response time, and has a low noise background level. Figure 7 is a schematic diagram of a typical hydrogen flame ionization type monitor. Recent investigations involving automated gas chromatographic separation of the air sample into three fractions has led to the development of a semicontinuous monitor for CO, CH_4, and total hydrocarbons. Measured volumes of ambient air are delivered periodically (4 to 12 times per hour) to a hydrogen flame ionization detector to measure its total hydrocarbon (THC) content. An aliquot of the same air sample is introduced into a stripper column which removes water, carbon dioxide, and hydrocarbons other than methane. Methane and carbon monoxide are passed gravitatively to a gas chromatographic column where they are separated. The methane is eluted first and is passed unchanged through a catalytic reduction

FIGURE 7: SCHEMATIC DIAGRAM OF A TYPICAL FLAME IONIZATION MONITOR

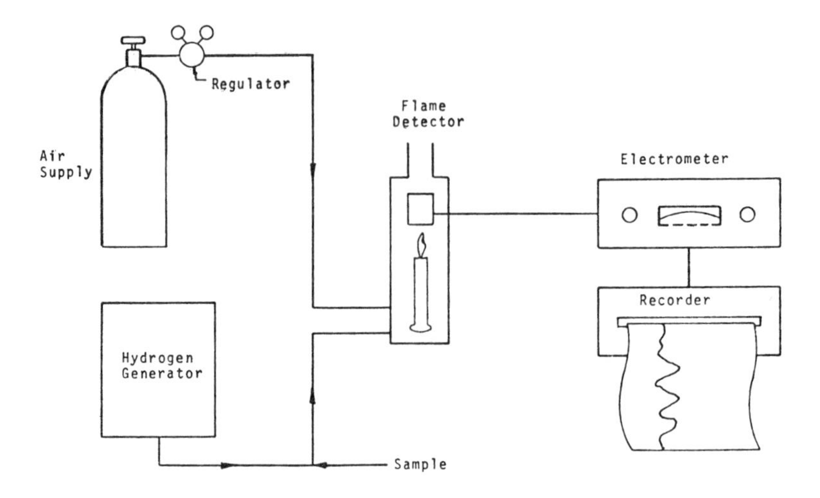

Source: Report PB 204,650, 1971

tube into the slave ionization detector. The carbon monoxide is eluted into the catalytic reduction tube where it is reduced to methane before passing through the flame ionization detector. Between analyses the stripper column is backflushed to prepare it for subsequent analysis. Hydrocarbon concentrations corrected for methane are determined by subtracting the methane value from the total hydrocarbon value.

The infrared absorption characteristics of several gases and vapors make possible their detection and analysis in continuous analyzers. Carbon monoxide as an air contaminant is uniquely suited to this method of analysis, as its absorption characteristics and typical concentrations make possible direct sampling.

A typical analyzer (Figure 8) consists of a sampling system, two infrared sources, sample and reference gas cells, detector, control unit and amplifier, and recorder. The reference cell contains a noninfrared-absorbing gas while the sample cell is continuously flushed with the sample atmosphere. The detector consists of a two-compartment gas cell (both filled with carbon monoxide under pressure) separated by a diaphragm whose movement causes a change of electrical capacitance in an external circuit, and ultimately an amplified electrical signal is suitable for input to a servo-type recorder.

During analyzer operation an optical chopper intermittently exposes the reference and sample cells to the infrared sources. At the frequency imposed by the chopper, a constant amount of infrared energy passes through the reference cell to one compartment of the detector cell while a varying amount of infrared energy, inversely proportional to the carbon monoxide concentration in the sample cell, reaches the other detector cell compartment. These unequal amounts of residual infrared energy reaching the two compartments of the detector cell cause unequal expansion of the detector gas. This unequal expansion causes variation in the detector cell diaphragm movement resulting in the electrical signal described earlier.

The output of this instrument is independent of sample flow rate. A reduction in the resolution of rapid changes in concentration can be made by introducing additional volume in the sample line ahead of the analyzer unit. Using the principles of reflectance and

FIGURE 8: SCHEMATIC DIAGRAM OF A TYPICAL NONDISPERSIVE INFRARED CO MONITOR

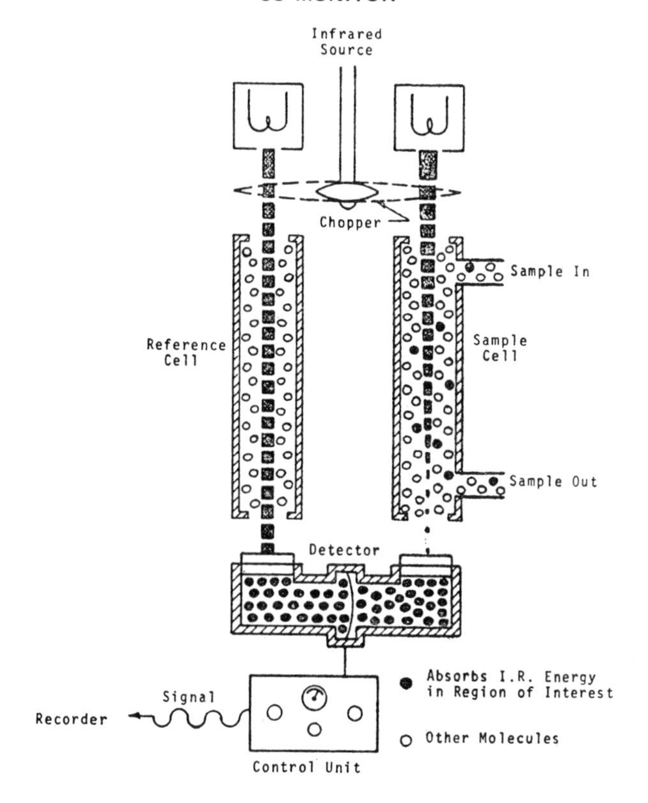

Source: Report PB 204,650, 1971

transmittance, the tape sampler is used to measure suspended particulates in ambient air. Air is drawn through a section of white filter paper. The most generally accepted sampling parameters for this measurement are: a 1 inch diameter spot, 7.07 l./min. (0.25 $ft.^3$/min.) flow rate and a two hour sampling period.

At the end of the sampling period, the tape advances automatically by means of a timing mechanism, placing a clean section of filter tape at the sampling port. The collected spot is automatically positioned under a photoelectric reflectance or transmittance head which evaluates the density of the spot by measuring the light reflected from or transmitted through the spot to the cell.

In general the greater the amount of particulate matter filtered out of the air, the darker the spot and the smaller the amount of light reflected or transmitted. The quantity of air sampled is expressed as linear feet and the final result reported in either of two units of measurement, the COH or the RUDS, per 1,000 linear feet of air. The units of 1,000 linear feet (LF) can be determined as follows:

$$LF = \frac{Qt}{1{,}000A}$$

where Q is sample air flow rate in cubic feet per minute; t is sampling time, minutes; and A is cross-sectional area of filter spot, square feet. The COH unit (coefficient of haze) can be defined as that quantity of particulate matter which produces an optical density of 0.01 when measured by light transmittance in the region of 375 to 450 nanometers. The transmittance of a clean filter is used as a reference and is set at 0.0 density (or 100% transmittance). The light transmitted through the filter tape is expressed as follows:

$$\text{Optical density} = \log \frac{I_0}{I}$$

where I_0 is the initial intensity of transmitted light through clean filter paper, and I is the transmittance observed through the soiled filter paper. The RUDS (reflectance unit of dirt shade) can be defined as that quantity of particulate matter which produces an optical reflectance of 0.01 when measured by reflectance through a green filter. The reflectance of a clean filter paper is set at 100% on the reflectance meter and is used as a reference. The reflectance of light from the filter tape is expressed as follows:

$$\text{Optical density} = \log \frac{R_0}{R}$$

where R_0 is the initial intensity of reflected light from clean filter tape and R is the reflectance observed from the soiled filter tape. This RUDS unit is also expressed in terms of 1,000 linear feet of sample collected. There is no general conversion factor which can be used between the COH and RUDS units: these measurements are dependent upon varying parameters such as the color of the particulates collected, their size distribution, depth of penetration in the filter tape, and the variation of paper tape thickness. There also is a general lack of uniform correlation between tape sampler measurements and high volume measurements. Figure 9 is a diagram of a typical tape sampler.

FIGURE 9: SCHEMATIC DIAGRAM OF A TYPICAL TAPE SAMPLER

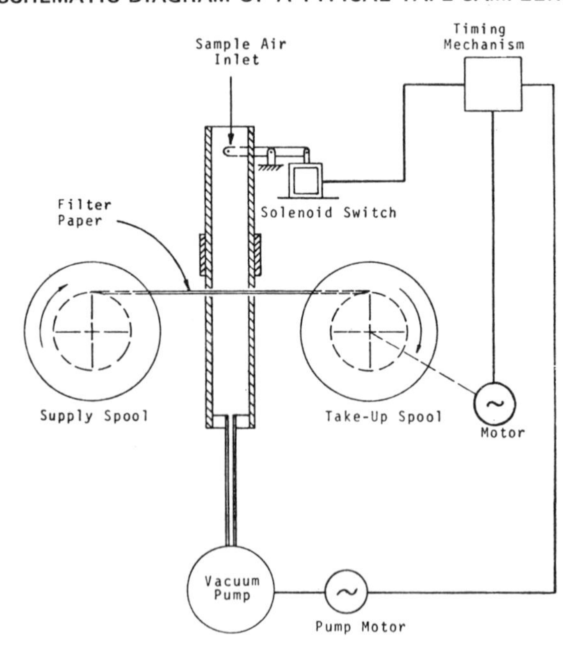

Source: Report PB 204,650, 1971

Chemiluminescent detection techniques have been employed for the measurement of atmospheric ozone. One method employs the reaction of rhodamine B that is impregnated on activated silica gel with ozone; the chemiluminescence produced is detected with a photomultiplier tube. The recommended method as described in the *Federal Register,* Vol. 30, No. 84, employs the gas-phase reaction of ethylene and ozone; the chemiluminescence is also detected by means of a photomultiplier tube. These specific ozone chemiluminescent techniques were recently evaluated and were found to be satisfactory as atmospheric ozone monitors. The gas-phase reaction is more advantageous for field use owing to its small size, simplicity of construction, and ease of operation. Figure 10 is a typical diagram of the gas-phase ozone monitor.

FIGURE 10: SCHEMATIC DIAGRAM OF A TYPICAL GAS-PHASE CHEMILUMINESCENT OZONE DETECTOR

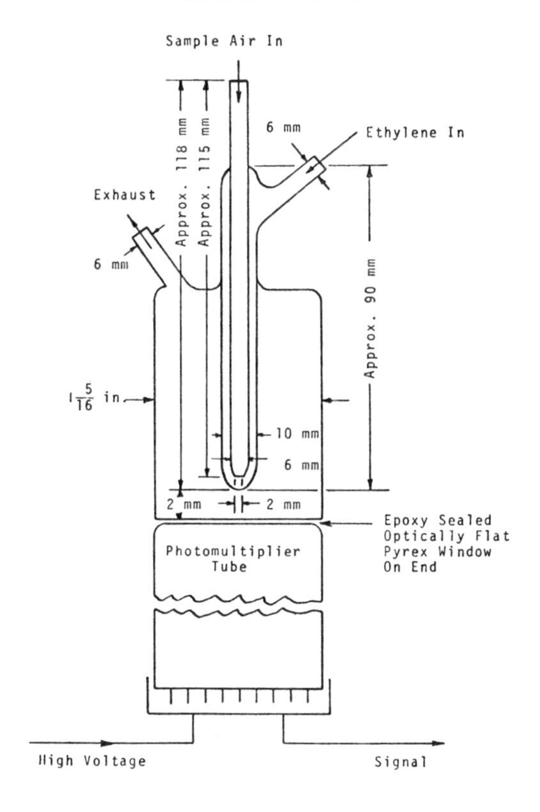

Source: Report PB 204,650, 1971

Nephelometry is based upon the principle that a common effect of particulate air pollution is the reduction of visibility. Small particles suspended in the air scatter light out of the line of vision, making distant objects appear less distinct. Visual range can be defined as that distance at which the difference in contrast between the background and the object being observed is too small to perceive. The air sample is drawn into the detection chamber where it is illuminated by a pulsed-flash lamp. The scattered light is measured over a range of scattering angles of 9° to 170° by means of a photomultiplier tube (See Figure 11). The

FIGURE 11: SCHEMATIC DIAGRAM OF A TYPICAL INTEGRATING NEPHELOMETER

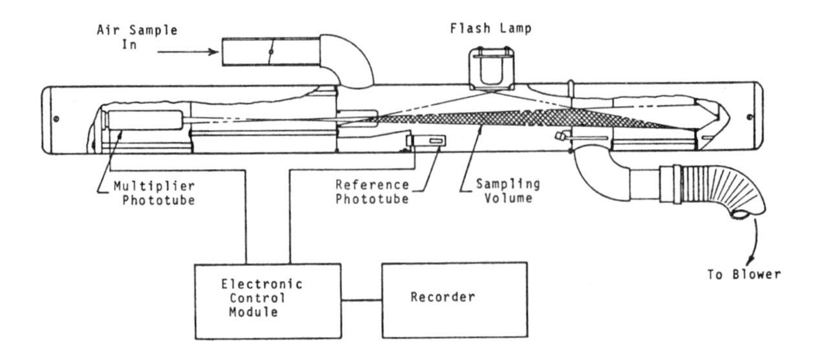

Source: Report PB 204,650, 1971

signal produced by the photomultiplier tube is averaged and compared with a reference voltage from another phototube illuminated by the flash lamp. Calibration is performed by introducing clean filtered air and Freon 12 as reference sources. The amount of light scattering is proportional to the mass concentration (μg./m.^{3}) of suspended particulates if the mass and size distribution of the particles is assumed constant, a good assumption owing to the inherent size-limitation constraints for well-mixed, aged aerosols. Measurements of suspended particulates with this device afford the user a continuous measurement in contrast with high volume sampling and weighing techniques which are time consuming and provide after-the-fact data.

Automated (Water): A general discussion of monitoring with respect to the determination of trace metal pollutants in water resources and stream sediment has been presented by C.R. Cothern of the University of Dayton (3). An excellent summary volume on instrumental analysis for water pollution control is that by Mancy (35b).

Prior to 1940, most metals were analzyed using volumetric or gravimetric procedures (35c). Since then, instrumental methods have achieved greater prominence because of increased sensitivity, selectivity, capability of simultaneous analysis, amenability of automation, and eventual lower cost per sample due, i.e., to a decrease in turnaround time.

Instrumentation for monitoring metals in water is classified as either laboratory or field types. Manual samplers are used to collect samples which are later analyzed in the laboratory, while automatic samplers are able to collect and analyze the sample continuously on-site. Instrumentation may be classified further as either selective or simultaneous metal analyzers. Selective analyzers determine one metal at a time, although they may have multielement capability. Simultaneous analzyers measure more than one metal in the same sample at the same time. Selective and simultaneous analzyers may be either manual or automatic sampling.

Many instrumental methods are available for measuring the metal content of water. Some of them are summarized in order of increasing sensitivity in Table 9. In principle, all these methods can be used on a particular sample by bringing the metal concentration within the sensitivity limits by appropriate pretreatments such as dilution or precipitation. Undesirable features of pretreatment include increased likelihood of analyst error, metal loss or sample contamination, longer analysis time and need for highly purified chemical reagents. For any method, the sensitivity will vary according to the particular metal.

TABLE 9: SENSITIVITIES OF INSTRUMENTAL METHODS FOR MONITORING METALS IN WATER

Instrumental Method	Sensitivity,ppb (μg/l)
Polarography	100 - 1000
Molecular Absorption	100 - 1000
Atomic Absorption (Flame)	100 - 1000
Atomic Emission	100 - 1000
Square Wave Polarography	10 - 100
Sweep Voltammetry	10 - 100
Emission Spectroscopy (Arc)	10 - 100
X-Ray Fluorescence	10 - 100
Pulse Polarography	1 - 10
Molecular Fluorescence	1 - 10
Spark Source Mass Spectrometry	1 - 10
Anodic Stripping	0.1 - 10
Neutron Activation	0.1 - 10
Atomic Absorption (Flameless)	0.1 - 10
Chemiluminescence	0.1 - 10
Emission Spectroscopy (Plasma)	0.1 - 10

Source: K.H. Mancy, *Instrumental Analysis for Water Pollution Control*, (1971)

Simultaneous analzyers are capable of measuring the concentrations of more than one metal in a sample during a single analysis. Most analyses using simultaneous analyzers are performed in the laboratory on manually obtained water samples.

Instrumental methods include emission spectroscopy, anodic stripping, x-ray fluorescence, activation analysis and spark source mass spectrometry. Plasma excitation and differential pulsed anodic stripping may be applied directly to dissolved metal analysis in fresh water. Neither method has been evaluated to the same extent as molecular absorption or atomic absorption methods; it is recommended that this be done. Differential pulsed anodic stripping is limited to some mercury soluble metals (e.g., Cd, Pb, Tl, In, Cu, and Bi). It does not have the wide range of multielement capability as emission spectroscopy. X-ray fluorescence methods should be field tested and evaluated for application to measuring metal concentrations in particulates, bottom sediments and sludges. The technique requires minimal sample handling and is essentially nondestructive so that the sample can be retained for subsequent testing.

Atomic absorption is by far the most widely used technique for measuring for metals in water analyses because it combines high specificity with high sensitivity. The method is applicable to most metals and some nonmetals at the ppb level, and can be used to measure some trace metals in fresh water without a sample preconcentration requirement. Atomic absorption measures one metal at a time. This is a disadvantage if several metals are to be determined because additional hollow cathode lamps must be purchased, one for each metal.

Atoms are capable of abosrbing light in the ultraviolet-visible region by interacting with a photon to promote an electronic transition from the ground state to one of the excited states of the atom. The physical laws which govern the relationship between the amount of light absorbed and the concentration of the absorbing atoms follows an exponential law. Over a concentration range, the variation in absorbance (logarithm of the ratio of the intensity of incident-to-transmitted radiation) is linear with metal concentration. This forms the basis for quantitative application of atomic absorption to trace metals analysis. Free atoms are necessary for atomic absorption; these are obtained only in the gas phase. The first requisite is therefore gasification of the metal from the sample matrix. Methods used to accomplish gasification include aspiration of the solution of the metal into a flame, use of a graphite furnace, and chemical reduction in solution followed by air bubbling.

Flame atomic absorption has been the usual gasification technique. Air-acetylene and nitrous oxide-acetylene mixtures are commonly used as the flame sources. The dissolved metal is aspirated as a fine mist into the flame wherein it undergoes a series of reactions eventuating in formation of free atoms. The flame method is very inefficient and wasteful of sample; losses occur because larger drops in the mist are not aspirated into the flame, or do not react as completely as smaller drops to form atoms because of the short residence time in the flame. Sensitivities vary according to the metal and fall usually in the 5 to 2,000 ppb range; a preconcentration step is needed for measurement of metals such as Hg, V, and Sn in natural waters. It is common practice to extract the metal into an organic layer of less volume to obtain a concentration enhancement.

An ultrasonic nebulizer increases the sensitivity by reducing the size of the droplets which form the aspirated mist. Benefits of smaller drop size include: (1) increased delivery efficiency through a decrease in condensation probability, (2) increased atomization efficiency because less time is required for evaporation of the droplet. An enhanced sensitivity for Ag, Cd, Mg, and Zn was found for an ultrasonic nebulizer. Detection limit of atomic absorption may be extended by a factor of 10 to 1,000 using a graphite furnace flameless technique. Direct analysis of water and biological samples is feasible, and premeasurement concentration is avoided so that procedures can be much simplified. A graphite oven is shown in Figure 12.

FIGURE 12: CROSS-SECTION OF THE PERKIN-ELMER HGA-2000

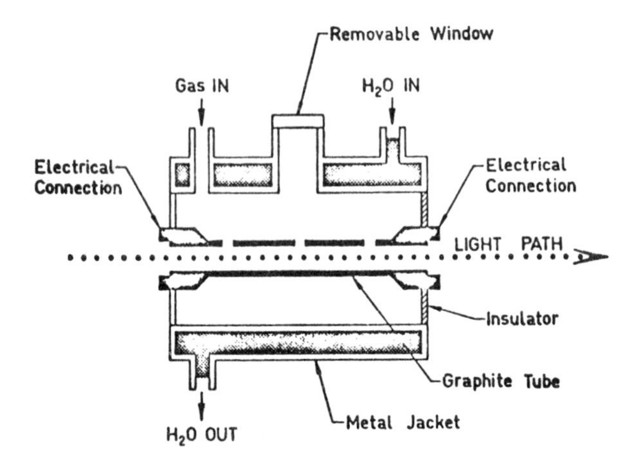

Source: Lawrence Berkeley Laboratory, *Instrumentation for Environmental Monitoring – Water,* (Feb. 1, 1973)

A third innovation in atomic absorption is the cold vapor flameless technique which is widely used to monitor mercury levels in water. Essentially, the system utilizes chemical reactions which convert all forms of mercury into mercuric form which is subsequently reduced to mercury metal. The mercury is swept out of the aqueous phase as the vapor by a gas stream and into the cell where measurement takes place. The method is applicable to Hg analyses at the sub-ppb level. Vapor methods have been similarly applied to As, Se and Sb by chemically converting each to the corresponding volatile hydrides. A gas stream subsequently sweeps the vapor into $Ar\text{-}H_2$ entrained air flame where the concentration is measured. Sensitivities for these vapor techniques are given below.

Element	Sensitivity, g.
Hg	2×10^{-9}
As	4×10^{-9}
Sb	5×10^{-8}
Se	5×10^{-9}

Atomic spectra have very narrow profiles so that lines are only about 0.2 nm. wide. Because monochromatic light from a conventional monochromator such as a hydrogen lamp is too wide for atomic absorption, no diminution in the incident light will be observed by the detector. Thus, hollow cathode tubes (composed of the analyzed metal) are used as the source of monochromatic radiation. In addition to the lines emitted by the metal vapor, there is a continuum from the inert carrier gas (Ar, He) which could be focussed on the exit slit/detector and swamp out the desired signal increase. A monochromator is placed in the system to disperse unwanted lines so that they do not fall on the detector. A monochromator is often a grating, mounted in the Czerny-Turner mode. A schematic is shown in Figure 13.

FIGURE 13: SCHEMATIC OF ATOMIC ABSORPTION FLAME PHOTOMETER

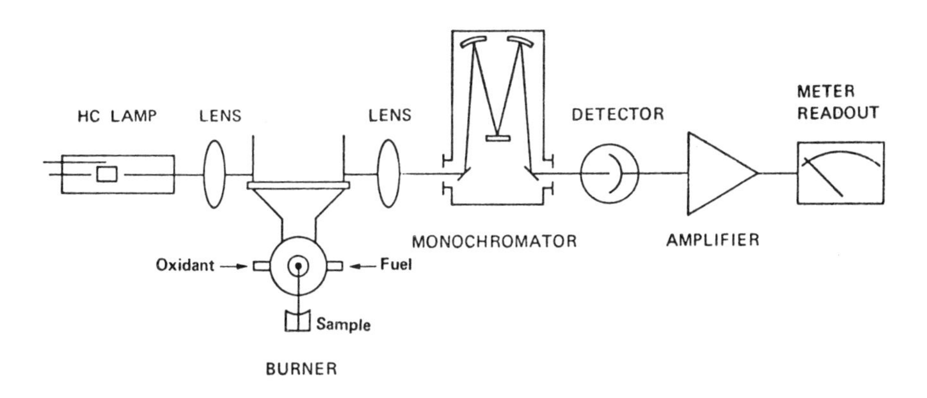

Source: Lawrence Berkeley Laboratory, *Instrumentation for Environmental Monitoring – Water,* (Feb. 1, 1973)

Molecular absorption (colorimetric) methods have been by and large replaced by atomic absorption techniques for analysis of metals in water. Sensitivity is in the ppb to ppm range. This can be extended to lower levels by a solvent extraction technique which serves to enhance the concentration by extracting into a smaller volume than the original sample. The excitation source is usually a tungsten lamp; a specified band width is selected by passing the light through a prism or grating monochromator. Detectors include phototube or photomultiplier tube.

Molecular absorption methods for metals usually involve forming colored species which absorb in the visible or near ultraviolet range of the spectrum. The colored metal ions or molecules absorb light of a specified bandwidth. The amount of light absorbed is compared with a previously obtained calibration plot, and is related to the metal concentration by the calibration data. For optimum color development, attention should be given to maintaining the correct pH, removal of interference, adjustment of oxidation state, masking interfering elements, or solvent extraction techniques. Usually, samples should be measured within a specified time to minimize problems associated with color fading. Molecular absorption methods are used in commercially available continuous analyzers such as the Hach analyzers for Si, Cu, Fe and hexavalent Cr. Arsenic is measured in the EPA method by evolution as arsine gas in a generator, followed by reaction with silver diethyldithiocarbamate to form a red complex. The detection limit is 10 ppb, as can be measured in most fresh waters where its mean value is 64 ppb. Selenium is also measured by molecular absorption in the EPA method, where selenite is reacted with diaminobenzidine to form the colored compound. The method was applied to synthetic samples containing 20 ppb Se.

In general, molecular absorption methods for metals in water lack specificity and sensitivity, and are time-consuming. To obtain suitable specificity solvent extraction techniques are commonly used which may involve two separate extractions. Sensitivity is obtained by extracting the metal from a large volume sample into a smaller volume. Extraction techniques involve use of a certain amount of laboratory glassware and several reagents. To minimize sample loss or contamination by handling, all glassware must be scrupulously clean, and all reagents of high purity. The main attraction of molecular absorption techniques for metals in water include the relatively low costs (less than $1,000 for some instruments), and the data available on reliability of the methods.

The increased emphasis on water pollution control as pointed out by Cothern (3) indicates the need for improved methods of analyzing water samples (36). Although a detailed discussion of all the analytical methods for determining the presence of trace metals in aquatic samples cannot be presented here, a few comments will be made regarding their relative merits. For further information see review articles by McCrone (37) and Woldseth et al (38).

The analytical tools with trace metal sensitivities comparable to or better than that of x-ray fluorescence (XRF) include atomic absorption (AA), neutron activation analysis (NAA), and emission spectroscopy (ES). In some special cases, e.g., using AA to measure the presence of Hg or using NAA to measure the presence of Si, a particular method is superior; but, in general all these methods are capable of measurement in the parts per million and in some cases the sub-parts per million range. Table 10 indicates some of the relative advantages of these methods.

TABLE 10: SOME ANALYTICAL METHODS FOR TRACE METAL DETERMINATION

Method	Data Analysis Simple	Nondestructive	Detect all Metals Simultaneously	Portability	Automated
Atomic absorption	X				X
Neutron activation analysis		X	X	X	
Optical emission spectroscopy	X				
X-ray fluorescence	X	X	X	X	X

Source: Report PB 213,369

Although none of these methods are portable at present, the future holds promise for XRF and NAA (it should be noted that portable XRF instruments are available now for determination of a single element) (39). NAA presently requires a large nuclear reactor (the minimum flux for best overall performance is around 10^{12} neutrons/sec./cm.^{2}) but the future availability of ^{252}Cf (40) should provide adequate portable activation sources. It is not unreasonable to expect that present solid state x-ray detectors used in XRF will become portable in the future (41). The only present difficulty in making these units portable is that they require cooling to liquid nitrogen temperature. A portable unit would weigh only 10 to 20 lbs. (42) and could rapidly determine in the field the presence of many metals in water samples.

Data analysis for XRF is less complicated than analyzing the gamma ray spectrum of NAA. The NAA energy spectrum contains ten or more peaks per element. Also the identification of elements in NAA spectra requires both intensity and half-life information. This requires the resolving and analysis of several complex spectra.

Only two of the above methods (XRF and AA) are instrumented in such a way that they can be automated. In the AA method the samples to be tested are poured into sample tubes and placed in a large rack, containing as many as 200 tubes. In XRF the filter paper on which the sample is collected is placed in a holder for immediate analysis. The complete analysis in both cases can be computerized. If the samples have to be shipped over large distances, the paper filters used in XRF have obvious advantages over the liquids used in AA.

One of the major problems with analysis of trace metals at present is the nonuniformity of results from both different techniques and different laboratories. In an effort to standardize the techniques and laboratories, the National Bureau of Standards is providing standardized materials through its Office of Reference Std. Matls. At present they provide orchard leaves which contain several trace metals of known quantities. In the near future, they will also provide samples of beef livers. The amount of trace metals in these standards are determined by at least three independent analytical techniques. In the need for independent methods x-ray fluorescence is a very valuable tool, since for many metals finding as many as three independent analytical methods is difficult.

That x-ray might be used in analytical techniques for the identification of elements was first discussed by Moseley in 1913. He determined that the x-rays given off from the target material in an x-ray tube were composed of several wavelengths known as the K and L series. He also showed that the wavelengths of each series were directly related to the atomic number of the element, the square root of the frequency increasing by an equal amount for each change of atomic number. The first general work that appeared concerning x-ray emission spectrography was *Chemical Analysis by X-rays and its Applications,* by Von Hevesy which appeared in 1932. During the following decade, x-ray emission spectrography began to be used for chemical analysis, usually when the spectra from emission spectrographs were too complex to analyze.

However, for a number of reasons, x-ray emission spectrography was not widely used; viz., x-rays were known to be hazardous, but these hazards were not well understood (as many of those who lost parts of their bodies can well testify), the apparatus was quite costly, the sample had to be somehow attached to the target in the tube and the heat generated there often destroyed the sample, the efficiency for excitation was very small compared to the broad background produced from the target and the analysis equipment was far from being very sensitive. Almost all analyses during this period were qualitative. One of the largest achievements of x-ray emission spectrography during this period was the discovery of hafnium by Von Hevesy and Coster by measurement of its x-ray emission spectrum.

The above method creates x-rays by electron bombardment in the x-ray tube itself. Some of the problems inherent in this approach can be eliminated by a modification. The x-rays produced by the x-ray tube can be used to excite the atoms whose characteristic x-ray energies are lower in energy than the energy of the exciting radiation. This process is similar to optical fluorescence and the x-rays produced are identical to those produced by electron bombardment. This secondary process is called x-ray fluorescence (XRF). X-ray fluorescence is superior to the earlier technique of x-ray emission in identifying elements because the background continuum is considerably reduced. In fact the background in XRF is due only to the backing material and the thickness of the sample. Thus it is important to prepare samples which are as thin as possible and whose backing material is as thin as possible.

Prior to 1945, XRF was not very widely used because it required a very high power x-ray tube and the analyzing techniques were not reliable enough. These techniques involved using a photographic plate or geiger tube to record the x-ray output as a function of the

Bragg angle of the analyzing crystal. After 1945, instrumentation became available which made XRF practical. In 1948 Friedman and Birks described a plane crystal x-ray spectrograph suitable for routine use. By 1954 the General Electric Company had produced a focusing spectrograph with a bent mica crystal for use in XRF. However, it was not until 1956 through the efforts of the Dutch Phillips Company that XRF began to be used routinely in analytical laboratories. In general, these devices were costly, bulky and by no means portable as noted by Cothern (3).

Two recent developments have greatly simplified XRF and may eventually lead to reasonably low cost instruments which are portable. The first development occurred shortly after World War II. The existence of large nuclear reactors made possible the fabrication of very intense radioactive sources which could be used to excite the x-ray fluorescence in place of the bulky x-ray tubes. At present, sources as large as 25 mc. are commercially available which are the size of a pin head and are relatively inexpensive (lower in cost than x-ray tubes). Although artificially produced radioactive sources have been available for several years they were not extensively used in XRF until a second new development.

In 1964, solid state detectors were developed by drifting lithium into germanium and silicon which revolutionized the spectrographic measurements of gamma rays and x-rays. These detectors were capable of resolutions roughly an order of magnitude better than the best of the previous detectors, namely the scintillation detectors (such as NaI and CsI). This development provided for the first time a multichannel detector capable of resolving x-rays for a large number of elements.

Probably the most important single part of the equipment for XRF is the solid state Si(Li) detector. The detector used in this study had an active area of 80 $mm.^2$ and a depth of 3 mm. Its operating voltage was 1,000 V. Since the detector was cooled to 77°K. by liquid nitrogen, it was enclosed in a vacuum cryostat to eliminate condensation (see Figure 14). The beryllium window on the front of the cryostat was 0.25 mm. thick and the gold dead layer on the front of the detector was 100 to 200 A. thick (0 to 2). An incident radiation penetrates this dead layer and creates electron-hole pairs (one pair per 2.9 eV) in the charge-free active area (A) or depletion region. These pairs are then swept away by means of the

FIGURE 14: GEOMETRICAL ARRANGEMENT OF EXCITING SOURCE, SAMPLE AND DETECTOR IN X-RAY FLUORESCENCE APPARATUS

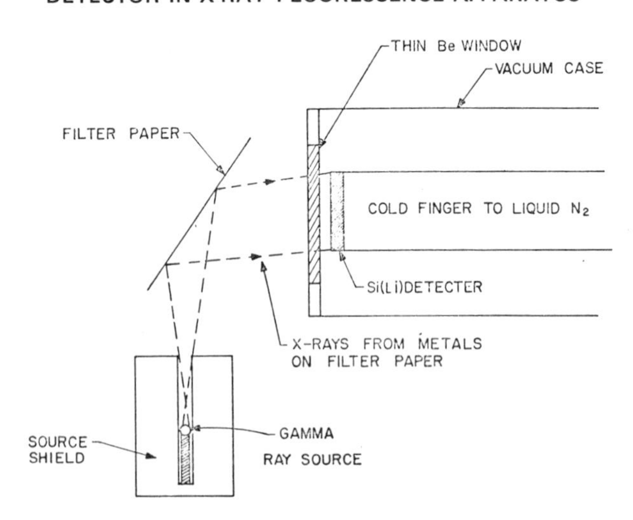

Source: Report PB 213,369, 1972

potential difference obtained through the applied bias voltage; hence, an amount of charge that is proportional to the energy deposited in the crystal is received at the positive bias lead. Some of the advantages of XRF compared to other techniques for trace metal determinations are; the method is nondestructive, that is, the sample is unaltered by the examination performed; the data analysis is simple and can be easily computerized; in addition the technique can be automated removing the need for a full-time technician to run the tests; this method determines the presence of a wide range of metals simultaneously and the method is potentially portable, that is in the near future field units weighing 30 lbs. can be constructed.

In a symposium on Traces of Heavy Metals in Water at Princeton University on Nov. 15 to 16, 1973, the various speakers compared monitoring techniques in a somewhat different way than in Table 9 as follows:

Technique	Pros	Cons
Flame AA	Fast, precise, and cheap.	Can not do many elements simultaneously.
Flameless AA	Much more sensitive than flame AA.	Slower than flame; limited to single elements. Interferences more of a problem.
Neutron activation	Sensitive, can do many elements simultaneously.	Most expensive.
Polarographic	Sensitive, precise, low in cost, and can do many elements at once.	Tedious, may give problems with solution equilibria.

At the same meeting data were presented on the detection limits of various techniques for various elements as shown in Table 11.

TABLE 11: DETECTION LIMITS IN PARTS PER BILLION FOR VARIOUS ANALYTICAL TECHNIQUES

	Flame AA				Neutron	
	Varian	Perkin-Elmer	EPA Data	Flameless AA	Activation	Polarographic
Al	--	40	--	--	--	--
Sb	--	60	--	--	--	--
As	250	600	50	20	5	20
Ba	--	60	--	--	--	--
Be	--	1	--	--	--	--
B	--	2,600	--	--	--	--
Cd	0.6	2	1	0.02	200	5
Ca	--	0.4	--	--	--	--
Cr	--	7	--	--	--	--
Co	--	16	--	--	--	--
Cu	3	3	5	1.4	4	5
Fe	5	3	4	0.6	50,000	100
Pb	20	2.5	10	1	40,000	5
Mg	--	0.3	--	--	--	--
Mn	3	3	5	0.1	0.1	100
Hg	200	--	--	20	5	--
Mo	--	14	--	--	--	--
Ni	--	6	--	--	--	--
Se	480	600	--	20	400	10
Si	--	80	--	--	--	--
Ag	3	2	10	0.04	2	0.1
Sn	30	120	--	12	10,000	5
Ti	--	60	--	--	--	--
V	--	2	--	--	--	--
Zn	2	2	5	0.016	400	5

Source: Conference on Traces of Heavy Metals in Water (Nov. 1973)

A device developed by F.J. Bogusz, (43) provides a water pollution monitoring system which may be coupled with data handling and processing equipment while the water pollution monitoring system is functioning. Varying levels of pollutants are detected by means of a variety of sensors and an electrical signal corresponding to the pollutant level is processed for recording and/or transmission. Means are provided for automatic periodic cleaning of the sensors so that a high level of accuracy can be maintained for an extended period of time. Additional means are provided for data reduction within the system to reduce massive volumes of detected information to that which is specifically of interest to the equipment operator. Still further means are provided for controlled suspension of the system in aqueous media to maximize system operation.

A further discussion of automated techniques for water monitoring has been presented by K.H. Mancy of the School of Public Health at the University of Michigan (9). Automated surveillance systems rely on the use of automatic, unattended measurement procedures operated on a continuous basis or intermittently at a determined frequency as pointed out by Mancy (9). Automatic water quality monitoring systems are being widely used in this country and abroad in a variety of water quality management programs. The main advantages in using automated systems lie in the ability of maintaining a continuous, instantaneous record on water quality in rivers, lakes and estuaries, industrial effluents, etc. Changes in water quality may occur suddenly as a result of intense storms, industrial spills, etc. In certain cases changes in water quality undergo rapid variation such as dissolved oxygen and salinity changes from natural tidal fluctuations.

Manual surveillance techniques seem to be inadequate and certainly not practical to monitor sudden changes in water quality. The main objectives of automatic water quality monitoring activity are to (a) detect violations of standards or other undesirable quality conditions so that remedial action can be taken quickly, (b) establish the water quality baseline and trends, (c) verify and provide data for predictive calculations on water quality including computerized mathematical models for the purpose of establishing design and operational criteria for water pollution control facilities, and (d) determine short-range water quality trends for program management and public information purposes. Selection of monitoring sites is usually made at places where (a) rapid fluctuations in water quality or quantity, (b) high potential for accidental spills of wastes, and (c) hydrologic or other conditions which permit continuous evaluation of the quality response of the stream system to waste discharges and other pollutional impacts occur.

There are two main approaches for water quality measurement in automatic monitoring systems. The first approach is based on in situ measurement by means of electrochemical transducers without altering the physicochemical characteristics of the test solution. The second approach is based on the use of automated repetitive wet analytical procedures in which chemicals are added to the test solution and measurements are made after suitable chemical reactions have taken place, e.g., the Technicon AutoAnalyzer.

Monitoring systems in which the measurement is done by transducer systems (without the addition of reagents) are being used widely for monitoring water quality of rivers, streams, estuaries and industrial and domestic waste effluents. These systems are usually capable of measuring temperature, electrical conductivity, dissolved oxygen, turbidity, pH, sunlight intensity, chlorides, oxidation-reduction potential (ORP) and alpha and beta radioactivity. Certain parameters are measured for specific applications, e.g., the monitoring of fluoride ions, residual chlorine, nitrates and hardness in drinking water supplies and monitoring cyanides, sulfides, copper ions in certain waste effluents. Electrochemical sensors used for water quality monitoring systems can be categorized based on the type of measurement, i.e., (a) conductometric, (b) potentiometric and (c) voltammetric sensors.

The second main type of monitoring systems is based on automating wet analytical techniques, e.g., the Technicon AutoAnalyzer. Such systems are essentially capable of automatic sampling, filtering, diluting, reagent addition, mixing, heating, digesting, and after appropriate delay for color development, colorimetric measurements are done. All these steps, which are usually done by an analyst, are automated and performed on a stream of samples moved by a fixed speed peristaltic pump. This technique finds its widest application in laboratory operations where large samples of waters are handled daily. There have been certain attempts however to use this technique for water quality monitoring where the AutoAnalyzer is kept in a trailer on a river bank (44). In other applications autoanalyzers were used for monitoring silicates and nitrates in seawater off the coast of California (45), and for monitoring nutrient concentrations in Lake Erie (46).

One of the main advantages of applying the AutoAnalyzer for water quality monitoring purposes lies in its ability for measurement of a large number of parameters. Table 12 shows a listing of water quality parameters that can be measured by such systems (44). The application of the AutoAnalyzer for water quality monitoring purposes has been faced with a number of limitations; (a) the effect of turbidity on colorimetric measurements, (b) these systems need more frequent attendance for servicing, and (c) AutoAnalyzer systems seem to have a higher capital cost and running expenses in comparison to the transducer-type monitoring systems. The parameters measured in the following table are from data from Technicon Corp., Tarrytown, N.Y. (1969).

TABLE 12: PARAMETERS MEASURED BY THE TECHNICON CSM 6

Parameter	Nominal Range* (ppm)	Detection Limit (ppm)
Phosphate	0 to 8	0.08
Chromium (hexavalent)	0 to 5	0.05
Copper	0 to 10	0.10
Iron	0 to 10	0.10
Ammonia	0 to 10	0.10
Methyl orange alkalinity**	0 to 500	5.0
Thymol blue alkalinity**	0 to 100	1.0
Hardness**	0 to 300	3.0
Sulfate***	0 to 500	5.0
Phenol	0 to 5	0.05
Cyanide**	0 to 3	0.03
Chemical oxygen demand	0 to 500	5.0
Chloride**	0 to 10	0.10
Nitrite	0 to 1.5	0.015
Nitrite + nitrate	0 to 2.0	0.02
Fluoride**	0 to 2.5	0.025
Orthophosphate	0 to 10	0.10
Silicate	0 to 15	0.15

*Alternate ranges optional.
**Deviates from linearity.
***Sulfate is only available on continuous monitoring applications.

Source: Technicon Corporation (1969)

Water quality monitoring should be carried out with specific objectives clearly in mind as pointed out by Sayers (47). Basic monitoring objectives include:

(1) the establishment of baseline water quality and identification of short- or long-term trends;
(2) evaluation of compliance or noncompliance with water quality standards;
(3) development of mathematical models for forecasting water quality under a variety of waste loadings and hydrological conditions; and
(4) obtaining data for input to existing models used in the day-to-day management of a water resource.

Monitoring carried out for the purpose of establishing baseline water quality and making trend evaluations will generally also serve to detect the emergence of adverse conditions before actual water quality standard violations occur. This permits action to be taken early to prevent a standards violation rather than later when the costs are greater. Thus, monitoring makes it possible for the water quality manager to spend his limited resources on ounces of prevention before damages occur rather than on pounds of cure after the damage is done. This is the essence of a true water quality management program. Maintaining compliance with water quality standards will require not only the existence of adequate waste treatment and control facilities but also the operation of these facilities day after day at or above design efficiency. A monitoring program sufficient to detect serious temporary standards violations will serve to discourage intentional waste bypassing or shoddy operation of the treatment and control facilities. Such a monitoring program is essential in achieving full implementation of the State-Federal standards developed over the past several years.

The huge expenditure required for waste treatment facilities across the nation in order to achieve and maintain compliance with water quality standards and the limited funds available for this endeavor demand that the funds be utilized in the most cost-effective manner. Accordingly, wastewater management planning efforts must be sufficient to identify maximum permissible waste loads and specific treatment requirements in successive reaches along each water course. Plans must also identify where joint treatment of wastes from several communities is less costly than individual treatment by each community. These plans must identify not only current treatment needs but more importantly, the treatment requirements of each community or region at various dates in the future. This information will serve to ensure that all treatment facilities are constructed and in operation by the time they are needed.

Preparation of these water quality management plans requires the development and use of mathematical models relating waste discharge loadings to water quality in the receiving body of water. The development of a mathematical model, in turn, requires sufficient monitoring to completely characterize water quality in the receiving body under a steady-state waste loading and hydrological condition.

In areas of dense population, complex waste sources, intensive water use, and close hydrologic interrelationships, mathematical models may well be used by basin managers to identify day-to-day adjustments needed in reservoir releases, storage and releases of treated effluents, waste treatment operating efficiencies, and other control measures in order to maintain optimal water quality conditions. Such a management system will require continuous monitoring at key points in the basin in order to provide the necessary model input data.

In the design of a monitoring program, the specific objectives to be served must be clearly in mind from the very beginning. Otherwise, the effort will probably not result in an efficient and meaningful program. The length of the survey, station selection, parameter coverage, and sampling frequency all depend on the specific objectives and the time-frame in which the objectives must be achieved. These variables also have a very significant impact on monitoring costs.

Monitoring for the purpose of establishing baseline water quality can generally be accomplished by intensive sampling over a one to two week period during each season of interest. If results are not needed for several years and the quality is not expected to change significantly during that period (e.g., in a wild river), then objectives can also be met by sampling at rather infrequent intervals over the entire period.

Monitoring for the purpose of mathematical model development requires intensive sampling of the waste sources and the receiving waters over a short time period. In that steady-state conditions are assumed to exist during a given sampling phase, it must be short enough so that variations in waste discharges and stream flow are minimal throughout the period. Monitoring for the purpose of long-term trend identification, evaluation of standards compliance, or river basin management must, of course, be a continuing program without end.

The number and spacing of monitoring stations also vary considerably, depending on the objectives to be achieved. Baseline water quality evaluation can be achieved with a minimal number of stations providing sparse coverage of a vast area. Mathematical model development, on the other hand, requires a relatively large number of stations giving complete coverage of a small area. The parameter coverage provided, likewise, varies with objectives. Baseline water quality and trend evaluations require measurement of all indices that are likely to be of interest in the future. Model development, conversely, requires evaluation of only those parameters being modeled plus the independent variables that influence the parameters of interest.

The sampling frequency necessary to characterize a body of water is related to the periods of cyclic phenomena which control quality and to random influence, mostly associated with meteorologic and hydrologic events. The number of time-dependent factors which must be considered depends on the type of water body in question, the wastes being discharged, and the particular parameter of interest. Generally, temporal variations in water quality are least frequent in large impoundments and most frequent in rivers.

The actual number of samples required over a given time period at each station to characterize water quality, in essence, depends on the variability of the parameters of interest, regardless of the number and kinds of influencing factors. In a statistical sense, there is no answer to the question of how many samples are needed, without foreknowledge of the variability of the parameter to be sampled and the precision desired. It can only be said that enough samples should be collected to define, at a specific level of significance, the water quality response caused by the imposed influencing factors (e.g., pollutant load, manipulation of stream flow, tidal conditions, and sunlight). The adequacy of the number of samples taken in a given period can be judged by the magnitude of the 95% confidence interval.

Often, too many stations are established with an insufficient number of samples collected at any one. A fewer number of stations with a sufficient number of samples to define results in terms of statistical significance is much more reliable than many stations with only a few samples at each. Sampling (for the purpose of describing a given condition) should not be spread over a long time interval during which the receiving water regimen is subject to a wide variety of conditions.

Attempts to assess all conditions by aimless sampling usually defines no condition and, in fact, may be misleading. Averaging of such noncomparable results is often a risky procedure. Instead, the sampling program should be designed so that each condition of interest is defined rather than attempting to define all the conditions (resulting from hydrologic, hydraulic, or hydrodynamic variations) by averaging them together.

Another factor which influences the number of samples required in a given time period is the precision of the test for the given parameter being measured. If, for instance, mean temperature and coliform density (MPN) were to be evaluated under steady-state conditions at some point in a small well-mixed stream, a greater number of coliform samples (minimum of 16) than temperature samples (minimum of 3) would be required for an equal degree of reliability in the results. This is because the MPN test procedure is subject to a much wider variation than the temperature test. Hence, when several parameters are being measured at a given site, the number of samples collected should be sufficient to satisfactorily evaluate that parameter tested by the least precise method.

Given a large number of samples collected at short time intervals, the results obtained can be analyzed to determine the optimum frequency for meeting a given set of objectives. Within the realm of practicability, however, this is seldom possible. Judgement must be used in arriving at a first approximation of the optimum frequency. Where the objective is to determine trends in water quality on a wild river, a sampling frequency that will require several years in order to accumulate sufficient data to make an evaluation may be satisfactory.

On the other hand, where the objective is to spot violations of standards in waters of marginal quality, a much greater frequency would be required, one that permits the accumulation of sufficient data to draw conclusions in a matter of hours or days, depending on the value of the resource affected. Evaluations to be made with the water quality data collected are generally not possible without some additional data on hydrological, hydrographic, and meteorological conditions during the period of the monitoring survey. If, for example, the evaluation is of a stream, information is needed on the stream discharge rate during the sampling period and the historical flows.

When the objective includes development of mathematical models for establishing cause-effect relationships, information on times-of-flow at a range of stream discharge rates is essential. This requires the measurement of velocities along a stream at a minimum of three different stream discharge rates. Dye tracers lend themselves very well to studies of this type. If reaeration rates (using the O'Connor-Dobbins formula) are to be included in a mathematical model, then stream depths must also be measured for a range of discharge rates. For most free-flowing streams, a plot on log-log paper of stream discharge rate versus velocity and depth will result in straight lines, thus permitting interpolation between actual measurements.

Studies of open waters (e.g., large impoundments, lakes, estuaries, and coastal waters) often require that several meteorological measurements be made. These include wind speed and direction, air temperature and humidity, solar radiation intensity, and tidal cycles. The meteorological parameters actually evaluated depend on the relationships that are to be established. If it is mixing and/or flow patterns, then wind speed and direction would be important; if it is the dissipation of heat resulting from a source of thermal pollution, then air temperature and humidity, in addition to wind speed, would be of interest.

Earliest water quality monitoring techniques employed the collection of grab samples using a bucket suspended from a rope with sample analysis carried out manually on the stream bank or in a nearby laboratory. This approach is still in wide use today. The reason for its continued use is not so much due to its outstanding advantages as it is to the fact that there is no alternative technique available for achieving the same objectives.

Early attempts to improve on the state-of-the-art of water quality monitoring centered largely around automation of the sample collection and laboratory analysis phases. There are numerous automatic sample collection devices on the market that can be timed to obtain discrete grab samples at predetermined intervals and to even composite the samples in proportion to wastewater flow rates. At the end of the sampling period, the station must be visited, the sample or samples picked up and then hand carried to the laboratory for analysis. These samplers are designed primarily for use on effluent lines rather than in open waters, however.

Many of the laboratory analyses that were formerly carried out manually have now been automated. Newer laboratory instrument designs have incorporated automatic sample handling, sequential analysis, and improved data presentation. Many instruments are capable of withdrawing an aliquot of sample, performing the measurement, and presenting the data as concentration of the constituent, in printed form. Connection of analytical instruments directly to a computer to eliminate all manual operation from sample aliquot withdrawal to final data storage is possible in some situations. This approach will very likely be employed to a greater and greater degree in the future. Some of the newer instruments found in routine use today in many laboratories include:

- Automatic Titrators
- Technicon AutoAnalyzers
- Total Carbon Analyzers
- Specific Ion Meters
- Gas Chromatographs
- Atomic Absorption Spectrophotometers
- Infrared Spectrophotometers

EPA and its predecessor agencies concerned with water analysis have long recognized the need for automatic field instrumentation capable of giving a direct and instantaneous readout of the levels of specific water quality indices. Work was begun in 1953 by the former U.S. Public Health Service Division of Water Supply and Pollution in cooperation with the Hays Corporation on the development of such a device for the continuous measurement of dissolved oxygen. Other work of a similar nature soon followed. Efforts centered on the development of sensing devices utilizing a direct electrical response mechanism. This work led to the preparation in November 1962 by A.F. Mentink of the first set of Public Health Service specifications for an integrated multiparameter water quality data acquisition system. Mr. Mentink of the Analytical Quality Control Laboratory in Cincinnati has now prepared the ninth edition of these specifications.

Of those parameters of general interest in water quality surveys, to date, eight of them can be reliably measured by continuously operated electronic sensors placed in the field. These are pH, temperature, dissolved oxygen, specific conductivity, chloride, turbidity, oxidation-reduction potential, and solar radiation intensity. Results can be recorded on site or telemetered to a central location. Many portable single parameter electronic sensors are also presently on the market. They are of considerable value in the conduct of short-term intensive surveys and for reconnaissance purposes.

Electronic monitors and aerial remote sensing techniques significantly increase the technical capability to monitor surface waters on a continuous or near-continuous basis. However, they are not a panacea. In terms of evaluating compliance with water quality standards, only one parameter of interest, temperature, can presently be measured quantitatively using remote or spectral sensing techniques. Electronic sensors provide a little more versatility in that they are capable of measuring six parameters likely to be found listed in water quality standards. This leaves 50 or so parameters contained in or alluded to by the standards for which no alternative for evaluation other than manual sample collection and separate laboratory analysis exists.

For those parameters where there is a choice between manual and automated monitoring techniques, the decision on which approach to pursue should be based on economic considerations. Once the number and location of stations, the parameter coverage, and the minimum necessary sampling frequency have all been identified, the cost of each alternative approach can then be determined. The least costly alternative should, of course, be the one implemented. When comparing the flexibility and relative costs of manual sampling and analysis versus electronic monitoring, the following points should be considered:

(1) Electronic monitors do not completely replace field personnel. Each monitoring site should be visited at least weekly for routine maintenance and recalibration. Equipment malfunctions may require more frequent visitations. Higher salaried personnel may be required for monitor maintenance than for manual sample collection.
(2) Automated monitoring system data handling costs will vary, depending on whether analog or digital output is used. In either case costs will generally be greater than if manual techniques were used. The type of output selected depends on the evaluation procedure and ultimate use of the data which must be fully considered in the design phase of the sampling program.

Generally the most important single factor in determining whether manual or automated techniques should be used is the monitoring frequency that is required. In the manual approach, monitoring costs vary almost directly with measurement frequency. At monitoring frequencies much greater than once per week, automated sensing will generally prove to be the more economical approach, provided that appropriate sensors exist.

Remote

Remote sensing is usually thought about in terms of remote sensing from satellites and/or aircraft and was the subject primarily covered by D.S. Barth and S.D. Shearer, Jr. of EPA (12). However, these authors said they would like to encourage those concerned to think of remote monitoring in a broader concept. They have listed four areas which the term remote sensing encompasses.

(a) Sensing of atmospheric pollutants from satellites and/or aircraft.

(b) Sensing of atmospheric pollutants from ground-based remote instruments with the telemetering of data to a central point.
(c) Sensing of emissions from stationary sources. Work currently underway in this area is being directed along three avenues:
 (1) sensing of emissions as they emerge from the stack,
 (2) sensing of emissions in the stack by outside-the-stack sensors, and
 (3) long-path sensing of emitted pollutants.
(d) Sensing of pollutant effects on ecology by remote means.

In addition, it is also necessary to carefully define and base remote sensing on needs and uses to which data obtained is to be put. Therefore, such sensing should probably be geared around the following purposes.

(a) Global and/or worldwide transport of pollutants.
(b) National pollutant trends and variations.
(c) Monitoring to determine changes in environmental pollution as a function of control measures and as a check on compliance with pollution control equipment.
(d) Monitoring to determine pollutant effects (human and ecological) in special localized areas.

The types, sophistication, and versatility of remote sensing instrumentation will vary widely depending on which of the above items are paramount. The sensing of air pollutants from aircraft and/or satellites pose many varied problems not encountered with traditional monitoring. The majority of techniques utilized to date with success have been restricted to short-range (several kilometers) sensing. Ludwig, Bartle, and Griggs (48) carried out a detailed study of several remote sensing techniques. In their study they considered in detail the applicability of the following instruments:

(a) The ordinary dispersion type spectrometers,
(b) filter wedge spectrometers,
(c) multiple detector spectrometers,
(d) scanning spectrometers,
(e) interferometer spectrometer (Michelson type),
(f) correlation spectrometer,
(g) matched filter spectrometer with multiple entrance slits,
(h) selective chopper radiometer, and
(i) the cross correlating spectrometer.

Table 13 presents a qualitative comparison of several spectroscopic instruments taken from the report by Ludwig. The conclusions of Ludwig, et al, regarding the detection of air pollutants by remote sensing from the ultraviolet (0.25 to 0.40 micron) to the microwave (greater than 500 micron) region were as follows:

(a) the ultraviolet visible region has more disadvantages,
(b) the infrared region (3.5 to 13μ) has more advantages than disadvantages,
(c) the microwave region (0.1 to 3 cm.) is not useful because of lack of specificity, and
(d) the quantitative measurements of aerosol content in the lower atmosphere probably cannot be made from a satellite.

Altshuller and McCormick (49) recently presented an article on remote sensing platforms for air pollution. This paper is an excellent review of the relative application of aircraft and satellites for remote sensing. Table 14 from their paper presents a summary of problem areas in remote sensing utilizing these two types of platforms.

"With so many limitations on satellite applications to air pollution measurement, it is clear that available resources should be used to exploit aircraft not satellite measurements. Measurements from aircraft can be made more rapidly, more effectively and at a lower overall cost for air pollution applications within urban areas or air quality regions."

TABLE 13: QUALITATIVE COMPARISON AMONG SPECTROSCOPIC INSTRUMENTS FOR THE DETECTION OF POLLUTANTS FROM SATELLITES

Instrument	Sensitivity	Specificity	Complexity Of Instrum.	Observ. Time	Information Content	Remarks
Radiometer (filter wheel or many single filters)	Medium	Low	Low	Short	Medium	Limit $\Delta\omega \sim 3\ cm^{-1}$; many channels required
Radiometer (polychromator)	Low Medium	High Low	Medium	Short Short	Medium	Assuming $\Delta\omega \sim < 1\ cm^{-1}$ Assuming $\Delta\omega > 1\ cm^{-1}$; Multiple detectors required
Scanning--Spectrometer	Low Medium	High Low	Medium	Long Medium	High	Assuming $\Delta\omega < cm^{-1}$ Assuming $\Delta\omega > cm^{-1}$
Optical Correlation Instrument	High	High	Medium	Short	Medium	Matched filter with multiple entrance slits and nondispersive
Interferometer--Spectrometer	High	High	High	Medium	High	Sampling time at least 10 sec; image motion compensation required; large information content for wide spectral interval

Source: EPA Workshop, Las Vegas, Nevada, November 30 to December 2, 1971

TABLE 14: REMOTE SENSOR SATELLITE/AIRCRAFT APPLICATIONS

PROBLEM AREA	SATELLITE APPLICATION	AIRCRAFT APPLICATION
Choice of instrument	Priority decision must be made well in advance	Various instruments may be flown separately or in combination as they are developed.
Instrument malfunction	If irreversible, complete failure results.	Instrument can be returned to ground for repair or modification.
Instrument	Appears limited to use infrared region.	Ultraviolet, visible, and infrared regions and point-sampling can be used.
Frequency of measurement	In a polar orbit a satellite is over a given urban area for only a very short time each day.	Repeared sweeps in a variety of patterns can be made to obtain diurnal variations in concentration.
Resolution	A 10 by 10 mile resolution is considered good, but even this resolution is marginal with respect to measurement output desired over urban areas. Resolution in third dimension or in time is unlikely or impossible.	Satisfactory resolution in time and space is possible.
Effect of temperature and pollutant profile	Quantitative measurements depend on concurrent measments of surface temperature, temperature profile, and estimate of pollutant profile	The two factors do not affect many types of measurements possible; temperature can be determined directly and concurrently without difficulty.
Clouds or overcast	Measurements lost. This is a very serious problem in terms of day by day measurement over specific urban areas.	Within energy limitations on instrument and flight minimums, measurements can be made below clouds or overcast.

Source: EPA Workshop, Las Vegas, Nevada, November 30 to December 2, 1971

TABLE 15: DETAILS ON DETECTION OF POLLUTANTS BY ABSORPTION OF LASER RADIATION

Pollutant	Source	Location of ir band centers (μ)	Laser line to be used	Concentration in parts per million to give 5% signal change over a 1 km path (estimated)	Remarks
CH_4	Marsh gas	3.35,7.7	He-Ne--3.39	0.03	Naturally in the air.
C_2H_2	Auto exhaust, combustion	13.7	Ne?	----	Very strong absorption
C_2H_4	Auto exhaust combustion	10.52	CO_2--10.53	0.02	Distinctive spectrum, strong Q-branch.
C_4H_{10}	Many, including auto exhaust	3.4	I_2---3.43	0.05	C-H band is a measure of total organics.
PAN	Atmospheric photochemistry	5.4,7.8	He-Ne?	0.05	Other bands are available
NO	Auto exhaust, combustion	5.3	I_2--5.5	0.5	Iodine line will penetrate humid atmosphere
NO_2	Auto exhaust combustion	3.4, 6.2	Rare gas?	----	A difficult case due to water interference.
CO	Auto exhaust, combustion	4.7	I_2 --4.86	2.0	Some other laser line could make detection more sensitive.
O_3	Atmospheric photochemistry	9.6	CO_2--9.5	0.05	One of the simplest cases.
SO_2	Burning of sulfur-cont. fuels	7.3,8.7, 18.5	Ne--7.4	0.03	Needs work to find better laser line.
NH_3	Organic wastes, industry	10.5	CO_2--10.7	0.02	Very strong lines at about 10.3 and 10.7
H_2S	Industry	2.6,7.7	Kr?	----	
HF	Burning of plastics	2.3	Kr?	----	
HCl	Burning of plastics	3.6	Xe?	----	
D_2O	Atomic energy installations	3.7	Xe	----	

Source: EPA Workshop, Las Vegas, Nevada, November 30 to December 2, 1971

Hanst (50) has presented a rather comprehensive discussion of infrared spectroscopy and infrared lasers in air pollution research and monitoring. Table 15 presents a summary given by Hanst of the detection of pollutants by absorption of laser radiation. For those interested in more technical details of remote sensing, the publications by Leonard (51) and McClintock, et al (52) and the recent book edited by Pitts and Metcalf (53) are particularly useful in addition to the ones mentioned previously. In the various workshops of this meeting much more detail will be given regarding the specific advantages and disadvantages of specific monitoring requirements.

It must be emphasized that the development and use of remote sensors must be tied very closely to the monitoring and surveillance needs in order to insure that valid and useful data result. Before significant resources are expended the monitoring requirements must be clearly defined for the various EPA program needs. In regard to the requirements of atmospheric monitoring the prime emphasis must be directed toward meeting the legal mandates of the 1970 Clean Air Act Amendments.

A system developed by R.T. Menzies (54) is one whereby the presence of selected atmospheric pollutants can be determined by transmitting an infrared laser beam of proper wavelength through the atmosphere, and detecting the reflections of the transmitted beam with a heterodyne radiometer transmitter-receiver using part of the laser beam as a local oscillator. The particular pollutant and its absorption line strength to be measured are selected by the laser beam wavelength. When the round-trip path for the light is known or measured, concentration can be determined. Since pressure (altitude) will affect the shape of the molecular absorption line of a pollutant, tuning the laser through a range of frequencies, which includes a part of the absorption line of the pollutant of interest, yields pollutant altitude data from which the altitude and altitude profile are determined.

Photography and remote sensing from airplanes are being used to gather information on (a) materials on the water surface, e.g., oil spills, (b) suspended matter in the water, and (c) certain soluble compounds. Measurement is based on the interactions of electromagnetic energy with matter floating, suspended or in solution. The electromagnetic energy source may be the sun, in which case sensors depending on solar reflections are referred to as passive systems, or the energy source may be an active source, e.g., laser (55). Applications of passive airborne systems, the measurements in the ultraviolet, visible and infrared regions of the electromagnetic spectrum have been recently reported in the literature (55)(56)(57).

A number of chemical pollutants can be detected by this technique based on their fluoresence properties. This includes a variety of compounds such as chlorophyll, phenol, lignin sulfonates, Rodamine WT and oils. Fluorometric techniques seem to be more applicable to airborne remote sensing than absorption techniques. Besides being more sensitive than absorption techniques, fluorometric procedures were found to be less affected by turbidity and quite useful in the analysis of mixtures (57). As low as 5 ppb of Rodamine WT were detected using airborne fluorometric measurement (57). In this case, the sun was used as the excitation source; hence, the procedure was limited to daylight hours.

Remote multispectral scanning techniques have been used for the measurement of various pertinent water quality parameters. Temperature measurements are usually made in the 8 to 13.5 nm. region throughout the day and night to an accuracy of 0.5°C. (55). This was used advantageously in thermal pollution measurement which provided the ability for a complete instantaneous mapping of large water areas. Detection of oil spills can be easily done with airborne remote sensing using IR and UV multispectral scanning (56). This is considered to be a very powerful technique because of its ability to follow the movement of the oil slicks over large areas of water surface.

The University of Michigan Willow Run Research Laboratory developed a unique multispectral system (55)(56) with the ability to examine a scene in 17 bands distributed between 0.3 and 1.35 nm. (i.e., a 17 channel spectrometer with electronic processing capabilities). This system was used for aerial surveying of oil spills and industrial waste discharge in lakes, rivers, etc. The system is capable of automatically determining the optimum wavelength, and in addition, the interactions of radiation with the effluent at other positions in the electromagnetic spectrum are also determined. Each effluent type and receiving body of water has its unique spectral characteristics which are used to differentiate it from other sources.

Aerial surveillance by multispectral remote sensing techniques seems to have a great potential in pollution monitoring. Such monitoring systems may not be able to provide information on the chemical composition of the system under investigation. Nevertheless, it is considered to be a valuable tool in the study of the dynamic conditions in aquatic environments and large scale water resources investigations.

Remote stack sensing techniques offer several advantages over the traditional methods of sampling through a probe introduced into a source of emissions. These advantages are as follows: (1) more representative sampling by virtue of spectral integration across the diameter of a stack plume, (2) no need for interfacing between stack and analyzer with probes and sample conditions, and (3) capability of measuring across an extended source such as an oil refinery.

Electrooptical techniques can be utilized remotely to characterize particulate and gaseous emissions. The use of Lidar systems for determining the opacity of plumes from power plants has received considerable attention (58)(59)(60)(61). A study of the optical properties of such plumes concluded that the optical transmission of the plume best characterizes the aerosol loading in emissions. A Lidar system has been fabricated as a research tool for field studies. More practical field equipment will probably utilize the signal backscattered from the plume aerosol rather than the signal backscattered from the ambient air beyond the plume. Such equipment would use a low powered laser. Mass loading measurements of aerosols in plumes also can be made by Lidar but interpretation of the signal would require measurements on other characteristics including particle size distribution.

The measurement of gases such as sulfur dioxide can be approached by use of ultraviolet transmission, infrared emission or Raman scattering. Field evaluations have been made utilizing correlation spectroscopy in the ultraviolet (62) and will be made with high resolution infrared emission spectroscopy. Nitric oxide presents a particularly difficult problem because of the overlap of the nitric oxide bands in the 5μ region by water vapor bands. Raman scattering provides a means of measuring nitric oxide with serious interference by other stack gas constituents. In a preliminary feasibility study insufficient signal to noise ratio was available (63). Improvements in (S/N) can be achieved by use of resonance Raman scatter and by fluorescent scatter techniques.

Detailed analysis of particulates in stationary source emissions for elements or compounds present many of the same problems as those already discussed for atmospheric analysis. Elemental analysis has been limited until the recent increase in concern about trace metals. However, available techniques already discussed for atmospheric particulate samples appear adequate for most stationary source applications. A reasonably extensive sampling and analysis program was conducted some years ago to analyze up to 10 polynuclear aromatic hydrocarbons in samples collected from a variety of combustion processes and industrial processes (64). Analyses were made by ultraviolet-visible spectrophotometry on the benzene soluble fraction of the samples following separations by column chromatography.

Another rather recent monitoring technique that may eventually have significant application in the evaluation of water quality is remote or spectral sensing from aircraft or satellite. It has the potential to provide rapid overall assessment of pertinent river basin characteristics, such as population distribution, land and water uses, and location of waste sources, as well as to provide information on the physical, chemical, biological, and hydrographic characteristics of surface waters. Spectral sensing in the infrared range is particularly useful in measuring variations in water surface temperature over wide areas. Natural color photography is also quite useful in that it provides a true bird's eye view of large water areas and permits a rapid qualitative evaluation of visible pollution problems. Locations where quantitative evaluations would be most meaningful are thus immediately identified.

Although the application of remote sensing technology to water resource management has really just gotten under way, the present state-of-the-art is already rather impressive. Remote sensing is expected to play an increasing role in future water quality management programs.

Public

In an attempt to increase the scope of the overall surveillance program certain states have initiated public surveillance programs, sometimes referred to as the water watchers. Similar to traffic safety patrols for school children, citizen groups are encouraged to watch for the safety of water resources and to telephone and report abnormal conditions of streams or lakes, e.g., oil or visually apparent chemical spills and fishkills.

SIZE OF THE MARKET FOR POLLUTION INSTRUMENTATION

A study of the market for air pollution measurement instrumentation has been made by R.R. Bertrand of Esso Research and Engineering Co. (65). Plans are underway to expand the network of ambient air monitors, to be run principally by federal, state and local government agencies. In addition to the ambient air monitoring network, each major pollution source will be monitored for its local contribution; these stationary source monitors will be run principally by local agencies and the pollution emitting industries themselves.

By 1974 there will be an estimated 790 ambient air monitoring stations, each containing a subsystem of 5 to 6 sensors, and an estimated 4,000 stationary source monitors (66). The total market for air pollution instrumentation for the decade 1970 to 1980 is estimated to reach $451 million (66). This figure includes an estimated $143 million for auto emissions monitoring. Table 16 gives an estimate of the air pollution instrumentation market total value 1970 to 1980.

TABLE 16: AIR POLLUTION INSTRUMENTATION MARKET

	Initial purchase	Replacement	Total
	(millions of dollars)		
Ambient level			
Agency (Automatic)	15	20	35
Agency (Non-automatic)	16	17	15
Industrial	10	--	10
Stationary source emission			
Industrial	140	86	226
Agency	12	10	22
Auto emissions			
Agency	33	--	33
Service station	110	--	110
		Total	451

Source: Report PB 204,174 (June 1971)

References

(1) Miller, S.S., *Envir. Sci. & Tech.*, 7, No. 11, 983 (Nov. 1973).
(2) Anon., *Chem. & Eng. News*, pp. 29-30,33 (July 19, 1971).
(3) Cothern, C.R., *Report PB 213,369; Springfield, Va.*, Natl. Tech. Information Service (1972).
(4) Stern, A.C., *Air Pollution*, Vol. I, "Air Pollution & Its Effects," New York, Academic Press (1968).

(5) Schroeder, H.A., *J. Chron. Dis.*, 18, 217 (1965).
(6) Davis, D.K., *Proceedings of the Third Annual Conference on Trace Substances in Environmental Health,* Columbia, Mo., p. 135 (1969).
(7) Pfitzer, E., Dept. of Environmental Health, College of Medicine, University of Cincinnati, quoted in Reference (3) above.
(8) Lawrence Berkeley Laboratory, *Report LBL-1,* Vol. I, Berkeley, Calif. (May 1, 1972).
(9) Mancy, K.H., in *NBS Publication 351,* "Analytical Chemistry: Key to Progress on National Problems," Wash., D.C., U.S. Government Printing Office (1972).
(10) Environmental Protection Agency, *Proceedings – Environmental Quality Sensor Workshop,* Las Vegas, Nevada, Nov. 30 to Dec. 2, 1971, Wash., D.C., EPA (1972).
(11) Holmes, D.C. in paper in Reference (10).
(12) Barth, D.S. and Shearer, S.D., Jr. in Reference (10).
(13) Amer. Soc. for Testing Materials, Part 23, Standard D-2928-71, Phila., Pa., ASTM (1971).
(14) Devorkin, H., *Air Pollution Source Testing Manual,* Los Angeles, Calif., Los Angeles Air Pollution District (Nov. 1963).
(15) Danielson, J.A., "Air Pollution Engineering Manual," *Publ. 999-AP-40,* Cincinnati, Ohio, U.S. Public Health Service (1967).
(16) Corbett, P.F., *J. Inst. Fuel,* 24, 237 (1961).
(17) Matty, R.E. and Diehl, E.K., *Power,* 101, 94 (1957).
(18) National Center for Air Pollution Control, *Publ. 999-AP-6,* "Methods of Measuring & Monitoring Atmospheric Sulfur Dioxide," Cincinnati, Ohio, U.S. Public Health Service (1964).
(19) Walden Research Corp., "Determination of SO_2 Emissions from Stationary Fossil Fuel Sources," *Publ. PB 209,267,* Springfield, Va., Nat. Tech. Info. Service (June 1971).
(20) Amer. Soc. for Testing Materials, Part 23, Standard D-1608-60, Phila., Pa., ASTM (1968).
(21) Smith, W.S., Shigehava, R.T. and Todd, W.F., paper on "A Method of Interpreting Stack Sampling Data," presented before 63rd. Annual Mtg. – Air Poll. Control Assoc., St. Louis, Mo. (June 1970).
(22) Smith, W.S., Martin, R.M., Durst, D.E., Hyland, R.G., Logan, T.L. and Haper, C.B., paper presented before 60th Annual Mtg. – Air Poll. Control Assoc., Cleveland, Ohio (June 1967).
(23) National Center for Air Pollution Control, "Specifications for Incinerator Testing at Federal Facilities," Cincinnati, Ohio, U.S. Public Health Service (1967).
(24) Hochheiser, S., Burmann, F.J. and Morgan, G.B., *Envir. Sci. & Tech.,* 5, No. 8, (Aug. 1971).
(25) Chand, R. and Marcote, R.V., *Development of Portable Electrochemical Transducers for Detection of SO_2 & NO_x,* Durham, N.C., Nat. Air Poll. Control Admin. (September 1970).
(26) Fontijn, A., Sabadell, A.J. and Ronco, R.J., *Anal. Chem.,* 42, 575 (1970).
(27) Fontijn, A., Sabadell, A.J. and Ronco, R.J., *Report on Contract CPA-22-69-11,* Durham, N.C., Nat. Air Poll. Control Admin. (September 1969).
(28) Snyder, A.D. and Wooten, G.W., *Report on Contract CPA-22-69-8,* Durham, N.C., Nat. Air Poll. Admin. (August 1969).
(29) Ronco, R.J. and Fontijn, A., *Report on Contract CPA-70-79,* Durham, N,C., Air Poll. Control Office (May 1971).
(30) Stevens, R.K., Mulik, J.D., O'Keefe, A.E. and Krost, K.J., *Anal. Chem.,* 43, 827 (1971).
(31) Walther, J.E. and Ambug, H.R., *TAPPI,* 50, 108A (1967).
(32) Anderson, K. and Borgstrom, J.G.T., *Svensk Papperstid (Stockholm),* 70, 805 (1967).
(33) Adams, D.F. and Koppe, R.K., *J. Air Pollution Control Assoc.,* 17, 161 (1961).
(34) Dorsey, J.A. and Kemnitz, D.A., *J. Air Pollution Control Assoc.,* 18, 12 (1968).
(35) Hodgeson, J.A., McClenny, W.A. and Hanst, P.L., *Science,* 182, 248-258 (October 19, 1973).
(35a) Elfers, L.A., "Field Operations Guide for Automatic Air Monitoring Equipment," *Report PB 204,650,* Springfield, Va., Nat. Tech. Info. Service (July 1971).
(35b) Mancy, K.H., *Instrumental Analysis for Water Pollution Control,* Ann Arbor, Mich., Ann Arbor Science Publishers (1971).
(35c) Lawrence Berkeley Laboratory, "Instrumentation for Environmental Monitoring – Water," *Publ. LBL-2,* Vol. 2, Berkeley Univ. of Calif. (February 1, 1973).
(36) American Chemical Society, "Cleaning Our Environment: The Chemical Basis for Action," Wash., D.C., Amer. Chem. Soc. (1969).
(37) McCrone, W.C., *American Laboratory,* 3, No. 4, 8 (1971).
(38) Woldseth, R., Porter, D.E. and Frankel, F.S., *Ind. Research,* 46 (February 1971).
(39) Berry, P.F., Paper before 16th National Symposium, Analysis Instrumentation Div., Instr. Soc. of America, Pittsburgh, Pa. (1970).
(40) U.S. Atomic Energy Commission, "Californium – Its Use & Market Potential," Wash., D.C., U.S.A.E.C. (1969 to 1971).
(41) Meyer, J.W., *Nuc. Insts. & Mat.,* 43, 55 (1966).
(42) Karttunen, J.O. and Henderson, J.S., *Anal. Chem.,* 37, 307 (1965).
(43) Bogusz, F.J., U.S. Patent 3,762,214; October 2, 1973.
(44) Technicon Corp., Environmental Sciences Div., Tarrytown, N.Y. (1969).

(45) Armstrong, F.A. and La Fond, E.C., *Limnol. Oceanogr.,* 11, 538 (1966).
(46) Allen, H.E., Research progress report on "Nutrient Dynamics in Lake Systems," Ann Arbor, Mich., Univ. of Michigan School of Public Health (1970).
(47) Sayers, W.T. of EPA in Reference (10).
(48) Ludwig, C.B., Bartle, R. and Griggs, M., *Report NASA-Cr-1380,* Wash., D.C., Natl. Aeronautics and Space Admin. (July 1969).
(49) Altshuller, A.P. and McCormick, R.A., *Journal of Remote Sensing* (March to April 1970).
(50) Hanst, P.L., *Appl. Spectroscopy,* 24, No. 2 (1970).
(51) Leonard, D.A., Avco Everett Research Laboratory Report (December 1970).
(52) McClintock, M. et al, *Studies of Techniques for Satellite Surveillance of Global Atmospheric Pollution,* Madison, Wisconsin, Univ. of Wisconsin (September 1970).
(53) Pitts, J.H. and Metcalf, R.L., Eds., *Advances in Environmental Science Technology,* New York, John Wiley & Sons (1971).
(54) Menzies, R.T., U.S. Patent 3,766,380; October 16, 1973; assigned to National Aeronautics & Space Admin.
(55) Wezernak, C.T. and Polcyn, F.C., *Proc. of 25th Purdue Industrial Waste Conference* (May 1970).
(56) Steward, S., Spellicy, R. and Polcyn, F.C., *Final Report 3340-4-F,* Ann Arbor, Mich., Willow Run Laboratories of Univ. of Michigan (1970).
(57) Hickman, G.D. and Moore, R.B., *Proc. of 13th Conference on Great Lakes Research,* Buffalo, N.Y. (1970).
(58) Hamilton, P.M., *Intern. J. Air & Water Poll.,* 10, 427 (1966).
(59) Johnson, W.B., Jr., *J. Air Pollution Control Assoc.,* 19, 176 (1969).
(60) Johnson, W.B., Jr. and Uthe, E.E., *Final Report on Contract PH-22-68-33,* Durham, N.C., Natl. Air Poll. Control Admin. (1971).
(61) Uthe, E.E., *Report on Contract CPA-70-173,* Durham, N.C., Natl. Air Poll. Control Admin. (1971).
(62) Barringer Research, *Report on Contract PH-22-68-44,* Durham, N.C., Natl. Air Poll. Control Admin. (December 1969).
(63) Leonard, D.A., *Final Report on Contract CPA-22-69-62,* Durham, N.C., Natl. Air Poll. Control Admin. (August 1970).
(64) Hangebrauck, R.P., Van Lehmden, D.J. and Meeker, J.E., *Publ. No. 999-AP-33,* Durham, N.C., Natl. Air Poll. Control Admin. (1967).
(65) Bertrand, R.R., *Report PB 204,174,* Springfield, Va., Natl. Tech. Information Service (June 1971).
(66) Bertrand, R.R., *J. Air Pollution Control Assoc.,* 20, 801 (1970).

SAMPLING TECHNIQUES

AIR AND GAS SAMPLING

Table 17 summarizes some common ambient air pollution sampling techniques as adapted from a paper by D.S. Barth and S.D. Shearer, Jr. (1).

A recent work on air sampling which was stated by a reviewer to be "...probably the best single guide to determining air pollution from industrial sources that has ever been published" is *Industrial Source Sampling* by D.L. Brenchley et al (1a).

TABLE 17: COMMON AMBIENT AIR POLLUTION SAMPLING TECHNIQUES

Type	Use	Specificity	Common Averaging Time	Relative Cost	Required Training of Personnel	Remarks
			Static			
Settleable particulates (dustfall)	Mapping and definition of special problem areas	Total settled particulates and general classes of pollutants	1 month	Collection, low; analysis, high	Collection low; analysis, moderate	Well equipped laboratory required for analysis only for definition of problem areas where a chemical analysis will pinpoint a particular source. Sensitive to temperature, wind, and humidity.
Sulfation devices	Mapping and general survey for sulfur dioxide	Responds to oxides of sulfur, hydrogen sulfide and sulfuric acid	1 month	Collection, low; analysis, high	Collection, low; analysis, moderate to high	
			Mechanized			
Hi-vol	Integrated quantification of suspended particulate	Total suspended particulate and multiple specific pollutants	24 hours	Moderate	Moderate	Detailed chemical analysis of Hi-vol and gas samples requires sophisticated laboratory, trained chemists; cost is high.
Gas sampler	Integrated quantification of gases	Sulfur dioxide nitrogen dioxide, mercury and other gases and vapors	24 hours	Moderate	High	
Spot tape sampler	Relative soiling index		2 hours	Low	Low	Provides only a rough, relative index of particulate soiling.

(continued)

TABLE 17: (continued)

Type	Use	Specificity	Common Averaging Time	Relative Cost	Required Training of Personnel	Remarks
			Automatic			
Gas	Continuous analysis of gaseous pollutants	Single gas or group of related gases	Continuous; sample integration usually 1 to 15 min.	Moderate to high	Moderate to high	Continuous measurements allow use of any desired averaging time by computation. Accuracy is generally much better than other methods. Calibration is simplified. Data are available instantaneously.
Particulate: soiling (automatic tape)	Continuous analysis of soiling rate	Unknown	Continuous; sample integration usually 1 to 15 min.	Moderate	Moderate	

Cost basis: low, 0 to $500; moderate $500 to $2,000; high, above $2,000.
Personnel training: low, maintenance level; moderate, technician; high, experienced technician or professional support staff.

Source: D.S. Barth and S.D. Shearer, Jr., EPA (1972)

A gas sampling probe for sampling particulate and moisture-laden gases has been developed by G.N. Thoen (2). The accurate sampling of flue gases or other industrial gases is rapidly becoming more and more important as stricter air pollution regulations are being passed by both State and Federal authorities. Identification and concentration of the gases being exhausted into the atmosphere on a continuous basis is becoming more and more important as industry attempts to comply with the enacted legislation. By continuous monitoring of the gas content of combustion operations, the combustion operation can be evaluated and corrections made rapidly if needed.

Industrial gases from power boilers, black liquor recovery furnaces, blast furnaces, lime kilns, etc. normally contain particulate matter and condensable materials, mainly water vapor. Obtaining a representative sample of the gaseous effluent from a stack is highly critical when the components making up the gaseous mixture are used as a criterion to optimize and control the then existing conditions of a combustion operation. To get an accurate identification and concentration of the gases contained in a typical gas sample the gas sample must be substantially free of moisture and particulate matter.

In addition the sampling probe used must be able to withdraw the gas sample and deliver it to the analytical apparatus without degradation or alteration of the compounds contained in the gas. If the gas sample is subjected to sudden temperature changes some degradation or alteration of the compounds contained therein is likely to occur, thereby causing erroneous results. Particulate matter such as dust, soot, lint, etc. tends to, over a period of time, plug the sampling probe inserted into the gas flow path.

This probe which is designed to overcome the above cited problems comprises a tubular shielding member having an open end in the gas flow path and a tubular sampling probe mounted concentrically within the shielding member. The sample probe is made of a low heat conductive substance permeable to moisture.

Particularly useful are ceramic materials. The probe allows moisture to evaporate through it into the atmosphere, cools the gas sample without degradation thereof, and is corrosion resistant. Particulate matter which deposits in the sampling probe is removed by periodically flushing the tubular probe with compressed air or other fluid. Valve means periodically and selectively connects the flushing fluid to the probe member. Figure 15 shows the construction of such a probe in sectional elevation.

FIGURE 15: SAMPLING PROBE FOR PARTICULATE AND MOISTURE-LADEN GASES

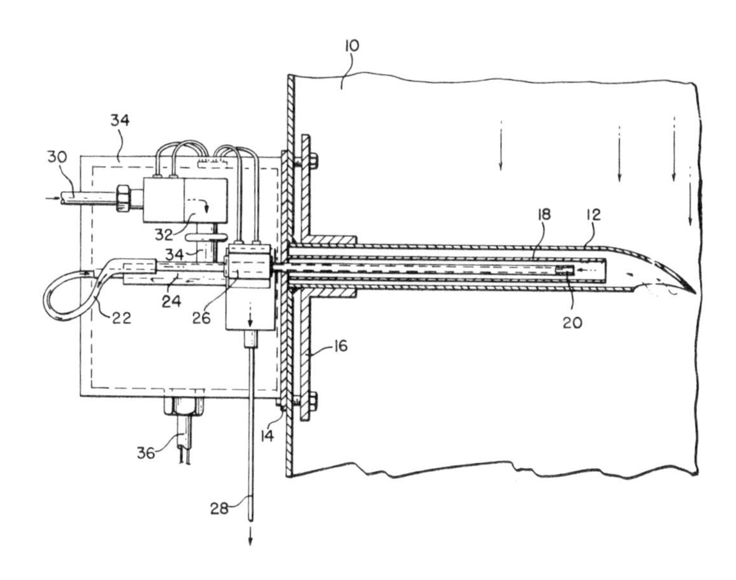

Source: G.N. Thoen; U.S. Patent 3,559,491; February 2, 1971

The probe is mounted in the wall of the furnace or in a stack leading from the furnace. An elongated tubular protective shield **12** of stainless steel or other corrosion resistant material is inserted through the stack or wall of the furnace into the gas flow path. The end of the tube in the gas flow path has a hood-like portion thereon to prevent particulate matter present in the gas from being drawn into the probe. As shown in the figure the opening in the end of the protective shield is opposite the direction of flow of the gas within the stack. The opposite end of the protective shield is welded in place to the wall of the furnace stack and is further secured by attachment to support plate **14** on the outside wall of the stack and plate **16** of the inside wall of the stack.

Within tubular shield member **12** are placed one or more tubular members or sampling probes **18** and **20** made of a material with low heat conductivity and permeability to condensable gases such as water vapor. Cast ceramic tubes, such as Coors AD 99, have been found to work satisfactorily as sampling probes. Other ceramic materials are equally applicable provided they are permeable to the flow of moisture therethrough and have fairly low heat conductivity.

As shown, tubular member **18** surrounds sampling probe **20**. If desired, tubular member **18** can be omitted although better results have been obtained by mounting probe **20** concentrically within ceramic tube **18** and the protective shield. The tubular ceramic probe **20** extends from the gas flow path inside the furnace stack through an opening in the stack wall. The end of probe **20** is connected to a flexible piece of tubing **22** and to conduit **24** leading by way of valve **26** to conduit **28** which is connected to a conventional gas analysis apparatus (not shown).

Valve **26** is solenoid operated and, in its normally open position, allows the gas sample withdrawn from the gas flow path to flow directly to the gas analysis apparatus. A vacuum created in conduit **28** by a vacuum pump continuously withdraws a gas sample from the gas flow path within the furnace or furnace stack. Over a period of time fine particulate matter, drawn into the sampling probe along with the gas sample, is deposited within the probe. If the particulate matter is not removed it will eventually plug the sample probe. To alleviate this problem the sample probe is periodically flushed with a flushing fluid flowing counter to the normal gas flow through the sample probe. Compressed air or other gaseous fluid under pressure is supplied through conduit **30**, solenoid operated valve **32** and conduit **34** to probe member **20**. Valve **32**, in its normally closed position, prevents the flushing fluid from entering probe member **20**.

Valves **26** and **32**, both solenoid operated, are connected to a cycle timer **34**. At predetermined intervals valve **26** is closed and valve **32** opened, thereby allowing a flushing fluid to blow probe member **20** free of any accumulated particulate matter. The timer is connected by means of wire **36** to a suitable source of electricity.

It has also been found that the probe member **20** can be kept free of particulate matter for longer periods of time by insertion of a filter means such as a ceramic frit in the sampling end of probe member **20**. The particulate matter is retained by the frit in the hot, higher than dew point zone. On flushing the probe member with compressed air the particulate matter is blown back into the furnace.

Flue gases from a typical black liquor recovery furnace issue at a temperature of approximately 700°F. and normally contain approximately 30% water by volume and 4 to 12 grains particulate matter per standard cubic feet per minute. Because the gas is corrosive by nature most metallic materials will not operate for long time periods in such an atmosphere. This ceramic sampling probe is corrosion resistant and capable of sampling the gas content continuously and efficiently. On withdrawal of a gas sample from the gas flow path of a furnace stack the ceramic probe, having low heat conductivity, does not thermally shock the gas sample.

In addition, substantially all of the moisture contained in the gas sample evaporates through the probe into the atmosphere with little or no degradation of the gas sample. Previously known probes were not capable of removing moisture from a gas sample without degradation thereof. Although the probe described here has particular utility in sampling flue gases in black liquor recovery furnaces, it is capable of being used in many other industrial combustion operations.

A sampler developed by J.C. Couchman and G.W. Applebay (3) is of the class of air sampling devices which experience high rates of airflow therethrough such as those used in conjunction with moving aircraft.

The use of aircraft mounted air samplers to obtain samples of the particulate matter contained in the atmosphere in order to identify the substances and their concentrations has become increasingly important. Reduced to simplest of terms, this type of sampler consists of a duct through which the sample of air passes and filter media positioned within the duct to retain the particulate matter. Collecting a representative sample of the particulate matter requires that no fractionation occur at the inlet of the system and that little or no plate-out loss occur in the duct leading to the filter media.

This type of sampling, isokinetic sampling, demands that the linear velocity of air entering the inlet of the sampling duct be equal to the linear velocity of the air in the free stream. When the inlet velocity is greater than the velocity of the free stream, the sample collected will be distorted by favoring the smaller particles in the free stream. When the inlet velocity is less than the velocity of the free stream the sample collected will be distorted by favoring the larger particles.

Achieving isokinetic conditions requires that all losses in head in the inlet duct and the pressure drop across the filter be compensated. The difficulty of completely compensating for such losses increases as the rate of airflow through the sampler increases. Of course, high rates of flow increase the size of the sample and are desirable when particulate materials of very low concentrations are of interest. It is desirable that the means utilized to compensate for the head losses and pressure drop permit adjustment for varying conditions since the ability to sample over a range of airspeeds is usually necessary. Some existing air samplers of the general type under consideration here have utilized vacuum pumps to provide a vacuum behind the filter in order to lower the exhaust pressure and achieve isokinetic conditions. However, when flow rates above about 50 cfm are of interest, the physical size (and weight) and the power requirements of a vacuum pump having sufficient capacity makes this arrangement impractical for use in aircraft in the size range normally considered for sampling operations.

The isokinetic device of this process does not require a vacuum pump. This sampler combines a sampling duct containing a particulate filter with a venturi shroud surrounding the duct. A turbulence condition which, it is found, tends to build up behind filters of relatively large diameter is minimized by providing means for delivering high velocity air, which is in addition to the air passing through the filter, to the region directly behind the filter.

This may be expeditiously accomplished by a plurality of tubes positioned so as to bleed air from the venturi into the outlet portion of the sampling duct. Control jets in the region of the venturi constriction enable the effect of the venturi to be varied by the injection of pressurized air or utilization of the siphon pumping action of the venturi.

A device developed by R.K. Riggs (4) permits obtaining samples of air in an aluminum reduction plant in order to determine the amount of dust and other particles carried therein. In aluminum reduction plants, as well as in manufacturing plants generally, it is necessary to sample the ambient air to test for dust or other foreign particle content, as well as to determine the degree of gaseous emissions emanating from the manufacturing operation. Should the deleterious matter in the air rise above acceptable levels, steps could be taken to remedy the situation before the health of the plant personnel is endangered or conditions become otherwise hazardous.

In the past, air sampling devices generally relied upon motor operated pumps to draw the ambient air into the device where particles or gas could collect on a filter or in a gas trap in order to be tested. In aluminum reduction plants, however, the magnetic fields generated as a result of the electrical current used in the reduction cells hindered the proper and reliable operation of electrically powered vacuum pumps that were used in collecting air to be sampled.

Because a proper analysis of the air sample depends to a large extent upon a constant and repeatable flow rate through the device, the effect of the magnetic fields on the electrically powered vacuum pumps rendered the test results unreliable. Moreover, in large plants the sampling sites are often located at great distances or heights from electrical outlets and thus require extensive lengths of electrical cord to reach the electrically operated sampling device. The setup of such test facilities is thus rendered more difficult, time-consuming, costly and hazardous.

In the device of this process, these problems are overcome by providing an air sampling device having a pipe that is adapted to be connected to any source of compressed air or other gas. A venturi tube is disposed within the pipe such that passage of the compressed air through the pipe will create a region of reduced pressure at the throat of the venturi. Air sampling chambers having membrane filters disposed therein are connected to the throat portion of the venturi such that the reduced pressure therein will draw ambient air into the sampling chambers whereby dust and other foreign particles may be trapped in the filters and thus tested to determine the purity of the air.

Alternatively, the air sampling chambers may be provided with a chemical solution trap or similar device whereby gaseous emissions present in the air may be filtered out and qualitatively and quantitatively tested.

A device developed by J.J. Quigley (5) permits continuously obtaining a representative multicomponent gas sample from a stream of furnace gas which is heavily laden with dust, such as gases from an open hearth furnace or a blast furnace.

Many types of apparatus have been devised for providing a gas sample to a gas analyzer, but none of the commercially available types are capable of continuously supplying a truly representative sample of a multicomponent high temperature furnace gas of the above type which contains a high proportion of solids and generally substantial amounts of moisture, carbon dioxide or carbon monoxide. Since modern multicomponent gas analyzers are very sensitive instruments, it is necessary to remove substantially all of the dust contained in the original gas sample without, however, removing any of the gaseous components which are to be analyzed.

Gas sampling systems which remove solids from a heavily dust ladened, moisture-containing gas sample by washing with water or condensing with steam also remove a significant amount of carbon dioxide which is often an important component of a furnace gas to be analyzed. Many other gas sampling systems are objectionable because of their inability to handle heavily dust ladened gas samples without becoming clogged with dust and requiring frequent attention for maintenance and repair.

The furnace gas sampling apparatus of this process employs in combination means for effecting first a demoisturizing treatment on the gas sample prior to specific treatment for removing dust particles and means for effecting a series of dry filter treatments of the sample to remove substantially all solids having a particle size in excess of about 1 micron.

To carry out the foregoing treatments continuously, a sample of a furnace gas atmosphere having a temperature between about 1000° and 3000°F., an average concentration of solid particulate matter of about 25 grains per standard cubic foot, and a moisture concentration as high as 50% is continuously withdrawn from the back uptakes of a furnace, such as an open hearth furnace, by means of a water-cooled gas-sampling probe which is preferably mounted on the end bulkhead so as to reach about the center of the uptake duct.

As shown in the diagram in Figure 16, the gas sample is conducted continuously through the probe **10** directly into a suitable heat exchanger **11** which preferably consists of a water-cooled pipe or coil **12** surrounded by an enclosure **13** through which cooling water flows continuously to effect the desired cooling and condensation of moisture in the gas. The exit temperature of the gas sample leaving the heat exchanger is approximately 65°F. when water is used as the cooling fluid. The exit temperature of the gas sample from the heat exchanger is, of course, dependent upon the temperature and flow of the cooling medium. If desired, one may use different cooling mediums to obtain predetermined gas temperature as the gas sample leaves the heat exchanger.

The heat exchanger serves essentially to condense moisture from the gas sample. The percentage of moisture removed by the heat exchanger through the moisture trap **14** can vary from 60 to 85%, depending upon whether or not the gas sample is supersaturated with moisture at the heat exchanger inlet. When a high percentage of moisture is present in the furnace gas, the gas sample is supersaturated with moisture at the heat exchanger inlet and an increased amount of moisture is removed from the gas stream in the heat exchanger.

Under normal conditions when the furnace or other process gas sample is simply saturated with moisture, the amount of moisture removal is approximately 65% when the gas sample leaving the heat exchanger has a temperature of about 65°F. The amount of moisture removed can, of course, be increased by lowering the temperature of the cooling medium of the heat exchanger.

If the dew point of the sample is so low that no condensation forms in the heat exchanger with a particular cooling fluid, the heat exchanger then simply cools the gas sample as the sample passes therethrough from a high temperature zone of the furnace. Also, if there is only a relatively small amount of moisture in the gas sample and no cooling of the sample is required, the gas sample can be conducted around the heat exchanger through the bypass piping **15**. In order to insure a continuous uniform flow of sample gas to the gas analyzer (not shown), a gas pump **17** is preferably used to control the flow of the gas sample through the probe **10** and the heat exchanger, and through the remaining mechanical

FIGURE 16: SAMPLER FOR OPEN HEARTH OR BLAST FURNACE EXIT GASES

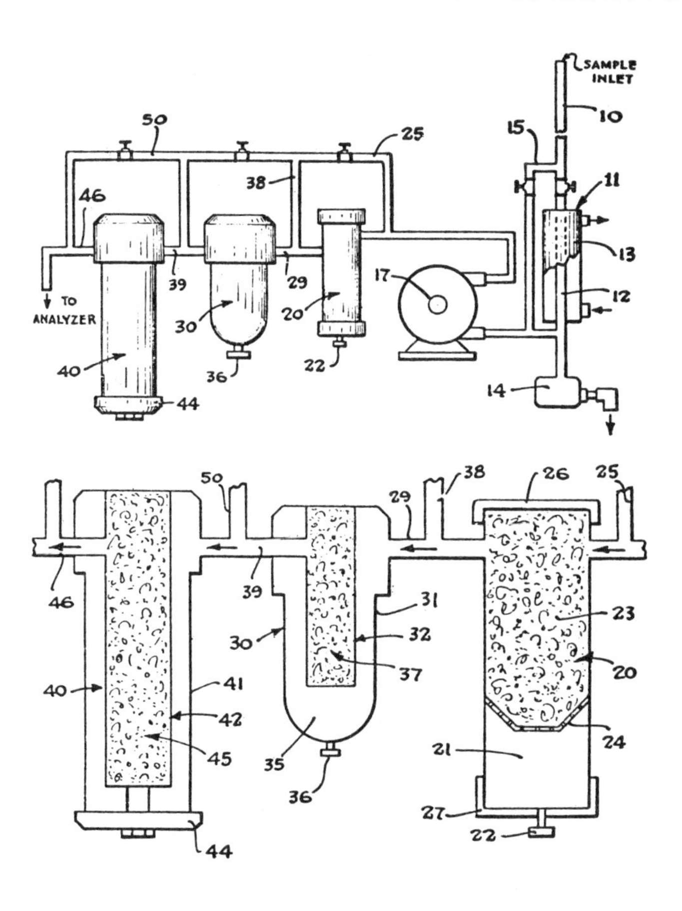

Source: J.J. Quigley; U.S. Patent 3,304,783; February 21,1967

filter units and gas analysis apparatus. The gas pump can be of any conventional design, including a flexible tubing type pump, with suitable drive means which can be used to regulate the speed thereof to control the rate of flow of gas sample. Additional moisture can be removed from the gas sample during the latter steps, if desired. In circumstances where conditions preclude the use of an electric motor drive, an air motor has been used and provides excellent results.

It should also be noted that a pump is not needed if the pressure of the atmosphere to be sampled is sufficiently high to provide a positive continuous flow of gas through the gas sampling apparatus and the gas analyzer. The need for and type of pump used will depend on considerations such as the gas sample pressure mentioned above, the distance from the sampling system to the analyzer, the pressure drop enroute to the analyzer, and the pressure and flow requirements of the gas analyzer. Since the pump flow and pressure of this apparatus are adjustable, the gas sampling system is able to meet a wide variety of gas analyzer requirements. Upon leaving the pump, the relatively dry gas sample is passed into a coarse mechanical filter unit **20** in which most of the large solid particles having a particle size larger than about 100 microns in the gas are removed, along with additional moisture which condenses and collects in the sump portion **21** at the bottom of the filter and which has a drain valve **22** for removing condensed moisture. The amount of moisture removed by the coarse filter can be up to about 25%. Part of the moisture removal in the coarse filter, of course, is due to a drop in temperature that occurs after the sample gas has been discharged from the pump.

In the event there is no coarse solid particulate material larger than about 100 microns in size suspended in the gas sample or when the filter unit must be shut down for repair, a by-pass tubing **25** is provided to convey the gas sample around the filter directly from the pump to a succeeding filter unit. The filter material **23** which is supported within the filter unit by a previous member **24** spaced from the lower end of the filter unit can be glass wool or any other inexpensive filter medium desired. For ease of maintenance, the filter unit is provided with removable top and bottom closures **26** and **27**, respectively.

The gas sample flows from the coarse filter unit through conduit **29** into the fine filter unit **30** which removes dust particles having a particle size down to about 75 microns. The body section **31** of the fine filter unit encloses and removably supports thereon a filter cartridge **32** in spaced relationship with the wall of the body section **31**. The filter material **37** in the cartridge **32** can be formed of glass wool or other filter material compressed therein to provide for the removal of predetermined particle size material.

If desired, the fine filter element can consist of a Monel screen with a transparent bottom section, such as a Norgren filter. The lower sump end **35** of the filter body section **31**, which is provided with a petcock or drain valve **36** for removing any condensed moisture which collects therein, is removable to facilitate replacement of the cartridge **32**, when required. It has been found that the fine filter removes up to 5% moisture from the gas sample. A by-pass conduit **38** is also provided to convey the gas sample around the fine filter unit directly to the ultrafine filter unit **40**, when required.

The gas sample is conveyed directly from the filter cartridge **32** of the fine filter unit through conduit **39** into the ultrafine filter unit wherein particles having a particle size larger than about 1 micron are removed from the gas sample. The body section **41** of the filter unit **40** encloses and removably supports therein in spaced relationship with the sides of the body section **41** a filter element **42**. The filtering material of this filter element can comprise packed glass wool **45**, if desired, but preferably is a commercial filter unit with a replaceable filter cartridge constructed to remove particles having a size larger than about 1 micron.

The glass wool or cartridge can be removed for replacement by removing the detachable end cap **44** at the lower end of the filter unit **40**. The substantially dry, dust-free gas sample after passing through the ultrafine filter element is carried by conduit **46** to the gas analysis apparatus (not shown) which forms no part of this process and can be of any desired type suitable for analyzing a multicomponent gas. The ultrafine filter unit can be by-passed by conduit **50**, if necessary.

Since the furnace gas sample is treated to remove therefrom a substantial proportion of the moisture before the gas sample enters the series of dry mechanical filters, there is substantially less tendency for the filters to become clogged than when the gas sample having a high concentration of both dust and moisture is initially subjected to a filtering treatment to remove dust and the like solids. While the gas sampling apparatus has been described as applied to sampling the gas from an open hearth furnace, it should be understood that other metallurgical furnace gases and process gases other than open hearth furnace gases, such as gases from a coke plant, can be continuously sampled by the apparatus of this process to provide a representative gas sample of a gaseous process stream to a continuous gas analyzer.

An apparatus developed by D.P. Manka (6) permits sampling a dirt and moisture-laden gas stream, such as blast furnace top gas. It employs parallel sampling probes and associated filter units to permit continuous alternate sampling and cleaning of the filters by back-flushing. The apparatus automatically ceases sampling whenever a foreign gas is introduced into the gas stream being sampled.

A sampling device developed by E.D. Neuberger and T.J. Junker (7) employs a steam eductor which utilizes steam as the force for moving the sample and at the same time for scrubbing the sample. Soluble and insoluble components of the sample are present in dissolved or suspended form in the condensate. The system is particularly useful to provide a continuous sample. A chemical may be added to the steam to enhance the tendency of a component of a gaseous sample to enter the liquid phase.

Figure 17 shows the essential elements of the device of this process. The stack or other conduit **1** contains waste gases and entrained liquids and solids which move in a vertical direction. A sample probe **2** is positioned in a prominent place in the stack. The probe is connected to a suitable tube **3** which passes through the wall of the stack. To induce the passage of the sample into the probe, a negative pressure is supplied by steam eductor **4**.

Steam passes through pipe **5** from a source thereof, not shown, and, after picking up the sample in the eductor, passes into condenser **6**. Pipe **12** may be used to introduce into the steam a chemical for enhancing the tendency of a particular component to enter the liquid phase, e.g., to dissolve in the steam or otherwise be entrained, conditioned or chemically combined for analysis as a component of the steam condensate.

In the condenser, the steam and condensable material therein are converted to a liquid which is drained or removed through conduit **7** to optional collector **8**. The collector should be equipped with a float valve **9** or other means to ensure that the liquid sample delivered to the analyzer through conduit **10** chronologically represents the composition of the gas sample, either continuously or intermittently. The collector has a gas escape line **13** on which is affixed a pressure gauge **42**, needle valve **44** and rotameter **41**, which contains an optional temperature and pressure indicator for calculating the absolute humidity in gas line **13**. A second rotameter **43** is placed in line **10**.

FIGURE 17: STEAM EDUCTOR SYSTEM FOR STACK GAS SAMPLING

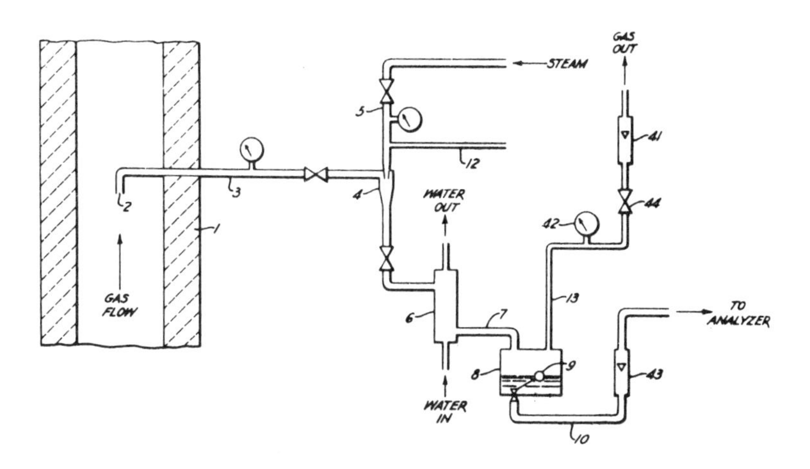

Source: E.D. Neuberger and T.J. Junker; U.S. Patent 3,641,821; February 15, 1972

A preferred eductor uses on the average about 30 lbs. of steam per hour, producing a maximum of about 225 cc of condensate per minute. When all factors such as temperature and pressure are controlled, this system provides a controlled, relatively small dilution of the sample. The dilution factor may be affected by the size of the nozzle in the eductor, the internal diameter of the sample probe, the flow rate of the steam in the eductor, and by opening or closing valve **44**. Most conveniently, this valve provided in exhaust line **13** will regulate the sample flow. This should be calibrated on installation.

The greater the negative pressure generated by the eductor, the larger will be the sample; however, the ratio of steam to sample should always be known. Accordingly, it is desirable to know at all times the steam and sample flow rates. The sample flow rate is influenced not only by the internal diameter of the conduit **3**, the degree of opening of valves thereon and on line **13** and the negative pressure in the eductor, but also by the velocity of the stack gas or other sample material. More or less ideal isokinetic conditions may be assumed, i.e., it can be assumed that the velocity of the sample in the sample probe is equal to the velocity of gas in the immediate vicinity of the probe nozzle.

Rotameter **41**, the optional pressure gauge **42** and the temperature indicator located in the rotameter, all on line **13**, will assist the operator in providing an analysis reported in units of analyzer constituent per unit of dry gas. The pressure and temperature in gas line **13** are relatively stable. The flow rates of the gas, measured by rotameter **41**, and the liquid sent to the analyzer, measured by rotameter **43**, are thus significant factors. An analyzer which measures a contaminant in terms of parts by weight of contaminant per million parts by weight of sample may be adjusted to read in absolute terms. That is, the reading may be taken as grams or pound of contaminant per unit of time.

By making a simple correction based on the absolute humidity (a function of the temperature and pressure) of the gas stream leaving the sampling system, it is possible to express the degree of contamination as units of contaminant per unit weight or volume of original dry gas sample. When the degree of contamination changes, the result is a corresponding, meaningful change in the concentration of contaminant in the liquid stream.

In calculating the influence of the water content of the stack gas or other material to be sampled, it should be observed that the exhaust gas emitted downstream of the condenser, as by outlet **13**, will contain some water vapor. If, for example, the sample in line **3** contains 5% water vapor and the exhaust line contains 3% water vapor, a correction factor may be computed relating these numbers to the respective flow rates. The correction factor is the ratio of noncondensables in the stack gas to the noncondensed material in the exhaust, or expressing these factors in percentages,

$$\frac{100 - 5}{100 - 3} = 0.976$$

A correction factor for temperature may be calculated by computing the ratio of the absolute temperature of the exhaust gas to the absolute temperature of the source or stack gas. The weight of the material analyzed, in parts by weight per million parts of dry sample, may be calculated by multipyling the weight of condensate per volume of dry gas sample (in pounds per cubic foot, for example) times the concentration of contaminant in the analyzer in parts per million of condensate. This answer may be converted easily to parts of contaminant per million parts by weight of dry emitted gas.

Where an alarm system or relatively rough reading may be used, the effect of water vapor may be ignored, or an empirical correction factor may be employed, as by computer. This will enable the user to read the final answer from the analyzer in absolute terms, i.e., pounds of contaminant emitted per hour through the stack. Any gas which is soluble in water will be collected by this system. In some cases, associated elements may be monitored if their relation to the undesirable gas or the efficiency of the process is known. Additives may be injected into the steam prior to, during or after, its passage through the eductor to increase the affinity of a gas, such as SO_2 or SO_3 for dissolution in the steam or the condensate, or otherwise enhance its tendency to enter the liquid phase. Illustrative examples of such additives are monoethanolamine, diethanolamine, and triethanolamine, and neutralizing amines such as cyclohexylamine and morpholine. No minimum amount is absolutely ineffective, i.e., a very small amount is effective to a correspondingly small degree. The maximum amount will be determined by economics and possible adverse effects on the condenser, etc.

The samples delivered by this system are particularly of use in continuous analyzers such as chloride, fluoride, sodium, conductivity, and colorimetric analyzers. It will be observed that the liquid collector **8** or elsewhere in the sample system may contain solid particles. Useful information may be obtained by measuring the turbidity of the sample, or by otherwise separating and analyzing the solid material. This system is applicable to the preparation of liquid samples as well as those derived from gas streams. The steam eductor functions in such cases as a pump and a diluter as well as a scrubber.

As pointed out by R.D. Keefer and J.S. Wyman, Jr. (8), gas sampling devices may have a means of passing a predetermined quantity of the gas to be sampled through a filter membrane. The filter membrane is then examined to determine what the contaminants were in the sample of gas. For example, the number of particles on the filter membrane might be counted or a beam of light and a photoelectric cell might be used to determine the change in translucency of the filter and hence the amount of contamination from dust, pollen particles or other debris in the gas.

The device might also be used to determine the radioactive fallout from nuclear tests or other sources in a gas sample, or a chemically treated filter tape might be used to identify the various gases of which a sample is composed. In gas samplers in which a filter material is in the form of a tape it is a problem to properly load and align the tape in the sampler and to provide a means for positively and uniformly feeding the filter tape through the sampler. This process is devoted to a detailed consideration of mechanical devices which may be used to overcome such difficulties.

A device developed by M.F. Scoggins (9) is a portable, low profile, drum-like structure of durable lightweight construction which can be easily and readily moved to different locations in order to provide comparative monitoring of atmospheric pollutants at different locations.

As shown in Figure 18, the sampling device, generally designated by the numeral **10**, consists of a shallow lower section **12** to which an upper section **14** is secured as by the bolt connections **16**. The lower section includes a flat circular base **20** and an arcuate, upstanding side wall **22** supported by the base adjacent its periphery and adapted to receive the substantially flat cover **24** of the upper section which is similar in size and shape to the base. The side wall of the lower section is provided with a single aperture or port **26** communicating with the interior of the drum-like sampling device and forming the atmosphere monitoring station thereof.

Rotatably mounted on the flat base within the lower section and spaced slightly therefrom is a sampler mounting magazine which takes the form of a flat disk **30** having a diameter only slightly smaller than the diameter defined by the arcuate side wall. Fixed securely to the disk **30** adjacent its periphery are a plurality of uniformly spaced mounting brackets **38** adapted to receive individual atmosphere monitoring samplers **40** which may be of the membrane filter or impinger type.

The number of mounting brackets carried by the disk may vary substantially depending upon such factors as the size and type of sampler employed, the desired programmed operation of the sampling device or the overall size of the device itself. In the form illustrated, provision is made on a 15 inch diameter disk for 12 uniformly spaced samplers which can be programmed for individual collection periods ranging from five minutes to 24 hours. In the form illustrated in the drawing, the thin, resilient mounting brackets **38** are provided with bifurcated or U-shaped sampler receiving portions **44** upstanding from the disk **30**

FIGURE 18: ATMOSPHERIC MONITORING STATION

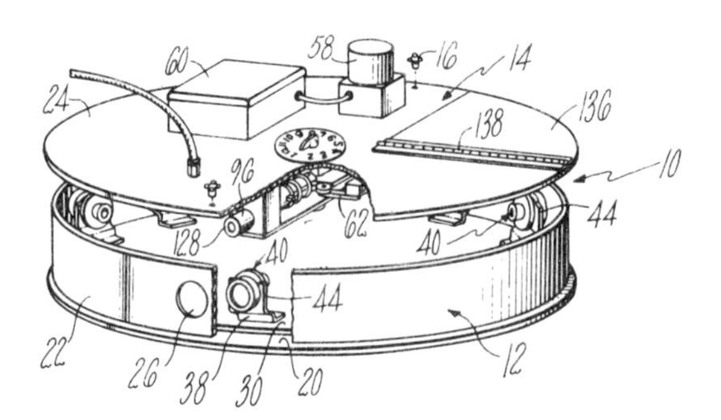

Source: M.F. Scoggins; U.S. Patent 3,540,261; November 17, 1970

and adapted to slidably receive the individual samplers **40**. In this manner the brackets **38** mount the samplers a short distance from and facing the side wall facilitating smooth radially outward movement thereof into the atmosphere monitoring station.

The flat cover which encloses the samplers **40** within the device fixedly supports a depending drive mechanism operable by a single motor **58** resting on the top of the cover. Operation of the motor is controlled by both a timing and switching mechanism **60**, also supported on the exterior of the device by the cover, and an interior microswitch **62** within the interior of the device.

The drive mechanism utilized to advance the individual samplers into registry with the monitoring station may also include provisions for moving the samplers toward the side wall and positioning them in the monitoring station, thereby fully exposing the sample collecting medium **52** to the atmosphere being monitored. To accomplish this the drive mechanism includes a reciprocating plunger head **96.**

As the plunger contacts the sampler a vacuum seal is created between the front face **128** of the plunger and the resilient rear wall of the sampler. The plunger continues its outward travel to drive the sampler into the side wall aperture **26** for exposure to the atmosphere while a separate source of vacuum draws the ambient atmosphere toward and through the sampler in order to collect the pollutants therein.

As illustrated, an extension of the center shaft may protrude through the top cover and be provided with an interior knob which will move with the center shaft and provide an immediate visual indication of which sampler is located at the monitoring station. Additionally, a portion **136** of the cover may be hingedly secured at **138** to provide ready access to individual samplers within the device prior to completion of the entire sampler collecting operation.

A remote operated sequential air sampling device has been described by D.M. Teel and J.H. Putnam, Jr. (10). The sampler is provided with a head in which are mounted a plurality of closed filter collecting devices using stacked filters, termed filter trains. An indexing table on which are mounted a pair of air cylinders is caused to rotate in a series of spaced steps. During each stationary interval of time the air cylinders are expanded at will exposing one of the filter trains. A cycling system is actuated to draw the gaseous medium through the filter train which is thereafter retracted. The indexing table is then indexed along its axis of rotation and the cycle is repeated for another filter train. An important aspect of the device is making the head containing the filter trains readily removable so that the samples can be collected with the removal of a minimum of equipment. In the laboratory, the filter trains can be removed later for examination. Another feature of this is that the sampler incorporates a unique approach for measuring the sample gas flow wet. A critical orifice is used, that is, when the downstream pressure is approximately one-half or less of the upstream pressure, sonic velocity will occur in the throat of the orifice. This velocity cannot be increased by further reducing the downstream pressure. While sonic velocity varies with the square root of the temperature of the gas flowing, it remains relatively constant over the temperature range of the sampler and is a known value for any given set of conditions.

Since the area of the throat is constant, the volume of the sample gas remains relatively constant and is also known for any given set of conditions. A mass flow meter in the system measures and records the mass of the sample gas thereby knowing the volume, mass, temperature, and pressure upstream of the orifice. Hence, a great deal is known about the sample gas flowing including all the measurements taken in the dry monitoring system used by other samplers. Sufficient is obtained this way so that a very good estimate of moisture trapped can be made if desired.

A device developed by G.S. Raynor (11) is an ambient air sampling arrangement which utilizes a container having a slidable hollow portion extending between a pair of end members to trap a fixed volume of air. The slidable portion is moved from a position away from the space between the end members to its position enclosing the fixed volume to trap the sample. A purging system removes the trapped volume and filters out the particulate.

When atmospheric air is sampled for the purpose of determining its particulate content, a major problem is that of obtaining a sample which contains a representative amount and distribution of particulate. Very often, at the entrance to the sampling apparatus, flow conditions are altered so that the particulate is distributed differently than under true ambient conditions. For example, when an intake tube is employed as the means of sampling the air, alignment difficulties under varying airflow direction may be encountered which place the results in question.

The device of this process overcomes the disadvantages of previous sampling devices by obtaining a sample of a gaseous fluid in a way which avoids the problems and difficulties associated with restricted entrance type sampling devices. In this device, provision is made to trap a fixed volume of a gaseous fluid to be sampled followed by purging to remove one or more constituents which may be particulate or aerosol. Trapping of the gas is done in such a way as to minimize the effect of altered local flow conditions. The apparatus includes a cycling arrangement so that samples of any size may be obtained.

Figure 19 is a diagram of this apparatus; the view at the lower left shows the collector in a partially open position. There is illustrated a collector **10** designed to sample ambient air. Collector **10** consists of a slidable hollow cylindrical body **12**, a lower stationary end cover **13** and an upper stationary end cover **14.**

The cylindrical body is supported by a member **15** which is attached to the outside of the body by way of a pair of arms **16** and **18.** The upper edge of the body is provided with an annular lip **12a** whose function is described below. Arms **16** and **18** are provided with holes (not shown) through which passes a stationary, vertically disposed rod **22** which is supported at its base and top as shown. It is thus seen that the body is slidable vertically as indicated by double arrow A.

Lower cover **13** is frustoconical in shape with a bottom end surface **24** through which passes an opening **25**. Cover **13** is supported by a plurality of legs **28** which may be spring loaded for reasons later explained. Along the upper edge of the cover is an annular ledge **29** against which the bottom edge of the body rests in its lowermost position. Upper cover **14** is similarly frustoconically shaped as shown with the outer diameter such as to permit collector body **12** to slide snugly though freely over the former as illustrated in the detail

FIGURE 19: AMBIENT AIR SAMPLING DEVICE

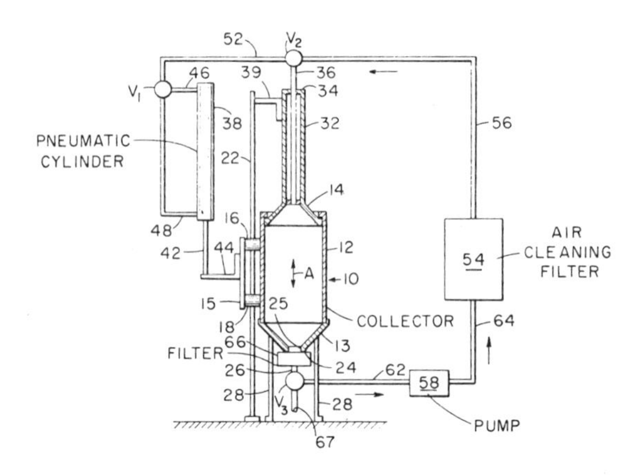

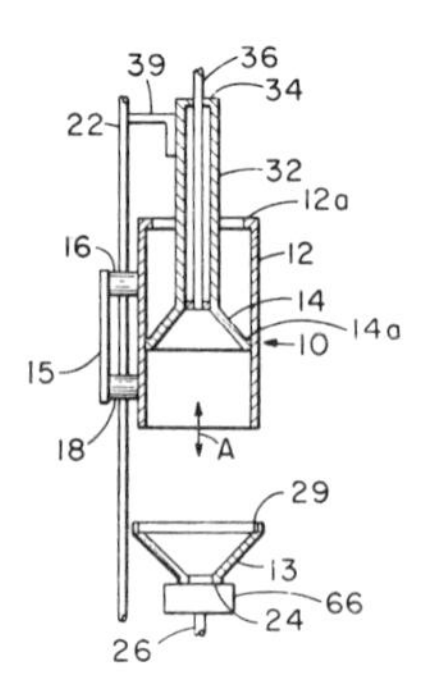

Source: G.S. Raynor; U.S. Patent 3,488,993; January 13, 1970

at the lower left of the figure. Cover **14** has an annular rim **14a**, which, when the body is in its lower position shown in the upper main view of the figure, contacts lip **12a** and thereby seals the collector at this edge. The body is slidable between a lowermost position wherein the collector is completely enclosed and an uppermost position where the lower edge of the body is adjacent to upper cover **14** so that the space between covers **13** and **14** is fully exposed to ambient air.

In the detail at the lower left of the figure, the body is in an intermediate position, either on the way down or on the way up. When the body is lowered, the bottom edge of the body contacts lower cover **13** before rim **14a** and lip **12a** contact each other so that the spring supports in legs **28** are depressed slightly insuring positive sealing with cover **13** when lip **12a** is pressed against rim **14a.**

Cover **14** is provided with a hollow extension **32** interconnected to rod **22** by way of a structural element **39** to insure rigidity of the assembly. It is understood, of course, that any convenient structural arrangement may be utilized. Mounted in convenient fashion near the collector is a pneumatic cylinder **38** from which extends rod **42** to one end of a link **44** whose other end is rigidly attached to member **15.** The upper end of the rod is connected to a piston (not shown) within cylinder **38,** which piston, as is understood in the art is moved upwardly or downwardly in accordance with any unbalance of fluid (in this case air) pressure on opposite sides thereof. Hence movement of the piston within cylinder **38** causes similar movement of collector body **12**. A pair of conduits **46** and **48** extending from a solenoid actuated valve V_1 supplies high pressure fluid to either the upper or lower ends, respectively, of pneumatic cylinder **38**. Pressurized air is supplied into either V_1 by way of a conduit **52** or cover **14** by way of conduit **36** from a solenoid actuated valve V_2 which is supplied with pressurized air from conduit **56** and air cleaning filter **54**. Fluid pressure for the system is provided by a pump **58** which receives its air from a conduit **62** and supplies the air under pressure to filter **54** by way of conduit **64**. Filter **66**, for collecting the particulate or aerosol constituents present in the trapped air is located against end surface **24** of cover **14**, its entry inlet being in communication with opening **25**. A solenoid actuated valve V_3, located at the interconnection of tube **26** from filter **66**, conduit **67** open to air, and conduit **62** permits either ambient air to enter the system by way of conduit **67** into conduit **62**, thereby bypassing filter **66**, or directing air from filter **66** to conduit **62**. It is understood that filter **66** may be any collecting device suitable for removing or measuring any of the constituents present in the trapped sample.

In the operation of the apparatus just described, initially, the collector body is in its raised position so that ambient air is flowing freely in the space between covers **13** and **14**. Solenoid valve V_3 communicates conduit **67** to conduit **62** to permit ambient air to bypass filter **66** and be supplied to pump **58** which operates throughout the whole cycle. Valve V_2 supplies the compressed air to valve V_1 which passes the air through conduit **48** to the bottom of cylinder **38** holding body **12** in its raised position. Valve V_1 is provided with a pressure relief valve (not shown) which discharges excess air being directed into either conduit **46** or **48**.

To start the sampling cycle, valve V_1 is actuated to direct compressed air to the upper portion of pneumatic cylinder **38** (by way of conduit **46**) which then causes the body to descend. When the collector becomes completely closed, thereby entrapping within a substantial body of ambient air, valve V_2 is actuated to direct the compressed air into the collector and valve V_3 is actuated to direct the discharge from filter **66** to conduit **62** so that for a period of time there is continuous closed cycle flow of air through collector **10**, filter **66**, pump **58** and cleaning filter **54**.

The flow stream of air from conduit **36** to opening **25** through the collector entrains air along the walls of the latter and gradually purges the collector of all the entrapped sample. The frustoconical shape of end covers **13** and **14** prevents or reduces the buildup of particulate within the collector.

After a predetermined period of such closed cycle flow, during which time several volumes of the collector may be passed through filter **66** in which the particulate or aerosol is collected, valve V_3 is actuated to direct flow from conduit **67** to conduit **62**, valve V_2 is actuated to direct flow to valve V_1 and the latter is actuated to direct flow into the lower entrance to cylinder **38**, thereby causing the body to rise and exposing the interior of the collector to ambient air.

The cycle is then repeated, and as many cycles as desired may be utilized to collect the desired sample. In order to obtain time discrimination in the sampling, if desired, filter **66** may employ a collection or sampling tape so that a distribution can be obtained with respect to time.

The purpose of air cleaning filter **54** is to provide clean air to the collector at all times so that all collections will have come from the ambient air sample trapped within the collector. It is understood that solenoid valves V_1, V_2 and V_3 may be actuated by hand, if desired, or automated through the use of limit switches, timers and relays as is well understood in the art.

An apparatus which has been constructed according to this process, employed a collector about 12 inches in height and 6 inches in diameter. A typical operating cycle is shown in the following table.

Step	Collector	V_1	V_2	V_3	Time (seconds)
1	Open	Open to 48	Open to V_1	Open	4
2	Closing	Open to 46	do	do	3
3	Closed	do	Open to collector	Open to filter 66	20
4	Opening	Open to 48	Open to V_1	Open	3

From the description it is seen that the apparatus is relatively simple in construction and operation and does not require elaborate procedures or testing to ensure representative sampling. It is dependable and any alteration of ambient airflow patterns that does occur is limited to a much smaller portion of the sample than in previous arrangements known in the art.

It is thus seen that there has been provided an improved arrangement for sampling gaseous fluids for particulates and aerosols. While only a preferred form of this process has been described, it is understood that many variations thereof are possible without departing from the principles of the process. For example, the collector may be used in pairs with staggered cycles so that sampling is continuous.

WATER SAMPLING

The topic of wastewater sampling has been discussed by K.H. Mancy (12). He has pointed out that the significance of a chemical analysis is no greater than that of the sampling program. Ideally speaking, a representative sample does not exist. Attempts are made however to come as close as possible to sample aquatic environment without disturbing its physicochemical and biochemical characteristics. There is no universal procedure for sampling applicable to all kinds of natural and wastewaters. The analyst has to design his sampling program based on a complete understanding of the purpose of analysis and the type of parameters to look for.

The most important requirements for a satisfactory sample are that it be both valid and representative. For a sample to be valid, it has to be one which has been collected by a process of random selection. Random selection is one of the most basic, yet most frequently violated, principles in development of a sampling program. Any method of sampling that sacrifices random selection will impair statistical evaluation of the analytical data. If nonrandom sampling procedures are contemplated, perhaps for significant reasons of convenience, it is highly desirable to first demonstrate that the results of the analysis check those which would be obtained by random sampling. This check would be essential prior to any statistical evaluation of the data.

A satisfactory sample is not only randomly drawn, but also is representative. This means that the composition of the sample should be identical to that of the water from which it was collected; the collected sample should have the same physicochemical characteristics as the sampled water at the time and site of sampling.

Planning for a sampling program should be guided by the overall objectives of analysis. Major factors of concern for any sampling program are: (a) frequency of sample collection; (b) total number of samples; (c) size of each sample; (d) sites of sample collection; (e) method of sample collection; (f) data to be collected with each sample; and (g) transportation and care of samples prior to analysis.

Frequency of sampling will depend to a large extent upon the frequency of variations in composition of the water to be sampled. There are two principal types of sampling procedures commonly used for analysis of natural and wastewaters. The first type is that which yields instantaneous spot or grab samples, while the second type yields integrated continuous or composite samples. A grab sample is a discrete portion of a wastewater taken at a given time; a series of grab samples reflects variations in constituents over a period of time. The size of such individual samples will depend on the objectives and methods of analysis, and on the required accuracy. The total number of grab samples should satisfy the statistical requirements of the sampling program. Composite samples are useful for determining average conditions, which when correlated with flow can be used for computing the material balance of a stream of wastewater over a period of time. A composite sample is essentially a weighted series of grab samples, the volume of each being proportional to the rate of flow of the waste stream at the time and site of sample collection. Samples may be composited over any time period such as 4, 8 or 24 hours, depending on the purposes of analysis. Selection of sampling sites should be made with great care. A field survey is often useful in planning for site selection. In the case of sampling of a stream, special consideration should be given to sources of waste discharge, dilution by tributaries, and changes in surrounding topography (13)(14).

Sampling of wastewater from pipes or conduits is more complicated than stream sampling, especially when the water to be sampled is under pressure. For example, in the case of a chemical treatment plant, selection of sampling sites may require extensive investigations and preliminary checking of samples from a number of effluent outlets. Proper positioning of the sampling outlet within the cross-section of a conduit is essential for obtaining a representative sample, particularly for conduits of large diameter.

The choice of a sampling site within the cross-section of a conduit is best done by examining and comparing samples drawn from several points along the vertical and horizontal diameters of the conduit. The cross-sectional area of the opening or inlet of the sampling line should be such that the flow of water in this line is proportional to the flow of the water in the conduit. An elaborate discussion on sampling of water from pipes and conduits can be found in the *ASTM Manual on Industrial Water and Industrial Wastewater* (15).

Wastes discharged by industry are of great variety, and sampling must be tailored to suit the particular characteristics of a given wastewater. Sampling procedures can be expected to vary widely from one wastewater to another. Special procedures have been reported for use with water sampled under reduced or elevated pressure and/or temperature (16). Procedures and equipment used for the sampling of waters containing dissolved gases and volatile constituents susceptible to loss upon aeration have been described.

A certain amount of precaution is sometimes required in sampling processes for reasons of safety. For example, strict precautionary measures (17) should be followed in taking samples from deep manholes to guard against accumulation of toxic and explosive gases and insufficiency of oxygen. Sampling can be accomplished by either manual or automatic means, again depending on the purpose of analysis and method of sampling. A grab sample is usually collected manually. When it is necessary to extend sampling over a considerable period of time, or when a continuous (repetitive) record of analysis at a given sampling point is required, automatic sampling equipment is commonly used.

Continuous sampling equipment, correctly designed and installed, will provide more frequent samples, tend to eliminate human errors, and in many cases be economically more feasible. A variety of automatic sampling equipment suitable for water-sampling under variable conditions and for different purposes, is presently available (16).

The maintenance of complete records regarding the source of the sample and the conditions under which it has been collected is an inherent part of a good sampling program. This is of particular importance in field, river, or in-plant surveys, where a great number of samples are collected from different sources and under variable conditions. For illustrative purposes, the U.S. Geological Survey has defined the minimum data required for samples of surface and groundwaters (17).

Surface Waters	Ground Waters
Name of water body	Geographical and legal locations
Location of station or site	Depth of well
Point of collection	Diameter of well
Data of collection	Length of casing and positions of screens

Surface Waters

Time of collection
Gage height or water discharge
Temperature of the water
Name of collector, and
Weather and other natural or other man-made factors that may assist in interpreting the chemical quality

Ground Waters

Method of collection
Point of collection
Water bearing formations
Water level
Yield of well in normal operations
Water temperature
Principal use of water
Name of collector
Date of collection
Appearance at time of collection, and
Weather or other natural or man-made factors that may assist in interpreting chemical quality

One of the most important aspects of the sampling process is the care and preservation of the sample prior to analysis. This point cannot be overemphasized. A water analysis is of limited value if the sample has undergone physicochemical or biochemical changes during transportation or storage. These changes are time dependent, but they usually proceed slowly. In general, the shorter the time that elapses between collection of a sample and its analysis, the more reliable will be the analytical results. Certain constituents may, however, require immediate analysis at the sample site.

Certain determinations are more sensitive than others to the method of handling of water samples before analysis. Changes in temperature and pressure may result in the escape of certain gaseous constituents (e.g., O_2, CO_2, H_2S, Cl_2, CH_4), or the dissolution of some atmospheric gases (e.g., O_2). It is recommended, therefore, that determinations for gases be done in the field, or, to fix such materials as O_2, Cl_2 and H_2S, the sample should be treated upon collection with stable oxidizing or reducing agents.

It is also recommended that the temperature and pH of the water be determined at the site of sampling. Changes in temperature and pH may cause changes in the solubility of dissolved gases and certain nonvolatile constituents, resulting in their separation from aqueous phase. Carbonic acid-bicarbonate-carbonate equilibria may be shifted to release gaseous CO_2, or to precipitate certain metal carbonates. Similarly, shifts in hydrogen sulfide-sulfide equilibria due to changes in pH and/or temperature may result in the escape of H_2S or the precipitation of metal sulfides.

Heavy metal ions may undergo a variety of physicochemical transformations during sample handling. It has been recommended that for analyses for Al, Cr, Cu, Fe, Mn and Zn, samples should be filtered at the site of collection and acidified to about pH 3.5 with glacial acetic acid (17). Acidification tends to minimize precipitation, as well as sorption on the wall of the sample container. Since acetic acid may stimulate growth of molds, it may be necessary to add a small quantity of formaldehyde to the sample solution as a preservative.

Another major point of interest for handling water samples is the effect of biological activity on the sample characteristics. Microbiological activity may be effective in changing the nitrate-nitrite-ammonia balance, in reducing sulfate to sulfide, in decreasing the dissolved oxygen content, biochemical oxygen demand (BOD), organophosphorous compounds and any readily degraded organic compound. Freezing of water samples is helpful in minimizing changes due to biological activity. Certain chemical preservatives, such as chloroform or formaldehyde, are sometimes added to water samples for this purpose.

That it is practically impossible to handle and process a water sample without changing its characteristics should be recognized. The best chance for error-free procedure lies in the use of in situ analyses. In the end, the dependability of even a well planned sampling program rests upon the experience and good judgment of the analyst.

A wastewater sampling device which has been developed by R.P. Farrell, Jr. (18) includes a grinder unit and a pump to deliver raw wastewater from the grinder into a sampling chamber. A sampler device which is operated automatically when the pump is running takes small samples of the wastewater flowing through the sampling chamber at fixed intervals and delivers them into a reservoir which may be refrigerated for subsequent analysis.

A wastewater sampling system developed by W.H. Merrill, Jr. (19) is one in which uniformly sized effluent samples are collected at the end of each period during which a predetermined volumetric flow of effluent has occurred.

A flow indicating device in response to the predetermined volume flow actuates a switch to operate the sample collecting device which utilizes an arm having a sampling cup at one end and a link at the other, the arm being reciprocated by a motor near its link end. In this way, a composite of these samples accurately reflects the composition of the effluent over the collection period.

A wastewater sampler developed by L.G. Lynn and D.A. Quadrini (20) is one in which effluent is discharged into a flume to flow past a selectively operable sampler device, and an adjacent probe develops and transmits to a remote control point a 4 to 20 ma. signal the amplitude of which is proportionate to the effluent flow rate.

At the control point the signal is applied to an integrator which produces an output voltage proportionate to the quantity (gallons) of effluent that has passed the probe in a preceding interval.

Each time this voltage reaches a predetermined value a threshold circuit resets the integrator and pulses a first register to record the quantity of effluent for a given period, and simultaneously pulses a presettable counter, which produces a sampler enabling signal every time the counter reaches zero and resets. This enabling signal momentarily energizes a solenoid in a remote sampler to cause it to pump a sample of wastewater from the flume to a sample receptacle.

References

(1) Barth, D.S. and Shearer, Jr., S.D. in paper before the Environmental Quality Sensor Workshop, Las Vegas, Nevada, November 30 to December 2, 1971, Washington, D.C., EPA (1972).
(1a) Brenchley, D.L., Turley, C.D. and Yarmac, R.F., *Industrial Source Sampling,* Ann Arbor, Mich., Ann Arbor Science Publishers (1973).
(2) Thoen, G.N.; U.S. Patent 3,559,491; February 2, 1971; assigned to Weyerhaeuser Company.
(3) Couchman, J.C. and Applebay, G.W.; U.S. Patent 3,710,557; January 16, 1973; assigned to the U.S. Atomic Energy Commission.
(4) Riggs, R.K.; U.S. Patent 3,765,247; October 16, 1973; assigned to National-Southwire Aluminum Company.
(5) Quigley, J.J.; U.S. Patent 3,304,783; February 21, 1967; assigned to Inland Steel Co.
(6) Manka, D.P.; U.S. Patent 3,748,906; July 31, 1973; assigned to Jones & Laughlin Steel Corporation.
(7) Neuberger, E.D. and Junker, T.J.; U.S. Patent 3,641,821; February 15, 1972; assigned to Calgon Corporation.
(8) Keefer, R.D. and Wyman, Jr., J.S.; U.S. Patent 3,654,801; April 11, 1972; assigned to The Bendix Corporation.
(9) Scoggins, M.F.; U.S. Patent 3,540,261; November 17, 1970; assigned to The Center for the Environment and Man, Inc.
(10) Teel, D.M. and Putnam, Jr., J.H.; U.S. Patent 3,657,920; April 25, 1972; assigned to the U.S. Atomic Energy Commission.
(11) Raynor, G.S.; U.S. Patent 3,488,993; January 13, 1970; assigned to the U.S. Atomic Energy Commission.
(12) Mancy, K.H., in *NBS Special Publication No. 351,* "Analytical Chemistry: Key to Progress on National Problems," Washington, D.C., U.S. Government Printing Office (1972).
(13) Velz, C.J., *Sewage Industrial Wastes* 22, 666 (1950).
(14) Haney, P.D. and Schmidt, J., *Technical Report W58-2,* Washington, D.C., U.S. Public Health Service (1958).
(15) American Society for Testing Materials, *Manual on Industrial Water and Industrial Wastewater,* Philadelphia, Pa.

(16) Rainwater, F.H. and Thatcher, L.L.; *Supply Paper 1454;* Washington, D.C., U.S. Geological Survey (1960).
(17) U.S. Department of Public Works, *Public Works* 85 (1954).
(18) Farrell, Jr., R.P.; U.S. Patent 3,478,596; November 18, 1969; assigned to General Electric Co.
(19) Merrill, Jr., W.H.; U.S. Patent 3,507,156; April 21, 1970.
(20) Lynn, L.G. and Quadrini, D.A.; U.S. Patent 3,719,081; March 6, 1973; assigned to Tri-Aid Sciences, Inc.

ACIDS AND ALKALIS

Industrial wastes of many types, and from a variety of processes, exhibit extreme pH values which can greatly influence the quality and aquatic life of receiving waters. Table 18 lists representative industries, and gives waste pH characteristics associated with each industry (1). Wastes may be released on a batch basis as acid or alkaline processing vats are dumped, or released as a continuous waste stream resulting from a specific industrial process (e.g., rinsing). The volume and composition of a waste stream containing acidic or basic compounds can be quite variable. Dickerson and Brooks (2) have described the waste stream from a cellulose production line of Hercules Powder Corporation as varying in volume from 3,000 to 9,000 gpm with acidity, measured as free sulfuric acid, ranging from 30 to 1,100 mg./l.

In addition to the free acids and bases (H_2SO_4, NaOH, etc.) typically encountered in industrial waste streams, some industries produce combined acidic or basic salts, which form weak acids or bases by hydrolysis upon dilution of the waste by receiving water. Typical of such industrial effluents are those originating from steel mills and other metal processing industries, where acid or alkaline cleaning solutions are employed. Control of effluent and stream pH requires treatment not only for the free acids and bases in the waste, but for the acidic and/or basic salts present as well. Waste treatment processes employed for the free acids and bases are effective in controlling acid and basic salts.

TABLE 18: pH CHARACTERISTICS OF INDUSTRIAL WASTES

Industrial Process	Waste pH Characteristics
Food and Drugs:	
Pickling	acidic or alkaline
Soft drinks	high pH
Apparel:	
Textile	highly alkaline
Leather processing	variable
Laundry	alkaline
Chemicals:	
Acids	low pH
Phosphate and Phosphorus	low pH
Materials:	
Pulp and paper	high or low pH
Photographic products	alkaline
Steel	mainly acid, some alkaline
Metal plating	acids
Oil	acids
Rubber stores	variable pH
Naval stores	acid
Energy:	
Coal mining	acid mine drainage
Coal processing	low pH

Source: Nemerow, N.L., *Theories and Practice of Industrial Waste Treatment,* (1963)

Measurement in Air

The reader is referred to the sections on specific acids such as Sulfuric Acid for information on acid determination in air.

Measurement in Water

An apparatus developed by J.P. McKaveney and C.J. Byrnes (3) involves mixing a known quantity of an acidic aqueous solution, the concentration of which is to be determined with a fluoride in ionic form permitting the release of the fluoride ion in the mixture. The fluoride is introduced in at least a stoichiometric amount with respect to the acid so that there are sufficient fluoride ions available for reaction with the available hydrogen ions of the acid to form hydrofluoric acid.

This mixture is placed in an electrolytic cell having a first electrode, which comprises the anode, of a semiconducting material, such as n-type silicon material. A second electrode, which is the cathode is additionally provided and is constructed of a conducting material substantially resistant to chemical attack by the mixture, such as stainless steel, platinum or titanium. An electrical potential is impressed between the electrodes, which results in the production of a limiting electrical current between the electrodes. The magnitude of electrical current has been found to be substantially, linearly proportional to the magnitude of the acid concentration in the acidic aqueous solution.

Figure 20 shows the apparatus which may be used in the conduct of this method. The electrolytic cell **10** consists of an electrolyte container, such as a standard beaker, designated as **12**, within which the electrolyte mixture **13**, as described above, is contained. Immersed within the electrolyte of the container is an electrode assembly consisting of a rod-shaped anode **14** of a semiconducting material surrounded by a cylindrical cathode **15**. In this manner, the cathode shields the anode to prevent damage thereto. The cathode **15** is constructed from stainless steel or other conducting materials suitable to resist chemical attack by the electrolyte. Both of the electrodes **14** and **15** are secured at their upper ends to a polyvinyl chloride cap, designated as **16**.

FIGURE 20: APPARATUS FOR DETERMINING ACID CONCENTRATIONS

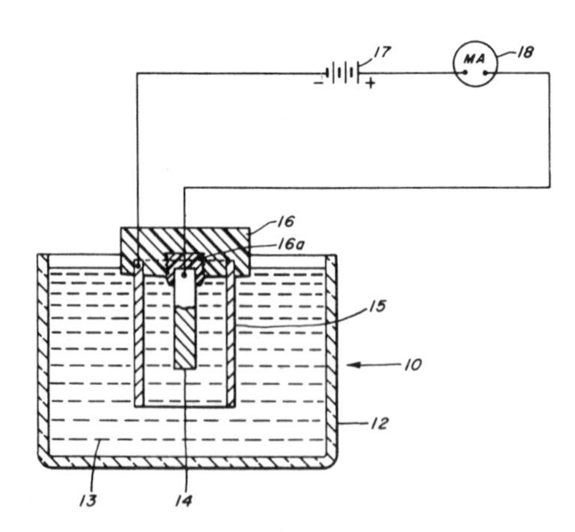

Source: J.P. McKaveney and C.J. Byrnes; U.S. Patent 3,528,778; September 15, 1970

The anode **14** is connected to cap **16** by a polyvinyl chloride nut **16a**, which is threaded into a tapped opening in the bottom of cap **16**. The end of the anode **14** is embedded within the nut **16a**. This arrangement is provided to permit easy replacement of the anode without necessarily requiring replacement of the cathode and vice versa. The anode **14** is connected to the positive pole of a battery **17**, and the cathode **15** is connected to the negative pole of the battery. Between the battery **17** and the anode **14**, there is electrically connected a conventional microammeter **18**.

Although the device shown is particularly adapted for use in determining the acid concentrations of steel pickling baths, it is also capable of many other uses. For example, by using relatively larger volume samples and relatively less water dilution, the method is applicable to very dilute acid solutions, such as metal-rinse waters and natural or other waters. This would be extremely useful in an application such as stream-pollution monitoring; this could conceivably be achieved on a substantially continuous basis.

A method developed by S.B. Dalgaard (4) involves monitoring the pH value in a continuously flowing liquid sample. The method comprises splitting a continuously flowing sample stream into two streams, passing one stream through a cation exchange resin column, and passing the other stream through an anion exchange resin column, measuring the specific electrical conductivity of the cation exchange resin column effluent, and obtaining the pH value of the sample by solving the equation

$$pH = 7 - \log \frac{Kcat}{1.65\ Kan}$$

where Kcat is the specific conductivity in $ohms^{-1}\ cm.^2$ of the cation exchange resin effluent, and Kan is the specific conductivity in $ohms^{-1}\ cm.^2$ of the anion exchange resin effluent.

Figure 21 is a schematic diagram showing the arrangement of apparatus for the conduct of the method. Starting at the top of the figure, a liquid sample of unknown pH value flows through a conduit **10** until it reaches a stream splitting joint connected to branch conduits **12** and **13** where the flow of the liquid sample is split into two streams, one stream going through conduit **12** into a cation exchange column **14** and the other stream going through conduit **13** into an anion exchange column **15**. The cation exchange column **14** is packed with cation exchange material selected from cation exchange materials available commercially and the anion exchange column **15** is packed with anion exchange material also selected from anion exchange materials available commercially.

In passing through the cation exchange column **14** the sample stream from conduit **12** exchanges cations for hydrogen, and in passing through the anion exchange column **15** the sample stream from conduit **13** exchanges anions for hydroxyl ions. Thus, the effluent from the cation exchange column **14** will contain primarily hydrochloric and sulfuric acid and the effluent from the anion exchange column **15** will contain primarily sodium, potassium, and ammonium hydroxides.

Positioned in the conduits **19** and **20** leading from the cation and anion exchange columns **14** and **15** respectively are conductivity cells **16** and **17** of conventional type including a pair of spaced electrodes immersed in the liquid flowing in the conduits **19** and **20**. The electrodes in the conductivity cells **16** and **17** are connected to conductivity monitors **16'** and **17'** respectively which produce voltage outputs corresponding to the conductivity, Kcat, of the cation exchange column effluent, and to the conductivity, Kan, of the anion exchange column effluent.

The conductivity cells **16** and **17** are preferably of the platinized, platinum conductivity cell type while the conductivity monitors **16'** and **17'** are selected from conventional types including solubridges, balanced bridges or other suitable electrical devices capable of registering the resistance between the pair of electrodes immersed in the effluent streams.

The effluents from the conductivity cells **16** and **17** may be discharged as acidic and alkaline wastes respectively.

FIGURE 21: APPARATUS FOR CONTINUOUS pH MONITORING

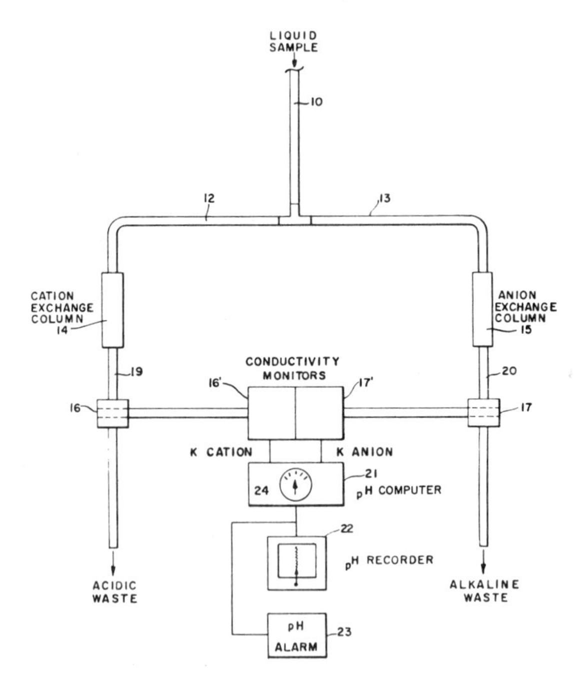

Source: S.B. Dalgaard; U.S. Patent 3,681,025; August 1, 1972

By providing a pH monitor **21** which is capable of electrically solving the equation

$$pH = 7 - \log \frac{Kcat}{1.65\ Kan}$$

and applying the voltage outputs from the conductivity monitors **16'** and **17'** as inputs to the pH monitor, a continuous computation of the pH value of the sample can be made. A voltage output from the pH monitor **21** corresponding to the computed pH value of the sample may be read directly from an indicator meter **24** calibrated in pH values, or the pH monitor output may be applied to a pH strip recorder **22**, and alarm **23** or to other devices which utilize the pH monitor output for useful purposes.

References

(1) Nemerow, N.L., *Theories and Practice of Industrial Waste Treatment,* Reading, Mass., Addison-Wesley Publ. Co. (1963).
(2) Dickerson, B.W. and Brooks, R.M., *Ind. Eng. Chem.,* 42, 599-605 (1950).
(3) McKaveney, J.P. and Byrnes, C.J., U.S. Patent 3,528,778; September 15, 1970; assigned to Hach Chemical Co.
(4) Dalgaard, S.B., U.S. Patent 3,681,025; August 1, 1972.

ALDEHYDES

Aldehydes are products of incomplete combustion of hydrocarbons and other organic materials. They are emitted into the atmosphere by exhaust from motor vehicles, incineration of wastes, and combustion of fuels (natural gas, fuel oils, and coal). Furthermore, aldehydes are formed from the photochemical reactions between nitrogen oxides and certain hydrocarbons, which are also emitted from the sources mentioned above. Thus, the ambient air is continually being polluted by aldehydes from emission sources and from atmospheric photochemical reactions. Moreover, aldehydes themselves can undergo photochemical reactions yielding oxidants (including ozone, peroxides, and peroxyacyl nitrate compounds) and carbon monoxide among the major products.

At low concentrations the principal effect of aldehydes on both humans and animals is irritation of the eyes and upper respiratory tract. This is particularly true for the lower molecular weight aldehydes. The unsaturated aldehydes are several times more toxic than the saturated aldehydes. In addition, aldehydes have been involved in plant damage. In some cases, the damage appears to be a result of oxidants produced by the photochemical reaction of aldehydes.

In the air pollution field major interest has been shown in two specific aldehydes: formaldehyde and acrolein. This is partly due to their effects on humans and to the fact that their concentrations are generally higher than those of other aldehydes present in the atmosphere. In addition, some reports indicate that formaldehyde and possibly acrolein may contribute to the odor and the eye irritation commonly experienced in polluted atmospheres.

Vehicle exhaust, particularly from automobiles, appears to be the major source of aldehydes. However, significant amounts may also be produced from other combustion sources such as open burning and incineration of solid waste materials, and the burning of fuels (gas, fuel, oil, coal). Another source of aldehyde emission is the thermal decomposition of hydrocarbons by pyrolysis in the presence of air or oxygen. Sources of these emissions include chemical manufacturing plants and industries that use drying or baking ovens to remove organic solvents in such processes as automobile painting and the manufacture of coated paper and metals.

Air sampling data indicate that plants manufacturing formaldehyde may be local sources of aldehyde pollution. However, the major amount of aldehyde pollution in some areas of the United States is from the photochemical reaction between nitrogen oxides and hydrocarbons. Hydrocarbons that yield formaldehyde are olefins, and to a lesser degree, other aldehydes and aromatic hydrocarbons. Diolefins produce most of the atmospheric acrolein. Some data indicate that in certain areas over two-thirds of the atmospheric aldehydes may have resulted from photochemical reactions. Of course, the sources that emit aldehyde pollutants are generally the same as those emitting hydrocarbons and nitrogen oxides.

In addition, aldehydes themselves may undergo photochemical reactions. They may produce, at low partial pressures in the presence of nitrogen oxides, other products such as carbon monoxide, lower aldehydes, nitrates, and oxidants. The oxidants produced include ozone, peroxyacyl nitrates, and alkyl hydroperoxides (hydrogen peroxide in the case of formaldehyde). No peroxyacids or diacetyl peroxides are found at low partial pressures of aldehydes.

The most characteristic and important effect of aldehydes, particularly of low molecular weight aldehydes, for both humans and animals is primary irritation of the eyes, upper respiratory tract, and skin. The observed symptoms in humans from inhalation of low concentrations of aldehydes include lacrimation, coughing, sneezing, headache, weakness, dyspnea, laryngitis, pharyngitis, bronchitis, and dermatitis. In most cases, the general and parenteral toxicities of these aldehydes appear to be related mainly to these irritant effects. The unsaturated aldehydes are several times more toxic than the corresponding aliphatic aldehydes. Also, the toxicity generally decreases with increasing molecular weight within the unsaturated and aliphatic aldehyde series. Sensitization has occurred from contact with formaldehyde solutions and other aldehydes, but sensitization of the pulmonary tract rarely is produced by inhalation of aldehydes. The anesthetic properties of aldehydes are generally overshadowed by the stronger irritant effects. Furthermore, concentrations that can be tolerated via inhalation can usually be metabolized so rapidly that systemic symptoms do not occur.

Formaldehyde concentrations as low as 600 μg./m.^{3} have been shown to cause cessation of the ciliary beat in rats. Animal experiments have shown that aldehydes can affect the responses of the respiratory system, causing such effects as an increase in flow resistance and in tidal volume and a decrease in the respiratory rate. Exposure of rats to 150 μg./m.^{3} of acrolein for 2 months caused a rise in the number of luminescent leukocytes in the blood. Exposure of animals to high concentrations of aldehydes has been shown by several investigators to produce edema and hemorrhages of the lungs and fluid in the pleural and peritoneal cavities. In a Russian study, formaldehyde was found to prolong the mean duration of pregnancy in rats and decrease the number of offspring. In addition, the weight of the lungs and liver of the offspring was less than that of the controls' offspring, but other organs exhibited an increase in weight.

Animal experiments also indicate possible synergistic effects between aldehydes and aerosols. Thus, acrolein and formaldehyde in the presence of certain inert aerosols appeared to be more toxic to mice than the pure compounds. Experiments with guinea pigs showed that formaldehyde with sodium chloride aerosols produced significant increases in the "respiratory work" compared with the effect of the pure vapor. In addition to the toxic effects, aldehydes may contribute to the annoyances of odor and eye irritation caused by polluted air. Aldehyde concentrations have been shown to correlate with the intensity of odor of diesel exhaust and the intensity of eye irritation during natural and chemically produced smogs. Data indicate that as little as 12 μg./m.^{3} of formaldehyde can cause human eye response.

Aldehyde air pollution may result in oxidant-type damage to plants, although atmospheric photochemically produced products from the aldehydes may actually cause the damage rather than direct attack by aldehydes. There are no data available to indicate the effect of aldehyde air pollution on materials. Air quality standards for various foreign countries spell out levels for formaldehyde, for example, as follows:

U.S.S.R.	0.029 ppm (single exposure)
	0.01 ppm (24-hour average)
Germany	0.02 ppm (long term exposure)
	0.06 ppm (only once in 4-hour period)

Measurement in Air

In 1967 the National Air Sampling Network began to report data for aliphatic aldehydes. The data for 1967 for several cities show that the average concentrations of aldehydes ranged from 3 to 79 μg./m.^{3} and that the maximum values ranged from 5 to 161 μg./m.^{3}. A Los Angeles area report indicates that the maximum values for two "smog" days in 1968 were 208 μg./m.^{3} for aliphatic aldehydes, 163 μg./m.^{3} for formaldehyde, and 27 μg./m.^{3} for acrolein. Generally, formaldehyde accounts for 50% or more of the total aldehydes, while acrolein accounts for about 5%.

There are numerous methods of analysis for aldehydes; too many to be covered thoroughly here. Only those methods that have been used for or appear applicable to determining formaldehyde, acrolein, or the aliphatic aldehydes in air samples or emission source samples will be discussed. Other methods of analysis for aldehydes can be found in the reviews of Altshuller (1), Sawicki (2), Altshuller et al (3), Farr (4) and Reynolds and Irwin (5).

Sampling Methods

Generally, common sampling methods employ bubblers or impingers containing a reactive reagent. In some cases the reactive reagent may result in a color product that may be used in the analysis procedure. Examples of the commonly used reagents are 3-methyl-2-benzothiazolone hydrazone (MBTH), (6)(7)(8)(9) sodium bisulfite, (10)(11)(12) and a mixture of sodium bisulfite and sodium tetrachloromercurate-(II) (13)(14) for "aldehydes"; chromotropic acid (3)(7)(9)(15) for formaldehyde; and 4-hexylresorcinol (7)(9) for acrolein. These reagents are preferred because they have high collection efficiencies (generally two bubblers in series yield 95% or better collection efficiencies) and produce fairly stable nonvolatile products, thus avoiding excess loss of aldehydes via evaporation or formation of undesirable by-products. In some cases, water is used as the collection medium (16) (17)(18) reportedly with high efficiency.

Qualitative Methods

The presence of aldehydes can be determined by infrared spectroscopy. The carbon-hydrogen stretch vibration of the aldehydic group absorbs as a doublet in the 3.5 to 3.7 μ region. Furthermore, the carbonyl of an aldehydic group has an absorption band in the 5.7 to 6.0 μ region, which, unlike the carbonyl bands of ketones and carboxylic acids, disappears when a chloroform solution is treated with phosphorus pentachloride (19). Many colorimetric methods applicable to formaldehyde, acrolein, and aldehydes have been used for spot tests or dectector tube methods. Some of the more common methods are summarized in Table 19.

TABLE 19: SUMMARY OF QUALITATIVE COLORIMETRIC DETERMINATION METHODS FOR ALDEHYDES

		Limits of Identification in Micrograms			
Reagent	Color	Aldehydes	Formaldehyde	Acrolein	References
Indole	Orange to red	~0.05 - 1	0.2	--	31
Fuchsin (Schiff method)	Violet to blue	~1 - 30	1	--	22
4-Phenylazo-phenyl-hydrazine sulfonic acid	Red to blue	0.2 - 0.4	0.25	0.2	22
2-Hydrazino-benzothiazole + p-nitrobenzenediazonium fluoborate	Blue to green	0.2 - 200	0.2	0.3	22,23
2-Hydrazino-benzothiazole (HBT)	Blue	0.01 - 3.0	0.01	--	24
3-Methyl-2-benzothiazolone hydrazone (MBTH)	Blue	0.1 - 80	0.1	--	25
(J-acid) 6-amino-1-naphthol-3-sulfonic acid	Blue	0.01 - 11	0.03	--	26

Source: Report PB 188,081

Quantitative Methods (Aldehydes in General)

Recently, one method has been used extensively to determine total water-soluble "aliphatic" aldehydes in atmospheric sampling (7)(27)(28)(29)(30). Since 1967 this method has been used by the National Air Sampling Network of the National Air Pollution Control Administration (31) according to the procedure described by Morgan et al (8). This method was first proposed by Sawicki et al (25) and refined by Hauser and Cummins (9)(32). The latter method used 3-methyl-2-benzothiazolone hydrazone (MBTH), with sulfuric acid added to avoid the dilution necessary in the earlier procedures.

The sensitivity is approximately 2.4 μg./m.^{3} (2 ppb), measured as formaldehyde. Aldehydes react with the MBTH to form a very stable product, which, upon oxidation with ferric chloride, produces a blue cationic dye that is measured at 628 mμ. Compounds which interfere with the analysis include aromatic amines, imino heterocyclics, carbazoles, azo dyes, stilbenes, Schiff bases, dinitrohydrazone (DNP) aldehyde derivatives, and compounds containing the p-hydroxy styryl group (9). Since most of these compounds are not gaseous or water-soluble, they will not generally interfere in analysis of atmospheric samples. Formaldehyde reacts in this procedure about 25% greater than the other aliphatic aldehydes and about 300% greater than branched-chain and unsaturated aldehydes. Altshuller and Leng (6) have suggested that a correction factor of 1.25 be used to take into account the various aldehyde responses to the method.

Most other colorimetric procedures that have been described in the literature show even larger response to formaldehyde in comparison with other aldehydes, and therefore, should not be used for quantitative determination of aldehydes (1). Such methods include chromotropic acid, J-acid, phenyl J-acid, Schiff's reagent, and phenylhydrazine reagent (Schryver's method).

A continuous monitor method for aldehydes (14) based on the method of Lyles et al (18), is a modified Schiff procedure using rosaniline and dichlorosulfuromercurate. The sensitivity is reported as 12 μg./m.^{3} (0.01 ppm), with a collection efficiency of greater than 90%. Nitrogen dioxide can cause interference in concentrations of 0.5 ppm or more and its response to formaldehyde is greater than to other aldehydes. Infrared spectroscopy has been used to determine aldehydes in irradiation chamber studies (33)(34). The carbon-hydrogen stretch vibration in the range of 3.5 to 3.7 μ was used.

The bisulfite method has been widely used for analysis of aldehydes in atmospheric samples (35)(36) auto exhaust (37), diesel exhaust (38)(39) and incinerator effluents (40). This method measures both aldehydes and ketones and thus really measures carbonyls (1). The sensitivity is not high, and the limit of applicability is approached with air sampling analysis. Therefore, this method is not very satisfactory for aldehyde analysis in air pollution. Numerous methods have been reported for the separation and identification of specific aldehydes by use of derivatives such as 2,4-dinitrohydrazone (2,4-DNP) (1)(2). Recently, gas chromatographic methods have been used for the separation and analysis of aldehydes that are collected as sodium bisulfite (11)(41)(42) or as 2,4-DNP derivatives (17).

A technique developed by L.H. Goodson and W.B. Jacobs (43) is one in which the presence of an organic vapor contaminant in air is determined by combining the air suspected of containing such vapor with a reactive halogen compound to convert the organic contaminant to the corresponding, volatile, covalent halide; then removing vaporous, ionically bound reaction by-products from the air; and then detecting the remaining covalent halide through the use of a halogen leak detector or other suitable halogen sensors. The process may proceed on a continuous or intermittent basis, and has general application with respect to the detection of alcohol, aldehyde or ketone vapor contaminants.

A technique developed by R.M. Neti and T.J. Kelly (44) involves determining aldehydes, unsaturated hydrocarbons and ketones in gases and liquids. A sample containing one or more of these constituents is photoexcited by ultraviolet radiation in the presence of oxygen to photooxidize the constituents thus forming peroxy acids. Means such as spectrophotometric techniques or reacting the acid with an aqueous halide solution to convert the halide ions to free halogen which can then be determined electrochemically or spectrophotometrically are provided for determining the peroxy acids as a function of the content of these constituents in the sample. The technique is particularly applicable to the monitoring of air for air pollution control.

Aldehydes, unsaturated hydrocarbons and ketones are common irritating components found in polluted air. For example, formaldehyde is a major source of eye irritation. In order that dangerous levels of these constituents may be known and corrective measures may be taken, means is required for practically and inexpensively determining the level of these constituents in air.

The only aldehyde analyzer which has been available to date is an elaborate wet chemical instrument in which the air being monitored is bubbled through a batch solution of mercuric chloride, potassium chloride and sodium or potassium sulfite to form a complex which is made to further react with p-rosaniline hydrochloride to form a highly colored product which, by the depth of its color, indicates the amount of aldehyde in the air sample. This colorimetric apparatus is expensive, complicated and requires complex equipment to translate the color into an electrical signal. The procedure is time consuming, temperature dependent and does not produce an output signal which is quantitatively related to the aldehyde content of the sample gas. This prior apparatus also has the disadvantage that its sensitivity to different aldehydes varies. Finally, the aforementioned apparatus has the disadvantage that it analyzes aldehyde samples on a batch basis rather than providing a continuous monitoring of the aldehyde level of a gas sample stream.

Figure 22 is a block diagram showing the arrangement of apparatus which may be used for the conduct of the technique which overcomes these difficulties.

FIGURE 22: MONITOR FOR ALDEHYDES IN AIR

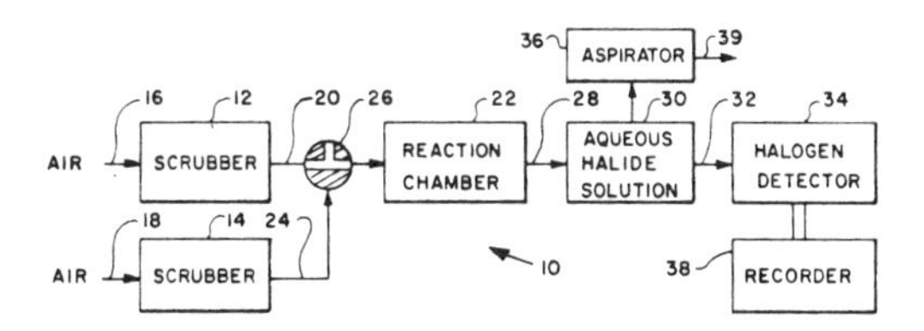

Source: R.M. Neti and T.J. Kelly; U.S. Patent 3,540,849; November 17, 1970

The apparatus **10** includes two scrubbers **12** and **14** each having a respective air inlet conduit **16** and **18**. The scrubber **12** is connected by a conduit **20** to a reaction chamber **22**. The reaction chamber **22** contains a source of radiation, not shown, for photoexciting the gas sample in the chamber. A suitable source of radiation is a double bore quartz lamp (known as Pen-Ray Lamp, Model No. 11SC-1). This lamp emits a narrow band of ultraviolet radiation centered about a wavelength of about 1849 A. at which formaldehyde is most efficiently photoexcited. However, other radiation sources could be employed, including those emitting radiation outside the UV spectrum, if they are capable of photoexciting the particular aldehyde molecules being determined and of disassociating molecular oxygen into atomic oxygen and, consequently, of photooxidizing the aldehydes to peroxy acid.

A conduit **24** connected to the outlet of the scrubber **14** joins the conduit **20** between the scrubber **12** and reaction chamber **22**. A three-way valve **26** is provided at the junction of the conduits **20** and **24** for controlling the flow of gases from the two scrubbers **12** and **14** to the reaction chamber.

The effluent from the reaction chamber **22** is conveyed by means of a conduit **28** to a chamber **30** containing aqueous halide solution. This chamber is connected by a conduit **32** to a halogen detector **34**. The air entering the inlet conduits **16** and **18** at the upstream end of the apparatus is drawn through the scrubbers **12** and **14** and reaction chamber **22**, bubbled through the aqueous halide solution in chamber **30** and vented to the atmosphere by means of an aspirator **36**. Obviously suitable pumping means could be provided at the inlet conduit **16** and **18** for conveying the air through the apparatus. The conduits preferably are tubes of chemically inert material such as polytetrafluoroalkane.

In the preferred form, the halogen detector **34** is a coulometric galvanic cell as disclosed in U.S. Patent 3,314,864 to Hersch. This cell is provided with an anode of active carbon or in some cases silver or mercury, and a cathode of inert conductive material such as platinum or graphite. The electrodes are joined by an aqueous halide electrolyte, preferably an iodide electrolyte although a bromide electrolyte could be used. The cell includes a reaction section into which the gas sample is initially delivered so that oxidizing species in the sample may convert the iodide ions of the electrolyte in the reaction section to free iodine.

This iodine is passed over the cathode of the cell where it is reduced back to iodide, and an electrical current flows in the external circuit connected to the electrodes which is a measure of the rate of supply of iodine and is related to this rate by Faraday's law. In the figure, the chamber **30** represents the reaction section of the Hersch cell while the detector **34** represents the section of the Hersch cell in which the electrodes are contained. It is understood that the halogen detector **34** includes some means for indicating the current output of the cell. This may be a microammeter or a suitable recorder **38**.

The scrubber **12** contains materials which remove from the air sample stream constituents other than aldehydes which will either oxidize iodide ions to produce iodine or consume iodine in the cell. Such constituents are generally referred to as interferents in the analysis procedure, and include mercaptans, hydrogen sulfide, ozone, nitrogen dioxide, sulfur dioxide and chlorine. A combination of $HgCl_2$, $Hg(SH)_2$, $Ca(OH)_2$ and Ag_2O has been found to efficiently remove these interferents and not affect the aldehyde content of the air sample stream, nor unsaturated hydrocarbons or ketones if they are present in the sample.

Thus, when a polluted air sample stream is drawn through the scrubber **12**, the effluent from the scrubber will consist essentially of a mixture of nitrogen, oxygen and aldehydes. When the valve **26** is positioned as shown, the effluent from the scrubber will pass to the reaction chamber **22** where it is subjected to ultraviolet radiation. The aldehydes in the sample stream are photooxidized to peroxy acid. The effluent from the chamber **22** containing the peroxy acid passes via the conduit **28** into the aqueous halide solution in chamber **30**. Peroxy acid is a powerful oxidizing agent. It oxidizes the iodide rapidly and liberates free iodine. The gas entering the chamber **30** exits therefrom via the aspirator **36** and a vent **39** while the aqueous solution containing free iodine passes to the electrode section of the Hersch cell, represented schematically by numeral **34**.

As discussed above, in the cell iodine is reduced at the cathode to iodide and a galvanic current is produced which is a measure of the rate of supply of iodine to the cell. This current is proportional to the rate of supply of both peroxy acid and ozone conveyed from the reaction chamber **22** to the chamber **30**. Thus, with the valve **26** positioned as shown, a first output signal is produced by the detector **34** which is proportional to the aldehyde and ozone content of the air sample stream.

In order to determine the aldehyde content of the air sample stream, it is necessary to determine that portion of the current output of the detector **34** which results from the ozone produced in the reaction chamber **22**. This is achieved by providing in the scrubber **14** materials which remove aldehydes as well as the abovementioned interferents from the air sample stream. A suitable combination of materials for this purpose is carbon and calcium hydroxide. Thus, when air is passed into the scrubber **14**, the effluent from the scrubber will comprise essentially a mixture of nitrogen and oxygen, and the oxygen content of the effluent will be equal to the oxygen content of the effluent from the scrubber **12**. The valve **26** is positioned to permit effluent from the scrubber **14** to pass through the conduit **24** and a conduit **20** to the reaction chamber **22**.

There the oxygen in the gas stream will be converted to ozone at the same rate that the oxygen in the effluent from scrubber **12** is converted to ozone. The effluent containing the ozone then passes from the reaction chamber **22** into the halide solution in chamber **30**. The solution is conveyed to the halogen detector **34** whereupon a second output signal is produced which is a function of the ozone produced in the reaction chamber **22**. It can be seen that by determining the difference between the first output signal resulting from the air entering conduit **16** and the second output signal resulting from the air entering inlet conduit **18**, a value results which corresponds to the aldehyde content of the sample.

In the practical operation of apparatus **10**, the valve **26** is positioned so that the air will pass through the scrubber **14** to provide a first output signal by the detector **34**. The signal level of the detector is then adjusted so that the first output signal corresponds to actual zero or the base line of the detector. Then the valve **22** is shifted to the position shown so that thereafter the signal produced by the detector provides a direct indication of the aldehyde content of the air sample stream.

The above described apparatus has been utilized for monitoring standard aldehyde and air mixtures with the result that the signal output of the Hersch cell corresponds quantitatively to the aldehyde content of the sample. The apparatus is capable of detecting levels of aldehydes as low as 10 ppb. The apparatus **10** has the advantage, besides being capable of continuously monitoring the aldehyde content of air, of being relatively inexpensive, fast responding, and is not temperature dependent. Moreover, the apparatus has equal sensitivity to different aldehydes.

Quantitative Methods (Formaldehyde)

The colorimetric method of determining formaldehyde with chromotropic acid (1,8-dihydroxynaphthalene-3,6-disulfonic acid) has had widespread use. Several variations have been described in the literature (3)(35)(45)(46)(47). The method proposed by Altshuller et al (45)(48) appears to be simple, rapid, and suitable for the analysis of effluents and air samples. The sensitivity of this method is approximately 12 μg./m.^{3} (0.01 ppm) (6). Nitrogen dioxide, most aldehydes and ketones, and straight-chain alcohols do not interfere significantly (45)(47). Aromatic hydrocarbons and olefins can cause serious interference, but the use of aqueous sodium bisulfite as the collection medium can reduce this interference.

Furthermore, compounds that are easily converted via hydrolysis or oxidation to formaldehyde in strong, warm sulfuric acid may also interfere. Compounds of this type, which include sugars, formaldehyde polymers, glyoxal, piperonal, and related compounds, have been discussed by Sawicki (2). The chromotropic acid method has been used in the analysis of air samples (7)(45), incinerator effluents (49)(50), automobile exhaust (3), and diesel exhaust (14)(15)(51)(52).

Many other colorimetric methods have been used or appear applicable to analysis of air samples. Sawicki et al (26)(53) found that 6-amino-1-naphthol-3-sulfonic acid (J-acid) and 6-anilino-1-naphthol-3-sulfonic acid (phenyl J-acid) have greater sensitivity than the chromotropic acid. However, these methods have not been used in analysis of effluents or air samples. A comparison of these methods and other spectrometric methods for determining formaldehyde has been made by Sawicki et al (53).

A continuous method for determining formaldehyde with a sensitivity of 12 μg./m.^{3} (0.01 ppm) has been reported (14)(18). The method is a modified Schiff method using para-rosaniline in sodium tetrachloromercurate(II) and sodium bisulfite. This method has been used for analysis of air samples. Only two aldehydes, acetaldehyde and propionaldehyde, gave positive reactions (18). Polarographic methods (54)(55) may also be applicable to analysis of air samples, but at present they need further study.

A microwave pollution monitor may be applied to formaldehyde as described by J.A. Hodgeson, W.A. McClenny and P.L. Hanst (56).

Quantitative Methods (Acrolein)

A highly sensitive spectrophotometric method for acrolein has been developed by Cohen and Altshuller (57), who based their method on a reagent first proposed by Rosenthaler and Vegezzi (58). Acrolein reacts with 4-hexylresorcinol in an ethanol-trichloroacetic acid solution to yield a blue-colored product with an absorption maximum at 605 mμ. The sensitivity is approximately 12.5 μg./m.^{3} (0.005 ppm). The method appears selective; no significant interferences were found from sulfur dioxide, nitrogen dioxide, ozone, aromatic compounds, ketones, olefins, and other unsaturated aldehydes (9)(57). Slight interferences are found with some dienes (9) and with malonaldehyde, which appears to form a similar blue product. This method has been used in analysis of automobile exhaust (3) (57), diesel exhaust (15)(38)(52) and atmospheric samples (7)(35).

Because colorimetric methods using tryptophan (59)(60) and phloroglucinol (61)(62) lack sufficient sensitivity and have appreciable interferences, they are not useful for analysis of air samples (57). A J-acid method can be used to determine acrolein with a sensitivity of 0.01 μg. but serious interference results with equal or higher amounts of formaldehyde (53). Polarographic (55), gas chromatographic (11)(63)(64), and paper chromatographic methods have been used in the analysis of acrolein from vehicle exhaust and air samples. However, these methods have not been generally applied because of the complexity of the techniques.

REFERENCES

(1) Altshuller, A.P. "Analysis of Aliphatic Oxygenated Compounds," in *Air Pollution,* vol. 2, A.C. Stern, Ed. (New York: Academic Press, p. 130, 1968).

(2) Sawicki, E., "Spot Test Detection and Spectrophotometric Determination of Microgram Amounts of Aldehydes and Aldehyde-Yielding Compounds – A Review," Preprint, Presented at International Symposium on Microchemical Techniques, University Park, Pa. (Aug. 13-18, 1961).

(3) Altshuller, A.P., et al, "Analysis of Aliphatic Aldehydes in Source Effluents and in the Atmosphere," *Anal. Chim. Acta* 25:101 (1961).

(4) Farr, J.P.G., "Analyses for the Industry," *Ind. Chemist* 389 (1956).

(5) Reynolds, J.G., and M. Irwin, "The Determination of Formaldehyde and Other Aldehydes," *Chem. Ind.* 419 (1948).

(6) Altshuller, A.P., and L.J. Leng, "Application of the 3-Methyl-2-Benzothiazolone Hydrazone Method for Atmospheric Analysis of Aliphatic Aldehydes," *Anal. Chem.* 35(10):1541 (1963).

(7) Altshuller, A.P., and S.P. McPherson, "Spectrophotometric Analysis of Aldehydes in the Los Angeles Atmosphere," *J. Air Pollution Control Assoc.* 13(3):109 (1963).

(8) Morgan, G.B., C. Golden, and E.C. Tabor, "New and Improved Procedures for Gas Sampling and Analysis in the National Air Sampling Network," *J. Air Pollution Control Assoc.* 17:300 (1967).

(9) *Selected Methods for the Measurement of Air Pollutants,* Public Health Service, Cincinnati, Ohio, Division of Air Pollution (May 1965).

(10) Devorkin, H., et al, *Air Pollution Source Testing Manual,* Air Pollution Control District, Los Angeles County, Calif. (1965).

(11) Ellis, C.F., R.F. Kendall, and B.H. Eccleston, "Identification of Some Oxygenates in Automobile Exhausts by Combined Gas Liquid Chromatography and Infrared Techniques," *Anal. Chem.* 37(4):511 (1965).

(12) George, R.E., and R.M. Burlin, "Air Pollution from Commercial Jet Aircraft in Los Angeles County, "Los Angeles County Air Pollution Control District, Calif. (Apr. 1960).

(13) Lahmann, E., and K. Jander, "Determination of Formaldehyde from Street Air," Text in German. *Gesundh. Ingr.* (Munich) 89(1):18 (1968).

(14) Yunghans, R.S., and W.A. Munroe, "Continuous Monitoring of Ambient Atmospheres with the Technicon Auto Analyzer," Presented at the Technicon Symposium, Automation in Analytical Chemistry, New York (Sept. 8, 1965).

(15) Linnell, R.H., and W.E. Scott, "Diesel Exhaust Analysis (Preliminary Results)," *Arch. Environ. Health* 5:616 (1962).

(16) Ellis, C.F., "Chemical Analyses of Automobile Exhaust Cases for Oxygenates," *Bur. Mines Rept. Invest. 5822* (1961).

(17) Fracchia, M.F., F.J. Schuette, and P.K. Mueller, "A Method for Sampling and Determination of Organic Carbonyl Compounds in Automobile Exhaust," *Environ. Sci. Technol.* 1(11):915 (1967).

(18) Lyles, G.R., F.B. Dowling, and V.J. Blanchard, "Quantitative Determination of Formaldehyde in the Parts Per Hundred Million Concentration Level," *J. Air Pollution Control Assoc.* 15(3):106 (1965).

(19) Sawicki, E., and T.R. Hauser, "Infrared Spectral Detection of Carboxylic Acids and Aldehydes in Air-Borne Particulates," *Anal. Chem.* 31:523 (1959).

(20) Stahl, Q.R., "Air Pollution Aspects of Aldehydes," *Report PB 188,081,* Springfield, Va., Natl. Tech. Information Service (Sept. 1969).

(21) Anger, V., and G. Fischer, "A New Spot Reaction of Aldehydes," *Mikrochim. Acta,* p. 592 (1960).
(22) Feigl, F., *Spot Tests in Organic Analysis* (New York: Elsevier, 1956).
(23) Sawicki, E., and T.W. Stanley, "Sensitive New Test for Aliphatic, Aromatic and Heterocyclic Aldehydes," *Mikrochim. Acta* p. 510 (1960).
(24) Sawicki, E., and T.R. Hauser, "Spot Test Detection and Colorimetric Determination of Aliphatic Aldehydes with 2-Hydrozinobenzothiazole: Application to Air Pollution," Presented at Air Pollution Symposium, 138th Meeting, American Chemical Society, New York (Sept. 1960).
(25) Sawicki, E., et al, "The 3-Methyl-2-benzothiazolone Hydrazone Test. Sensitive New Methods for the Detection, Rapid Estimation, and Determination of Aliphatic Aldehydes," *Anal. Chem.* 33:92 (1961).
(26) Sawicki, E., T. R. Hauser, and S. McPherson, "Spectrophotometric Determination of Formaldehyde and Formaldehyde-Releasing Compounds with Chromotropic Acid, 6-Amino-1-naphthol-3-sulfonic Acid (J Acid), and 6-Anilino-1-naphthol-3-sulfonic Acid (Phenyl J Acid)," *Anal. Chem.* 34:1460 (1962).
(27) Basbagill, W.J., "Air Contaminant Measurements at Roosevelt Field, Nassau County, New York (January-February 1964)," Public Health Service, Cincinnati, Ohio, Division of Air Pollution (July 1965).
(28) Basbagill, W.J., and J.L. Dallas, "Air Quality in Boston, Massachusetts (November-December 1963)," Public Health Service, Cincinnati, Ohio, Division of Air Pollution (Nov. 1964).
(29) Hochheiser, S., M. Burchett, and H.J. Dunsmore, "Air Pollution Measurements in Pittsburgh (January-February 1963)," Public Health Service, Cincinnati, Ohio, Division of Air Pollution and Alleghany County Health Dept., Pittsburgh, Pa., Bureau of Air Pollution Control (Nov. 1963).
(30) Hochheiser, S., M. Nolan, and H.J. Dunsmore, "Air Pollution Measurements in Duquesne, Pennsylvania (September-October 1963)," Public Health Service, Cincinnati, Ohio, Division of Air Pollution and Allegheny County Health Dept., Duquesne, Pa., Bureau of Air Pollution Control (Oct. 1964).
(31) McMullen, T.B., Technical Reports Unit, Division of Air Quality and Emission Data, National Air Pollution Control Administration (May 1969).
(32) Hauser, T.R., and R.L. Cummins, "Increasing Sensitivity of 3-Methyl-2-Benzothiazolone Hydrazone Test for Analysis of Aliphatic Aldehydes in Air," *Anal. Chem.* 36(3):679 (1964).
(33) Altschuller, A.P. and J.J. Bufalini, "Photochemical Aspects of Air Pollution: A Review," *Photochem. Photobiol.* 4:97 (1965).
(34) Schuck, E.A., H.W. Ford, and E.R. Stephens, *Air Pollution Effects of Irradiated Automobile Exhaust As Related to Fuel Composition,* Report No. 26, Air Pollution Foundation, San Marino, Calif. (Oct. 1958).
(35) Renzetti, N.A., and R.J. Bryan, "Atmospheric Sampling for Aldehydes and Eye Irritation in Los Angeles Smog – 1960," *J. Air Pollution Control Assoc.* 11(9):421 (1961).
(36) Renzetti, N. A., L.H. Rogers, and R. Tice, in *An Aerometric Survey in the Los Angeles Basin, August-November, 1954,* N.A. Renzetti, ed., Report No. 9, Air Pollution Foundation, Los Angeles, Calif. (1955).
(37) Larson, G.P., J.C. Chipman, and E.K. Kauper, "Study of the Distribution and Effects of Auto Exhaust Gas," *J. Air Pollution Control Assoc.* 5:84 (1955).
(38) Battigelli, M.C., "Air Pollution from Diesel Exhaust," *J. Occupational Med.* 5:54 (1963).
(39) Schumann, C.E., and C.W. Gruber, "Motorist Exposures to Aldehydes from Diesel-Powered Buses," *J. Air Pollution Control Assoc.* 14:53 (1964).
(40) Yocum, J.E., G.M. Hein, and H.W. Nelson, "Effluents from Backyard Incinerators," *J. Air Pollution Control Assoc.* 6:84 (1956).
(41) Levaggi, D.A., and M. Feldstein, "Gas Chromatographic Determination of Aldehydes, Bay Area Air Pollution Control District, San Francisco, Calif." Presented at State Health Dept. Methods Conference, Pasadena, Calif. (Feb. 7-9, 1968).
(42) Levaggi, D.A., and M. Feldstein, "The Collection and Analysis of Low Molecular Weight Carbonyl Compounds from Source Effluents," *J. Air Pollution Control Assoc.* 19:43 (1969).
(43) Goodson, L.H. and Jacobs, W.B., U.S. Patent 3,711,251; Jan. 16, 1973; assigned to the Insurance Institute for Highway Safety.
(44) Neti, R.M. and Kelly, T.J.; U.S. Patent 3,540,849; November 17, 1970; assigned to Beckman Instruments, Inc.
(45) Altshuller, A.P., D.L. Miller, and S.F. Sleva, "Determination of Formaldehyde in Gas Mixtures by the Chromotropic Acid Method," *Anal. Chem.* 33(4):622 (1961).
(46) Bricker, C.E., and A.H. Vail, "Microdetermination of Formaldehyde With Chromotropic Acid," *Anal. Chem.* 22:720 (1950).
(47) West, P.W., and B. Sen, "Spectrophotometric Determination of Traces of Formaldehyde," *Z. Anal. Chem.* 153:177 (1956).
(48) Altshuller, A.P., L.J. Leng, and A.F. Wartburg, Jr., "Source and Atmospheric Analyses for Formaldehyde by Chromotropic Acid Procedures," *Intern. J. Air Water Pollution* 6:381 (1962).

(49) Stenburg, R.L., et al, "Effects of High Volatile Fuel on Incinerator Effluents," *J. Air Pollution Control Assoc.* 11:376 (1961).
(50) Stenbury, R.L., et al, "Field Evaluation of Combustion Air Effects on Atmospheric Emissions from Municipal Incinerators," *J. Air Pollution Control Assoc.* 12:83 (1962).
(51) Linnell, R.H., and W.E. Scott, "Diesel Exhaust Composition and Odor Studies," *J. Air Pollution Control Assoc.* 12:510 (1962).
(52) Reckner, L. R., W.E. Scott, and W.F. Biller, "The Composition and Odor of Diesel Exhaust," *Proc. Am. Petrol. Inst.* 45:133 (1965).
(53) Sawicki, E., T.W. Stanley, and J. Pfaff, "Spectrophotofluorimetric Determination of Formaldehyde and Acrolein with J. Acid, Comparison with Other Methods," *Anal. Chim. Acta* 28 (1963).
(54) Barnes, E.C., and H.W. Speicher, "The Determination of Formaldehyde in Air," *J. Ind. Hyg. Toxicol. 24:9 (1942).*
(55) Coulson, D.M., "Polarographic Determination of Semicarbazones," *Anal. Chim. Acta* 19:284 (1958).
(56) Hodgeson, J.A., McClenny, W.A., and Hanst, P.L., *Science* 182, 248-258 (1973).
(57) Cohen, I.R., and A.P. Altshuller, "A New Spectrophotometric Method for the Determination of Acrolein in Combustion Gases and in the Atmosphere," *Anal. Chem.* 33(6):726 (1961).
(58) Rosenthaler, L., and G. Vegezzi, "Detection and Determination of Acrolein in Alcoholic Liquors," *Lebensm-Untersuch U. Forsch* 99:352 (1954).
(59) Plotnikova, M.M., "Acrolein As an Atmospheric Air Pollutant," *Gigiena i Sanit,* 22:10 (1957); Translated by B.S. Levine, *U.S.S.R. Literature on Air Pollution and Related Occupational Diseases* 3:188 (May 1960).
(60) Senderikhina, D.P., "Determination of Acrolein in the Air," *Limits of Allowable Concentrations of Atmospheric Pollutants,* Book 3, (1957).
(61) Powick, W.C., "A New Test for Acrolein and Its Bearing on Rancidity in Fats," *Ind. Eng. Chem.* 15:66 (1923).
(62) Uzdina, I.L., "Determination of Acrolein and of Formaldehyde in the Air," *Hig. Truda* 15:63 (1937).
(63) Hughes, K.J., and R.W. Hurn, "A Preliminary Study of Hydrocarbon-Derived Oxygenated Material in Automobile Exhaust Gases," *J. Air Pollution Control Assoc.* 10:367 (1960).
(64) Mueller, P.K., M.F. Fracchia, and F.J. Schuette, 152nd National Meeting, Am. Chem. Soc., New York (1966).

ALUMINUM

Finely divided metallic aluminum has been reported by Sax (1) to produce lung damage when inhaled. Aluminum oxide smoke containing free silica produces an illness known as "Shaver's Disease." Thus, personnel exposed to high concentrations of aluminum powder or aluminum oxide smoke should wear respirators to avoid excessive inhalation.

The trace element tolerance for aluminum in agricultural water supplies has been given as 1.0 ppm for continuous use and 20 ppm for short term use in fine textured soil as noted earlier in Table 5.

Measurement in Water

Aluminum may be monitored in water by spark source mass spectrometry and by atomic absorption spectrometry. The detection limit has been cited as 40 ppb for atomic absorption as shown earlier in Table 11. Aluminum may also be determined by molecular fluorescence spectrophotometry as described by K.H. Mancy (2). The sensitivity of the method is given as 0.007 μg./ml. using Alizarin Garnet R as a reagent. A large number of ions interfere with this method, however.

References

(1) Sax, N.I., *Handbook of Dangerous Materials,* New York, Reinhold Publ. Corp. (1951).
(2) Mancy, K.H., *Instrumental Analysis for Water Pollution Control,* Ann Arbor, Mich., Ann Arbor Science Publishers (1971).

AMMONIA

Atmospheric concentrations of ammonia are generally below the level at which health hazards and deleterious effects to humans, animals, plants, and materials are known to occur, according to Miner (1). High concentrations, usually caused by accidental spillage, can result in corrosive action to mucous membranes, permanent injury to the cornea, damage to the throat and upper respiratory tract, chronic bronchial catarrh, and edema. In sufficiently high doses (1,700,000 μg./m.^{3} to 4,500,000 μg./m.^{3}), ammonia acts as an asphyxiant. Ammonia combines with sulfur oxides and metallic materials in the atmosphere to form ammonium sulfate and zinc ammonium sulfate aerosols. These aerosols are thought to be in part responsible for the irritant effects of the air during the Donora Smog Episode in 1948. Severe irritation was produced by these salts in guinea pigs.

Ammonia has been shown to be toxic to most plant life, producing injury to leaf and stem tissue, and reducing or delaying germination of seeds at concentrations of 700,000 μg./m.^{3}. It also has a corrosive effect on certain materials.

The bulk of the ammonia found throughout the world in the atmosphere is produced by natural biological processes which release this substance to the air. The background concentration of ammonia is 6 μg./m.^{3} in the mid-latitudes and 140 μg./m.^{3} near the equator. A secondary source of atmospheric ammonia is urban-produced ammonia, which may be an important source in localized situations. The average concentration of ammonia in the urban atmosphere is 20 μg./m.^{3}, although measurements as high as 7,200 μg./m.^{3} have been recorded in areas adjacent to chemical manufacturing complexes.

The combustion process, the main source of urban-produced ammonia in the atmosphere, includes combustion of fuels, incineration of wastes, and use of the internal combustion engine. Industrial sources consist of chemical plants, coke ovens, and refineries. Ammonia air pollution also results from biological degradation in piggeries, stockyards, and similar installations.

Equipment such as wet scrubbers, and bag filters added to control systems to reduce other combustion-product emissions will also reduce ammonia emissions. In coke ovens, ammonia emission can be reduced along with other gases and dust by collecting the coke-oven gases during charging and discharging. In chemical plants, ammonia emissions are generally kept to a minimum due to economic considerations; however, wet scrubbers can be used to reduce emissions even more effectively. In places where animals are kept, impregnated charcoal has been used effectively to control ammonia pollution.

The economic impact of ammonia pollution on humans, animals, plants, and materials is believed to be minimal, due to the low natural concentrations of ammonia and the absence of harmful effects at these levels. Data on economic losses from ammonia are not available. Localized accidental emissions could, of course, create loss of life and serious damage to plants. Air quality standards in the U.S.S.R. call for a maximum of 0.28 ppm ammonia in air on either a single exposure or 24-hour average basis. Czechoslovak standards specify 0.42 ppm on a single exposure basis and 0.14 ppm on a 24-hour average basis. Ontario (Canada) on the other hand, specifies 5.0 ppm of ammonia on a single exposure basis.

Measurement in Air

The primary method used in air pollution for analyzing for ammonia in air is the Nessler colorimetric method (2)(3)(4). The sample is collected by passing the air through a standard impinger containing 0.1N sulfuric acid. The collected sample is then contacted with Nessler's reagent and examined in a colorimeter. If a cloudy solution forms after the addition of Nessler's reagent, alkaline Rochelle salt is added to clear it up (5).

To obtain more accurate results prior to Nesslerization, the acidic sample may be made alkaline and the ammonia distilled into a receiver containing 0.02N sulfuric acid (6).

Nessler's reagent is then added, and the sample is analyzed colorimetrically. The Nessler colorimetric method of analysis gives the total ammonia content of the air: i.e., both gaseous and particulate components. Equipment based on Nessler's method has been developed for automatic analysis (7).

Another method utilized for analyzing for ammonia is the indophenol blue technique. The sample is collected as outlined above. It is then contacted with alkaline phenol and sodium hypochlorite, which turns it blue-green. The sample color is then read on a colorimeter. The ammonia determined is the total ammonia and ammonium in the sample. This method was developed for controlled atmosphere applications but can be applied to air pollution work. The indophenol blue technique has also been adapted for use in automatic ammonia analyzers (8).

Smolczyk (9) showed that paper impregnated with phenolphthalein will change color in air in the presence of 10 to 100 ppm (7,000 to 70,000 $\mu g./m.^3$) of ammonia gas. Korenman (10) used impregnated diazotized alpha or beta-naphthylamine to test for ammonia. Cambi (2) indicated that ammonia samples with as little as 0.01 ppm (7 $\mu g./m.^3$) ammonia can be analyzed by titrating directly with standard solutions of sodium hydroxide and sulfuric acid. In addition, industrial methods based on infrared analysis and colorimetric techniques (3) are used for ammonia analysis.

A device developed by J.L. Radawski for visual detection of the leakage of ammonia fumes comprises polyvinyl alcohol, water and a suitable indicator with or without the addition of silica gel and glycerine. Suitable indicators include phenolphthalein and an equal mixture of cresol red and thymol blue. The composition is a hygroscopic, water-soluble emulsion which can be applied to any surface in the same manner as paint by spraying, brushing or dipping. In order to be operative, the detector does not require water beyond atmospheric or chemically bound moisture and is regenerative to its original color even after exposure to ammonia fumes. Upon application to a suitable surface, the composition assumes a yellow color in an ammonia-free atmosphere, but changes color to red with a phenolphthalein indicator and to blue with a cresol red-thymol blue indicator upon exposure to ammonia fumes.

An apparatus developed by G.F. Skala (12) permits detecting the presence of small amounts of ammonia in a gas such as air. The gas-ammonia mixture is treated with a corona discharge before it is exposed to acid vapor to form ammonium salt condensation nuclei which are then detected and measured. Alternatively, an ammonia-free gas may be treated with a corona discharge and then mixed with the ammonia-containing gas before exposure to the acid vapor. The use of the corona treatment increases the sensitivity of the instrument by a factor of 10 or more.

According to F.W. Van Luik, Jr. (13) if a gas such as air containing small amounts of ammonia is exposed to ultraviolet radiation prior to being contacted with an acid to form ammonia compound condensation nuclei, about 1,000 times more nuclei are formed than if such radiation is omitted.

If, however, the gas also contains sulfur dioxide, the ultraviolet light produced by the corona discharge converts the sulfur dioxide to sulfur trioxide. The sulfur trioxide will combine with water vapor to form droplets of sulfuric acid which are detectable as particles by the condensation nuclei counter and give a spurious reading. It would therefore be desirable to provide a gas conversion apparatus for detecting and measuring the ammonia concentration in a gas such as air which would be insensitive to sulfur dioxide in one mode of operation and in another mode of operation, detect and measure both the ammonia and sulfur dioxide concentrations so that the sulfur dioxide concentration could be determined by difference.

Thus, an apparatus has been developed by P.E. Coffey which is a gas conversion apparatus for use in conjunction with a condensation nuclei counter in the detection and measurement of ammonia and sulfur dioxide in a gas such as air.

The total condensation nuclei consisting of ammonium salt particles and sulfuric acid droplets are counted. Ultraviolet radiation is then screened off and the process repeated with another sample of the gas substantially identical to the first sample. Under these conditions, any sulfur dioxide present is not changed to sulfur trioxide and only the ammonia is converted to condensation nuclei, which are counted by the condensation nuclei counter. If sulfur dioxide is present in gas in addition to ammonia, the two counting results will be different, with the difference between the results of the two measurements reflecting the concentration of sulfur dioxide in the carrying gas.

Measurement in Water

The permissible amount of ammonia in domestic water supplies is 0.5 ppm and the desirable concentration in domestic water supplies is less than 0.01 ppm as noted earlier in Table 3. Effluents to storm sewers or streams should have ammonia contents below 2.5 ppm. Ammonia may be monitored in water by the following instrumental techniques (15): Activation analysis; automated colorimetric; colorimetric; and ion selective electrode.

As noted earlier in Table 12, the Technicon CSM-6 may be used to measure ammonia in a normal range of 0 to 10 ppm with a detection limit of 0.10 ppm. As described by K.M. Mancy (16), most automated methods for the determination of ammonia use the alkaline phenol-hypochlorite reaction. It is believed that the colored product is an indophenol compound.

Also, as pointed out by Mancy (16), the most widely used conventional method used for analysis of ammonia is the Nesslerization method based on the development of a yellow-brown colloidal color on addition of Nessler's reagent to an ammonia solution. The ASTM reference test uses this reaction.

References

(1) Miner, S., "Air Pollution Aspects of Ammonia," *Report PB 188,082,* Springfield, Va., Nat. Tech. Information Service (Sept. 1969).
(2) Cambi, F., "Sampling, Analysis, and Instrumentation in the Field of Air Pollution," *World Health Organization Monograph Series No. 46,* Geneva (1961).
(3) Louw, C.W., "Atmospheric Pollutants and Chemical Analysis," *CIR Special Report SM 062,* UDC 614.71: 543.27, Pretoria, South Africa (1966).
(4) Stern, A.C., *Air Pollution, Vol. II - Analysis, Monitoring, and Surveying* (New York: Academic Press, 1968).
(5) Jacobs, M.B., *The Chemical Analysis of Air Pollutants* (New York: Interstate Publishers, 1960).
(6) "Atmospheric Pollution in the Great Kanawha River Valley Industrial Area," West Virginia Department of Health, Bureau of Industrial Hygiene (1952).
(7) "Air Quality Data from the National Air Sampling Networks and Contributing State and Local Networks, 1964-1965," U.S. Department of Health, Education, and Welfare, Public Health Service, Division of Air Pollution, Cincinnati, Ohio (1966).
(8) Kawasaki, E.H., et al, "Application of the Autoanalyzer for Atmospheric Trace Contamination Analysis in Close Environmental Systems," Presented at the Technician Symposium, Automation in Analytical Chemistry, New York (Oct. 1967).
(9) Smolczyk, E., and H. Cabler, "Chemical Detection of Respiratory Poisons," *Wasser Abwasser* 28:95 (1930).
(10) Korenman, I.M., "Detection of Ammonia in the Air," *Z. Analy. Chem.* 90:115 (1932).
(11) Radawski, J.L., U.S. Patent 3,528,780; September 15, 1970; assigned to Secretary of the Air Force.
(12) Skala, G.F., U.S. Patent 3,503,711; Mar. 31, 1970; assigned to General Electric Co.
(13) Van Luik, F.W., Jr., U.S. Patent 3,578,410; May 11, 1971; assigned to General Electric Co.
(14) Coffey, P.E., U.S. Patent 3,562,128; Feb. 9, 1971; assigned to General Electric Co.
(15) Lawrence Berkeley Laboratory, "Instrumentation for Environmental Monitoring Water," *Publication LBL-1,* Vol. 2, Berkely, Univ. of Calif. (Feb. 1, 1973).
(16) Mancy, K.H., *Instrumental Analysis for Water Pollution Control,* Ann Arbor, Mich., Ann Arbor Science Publishers (1971).

ANTIMONY

Antimony is introduced to the environment from various industrial sources. As to the health effect of antimony, it is said to shorten the life span in rats (1). In the case of humans, antimony has been reported to cause dermatitis and gastrointestinal disturbances. Poisoning from occupational exposure has been very rare however; the gastrointestinal upset is usually acute rather than chronic in character.

British emission standards for antimony and its compounds in air, call for a maximum of 115 mg./cu. m. at STP at flows of less than 5,000 cfm and 46 mg./cu. m. at flows above 5,000 cfm. On the other hand, Australian emission standards call for a maximum of 23 mg./cu. m. at STP. Antimony does not seem to be cited in air quality standards of various countries summarized by A.C. Stern (2).

Measurement in Air

Antimony may be monitored in air by a colorimetric instrumental technique (3). A tentative method for antimony analysis of the atmosphere involving the reaction of pentavalent antimony with Rhodamine B to form a colored complex has been outlined (4).

Measurement in Water

Antimony can be measured in water by atomic absorption spectrometry to a detection limit of about 60 ppb as noted earlier in Table 11.

References

(1) Anon., *Chem. & Eng. News,* pp. 29-30, 33 (July 19, 1971).
(2) Stern, A.C., in *Industrial Pollution Control Handbook,* H.F. Lund, Ed., New York, McGraw-Hill Book Co., (1971).
(3) Lawrence Berkeley Laboratory, "Instrumentation for Environmental Monitoring – Air," *Publication LBL-1,* Vol. 1, Berkeley, Univ. of California (Feb. 1, 1973).
(4) American Public Health Assoc., *Methods of Air Sampling and Analysis,* Wash., D.C. (1972).

ARSENIC

Arsenic is toxic to some degree in most chemical forms. Arsenical compounds may be ingested, inhaled, or absorbed through the skin. Industrial exposure to arsenic has shown that it can produce dermatitis, mild bronchitis, and other upper respiratory tract irritations including perforation of the nasal septum. However, because of the irritant qualities of arsenic, it is doubtful that one could inhale sufficient amounts to produce systemic poisoning, according to Sullivan (1).

Skin cancer can result from prolonged therapeutic administration of arsenic. Similar cancers have not been observed among industrial workers. Moreover, lung tumors which resulted from inhaling mixed industrial dusts were often thought to be the result of inhaling arsenic. Recently, this relationship has been questioned because animal experiments have failed to demonstrate that arsenic is a carcinogen. Therefore, the causal relationship between cancer and arsenic is disputed.

Arsenic is poisonous to both animals and plants, but no damage to materials was found. Two air pollution episodes in the United States have shown that there is an arsenical air pollution potential at every smelter which refines arsenical ores. Arsenical compounds are used as insecticides and herbicides. Although the use of arsenical pesticides declined sharply after the appearance of DDT and 2,4-D, arsenical compounds are still used as desiccants, herbicides, and sterilants. Some undetermined amounts of air pollution take place during spraying and dusting operations with arsenical pesticides. Pollution from cotton gins and cotton trash burning has been cited as an important source of agricultural pollution. While the emission rates from cotton trash burning have not been determined, as much as 1,258,000 μg./m.^{3} of exhaust air (580,000 μg./min.) may be emitted during the ginning operation. This produced concentrations of only 0.14 μg./m.^{3} of arsenic in the air 150 feet from the gin.

Arsenic is found to the extent of approximately 5 μg./g. in coal. Therefore, the air of cities which burn coal contains some arsenic. Air quality data from 133 sites monitored by the National Air Sampling Network showed an average daily arsenic concentration of 0.02 μg./m.^{3} in 1964. Control of arsenic emissions requires special attention to the temperature of exhaust gases since arsenic trioxide sublimes at 192°C. For this reason exhaust fumes must be cooled to approximately 100°C. prior to removing them as particulates.

In general, the most stable natural valence of an element, that normally found in soil and water, is the least toxic, for example chromium(III), manganese(II) and arsenic(V) as opposed to the very toxic chromium(VI), manganese(VII) and arsenic(III), as noted by Cothern (1a). Sodium arsenate is nontoxic in amounts 100 times as large as the LD (lethal dose) of trivalent arsenic. Commercial arsenic is almost entirely the trivalent oxide (2). Adding 12 to 15 ppm of arsenate counteracts selenium poisoning in cattle, dogs and chickens. Metallic arsenates, potassium, calcium and lead are toxic to insects (2). However, Warnick and Bell (3) say that less than 20 ppm of arsenic is not toxic to aquatic insects.

Industrial exposures of arsenic can give rise to keratosis of the skin that occasionally become malignant. However, most attempts to produce cancer with arsenic have been unsuccessful (2). Arsenite given in a dose of 5 micrograms per milliliter of water significantly decreased the incidence of all tumors, except malignant ones where no effect was noted (4). Intravenous injection of 20 mg./kg. of sodium arsenate in pregnant golden hamsters led to a high incidence of malformations of the embryos (5). Buechley (6) made the suggestion that the epidemiological chain for cancer leads from arsenic high mines and smelters, through arsenical insecticides to arsenic-sprayed tobaccos used in cigarettes. Along this line, lung cancer appears to be proportional to the amount of inhaled arsenic.

If arsenic is an essential trace element for mammals, the requirements are of the order of 1 microgram per day or less. (1a). Trivalent arsenic given at a dose of 5 ppm in drinking water for the life of a mouse showed a slight tendency to accumulate in the heart and lung and was slightly toxic in terms of medium life span, longevity and survival, but not in terms of growth (7). Arsenic trioxide fed in drinking water in concentration of 0.01% reduced the number of induced cutaneous malignant skin tumors. It is still questionable whether arsenic is carcinogenic (8). Arsenite given in 5 ppm in drinking water caused the accumulation of arsenic in the aorta and red blood cells with no signs of toxicity (9).

Chronic arsenic intoxication by ingestion is usually characterized by weakness, loss of appetite, gastrointestinal disturbances, peripheral neuritis, occasionally hepatitis and skin disorders such as keratosis and pigmentation (10). The common gray form of elemental arsenic is essentially nonpoisonous by ingestion and most of it is eliminated unchanged. The oral LD of arsenic trioxide in man is between 70 and 180 mg./kg. Thus, although arsenic is not a serious health problem at the ppm level, it is an undesirable trace metal in water supplies.

Industrial sources of arsenic have been reviewed by Patterson and Minear (11). Arsenic and arsenical compounds have been reported in wastewaters from the metallurgical industry, glassware and ceramic production, tannery operations, dye, and pesticide manufacture (12). Other industrial sources include the organic and inorganic chemicals and petroleum refining industries (13), and the rare-earth industry (14). The manufacture of Paris green and calcium meta-arsenate, both insecticides, was reported to produce wastewaters containing 362 mg./l. of arsenious oxide (15). Although arsenic is widely associated with the manufacture of herbicides and pesticides (16), a recent study concluded that the inorganic pesticides industry (including arsenic based compounds) is fading (17). Arsenical wastes from these sources would be expected to assume less importance in that event. Arsenic wastes from nonferrous smelting have been reported in the literature (13) (18)(19), although the main form appears to be as extremely small (1 to 3 micron diameter) particles from which arsenic can be profitably recovered (18). Very little current information is available on arsenic waste treatment in the smelting industry, perhaps because of the highly efficient arsenic recovery processes employed, as suggested by Swain (18).

High arsenic levels have been encountered in raw municipal water supplies, necessitating arsenic removal by the water treatment industry. Groundwater supplies in Central and South America are frequently reported to contain excessive arsenic levels (20)(21). Shen and Chen have described deep well waters in Taiwan, China which contain up to 1.1 mg./l. arsenic, and are believed to be responsible for an endemic illness in the area called "black-foot" desease (22). Deep well waters in many desert areas of the southwest United States contain arsenic at levels exceeding 0.1 mg./l. (23).

Limited information was encountered on current levels of arsenic in industrial wastes, and on current treatment processes and removals obtained (14)(24)(25). Much of the literature describing industrial sources and treatment of arsenic is 30 or more years old (15)(18). Recent industrial literature (13)(16)(17)(26), while referring to arsenical wastes and to the severe pollution resulting from their discharge, presents no specific treatment processes or industrial waste values. More information is available on removal of arsenic from drinking water and methods for arsenic treatment of drinking water and industrial wastes are similar.

Measurement in Air

Sampling Methods: Dusts and fumes of arsenic compounds may be collected by any method suitable for collection of other dusts and fumes; the impinger, electrostatic precipitator, and filters are commonly used. The National Air Sampling Network uses a high-volume filtration sampler (27).

Quantitative Methods: Several methods are available for detecting trace amounts of arsenic in dusts; however, only a few of these are quantitative. The chemical methods generally rely on the reduction of arsenical compounds to arsine. The arsine is transported as a gas from the reaction vessel to a second reaction chamber, where it reacts with copper foil (Reinsch's method) or is heated to produce metallic arsenic (Marsh's test), silver nitrate or mercuric chloride (Gutzeit's test), and silver diethyldithiocarbamate (ACGIH tests) (10) (28). The National Air Pollution Control Administration uses silver diethyldithiocarbamate in the second reaction vessel. Neutron activation methods are both quantitative and extremely sensitive, but they require a neutron source. They are sensitive to approximately 0.1 μg. of arsenic, corresponding to 0.24 μg./m.^{3} in a 30 cu. ft. air sample (28).

Thompson et al (27) have reported that the National Air Pollution Control Administration uses atomic absorption to supplement analyses obtained by the Gutzeit method. The method has a minimum detectable limit of 0.2 μg./m.^{3} based on a 2,000 cubic meter air sample.

Arsenic may be monitored in air by the following instrumental techniques (29): colorimetric, neutron activation, atomic absorption. The U.S.S.R. air quality standards for arsenic in air is 0.003 mg./cu. m. on a 24-hour average basis. It should be pointed out that the Soviet air quality standards represent the lowest pollutant concentrations at which a response to a standardized battery of tests was evoked in man or in test animal. It is assumed that such a response is adverse and is therefore to be avoided. This leads to much lower air quality standards in the U.S.S.R. than in other countries where more criteria may be considered and where the definition of adverse response differs as noted by A.C. Stern (30). British emission standards are 115 mg./cu. m. for flows of effluent less than 5,000 cu. ft. per minute and 46 mg./cu. meter for flows of effluent above 5,000 cu. ft. per min. The Australian emission standard is 23 mg./cu. meter.

Measurement in Water

The measurement of arsenic in potable desert groundwater has been discussed by G.C. Whitnack and H.H. Martens (31). The permissible concentration of arsenic in domestic water supplies was 0.05 ppm as shown earlier in Table 3; the desirable concentration is zero. The Safe Drinking Water Act of 1973 has raised the arsenic tolerance level to 0.10 mg./l. The arsenic tolerance in irrigation water has been given in Table 5 as 1.0 ppm for continuous use and 10.0 ppm for short term use in fine textured soil. Aqueous discharge to storm sewers or streams should contain a maximum of 1.0 ppm of arsenic. As shown earlier in Table 11, arsenic may be measured down to detection limits of 20 to 50 ppb by polarography and atomic absorption; neutron activation has shown a detection limit of 5 ppb for arsenic.

A device developed by Stroterhoff (32) provides an economical means and method which can be operated by unskilled personnel to detect low concentrations of arsenical materials in water. A sample of water is taken into a clear and flexible plastic tube; a cap having a detection paper impregnated with a mercuric salt mounted in the cap is inserted onto the plastic tube to retain the water sample therein; a sealed thin wall glass ampule fixedly retained in the plastic tube is broken by squeezing the plastic tube at the ampule location and with the ampule between the fingers releasing to the water sample a mixture of potassium bisulfate and cupric sulfate contained in the ampule to react with the water sample and any arsenical material therein and a zinc strip within the plastic tube.

Any arsenical material present in the water is detected by a yellow to brown color imparted to the detection paper upon reaction of the mercuric salt therein with arsine produced upon release of the ampule contents to the water sample.

References

(1) Sullivan R.J. "Air Pollution Aspects of Arsenic and Its Compounds," *Report PB 188,071,* Springfield, Va., Nat. Tech. Info. Service (Sept. 1969).

(1a) Cothern, C.R., "Determination of Trace Metal Pollutants in Water Resources and Stream Sediments," *Report PB 213,369,* Springfield, Va., Nat. Tech. Information Serivce (1972).

(2) Schroeder, H.A. and Balassa, J.J., *J. Chron. Dis.* 19, 85 (1966).

(3) Warnick, S.L. and Bell, H.L., *J. Water Pollution Control Fed.* 41, 280 (1969).

(4) Konisawa, M. and Schroeder, H.A., *Cancer Res.* 27, 1192 (1967).

(5) Ferm, V.H. and Carpenter, S.J., *Reprod. Fert.* 19, 199 (1968).

(6) Buechley, R.W., *Am. J. Pub. Health,* 53, 1229 (1963).

(7) Schroeder, H.A. and Balassa, J.J., *J. Nutrition* 92, 245 (1967).

(8) Milner, S.E., *Arch. Envir. Health* 18, 7 (1960).

(9) Schroeder, H.A., Kanisawa, M., Frost, D.V. and Mitchener, M.J., *J. Nutrition* 96, 37 (1968).

(10) "Hygiene Guide Series," *Am. Ind. Hyg. Assoc. J.* 25 Nov. - Dec. 1964), p. 610.

(11) Patterson, J.W. and Minear, R.A., *Wastewater Treatment Technology,* 2nd Ed., Report PB 216,162, Springfield, Va., Nat. Tech. Information Service (Feb. 1973).

(12) J.E. McKee and H.W. Wolf, *Water Quality Criteria,* 2nd ed., California State Water Quality Control Board Publication No. 3-A, 1963.

(13) J.G. Dean, F.L. Bosqui and K.H. Lanouette, "Removing Heavy Metals from Wastewater," *Environ. Sci. Tech.,* 6:518-522, 1972.

(14) T. Skripach, V. Kagan, M. Romanov, L. Kamer and A. Semina, "Removal of Fluorine and Arsenic from the Wastewater of the Rare-Earth Industry," in *Proc. 5th Internat. Conf. Wat. Poll. Research,* 2:III-34, Pergamon Press, New York, 1971.

(15) S.N. Cherkinski and F.T. Ginzburg, "Purification of Arsenious Wastewaters," *Water Poll. Abst.,* 14:315-316, 1941.

(16) H.F. Lund, *Industrial Pollution Control Handbook,* McGraw-Hill Book Co., New York, 1971.

(17) Anonymous, *The Economics of Clean Water, Vol. III, Inorganic Chemicals Industry Profile,* U.S. Dept. Interior, Washington, 1970.

(18) R.E. Swain, "Waste Problems in the Nonferrous Smelting Industry," *Indust. Engng. Chem.,* 31:1358-1361, 1939.

(19) H. Stooff and L.W. Haase, "Occurrence and Removal of Arsenic in Drinking Waters," *Chem. Abst.,* 32:6370 (4), 1938).

(20) R.A. Trelles and F.D. Amato, "Treatment of Arsenical Waters with Lime," *Water Poll. Abst.,* 23:125, 1950.

(21) G. Viniegra and R.E. Marquez, "Chronic Arsenic Poisoning in the Lake Region: Section 4. Treatment of Drinking Water," *Water Poll. Abst.,* 38:430-431, 1965.
(22) Y.S. Shen and C.S. Chen, "Relation Between Black-foot Disease and the Pollution of Drinking Water by Arsenic in Taiwan," in *Proc. 2nd Internat. Conf. Wat. Poll. Research, Tokyo,* 1:173-190, Pergamon Press, New York, 1964.
(23) E. Bellack, "Arsenic Removal from Potable Water," *Jour. Amer. Water Works Assoc.,* 63:454-458, 1972.
(24) N.A. Curry, "Philosophy and Methodology of Metallic Waste Treatment," presented at 27th Purdue Indust. Waste Conf., Purdue Univ., 1972.
(25) T. Maruyama, S.A. Hannah and J.M. Cohen, "Removal of Heavy Metals by Physical and Chemical Treatment Processes," presented at 45th Ann. Conf., Wat. Poll. Control Fed., Atlanta, Georgia 1972.
(26) Electrolytic Metal Corp., "Recovery of Manganese from Solutions Containing Iron, Manganese, Nickel, and Cobalt – Patent," *Chem. Abst.,* 54:12960 (c), 1960.
(27) Kanisawa, M., and H.A. Schroeder, "Life Term Studies on the Effects of Arsenic, Germanium, Tin and Vanadium," *Cancer Research* 27:1192 (1967).
(28). Buchanan, W.D., *Toxicity of Arsenic Compounds* (New Jersey: Van Nostrand, 1962).
(29) Lawrence Berkeley Laboratory, "Instrumentation for Environmental Monitoring – Air," *Publication LBL-1,* Vol. 1, Berkeley, Univ. of California (Feb 1, 1973).
(30) Stern, A.C., in *Industrial Pollution Control Handbook,* H.F. Lund, Ed., New York, McGraw-Hill Book Co. (1971).
(31) Whitnack, G.C. and Martens, H.H., *Science* 171, No. 3969, 383-85 (Jan. 29, 1971).
(32) Stroterhoff, H.L. U.S. Patent 3,741,727; June 26, 1973; assigned to Secretary of the Army.

ASBESTOS

Asbestos is an air pollutant which carries with it the potential for a national or worldwide epidemic of lung cancer or mesothelioma of the pleura or peritoneum, according to Sullivan et al (1). Asbestos bodies have been observed in random autopsies of one-fourth to one-half of the population of Pittsburgh, Miami, and San Francisco and will probably be found in the people of every large city. Although asbestos has been shown to produce asbestosis, lung cancer, and mesothelioma in asbestos workers, the relationship between "asbestos bodies" and cancer or asbestosis has not been determined.

The latent period required to develop asbestosis, lung cancer, or mesothelioma is 20 to 40 years, and the exposure required to cause asbestosis has been estimated to be 50 to 60 mppcf-years. No such exposure relationship has been established between asbestos and lung cancer or mesothelioma. Asbestosis, lung cancer, and mesothelioma are all diseases which, once established, progress even after exposure to dust ceases.

Experiments with animals have shown that animals may develop asbestosis or cancer after inhaling asbestos. No information has been found on the effects of asbestos air pollution on either plants or materials. The likely sources of air pollution appear to be the vast number of asbestos products used in our modern society, particularly in the building industry. Mines, factories, and shipping yards may also constitute pollution sources.

The world production of asbestos has approximately doubled over the past 10 to 12 years, whereas the domestic consumption has apparently remained relatively constant. These results indicate that while the asbestos air pollution problem of the world may be increasing, the air pollution potential in the United States has remained relatively unchanged. The increase in the number of lung cancer or mesothelioma cases reported in the the current literature may be due to increases in asbestos use that occurred 20 to 40 years ago. Moreover, the effects of the asbestos being inhaled today may not be reflected in the general health of the population until the 1990's or the next century.

The number of people exposed to asbestos has been estimated to be 100,000 asbestos workers, 3.5 million people working in areas where asbestos is handled in ways which emit small quantities of dust, and 50 to 100 million people who have breathed or will breathe enough fibers to show positive asbestos bodies at autopsy. No measurements have been made of the concentration of asbestos in urban air. A single estimate of 600 to 6,000 particles per cubic meter has been reported.

Bag filters are used to control asbestos in the exhaust gases from asbestos factories. Wetting the asbestos or its products has also been used to keep the dust from becoming airborne. The cost of abatement equipment in a British factory amounted to 27.5% of the total factory cost and approximately 7% of the operating cost. Similar data on the costs of abatement or economic losses due to asbestos air pollution in the United States were not found.

Some of the problems of asbestos in air and its effects on workmen have been described graphically in the magazine, *The New Yorker* (1a). A threshold limit value of 5 million particles per cubic foot of air is cited there as propounded in 1946 by the American Conference of Governmental Industrial Hygienists for all particulate matter. Subsequently, a limit of a particular number of asbestos fibers per cubic centimeter of air was set. As of Oct. 1969 this limit was 2 fibers per cc as set by British occupational health authorities. However, the U.S. Bureau of Occupational Safety and Health had a 12 fibers/cc limit as of that same time. It should be noted that the old standard of 5 million particles per cubic foot is equivalent to about 30 fibers per cc.

The problems of asbestos in water have recently been recognized. At a conference called by the National Institute of Environmental Health Sciences at Durham, North Carolina in Nov. 1973, the problem was discussed. The conference was prompted largely by the discovery earlier in 1973 that the drinking water of Duluth, Minn., and other cities on Lake Superior was heavily contaminated with asbestos, the presumed result of pollution by a mining company that has been dumping 67,000 tons of rocky waste into the lake each day for 17 years. As pointed out in the introductory chapter of this volume, aqueous effluent standards for asbestos processing industries are being established by the Environmental Protection Agency.

Measurement in Air

Of the methods presently being used (2)(3)(4) to count dust samples in the asbestos industry, none is applicable to atmospheric asbestos air pollution. In all the asbestos monitoring methods used, microscopic counting of the fibrous particles is necessary to determine the proportion of fibrous material, and even then it is not known what fraction of the fibers are asbestos. Counting of fibers by eye under the microscope is tedious and difficult. If the number of fibers is less than 1% (less than 5 weight percent) of total dust, the other dust masks the fibers, and quantitative results cannot be obtained.

In parts of the asbestos industry where the asbestos-to-dust ratio is high (greater than 5 wt. percent), it is often possible to determine the asbestos content indirectly (5). For example, if the proportion of asbestos in the airborne dust was known by microscopic count for a given sampling location, the concentration (at least the order of magnitude) could then be inferred from a simple measurement of the concentration of the total dust.

There are at present no proven satisfactory methods for the collection, detection, and identification of asbestos fibers in the 0.1 to 5.0 μ range in ambient air. Satisfactory sampling can probably be accomplished by use of a membrane filter-pump system. The major difficulty lies in the problem of identifying a very few asbestos fibers in the presence of relatively large numbers of a wide variety of other inorganic particulate matter found in the same air. Attempts to determine the asbestos content of urban air have revealed the need for development of new methods.

Modern analytical methods and instrumentation used in the asbestos industry are listed on the following page.

Microscopic particle counting of samples on membrane filters (2)(3)(5)(6)(7)(8)(9)(10)
Thermal precipitators (2)(3)(5)(9)(10)
Impingers (2)(3)(5)(6)(7)(10)
Royco particle counter (2)(3)(5)(10)
Mass concentration methods (2)(5)(6)(10)
Microsieving (11)
Digestion (11)
Column chromatography of organics adsorbed on the surface (11)
X-ray diffraction (6)(11)(12)(13)
Low-temperature ashing (11)
Atomic absorption spectrophotometry (6)(11)
Electron microprobe (11)
Neutron activation (11)
Owens jet counter (2)(3)
Konimeter (2)(3)

Asbestos may be monitored in air by electron microscopy (14). The results of the Battelle Columbus Laboratories attempt to develop a method for asbestos determination in air have been summarized by R.E. Heffelfinger, C.W. Melton, D.L. Kiefer and W.M. Henry (15). The method developed included sampling, beneficiation of asbestos fiber and fibril, and determination of total asbestos by a transmission electron microscopic technique.

References

(1) Sullivan, R.J. and Athanassiadis, Y.C., "Air Pollution Aspects of Asbestos," *Report PB 188,080,* Springfield, Va., National Tech. Information Service (Sept., 1969).
(1a) Brodeur, P., *The New Yorker,* p. 41 ff., (October 29, 1973).
(2) Addingley, C.G., "Dust Measurement and Monitoring in the Asbestos Industry," *Ann. N.Y. Acad. Sci.* 132:298 (1965).
(3) Addingley, C.G., "Asbestos Dust and Its Measurement," *Ann. Occup. Hyg.* (London) 9:73 (1966).
(4) Smith, R. and Tabor, E. (Sept. 1968) in Reference (1) above.
(5) Lane, R.E., et al, "Hygiene Standards for Chrysotile Asbestos Dust," *Ann. Occup. Hyg.* II(2):47 (1968).
(6) Ayer, H.E., and J.R. Lynch, "Motes and Fibers in the Air of Asbestos Processing Plants and Hygienic Criteria for Airborne Asbestos," *Proc. Intern. Symp. Inhaled Particles Vapours, II,* Cambridge, Eng., p. 511 (1967).
(7) Ayer, H.E., J.R. Lynch, and J.H. Fanney, "A Comparison of Impinger and Membrane Filter Techniques for Evaluating Air Samples in Asbestos Plant," *Ann. N.Y. Acad. Sci.* 132:274 (1965).
(8) Edwards, G.H., and J.R. Lynch, "The Method Used by the U.S. Public Health Service for Enumeration of Asbestos Dust on Membrane Filters," *Ann. Occup. Hyg.* II(1):1 (1968).
(9) Holmes, S., "Developments in Dust Sampling and Counting Techniques in the Asbestos Industry," *Ann. N.Y. Acad. Sci.* 132:288 (1965).
(10) Roach, S.A., "Measurement of Airborne Asbestos Dust by Instruments Measuring Different Parameters," *Ann. N.Y. Acad. Sci.* 132:306 (1965).
(11) Keenan, R.G., and R.E. Kupel, "Modern Techniques for Evaluating Mixed Environmental Exposures to Fibrous and Particulate Dusts in the Asbestos Industry," Preprint. U.S. Public Health Service, National Center for Urban and Industrial Health, Cincinnati, Ohio (1958).
(12) Crable, J.V., "Quantitative Determination of Chrysotile, Amosite and Crocidolite by X-ray Diffraction," *Am. Ind. Hyg. Assoc. J.,* 27:293 (1966).
(13) Crable, J.V., and M.J. Knott, "Quantitative X-ray Diffraction Analysis of Crocidolite and Amosite in Bulk or Settled Dust Samples," *Am. Ind. Hyg. Assoc. J.* 27(5):449 (1966).
(14) Lawrence Berkeley Laboratory, "Instrumentation for Environmental Monitoring – Air," *Publication LBL-1,* Vol. 1, Berkeley, Univ. of California (Feb. 1, 1973).
(15) Heffelfinger, R.E., Melton, C.W., and Kiefer, D.L., *Report PB 209,477;* Springfield, Va., Nat. Tech. Information Service (Feb. 29, 1972).

AUTOMOTIVE EXHAUST

The Clean Air Act authorized the promulgation of Federal emission standards for motor vehicle exhausts; it did not authorize the setting of Federal emission standards for any other class of sources as pointed out by A.C. Stern (1).

In view of the massive impact of the automobile industry on the U.S. economy as regards consumption of glass, steel, rubber, etc., the consequences of such regulations are bound to make waves, however. As reviewed recently (1a), both emission standards and test methods have been subjected to many questions.

Research measurements of the detailed composition of emissions of motor vehicles have been made for at least 15 years although emission standards for hydrocarbons in blowby gases were not promulgated in California until 1963 and for exhaust hydrocarbons and carbon monoxide in 1966 in California and in 1968 for the U.S. Subsequently, nitrogen oxide standards have been established. Much more restrictive standards for hydrocarbons, carbon monoxide and nitrogen oxides will apply to 1975 and 1976 model year vehicles as noted by Altshuller (1b).

Evaporative loss controls also have been established. The limiting factor on implementing standards has not been analytical methods, but control technology applicable to mass production vehicles. The original techniques used for hydrocarbons, carbon monoxide and nitrogen oxides involved use of nondispersive infrared analyzers. The infrared analyzers for nitrogen oxides have suffered from water vapor interference as well as limitation in sensitivity.

Many techniques have been applied for detailed analysis of hydrocarbon emissions including mass spectroscopy, dispersive infrared analysis, and coulometric and colorimetric methods (2). Gas chromatography was applied to automobile emissions shortly after its first use in the U.S. However, thermal conductivity detectors were limited in sensitivity and suffered from water vapor and carbon dioxide interference. Chemical pretreatment to remove interferences and concentration techniques greatly complicated the practical application of gas chromatography to automobile exhaust emissions in the late 1950's.

In the early 1960's, the application of flame ionization detectors eliminated these earlier limitations and accelerated the use of gas chromatography to measure the detailed hydrocarbon composition of automotive emissions (3)(4). With gas chromatographic capability well established other analytical methods have not been utilized in recent years. The flame ionization analyzer has replaced the nondispersive infrared analyzer as the motor vehicle certification technique for hydrocarbons. The nondispersive infrared technique was of somewhat limited sensitivity for the more restrictive standards and its response depended on composition in an undesirable manner.

Adsorptive columns have been used for the analysis of paraffinic, olefinic, and aromatic hydrocarbons in vehicle emissions using a flame ionization analyzer (5). This technique provides a more rapid approach to class analysis of emission than does gas chromatography.

The use of open-tubular columns combined with solid absorbent or packed columns combined with temperature programming makes it possible to make about any analysis for individual hydrocarbons in emissions desired. Considerable work has been done applying gas chromatographic techniques to organic oxygenates in exhaust. Recently, a chemical ionization mass spectrometer has been demonstrated to be capable of measuring a number of aldehydes and ketones in exhausts. Work also has been done on an electrochemical approach to aldehyde analysis, but this technique is not far advanced.

Nitrogen oxide analysis primarily for nitric oxide has been done by several other analytical techniques in addition to the nondispersive infrared analyzer. These other approaches included use of an oxidizing step prior to colorimetric analysis or use of an ultraviolet analyzer. A mass spectrometric analyzer for nitric oxide also was developed.

More recently the need to measure the lower concentrations of nitric oxide required by future standards with use of constant total volume samples has stimulated development of more sensitive analyzers. Electrochemical and particularly chemiluminescent types of analyzers have received recent evaluation.

The chemiluminescent analyzers involving gas titration of nitric oxide with ozone are highly specific, very sensitive and have very rapid response times. A thermal decomposition stage has been used to decompose any nitrogen dioxide to nitric oxide. A closely related technique involves gas titration with atomic oxygen which results in equal responses for nitric oxide and nitrogen dioxide. The nondispersive infrared analyzer for carbon monoxide has had adequate sensitivity, specificity and speed of response for vehicle emission applications. Therefore, there has not been much incentive to develop other analyzers. Gas chromatographic analysis for carbon monoxide and the recent fluorescent infrared techniques provide more sensitivity if needed.

All of the instruments or methods discussed previously have been developed either for research, certification or surveillance needs. Inspection of motor vehicles or production line testing requres simple and inexpensive instruments. Some of the instruments discussed are useable for these purposes although more expensive than desired. Catalytic techniques have been shown to have potential. A simple, rapid response optical instrument of moderate cost also has much appeal. Considerable effort has gone into analysis of oxygenated hydrocarbons particularly aldehydes and ketones. In internal combustion engine exhaust (2)(3)(6)(7), gas chromatographic and colorimetric methods have been emphasized. Phenols also have received attention as products in automobile exhaust (8)(9) by chromatography and colorimetric techniques.

Polynuclear aromatic substances particularly the hydrocarbons have been the subject of several measurement projects on automobile exhaust (9). The work cited on polynuclear aromatic hydrocarbons for stationary sources also included analyses for passenger cars and trucks (10). Recently, additional work has been done associating polynuclear aromatic substances and phenols with fuels and fuel additives and engine variables (11). The presence of aza heterocyclic hydrocarbons in automobile exhaust has been demonstrated (12).

Earlier work on particulates in automobile exhaust was concerned with amounts of lead-containing particles (13). During the same period lead was determined polarographically on each of a series of particle sized fractions from the Anderson Sampler, the Goetz spectrometer and other devices from auto exhaust (14). Total particulate and benzene-soluble particulates have been measured in two investigations (15)(16) on exhaust from a number of passenger vehicles and trucks. Auto exhaust particulates after size fractionation in an Anderson Sampler have been analyzed for lead, nitrate, sulfate and chloride by atomic absorption and by nephelometric or colorimetric procedures for the anions (17).

The ratio of water-soluble to water-insoluble lead also was compared. An Anderson Sampler as well as a constant volume sampler were used to obtain total particulate and size-fraction atom with analysis for lead, iron and zinc by atomic absorption and bromine by neutron activation (18). Both leaded and nonleaded fuels were utilized. A tunnel type sampling system capable of sampling auto exhaust for particulate matter under realistic operating conditions has been developed (19). This system was utilized with Andersen and Monsanto impactors to do light particle size distributions as a function of vehicle operation conditions and driving history (19).

A similar sampling system was used in a more comprehensive study of composition of particulate and gases with fuels with various tetraethyl lead contents (20). Total particulate weight, particle size distribution of particulate, metal analysis on exhaust particulate as well as fuel and engine oil by optical emission spectroscopy, chlorine and bromine by neutron activation, particle characterization by electron and light microscopy, and organics by mass spectrometric and ultraviolet fluorescent analysis were the measuring techniques applied to the particulate fraction of the auto exhaust (20).

Diesel exhaust emissions usually are considered of concern because of smoke and odor problems in the vicinity of individual diesel vehicles. Levels of carbon monoxide are very low from diesels (21). Hydrocarbon concentration levels can vary widely (21). Because of the higher molecular weight of diesel fuels the combined emissions of fuel components, low molecular cracked hydrocarbons as well as partial oxygenated organic products presents a substantial analytical problem (3). A portion of these organic components is responsible for the odors associated with diesel exhaust emissions (22)(23)(24). Nitrogen oxides also can vary considerably in diesel emissions overlapping the concentration levels produced by vehicles with spark ignition internal combustion engines (21). Polynuclear aromatic hydrocarbons also have been measured at substantial concentration levels in diesel exhaust emissions (25).

Measurement of the concentration levels of a number of components such as carbon monoxide, nitrogen oxides, low molecular weight hydrocarbons and polynuclear aromatic hydrocarbons do not present substantially different analytical requirements from those well established for spark ignition internal combustion engines. Sampling and analysis of fuel hydrocarbons and their partial oxygenated products still present opportunities for analytical activity (3). Although substantial progress has been made in recent years on identification of the odorous components of diesel exhaust much more analytical work is still needed (24).

Considerable use was made in earlier work by Scott research laboratories of dispersive infrared measurements for carbon monoxide, carbon dioxide, nitrogen oxides and some hydrocarbons (22). Nondispersive infrared instruments were used in Bureau of Mines investigations for carbon monoxide and nitric oxide while nitrogen dioxide was measured with a nondispersive ultraviolet analyzer (21). Colorimetric methods have been used for nitrogen oxides, formaldehyde, acrolein and total aliphatic aldehydes (22).

Gas chromatography has been used for analysis of hydrocarbons in combination with column chromatography with identification by mass spectrometry for odor components (24). Considerable progress has been made by this combination of techniques when applied to the oily-kerosine odor fraction and similar techniques are being applied to the smoky-burnt fraction. Column chromatography combined with fluorescence spectroscopy has been used to identify and measure a number of polynuclear aromatic hydrocarbons in diesel exhaust emissions (25).

The use of high molecular weight fuels presents a special problem with respect to the actual form of the higher molecular weight products upon emission. These materials may be present in the atmosphere as vapors, as finely divided organic aerosols, or as large droplets that settle to the ground rapidly. Additional investigation is needed to properly define this situation under realistic operating conditions.

While smoke and odor have been the main concern from the standpoint of concern about regulations, other emissions may well have to be included in emission regulations in the future. If emissions from the passenger vehicle equipped with spark ignition internal combustion engines are very effectively controlled by the middle 1970's, the residue contributions from other propulsions systems such as the diesel engine may well become of greater concern. The timing and degree of control of automotive emissions continues to be the subject of controversy between environmentalists and the automotive industry, however.

In the case of nitrogen oxides, for example, the Clean Air Act of 1970 set the nitrogen-oxide limit for cars at 0.4 grams a mile. This was to go into effect on 1976 models. But in July 1973, the EPA authorized the year's delay it was allowed to grant under the law. Then in November 1973, a proposal was made to set the standard from 1977 through 1981 at 2 grams a mile. Then it would go to 1 gram a mile through 1989, and to the original 0.4 gram a mile starting in 1990. The interim standard that EPA set in July 1973 for the 1976 to 1977 grace years was 2 grams a mile.

The large scale monitoring of automobile exhaust particulates has been discussed by

L.S. Birks, J.V. Gilfrich and D.J. Nagel of the Naval Research Laboratory (26). A sampler for auto exhaust emissions has been developed by W.F. Kaufman (27).

A centrifugal type extractor pump is used to draw gas through a critical flow venturi of predetermined flow rate. Exhaust emissions are diluted with air and the critical flow venturi acts as a constant volume flow device serving a metering function. Samples are taken and collected continuously during the test run, and are available to be analyzed for contaminants such for example as hydrocarbon, carbon monoxide and oxides of nitrogen.

A sampling process for auto exhaust gases developed by A. Cleveland (28) includes the steps of introducing a continuous stream of ambient air into a conduit system, reducing the temperature of the ambient air to not more than about 40°F., subsequently heating the air to about 130°F., introducing exhaust gases into the system downstream from the heated air and mixing it with the heated air, passing the mixture of gases through a heat sink to obtain a uniform temperature, exhausting the gases from the heat sink at a constant volume, and drawing off a quantity of gas into a sample tube.

This method provides a sample of constant mass in terms of temperature, volume, and moisture content which can then be inserted into a measuring instrument reading in percentage, part per million or similar proportional measurement to determine the mass per unit of time of nitrogen oxides present in the sample.

A sample handling system for auto exhausts has been developed by L.D. Gemache (29). This system provides pumping means to draw the exhaust through the detector at elevated temperatures. Since auto exhaust has very little positive pressure, it requires boosting before it can adequately flow through an instrument to be analyzed. Further, auto exhaust must be maintained hot, above 150°C., to prevent changes in its composition. Booster pumps then encounter severe operating conditions and standard pumps cannot meet these high temperature requirements.

Accordingly, it is an object of this development to provide a sample handling system for auto exhaust analyzers which will reduce the temperature of the pump and maintain the pressure going into the pump at a constant level. Another object is to provide a vacuum barrier around the detector to prevent atmospheric contamination from affecting the output of the detector. These and other objects are achieved by providing a sample handling system for an auto exhaust analyzer including an analyzer through which the exhaust is drawn by means of a sample pump. An auxiliary air supply is drawn through a pressure reduction regulator to the point in the stream between the analyzer and the pump to maintain the pressure at that point constant and reduce the pump temperature.

A sampler for auto exhausts has been developed by H. List and E. Schreiber (30). It is said to be an improved and simplified device capable of supplying representative and readily analyzable waste gas samples. It provides for an air meter connected to the intake of the internal combustion engine with the interposition of a surge tank and for a synchronizer located between the air meter and the pump motor and serving to control the speed of the pump motor in proportion of the speed of the air meter.

With such an arrangement the pump always takes waste gas in an amount proportionate to the quantity of air drawn in irrespective of the load and speed of the internal combustion engine at any given moment. It is thus possible to determine the quantity of waste gas produced during any given period of time and the total of noxious components contained therein technically with a high degree of precision. At the same time, the space required for collecting the samples is much less than with conventional methods and the work of the analyst is greatly simplified and expedited. Speedier completion of both the sampling and analyzing procedures makes it possible to use a single analyzer for a number of test stands, thereby reducing plant investment.

Figure 23 is a schematic drawing showing the essential components of the sampler. The apparatus shown comprises an air meter 1 of conventional design, connected to the suction pipe **3** of the internal combustion engine **4** whose waste gases are to be analyzed, with the interposition of an equalizing tank **2**. The exhaust pipe **5** of the internal combustion engine **4** terminates in a chamber **6** from where a sampling pipe **7** branches off. The sampling pipe **7** extends through a waste gas cooler **8**, its other end being connected to the suction side of a motor driven pump **10**.

FIGURE 23: AUTOMOTIVE EXHAUST GAS SAMPLER

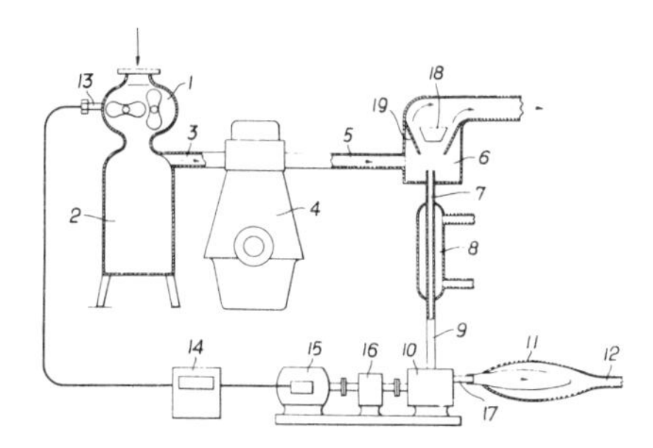

Source: H. List and E. Schreiber; U.S. Patent 3,610,047; October 5, 1971

A temperature feeler **9** provided in the sampling pipe **7** between the waste gas cooler **8** and the pump **10** serves to monitor the temperature of the waste gas sample and is also used, if necessary, for controlling the temperature of the waste gas cooler **8**. At its delivery side, the pump **10** carries a pipe connection **17** to which a double-walled plastic bag **11** receiving the waste gas sample can be attached. In order to ensure complete evacuation of the waste gas sample during the connection of the plastic bag to the analyzer (not shown in the drawing), the outer envelope of the plastic bag **11** has an opening **12** through which compressed air can be introduced.

Associated with the air meter **1** is a pulse generator **13** which is excited by means of an indicator element (not shown) rotating at the same speed as the air meter. The pulses produced by the pulse generator **13** are transmitted to a speed regulator **14** associated with the driving motor **15** of the pump **10**. The motor **15** drives the pump **10** via an interposed governor gearing **16**. The apparatus also comprises a device for maintaining the waste gas pressure in the chamber **6** on a constant level. This device consists of a floating body **18** capable of freely floating above a conical insert **19** of the chamber **6** and controlling the exhaust aperture for the waste gases in the chamber **6** in such a manner that the waste gas pressure in the chamber **6** will remain at least approximately on a constant level whatever the load and speed of the internal combustion engine **4**.

The operation of the device is as follows: With the internal combustion engine **4** running, the speed of the air meter **1** is the measure for the rate of air flow in the engine at a given moment. The pulse generator **13** delivers a number of pulses the number per unit of time of which is proportional to the speed of the air meter **1** and consequently, to the rate of air flow. The transmitter pulses act upon the speed regulator **14** in such a manner that the driving motor **15** of the pump **10** will rotate at a speed which is proportionate to the rate of air flow in the internal combustion engine. Consequently, the pump **10** extracts from the chamber **6** a quantity of waste gas which is proportionate to the amount of air drawn in at the same time, irrespectively of the given speed and load of the internal combustion engine **4**.

The relationship between the amount of waste gas on the one hand, and the amount of air drawn in on the other hand, can be predetermined by means of the governor gearing **16**.

It is thus possible to obtain also over a very short period of operation a representative waste gas sample of a very small volume to be accommodated in a plastic bag **11** of a minimum size.

A process has been developed by G.J. Fergusson (31) for determining hydrocarbon and carbon monoxide concentrations in the gases produced by the combustion of hydrocarbon fuels. A sample of gas is passed through a first stage wherein the free hydrogen present is selectively oxidized in a catalytic oxidation tube without oxidizing any of the hydrocarbons or carbon monoxide present therein. The sample is then passed through a drying tube wherein any water present is substantially completely removed. The dried gas is then passed through a second oxidation tube wherein the hydrocarbons and carbon monoxide are oxidized to water and carbon dioxide. The thermal conductivity change caused by the oxidation of the hydrocarbons and carbon monoxide is measured and a first signal proportional thereto is produced.

The water concentration is also measured and a signal produced which is indicative of the hydrocarbon concentration. A portion of this hydrocarbon signal is subtracted from the first signal in order to generate an output indicative of the carbon monoxide content.

An analyzer for exhaust emissions developed by H.R. Geul (32) combines the known measure of exhaust gases based upon the thermal conductivity of the exhaust gases and the known measure based upon the catalytical burning of the exhaust gases into a system, according to which the output signals obtained by these two measurements, made simultaneously, are subtracted from each other. This results in an analyzer which has an unequivocal indication in the region below and above the stoichiometrical point.

A device developed by A.J. Andreatch (33) utilizes a filament having a high thermal coefficient of conductivity which is coated with an oxide catalyst and positioned in a gas stream to be analyzed. A heater supplies heat to the stream as it is passed to the catalyst coated filament and maintains the stream at temperatures at which the components of the stream to be detected react in the presence of the catalyst.

An apparatus for exhaust gas analysis is described by M. Dodson and R.J. Gomez (34). The exhaust gas is conveyed by a pump to an air cooled condenser to lower the temperature of the gas to essentially ambient temperature and thus condense water vapor therein. The gas is then heated to a temperature above the ambient temperature while being conveyed from the condenser through a conduit to a sample cell of a radiant energy analyzer so that water vapor in the gas passing through the conduit and cell will not condense upon the walls thereof.

The cell is positioned above a water trap in the condenser a distance sufficiently great so that no water in the trap will be carried into the cell by the flow of gas passing through the apparatus. The gas is preferably conveyed through the apparatus by means of a jet pump connected to an outlet port on the cell. The jet pump embodies a restriction at its outer end to provide automatic pressure regulation in the apparatus so that the gas pressure in the sample cell will remain relatively constant regardless of pressure fluctuations in the air supply that drives the pump.

A device has been developed by W.R. Freeman and K.J. Bombaugh (35) for the determination of reactive hydrocarbons in automotive exhaust gases. Air pollution workers believe that photochemical smogs are caused, at least in part, by photolytic ozonolysis of reactive hydrocarbons. They are therefore concerned with the removal of those hydrocarbons from automotive exhaust gases, at least to below an amount responsible for smog formation. Consequently, there are moves at both Federal and State levels not only seeking to limit the amount of RH emitted by automotive vehicles, but also to establish actual standards setting maximum permissible amounts. However, control of the allowable quantity of RH in exhaust gases is at present hampered by the lack of a simple method of measuring the amount of those hydrocarbons in the exhaust gases.

In this device, reactive hydrocarbons (RH) in automotive exhaust gases are determined by the difference in heat effect of combustion upon the gases with and without their content of RH. Preferably, this is accomplished by subjecting a stream of the exhaust gases to combustion then removing the RH from the stream and again subjecting it to combustion, the content of reactive hydrocarbons being determined as a function of the difference in the heats of combustion before and after removal of the RH. The RH removal is effected by passage of the gases through a body of acid-resistant granular material impregnated with an anhydrous solution of sulfuric acid and an alkali metal dichromate.

With increasing public awareness of the problem of air pollution caused by emissions from internal combustion engines of vehicles, various state legislatures and agencies are developing and requiring adherence to higher standards for vehicle emissions. In the past, it has been necessary only to determine the concentration, or percentage, of a given pollutant or constituent in the exhaust gas being emitted from the engine of a vehicle without regard to the volume of exhaust or the rate at which it is being emitted. However, there is a developing concern that vehicles with large engines will expel a greater amount of exhaust gas and hence will emit a greater amount of pollutant than would a car having only a small engine. It is therefore desirable to be able to measure the total quantity of emissions per mile travelled for a vehicle to obtain a true indication of the polluting effect of such vehicle.

An apparatus and method for the determination of mass emissions from vehicle exhausts has been developed by C.W. Morris and J.R. Ulyate (36). It is therefore the primary object of the technique to disclose and provide methods and apparatus for maintaining a constant mass or weight flow of diluted exhaust gases through a test apparatus, including a flow-through duct or conduit, during a test period and to obtain a true proportional sample thereof during such test period so that the total mass of a given pollutant or constituent in the exhaust gases can be computed by multiplying the concentration of such pollutant found in the sample obtained times the total mass or weight or exhaust gases and air passing through the duct during the test, the latter value being determined from an initially established volumetric flow rate established for conducting the test.

A device developed by M. Steinberg (37) is one in which automotive exhaust gas to be monitored is passed through a plurality of adjacently positioned serially connected absorption cells of varied length. A chopper wheel carrying a plurality of spectral filters sequentially passes radiation from a source at discrete wavelengths to each cell in sequence. The amount of radiation absorbed by the gas is detected to determine the concentrations of certain components in the gas.

References

(1) Stern, A.C. in *Industrial Pollution Control Handbook,* H.F. Lund, Ed., New York, McGraw-Hill Book Co. (1971).

(1a) Anon., "More Auto Emission Muddles," *Chem. Eng.* pp. 54, 56 (May 28, 1973).

(1b) Altschuller, A.P., *Proceedings Environmental Quality Sensor Workshop,* (Nov. 30 - Dec. 2, 1971, Las Vegas, Nevada) Wash. D.C., U.S. Govt. Printing Office (1972).

(2) Altshuller, A.P., "Analysis of Gaseous Organic Pollutants" in *Air Pollution,* A.C. Stern, Ed., New York, Academic Press, (1968).

(3) Dimitriades, B., Ellis, C.F., and Seizinger, D.E., "Gas Chromatographic Analysis of Vehicular Exhaust Emissions," *Advances in Chromatography,* Ed. J.C. Giddings and D.A. Keller, New York, Marcel Dekker Publisher (1969).

(4) Dimitriades, B., and Seizinger, D.E., *Envir. Sci. and Tech.* 5, 222 (1971).

(5) Klosterman, D.L. and Sigsby, J.E., Jr., *Envir. Sci. and Tech.* 1, 309 (1967).

(6) Fracchia, M.F., Schutte, F.J. and Mueller, P.L. *Envir. Sci. and Tech.* 1, 915 (1967).

(7) Bellar, T.A. and Sigsby, J.E., Jr., *Envir. Sci. and Tech.* 4, 150 (1970).

(8) Smith, R.G., MacEwen, J.D. and Barrow, R.E., *J. Amer. Ind. Hyg. Assoc.* 20, 149 (1959).

(9) Hoffman, D. and Wynder, E.L., "Organic Particulate Pollutants," in *Air Pollution,* Stern, A.C., Ed., New York, Academic Press (1968).

(10) Hangebrauck, R.P., Von Lehmden, D.J. and Meeker, J.E., "Sources of Polynuclear Hydrocarbons in the Atmosphere," *Publ. No. 999-AP-33,* Durham, N.C., U.S. Public Health Service (1967).

(11) Esso Research and Eng. Co., "Gasoline Composition and Vehicle Exhaust Gas Polynuclear Aromatic Content," *Report on Contract CPA-22-69-56,* Durham, N.C., U.S. Public Health Service (Feb. 1970).
(12) Sawicki, E., Meeker, J.E., and Morgan, M.J., *Arch. Envir. Health* 11, 773 (1965).
(13) Hirschler, D.A., and Gilbert, L.F., *Arch. Envir. Health* 8, 109 (1964).
(14) Mueller, P.K., Helwig, H.L., Alcocer, A.E., Cong, W.K. and Jones, E.E. in *Publ. 352,* Phila., Pa., Amer. Soc. for Testing Matls. (1964).
(15) McKee, H.E. and McMahou, W.A., Jr., *J. Air Poll. Control Assoc.* 10, 456 (1960).
(16) Hangebrauck, R.P., Lauch, R.P. and Meeker, J.E., *J. Amer. Ind. Hygiene Assoc.* 27, 47 (1966).
(17) Lee, R.E., Jr., Patterson, R.K., Crider, W.L. and Wagman, J., *Atm. Environ.* 5, 225-237 (1971).
(18) Nimomjii, J.W., Bergman, W. and Simpson, B.H., "Automobile Particulate Emissions," Second International Clean Air Congress, Wash., D.C. (Dec. 6 - 11, 1970).
(19) Habibi, I., *Envir. Sci. and Tech.* 4, 239 (1970).
(20) Moran, J.B. and Manery, O.J., *Report on Contract CPA-22-69-145,* Durham, N.C., U.S. Public Health Service (July 1970).
(21) Marshall, W.F. and Fleming, R.D., Report of Investigation, "Diesel Emissions Reinventoried," Wash. D.C. U.S. Bureau of Mines (1971).
(22) Linnell, R.H. and Scott, R.H., *J. Air Pollution Control Assoc.* 12, 510 (1962).
(23) Vogh, J.W., *J. Air Pollution Control Assoc.* 19, 773 (1969).
(24) Arthur D. Little, Inc., "Chemical Identification of the Odor Components in Diesel Engine Exhaust," *Final Report on Contract CPA-22-69-63,* Durham. N.C., U.S. Public Health Service (June 1970).
(25) Reckner, L.R., Scott, W.E. and Biller, W.F., "The Composition and Odor of Diesel Exhaust," before 30th Midyear Meeting of Amer. Petroleum Inst., Montreal, Canada (May, 1965).
(26) Birks, L.S., Gilfrich, J.V. and Nagel, D.J., *Report AD-738,801,* Springfield, Va., Nat. Tech. Infor. Service (Oct. 1971).
(27) Kaufman, W.F., U.S. Patent 3,699,814; October 24, 1972; assigned to Philco-Ford Corp.
(28) Cleveland, A., U.S. Patent 3,611,812; Oct. 12, 1971; assigned to Olson Laboratories, Inc.
(29) Gamache, L.D., U.S. Patent 3,537,296; Nov. 3, 1970; assigned to Beckman Instruments, Inc.
(30) List, H. and Schreiber, E., U.S. Patent 3,620,047; Oct. 5, 1971.
(31) Fergusson, G.J., U.S. Patent 3,549,327; Dec. 22, 1970; assigned to Scientific Research Instruments Corp.
(32) Geul, H.R., U.S. Patent 3,674,436; July 4, 1972.
(33) Andreatch, A.J., U.S. Patent 3,595,621; July 27, 1971.
(34) Dodson, M. and Gomez, R.J., U.S. Patent 3,593,023; July 13, 1971; assigned to Beckman Instruments, Inc.
(35) Freeman, W.R. and Bombaugh, K.J., U.S. Patent 3,558,283; Jan. 26, 1971; assigned to Mine Safety Appliances Co.
(36) Morris, C.W. and Ulyate, J.R., U.S. Patent 3,603,155; Sept. 7, 1971; assigned to Chromalloy American Corp.
(37) Steinberg, M., U.S. Patent 3,743,426; July 3, 1973; assigned to General Motors Corp.

BACTERIA

Biological aerosols, suspensions of microorganisms in the air, can cause diseases of humans, animals, and plants, and degradation of inanimate materials. The microorganisms generally involved are the bacteria, fungi (yeast and molds), and viruses. Bacterial and viral aerosols are detrimentally affected by the atmospheric environment and, therefore, air-borne transmission of such diseases is limited to short distances and crowded conditions. Fungi are better adapted to aerial dissemination and are known to have been transmitted hundreds of miles from their source.

Generally, the symptoms produced by air-borne infectious organisms in humans and animals are those of a respiratory disease. The human diseases in this category include tuberculosis, pneumonia, aspergillosis, influenza, the common cold, and others. As more data are gathered, there is increasing evidence that biological and nonbiological air pollutants are capable of producing synergistic effects. An increase in the incidence of respiratory diseases has been reported in metropolitan areas during occasions of excessively high air pollution. This potential effect has been confirmed through the use of experimental animals in the laboratory. For example, mice have been found to exhibit a higher mortality rate after a controlled dosage of *Klebsiella pneumoniae* when preceded by exposure to ozone or nitrogen dioxide. Compared to humans, relatively few diseases of animals are spread by airborne transmission. Those that are include tuberculosis, glanders, aspergillosis, hog cholera, and Newcastle disease.

Plants are susceptible both to specific plant pathogens and to the indigenous saprophytic decay produced by microorganisms present in the soil. Of the plant pathogens, fungi are the most commonly transmitted by air and in the past have been the agents for devastating epidemics. For example, wheat rust destroyed an estimated 135 million bushels of wheat in 1935. Saprophytic microorganisms are ubiquitous in nature. Consequently, surfaces of material in contact with a humid environment often show microbial, especially fungal, growth. There are no environmental standards applicable to biological aerosols at the present time, according to Finkelstein (1).

Sewage treatment plants have been investigated as a source of hazardous biological aerosols. Although potentially pathogenic microorganisms have been isolated downwind of sewage tanks, the full significance of this condition is not as yet known. Industrial fermentations with microorganisms produce a number of economically important materials, such as organic solvents, vitamins, and antibiotics, but no instance has yet been reported of a disease being transmitted to the general population as a result of any of these processes.

It is not valid to present any one value for the aerial microbial concentration of a given area. Any count is influenced by the temperature, meteorological conditions, vegetation, human and animal population, and time of day, as well as by the inability to determine all types of microorganisms by any one sampling procedure. However, some data have been presented as indicative of certain areas under noted environmental conditions and sampling procedures.

The problem of abatement and control of biological aerosols is exceedingly difficult and complex. Attempts have been made to control airborne infections indoors by ultraviolet light irradiation. Dust control, treatment of carriers with antibiotics, washing with disinfectant soaps, and the use of disinfectants and surgical masks have reduced significantly the spread of airborne disease in hospitals. Within recent years, a number of air filtration devices have been made commercially available that are capable of removing extremely small particles, including microorganisms. These devices have been produced in different sizes and efficiencies and can be used in air-conditioning systems. Their full potential in the control of biological aerosols has not as yet been realized.

The control of outdoor airborne infections has been limited essentially to dust control and location and elimination of sources for specific outbreaks of certain diseases. Progress in the area has been hindered by lack of knowledge concerning the outdoor transmission of airborne disease.

Measurement in Air

The problems of obtaining representative samples for analysis of airborne particles covering the wide range of atmospheric conditions, biological types, and particle size are such that no single procedure is adequate for all. Therefore, the methods of analysis tend to be specialized for relatively narrow fields of study; consequently, many different individual sampling devices have been used. The best reviews of the subject are by Wolf et al (1a), Gregory (2), Noble (3) and May (4).

The methods used for sampling biological aerosols are basically the same as the methods used to sample dust and other airborne particulates. However, since the objective is generally to determine the viability of collected particles, following collection the samples must undergo an additional step: growth in a suitable nutrient under proper environmental conditions, followed by observation of the growth and evaluation of the results. Since no one method of analysis will yield information concerning all parameters of a sample, procedures should be chosen which will yield the information that is of greatest concern. The The basic methods are these:

1. Sedimentation (5): In this method, particulates suspended in the air are allowed to settle either on plain surfaces or on surfaces coated with a nutrient medium. This method can yield information on the

number of viable particles that have settled during the sampling time, and the total number and size of all particles that settle in a given time. Results will be influenced greatly by air movement and diameter of the aerosol particles.

2. Impingement into liquids (6)(7)(8)(9): Air is drawn through a small jet and is directed against a liquid surface, and the suspended particles are collected in the liquid. Due to the agitation of the particles in the collecting liquid, aggregates are likely to be broken up. Therefore, the counts obtained by this method tend to reflect the total number of individual organisms in the air and are higher than the values obtained by other methods.
3. Impaction onto solid surfaces (10)(11): Air is drawn through a small jet(s), and the particles are deposited on dry or coated solid surfaces or on an agar nutrient. Samples taken by this method have been used to determine total numbers, size, viable numbers, and variation in numbers per unit of time during a long sampling period.
4. Filtration (12)(13)(14)(15): The particulates are collected by passing the air through a filter, which can be made of cellulose-asbestos paper, glass wool, cotton, alginate wool, gelatin foam, or membrane material. The particulates are washed from the filters and assayed by appropriate microbiological techniques. In this method, the viability of organisms can be detrimentally affected by dehydration in the air stream and the results thereby biased.
5. Centrifugation (16)(17): The particulates are propelled by centrifugal force onto the collecting surface, which can be glass or an agar nutrient. Size and number information can be obtained by this method.
6. Electrostatic precipitation (18): Particles are collected by drawing air at a measured rate over an electrically charged surface of glass, liquid, or agar. The total number of particles or viable number is then determined.
7. Thermal precipitation (19): Particles are collected by means of thermal gradients. The design is based on the principle that airborne particles are repelled by hot surfaces and are deposited on colder surfaces by forces proportional to the temperature gradient. The particle size distribution can be determined.

Because of the great number of different aerosol samples used by investigators, general agreement was reached at the International Aerobiology Symposium (sponsored by the Office of Naval Research and the University of California in October 1963) that data obtained with any specialized sampler should be correlated with at least some results obtained with a standard reference sampler (20). The participants at the Symposium also agreed that the United States Army Chemical Corps all-glass impinger (AGI 30 Impinger)(1a) be recommended as the standard liquid impinger, and that the Anderson Stacked Sieve sampler (10) be recommended as the standard apparatus for impaction on solid surfaces.

A device for the detection of individual bacteria in a constantly flowing stream of air has been developed by T.A. Rich, J.N. Groves and G.F. Skala (21). Any bacteria present in the air, being composed of protein, are pyrolyzed to produce a bubble-like volume of ammonia in the air stream which is then converted to condensation nuclei which are detected. By maintaining streamline flow of the air stream, the bubble form of the ammonia gas is maintained and when converted forms a cluster or cloud of particles.

A schematic perspective view of the apparatus employed in the process is shown in Figure 24. Referring to the figure, there is shown an air inlet **1** through which flows the volume of air under inspection and which may possibly contain bacteria. The air flow then strikes an impactor **2** which includes a plate **3** which blocks large high inertia airborne particles such as pollen and which allows smaller lower inertia particles to flow around its corners following paths such as indicated by the arrows. The impactor **2** is conventional and a detailed description of it and its operation will be unnecessary for those skilled in the art. The volume of air containing the bacteria then enters a pyrolyzer tube **4** in which the bacteria is converted to ammonia gas by the action of heat supplied by an electric radiant heater **5** which may be energized from any convenient source not shown. It is well-known that the protein in bacteria, when subjected to heat, converts into ammonia gas.

FIGURE 24: DETECTOR FOR BACTERIA IN AIR

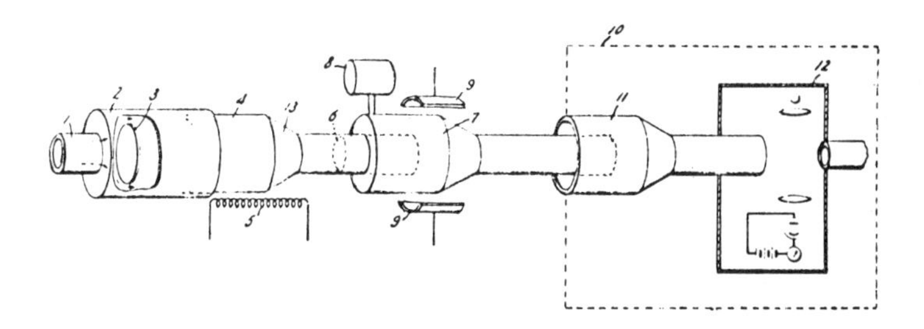

Source: T.A. Rich, J.N. Groves and G.F. Skala; U.S. Patent 3,458,284; July 29, 1969

The sensitivity of a bacteria detector can be increased such that individual particles can be discerned if the air flow is kept continuous so that little or no mixing of the ammonia vapor with the air occurs after pyrolysis. When this is done the ammonia remains as a discrete "bubble" of ammonia gas which can be detected by means which will become apparent. Thus, it is desirable to insure that the air flow through the pyrolysis device is kept continuous without the occurrence of turbulence or any other event which may cause mixing.

After pyrolysis the air containing the ammonia bubble is filtered through a filtering device such as a thin porous disc **6** which removes any solid particles which may have resulted from combustion of nonprotein containing substances in the pyrolysis device. A thin porous disc, if used, will not cause mixing of the air flow up but rather will pass the ammonia bubble in a streamlined flow of air.

The filtered air and ammonia bubble then enters chamber **7** where it is subjected to a saturated hydrogen chloride and water vapor gas mixture supplied from a source **8**. A corona discharge, supplied by a pair of electrodes **9**, converts the ammonia gas bubble to a cloud of extremely small ammonium chloride solid particles. The conversion technique utilizing the corona discharge has been found desirable but any means by which the HCl gas reacts with the ammonia gas to form ammonium chloride particles would be acceptable. The only criterion is that the reaction does not unduly disperse the NH_4Cl particles thus formed. The resulting NH_4Cl particles are so small that they are invisible but, if the flow of air is controlled properly, the particles will not disperse but will remain clustered together in a sort of cloud.

The flow of saturated air containing the cloud of ammonium chloride particles and water vapor then enters a condensation nuclei detector **10** including a chamber **11** in which ambient air at 100% humidity enters. Since the temperature of the ambient air is much lower than the temperature of the air emerging from the chamber **7** due to the heat of the pyrolysis, a supersaturated air mass results which has a humidity greater than 100%. Since this supersaturated air mass has been obtained while maintaining a continuous air flow, unlike conventional methods which obtain supersaturation by a cyclic expansion, the ammonium chloride particles remain in a discrete bundle.

In such conditions, water vapor begins to condense out of the air around any solid particles existing therein. Since the only solids present in this air mass are the cluster of small ammonium chloride particles, water vapor begins to condense around these particles using them as a nucleus. As the amount of water which condenses around the ammonium chloride nucleus increases, the size of the resultant particle increases until its size is sufficient so that it can be detected with any suitable detecting device such as a scattered light measuring device **12**.

Another apparatus which may be used for the detection of bacteria has been developed by S. Witz, L.T. Carleton, H.H. Anderson, R.H. Moyer and H.A. Neufeld (22). The apparatus for detecting biological agents such as vegetative bacteria, spores and viruses is capable of operating satisfactorily when supplied with minute samples of material to be tested, even when present in a continuous background of matter similar in nature. The equipment utilizes the phenomenon of chemiluminescence and, more particularly, provides the proper conditions for chemiluminescence of luminol by hydrogen peroxide, operating in an intermittent flow system supplied with the agents by an aerosol particle collector, and in which detection of the chemiluminescence is by a photomultiplier tube the output of which is monitored. Photomultiplier output could be recorded on a chart, magnetic tape or merely designed to set off an alarm when values exceed a prescribed threshold.

Measurement in Water

Bacteria and viruses can be monitored in water by the following instrumental techniques (23): Colony counters; fluorescent antibody technique.

References

(1) Finkelstein, H., "Air Pollution Aspects of Biological Aerosols (Microorganisms)," *Report No. PB 188,084,* Springfield, Va., Nat. Tech. Information Service (Sept. 1969).
(1a) Wolf, H.W., et al, "Sampling Microbiological Aerosols," *U.S. Public Health Monograph No. 60,* Washington, D.C. (1959).
(2) Gregory, P.H., *The Microbiology of the Atmosphere,* (New York: Interscience, 1961).
(3) Noble, W.C., "Sampling Airborne Microbes – Handling the Catch," in *Airborne Microbes, Seventeenth Symposium of the Society for General Microbiology,* P.H. Gregory and J.L. Monteith, Eds. (Cambridge, Engl.: University Press, 1967).
(4) May, K.R., "Physical Aspects of Sampling Airborne Microbes," in *Airborne Microbes, Seventeenth Symposium of the Society for General Microbiology,* P.H. Gregory and J.L. Monteith, Eds. (Cambridge, Engl.: University Press, 1967).
(5) Richardson, J.F., and E.R. Wooding, "The Use of Sedimentation Cell in the Sampling of Aerosols," *Chem. Eng. Sci.* 4:26 (1955).
(6) Cown, W.B., T.W. Kethley, and E.L. Fincher, "The Critical-Orifice Liquid Impinger as a Sampler for Bacterial Aerosols," *Appl. Microbiol.* 5:118 (1957).
(7) Ferry, R.M., L.E. Farr, and M.G. Hartman, "The Preparation and Measurement of the Concentration of Dilute Bacterial Aerosols," *Chem. Rev.* 44:389 (1949).
(8) Greenburg, L., and J.J. Bloomfield, "The Impinger Dust Sampling Apparatus as Used by the United States Public Health Service," *Public Health Repts. (U.S.)* 47:654 (1932).
(9) Moulton, S., T.T. Puck, and H.M. Lemon, "An Apparatus for Determination of the Bacterial Content of Air," *Science* 97:51 (1943).
(10) Andersen, A.A., "New Sampler for the Collection, Sizing, and Enumeration of Viable Airborne Particles," *J. Bacteriol.* 76:471 (1958).
(11) DuBuy, H.G., A. Hollaender, and M.D. Lackey, "A Comparative Study of Sampling Devices for Airborne Microorganisms," *Public Health Repts. Suppl. No. 184,* Washington, D.C. (1945).
(12) Mitchell, R.B., J.D. Fulton, and H.V. Ellingston, "A Soluble Gelatin Foam Sampler for Airborne Microorganisms at Surface Levels," *Am. J. Public Health* 44:1334 (1954).
(13) Noller, E., and J.C. Spendlove, "An Appraisal of the Soluble Gelatin Foam Filter as a Sampler for Bacterial Aerosols," *Appl. Microbiol.* 4:300 (1956).
(14) Sehl, F.W., and B.J. Havens, Jr., "A Modified Air Sampler Employing Fiber Glass," *A.M.A. Arch. Indust. Hyg. Occupational Med.* 3:98 (1951).
(15) Thomas, D.J., "Fibrous Filters for Fine Particle Filtration," *J. Inst. Heating Ventilating Engrs.* 20:35 (1952).
(16) Sawyer, K.F., and W.H. Walton, "The Conifuge – A Size Separating Sampling Device for Airborne Particle," *J. Sci. Instr.* 27:272 (1950).
(17) Wells, W.F., "Apparatus for Study of Bacterial Behavior of Air," *Am. J. Public Health* 23:58 (1933).
(18) Kraemer, H.F., and H.F. Johnstone, "Collection of Aerosol Particles in Presence of Electrostatic Fields," *Ind. Eng. Chem.* 47:2426 (1955).
(19) Kethley, T.W., M.T. Gordon, and C. Orr, Jr., "A Thermal Precipitator for Aerobacteriology," *Science* 116:368 (1946).
(20) Brachman, P.S. et al, "Standard Sampler for Assay of Airborne Microorganisms," *Science* 144:1295 (1964).
(21) Rich, T.A., Groves, J.N. and Skala, G.F., U.S. Patent 3,458,284; July 29, 1969; assigned to General Electric Co.
(22) Witz, S., Carleton, L.T., Anderson, H.H., Moyer, R.H. and Neufeld, H.A.; U.S. Patent 3,690,837; Sept. 12, 1972; assigned to Aerojet-General Corp. and Secretary of the Army.

(23) Lawrence Berkeley Laboratory, "Instrumentation for Environmental Monitoring – Water," *Publ. LBL-1,* Vol. 2, Berkeley, Univ. of Calif. (Feb. 1, 1973).

BARIUM

Metallic barium is highly reactive, and if released to the atmosphere, quickly becomes converted to a barium salt. Soluble barium salts are very toxic, and if ingested, have a strong stimulating effect on all muscles, including the heart. Symptoms of barium poisoning are excessive salivation; vomiting; colic; diarrhea; convulsive tremors; slow, hard pulse; and increased blood pressure. There have been very few reported deaths due to barium poisoning. Insoluble barium compounds, such as common barium sulfate, are generally nontoxic; Inhaled barium compounds are known to cause a benign pneumoconiosis called baritosis, which does not cause symptoms of emphysema, bronchitis, or reduced respiratory capacity according to Miner (1).

Tests on animals with exhaust solids from a diesel engine using fuels with barium-base smoke-suppressant additives indicated that the LD_{50} of these exhaust solids was in excess of 10 grams/kg. of body weight. Very limited animal studies with these exhaust solids do not appear to have demonstrated any acute health effects from the concentrations used; no information is available on chronic effects. No information was found on the effect of barium and its compounds on plants or materials.

Barium occurs naturally mainly as barite (barium sulfate) and witherite (barium carbonate). Approximately 90% of the barite produced is used as well-drilling mud. Lithopone and other barium compounds are produced with the remainder. No environmental emission data were found for the mines or refineries producing barite, lithopone, or any of the other barium compounds.

A number of proprietary barium-base organometallic compounds used as additives in diesel fuel have been found to be very effective in reducing smoke emissions from diesel engines. Tests with diesel engines with and without the smoke-suppressant additives showed very little change in gaseous emissions. Analysis of solids emitted from these engines revealed the presence of barium sulfate and barium carbonate. The soluble barium carbonate content varied from under 1% to about 25% of the total barium content of the exhaust solids. Based on the 25% soluble salt content, it was estimated that the maximum emission of barium from a diesel engine would be on the order of 48,000 $\mu g./m.^3$ of exhaust. Sufficient data are not available to give detailed physical and chemical properties of these barium-containing diesel engine exhaust solids.

Although no quantitative measurements have been made of the environmental air concentrations of barium, qualitative measurements in some major U.S. cities have shown the presence of barium. Additional data on atmospheric emissions of barium compounds has been presented by W.E. Davis and Associates (2). No information has been found on techniques for control of emissions of barium and its compounds. However, for industrial processes, control equipment such as filters, electrostatic precipitators, and wet scrubbers could be used. No information has been found on the economic costs of barium air pollution or on the costs of its abatement.

Very little information is available in the literature on levels of barium in industrial wastewaters, on existing methods of treatment of levels of barium removal achievable, or on costs associated with such methods of treatment according to Patterson and Minear (3). Several reports mention barium as an industrial type of waste, or discuss its toxicity. Barium as barite (barium sulfate), is used as a white pigment for paint (4). Barium is employed in metallurgy, glass, ceramics and dyes manufacture, and in vulcanizing of rubber (5). Barium has been also reported present in explosives manufacturing wastewater (6).

The permissible concentration of barium in domestic water supplies is 1.0 ppm as noted earlier in Table 3. Also, effluents to storm sewers or streams should contain a maximum of only 1.0 ppm of barium.

Measurement in Air

Sampling Methods: Dusts and fumes of barium compounds may be collected by any of the usual methods for collection of particulate matter. For concentrations of barium encountered in ambient air, sampling has to be done with membrane filters (7). Barium solids from diesel engine exhaust samples have been collected on analytical-grade filter paper (8).

Quantitative Methods: Any barium sample soluble in hydrochloric acid can be analyzed by spectrographic methods (9). Low concentrations of barium can be quantitatively determined by emission spectrography or atomic absorption (7). Thompson et al (10) reported that when barium is analyzed by atomic absorption, the minimum detectable limit is 0.02 $\mu g./m.^3$ based on an air sample of 2,000 $m.^3$.

To convert all the barium collected on a filter to a soluble form for spectrographic analysis, the filter is ashed and the ash is dissolved in a hydrochloric acid and nitric acid mixture (7). Golothan (8) separated the soluble and insoluble barium compounds by digesting the solids on the filter paper with hydrochloric acid. This acid solution was analyzed directly for the soluble barium compounds. The solid residue (remaining after the acid treatment) was ashed and treated with acid and subsequently analyzed to determine the amount of insoluble barium compounds in the original sample.

Industrially, barium is usually determined by precipitating it as insoluble barium sulfate, separating the precipitate, and weighing it (11)(12). When barium is associated with other alkaline earths such as calcium and strontium, the calcium is separated as calcium nitrate by dissolving it in a mixture of alcohol and absolute ether. The barium is then precipitated as barium chromate from the slightly acidified (acetic acid) mixture (12). Barium may be monitored in air by emission spectroscopy (13).

Measurement in Water

Barium may be monitored in water down to a detection limit of 60 ppb by atomic absorption spectroscopy as noted earlier in Table 11.

References

(1) Miner, S., "Air Pollution Aspects of Barium and Its Compounds," *Report PB 188,083;* Springfield, Va., Nat. Tech. Info. Service (Sept. 1969).
(2) Davis, W.E. and Associates, *Report PB 210,676,* Springfield, Va., Nat. Tech. Info. Service (May (1972).
(3) Patterson, J.W. and Minear, R.A., "Wastewater Treatment Technology," *Report PB 216,162,* Springfield, Va., Nat. Tech. Info. Service (Feb. 1973).
(4) Anonymous, *The Economics of Clean Water, Vol. III, Inorganic Chemicals Industry Profile,* U.S. Dept. Interior, Washington, 1970.
(5) J.E. McKee and H.W. Wolf, *Water Quality Criteria,* 2nd ed., California State Water Quality Board Publication No. 3-A, 1963.
(6) Anonymous, "Taming Explosives," *Indust. Engng. Chem.* 46 (12):20A, 1954.
(7) Morgan, G., Director, Division of Air Quality and Emissions Data, National Air Pollution Control Administration, (May 1969) quoted in Reference (1).
(8) Golothan, D.W., "Diesel Engine Exhaust Smoke. The Influence of Fuel Properties and the Effects of Using Barium-Containing Fuel Additives," *SAE-670092,* Society of Automotive Engineers, (1967).
(9) Abernethy, R.F., and F.H. Gibson, "Rare Elements in Coal," *U.S. Bur. Mines, Inform. Circ. 8163,* (1963).
(10) Thompson, R.J., G.B. Morgan, and L.J. Purdue, "Analyses of Selected Elements in Atmospheric Particulate Matter by Atomic Absorption," Preprint, presented at the Instrument Society of America Symposium, New Orleans, La. (May 5-7, 1969).

(11) "Barium," in *Encyclopedia of Science and Technology,* vol. 2 (New York: McGraw-Hill, p. 94, 1960).
(12) "Barium" and "Barium Compounds," in *Kirk-Othmer Encyclopedia of Chemical Technology,* vol. 3 (New York: Wiley, pp. 77, 80, 1964).
(13) Lawrence Berkeley Laboratory, "Instrumentation for Environmental Monitoring – Air," *Publ. LBL-1,* vol. 1, Berkeley, Univ. of Calif. (Feb. 1, 1973).

BERYLLIUM

Inhalation of beryllium or its compounds is highly toxic to humans and animals, producing body-wide systemic disease commonly known as beryllium disease. Both acute and chronic manifestations of the disease are known. The effects of beryllium intoxication can be mild, moderate, or severe, and can prove fatal, depending on the duration and intensity of exposure, according to Durocher (1). Acute beryllium disease is manifested by a chemical pneumonitis ranging from transient pharyngitis or tracheobronchitis to severe pulmonary reaction. As of June 1966, 215 acute cases had been recorded in the Beryllium Case Registry.

Chronic beryllium disease generally occurs as lesions in the lung, producing serious respiratory damage and even death. However, every organ system may be involved in response to beryllium exposure, except for the organs in the pelvic area. The chronic form is characterized by a delay in onset of disease, which may occur weeks or even years after exposure. In June 1966, 498 chronic cases had been recorded, plus 47 acute-to-chronic cases. Of the total 760 cases recorded in the Beryllium Case Registry, 210 fatalities, or 27.5%, had occurred by June 1966.

Cancer has been produced experimentally in animals and 20 cases of cancer have been found (as of 1966) in humans afflicted with beryllium disease. However, insufficient information exists at this time to causally relate beryllium poisoning to development of cancer in humans. Beryllium and its compounds can produce dermatitis, conjunctivitis, and other contact effects; however, these manifestations are rare. There is some evidence that beryllium in soils is toxic to plant life; no evidence was found on the effects of atmospheric beryllium on plants or on materials.

The major potential sources of beryllium in the atmosphere are industrial. The processes of extraction, refining, machining, and alloying of the metal produce toxic quantities of beryllium, beryllium oxide, and beryllium chloride, which if allowed to escape into the atmosphere would cause serious contamination. Recognition of the serious hazards to health from these sources has led to adaptation of control procedures minimizing this potential. However, beryllium in limited quantity is emitted from these industrial processes, and danger also exists from accidental discharges. One major source of beryllium contamination, the use of beryllium in fluorescent light tubes, was discontinued in 1949.

Other sources could be the use of metallic beryllium in rocket fuels, and the combustion of coals. Rocket fuels could present a hazard in the handling and storage of the powdered metallic beryllium used as an additive in the fuels. Also, the exhaust fumes, which contain oxidized beryllium as well as other compounds of beryllium would be of significance in local soil and air pollution if not contained. As beryllium is a normal constituent (above 2 ppm) of coals, the combustion of coal may add a significant quantity of beryllium to the atmosphere.

Measurements are made of the beryllium concentration at 100 stations in the United States. The average 24-hour concentration is less than 0.0005 μg./m.^{3}; the maximum value recorded during the 1964-65 period was 0.008 μg./m.^{3}. Abatement measures have been implemented industry-wide, with a very high degree of success. Conventional air-cleaning procedures have been employed, including the use of electrostatic precipitators, bag houses, scrubbers, etc. These procedures have enabled the beryllium industry to meet the industrial hygiene standards established for beryllium. Data on the economic losses resulting from beryllium air pollution are not available. Court cases are pending in the State of Pennsylvania, however, which may provide data on the economic values of impairment to health resulting from exposure to beryllium. Only one analysis of the costs for abatement was found. This study indicated that the added costs for control amounted in 1952, to approximately 20% of the normal cost of operation for the particular plant analyzed.

Ontario (Canada) has set an air quality standard for beryllium of only 0.00001 mg./cu. m. at STP for either single exposure or 24-hour average exposure. New York State, Montana, Pennsylvania and Texas have similar standards. The tolerance for beryllium in irrigation waters has been given in Table 5 as 0.5 ppm for continuous water use and 1.0 ppm for short term water use in fine-textured soils.

Measurement in Air

Sampling Methods: The most common method of sampling beryllium concentrations in air employs a high-volume sampler which draws air to be analyzed through a filter for the appropriate sampling period (2). The Hi-Vol sampler, which samples at a rate of 1.5 to 1.7 $m.^3$/min. when operated 24 hours, will provide an adequate sample for the measurement of beryllium in ambient air in concentrations as low as 0.002 μg./$m.^3$. Cellulose-fiber filter paper with very low ash content is widely used; cellulose ester and fiber glass papers are also used, and are particularly suitable for use when collecting contaminants are of very small size.

Quantitative Methods: Atmospheric beryllium may be analyzed by colorimetric, fluorimetric, or spectrographic procedures (3). These methods are accurate and sensitive but do not discriminate between the various compounds of beryllium.

Morin Fluorescent Method – The morin method is suitable for measuring quantities of beryllium in the air surrounding beryllium plants; its lower limit of detection is approximately 0.01 μg. of beryllium (3). In this procedure, the beryllium is removed from the filter, processed to remove interfering elements and to form an alkaline solution, and fluorescent morin solution ($C_{15}H_{10}O_7 \cdot H_2O$) is added. Beryllium acts to reduce the fluorescent characteristics of the morin solution, as a direct function of the quantity of beryllium present. Measurement of the fluorescence by fluorimeter against standard samples indicates the quantity of beryllium. (Note: Keenan (4) states that the procedure is sensitive to 0.00002 μg./ml. of the solution.

Colorimetric Method – This method, commonly called the Zenia method, is suitable for large concentrations from which at least 2 μg. of beryllium can be obtained, and is used in monitoring the atmospheres of industrial plants. The procedure involves removal from the filter paper, removal of interfering elements, and addition of the Zenia (p-nitrobenzene-azoorcinol) solution. Zenia solution added to a beryllium solution forms a reddish-brown solution, whose absorption of light is a function of its beryllium content. Measurements can be made subjectively, to give a quick spot-check (5) or by means of a spectrophotometer. Sensitivity to 0.5 μg./ml. of solution is claimed (4).

Spectrographic Method – This spectrographic method is considered suitable for measuring concentrations in the general atmosphere, as it permits detection of trace quantities down to 0.003 μg. Also, automatic detectors are available that are capable of handling large numbers of samples rapidly and accurately. This method adds a thallium standard to the beryllium-Zenia solution produced in the colorimetric method, concentrates the solution, subjects it to an electric arc, and photographs its spectra. The ratio of the intensity of the beryllium line at 2348.6 A. to that of the thallium line at 2379.7 A. is determined, giving a measure of the concentration of beryllium.

McCloskey (6) reported on a variation of this method using aluminon reagent in place of the Zenia solution that was said to be more sensitive than the normal spectrographic procedure.

Other Methods – Another method for automatic monitoring of beryllium in air, under development by the U.S. Air Force, was reported on by Braman (7) of the ITT Research Institute. This device attempted to use the nuclear reaction of beryllium for the detection of beryllium. Design goals called for an alarm signal at 25, 2 and 0.01 $\mu g./m.^3$ of beryllium over sampling periods of 30 seconds, 1 hour, and 24 hours respectively.

Emission spectroscopy has been used by the National Air Pollution Control Administration for beryllium analysis of samples from the National Air Sampling Network (8). The samples are ashed and extracted to eliminate interfering elements. The minimum detectable beryllium concentration by emission spectroscopy is 0.0008 $\mu g./m.^3$ for urban samples and 0.00016 $\mu g./m.^3$ for nonurban samples. The different sensitivities result from the different extraction procedures required for urban samples. (9).

The National Air Pollution Control Administration uses atomic absorption to supplement analyses obtained by emission spectroscopy. The method has a minimum detectable limit of 0.01 $\mu g./m.^3$ based on a 2,000 $m.^3$ air sample (8).

Measurement in Water

Beryllium may be measured in wastewaters by atomic absorption spectroscopy down to a detection limit of 1.0 ppb. Beryllium may also be measured in water by spectrofluorometric methods as described by K.H. Mancy (10). Using 1-amino-4-hydroxyanthraquinone as the reagent, a sensitivity of 0.02 $\mu g./ml.$ can be attained with only chromate and lithium ions interfering. Using 8-hydroxyquinoline as the reagent, a sensitivity of 0.001 $\mu g./ml.$ can be attained but many more ions interfere. Using 2,3-hydroxy-naphthoic acid, a sensitivity of 0.0002 $\mu g./ml.$ can be attained but an even greater number of other ions interfere as noted by Mancy (10).

References

(1) Durocher, N.L., "Air Pollution Aspects of Beryllium and Its Compounds," *Report PB 188,078,* Springfield, Va., Nat. Tech. Info. Service (Sept. 1969).

(2) "Community Air Quality Guide for Beryllium," *Am. Ind. Hyg. Assoc. J.* 29 (1968).

(3) Cholak, J., R.A. Kehoe, L.H. Miller, F. Princi, and L.J. Schafer, "Toxicity of Beryllium," *U.S. Air Force Systems Command Report ASD-TR-62-7-665,* (April 1962).

(4) Keenan, R.G., "Analytical Determination of Beryllium," Chapter 5 in *Beryllium, Its Industrial Aspects,* H.E. Stokinger, Ed. (New York: Academic Press, 1966).

(5) Donaldson, H.M., R.A. Hiser, and C.W. Schwenzfeier, "A Rapid Method for Determination of Beryllium in Air Samples," *Am. Ind. Hyg. Assoc. J.* 22:280 (1961).

(6) McCloskey, J.P., "Spectrophotometric Determination of Beryllium in Airborne Dust Samples," *Microchem. J.* 12:401 (1967).

(7) Braman, R.S., "Research and Development of an Automatic Beryllium-in-Air Monitor," *U.S. Air Force Systems Journal Report RTD-TDR-63-1112* (1963).

(8) Thompson, R.J., G.B. Morgan, and L.J. Purdue, "Analyses of Selected Elements in Atmospheric Particulate Matter by Atomic Absorption," Preprint, presented at the Instrument Society of America Symposium, New Orleans, La. (May 5-7, 1969).

(9) "Beryllium, Actual and Potential Resources, Toxicity, and Properties in Relation to its Use in Propellants and Explosives," Naval Ordinance Laboratory, *Bureau of Naval Weapons Report 7346* March (1961).

(10) Mancy, K.H., "Instrumental Analysis for Water Pollution Control," Ann Arbor, Mich., Ann Arbor Science Publishers (1971).

BISMUTH

Bismuth is considered one of the less toxic of the heavy metals. Bismuth and its salts can cause kidney damage, although the degree of such damage is usually mild.

Measurement in Water

Bismuth may be measured in the range of 0.01 ppm after extraction with dithizone and carbon tetrachlorides as described by K. Heller et al (1). The detection limit for bismuth determination by atomic absorption spectrophotometry is 0.05 ppm according to K.H. Mancy (2).

References

(1) Heller, K., Kuhla, G. and Machek, F., *Mikrochemie* 18, 193 (1935).
(2) Mancy, K.H., *Instrumental Analysis for Water Pollution Control,* Ann Arbor, Mich., Ann Arbor Science Publishers (1971).

BORON

Boron compounds, in general, are toxic to humans and animals when ingested. Most boron compounds are moderately toxic by inhalation, while certain compounds, such as the high-energy borane fuels, are highly toxic when inhaled. Dusts of some boron compounds are considered merely nuisance dusts; others, such as dusts of boric acid and boron oxide, are moderately toxic, according to Durocher (1). Borane poisoning in animals has produced listlessness, incoordination, convulsions, coma, and death. Similar symptoms have been observed in humans exposed to the boranes, although none have died. The incidence of acute borane intoxication in humans has been low, and adequate data on exposure, concentrations, and reactions are lacking.

Environmental problems concerning boron and its compounds are generally limited to the particular areas where the production of boron compounds emits dust particles to the atmosphere. However, the highly toxic nature of the borane fuels poses a constant threat of environmental pollution wherever they are produced, stored, or used. Recognition of this hazard has resulted in strict safety measures which have minimized accidental exposures.

The use of boron additives in petroleum fuels and the presence of boron in coals, possibly contribute quantities of boron to the atmosphere, but no assessment has been made of the amount of these contributions or of their impact as air pollutants.

Data are not available on the concentration of boron or its compounds in the atmosphere. This absence of recorded data indicating qualitatively or quantitatively the presence of boron compounds suggests that their presence in the atmosphere may not be fully understood, and the exact nature of boron as an air pollutant may not be fully appreciated. Environmental monitoring of boron concentrations should be considered, particularly near industrial plants engaged in boron production or use and near borane fuel sites.

Normal entrapment and precipitation procedures applicable to most particulate pollutants should be effective in controlling or reducing the amount of boron compound dusts emitted into the atmosphere. No specific methods for abatement of atmospheric borane compounds have been identified, but strict regulatory practices are used to prevent accidental emissions. No information has been found on the economic costs of boron air pollution or on the costs of its abatement. Additional information on atmospheric emissions of boron is available in a report by W.E. Davis and Associates (2).

Despite its widespread industrial use (see Table 20), information on industrial waste levels of boron, and on treatment processes for removal of boron from industrial wastewaters is almost nonexistant in the literature surveyed (3). Waggott (4) has reported boron levels up to 5.5 mg./l. in raw domestic sewage. A strong correlation between anionic detergent and boron concentrations was observed, indicating that the major source of the boron in this wastewater was sodium perborate, used as bleach in household washing powders. Fluoborate solutions are sometimes used for plating of cadmium, copper, lead, nickel, tin, zinc and a number of alloys (5). They are used only for special purposes such as deposition of lead on battery bolts and terminals, plating of tin-lead alloys on bearings, or in cases where high-speed deposition is required for continuous wire or strip plating. Lowe has given fluoborate plating solution concentrations (as BF_4) of 125,000 to 265,000 milligrams per liter (5). This is equivalent to boron concentrations of 15,800 to 33,500 milligrams per liter.

The permissible boron concentration in domestic water supplies is 1.0 ppm as shown earlier in Table 11. The boron tolerance in irrigation waters is 0.75 ppm for continuous rise and 2.0 ppm for short term use in fine textured soil, as shown earlier in Table 5.

TABLE 20: INDUSTRIAL USES OF BORON

Nuclear Reactor Core Solution
Detergent Manufacture
Weatherproofing Wood
Fireproofing Fabrics
Manufacture of Glassware and Porcelain
Production of Leather and Carpets
Cosmetics Manufacture
Photographic Supplies
Rocket Fuels (boron hydrides)
Wire Drawing
Fertilizers (boron deficient soils)
Disinfectants
Weed Control
Welding and Brazing
Cutting Fluids
Plating
Metallurgy (impact-resistant steels)

Source: Report PB 216,162

Measurement in Air

Several devices for detecting pentaborane in air are available, but they are nonspecific (i.e., they detect other compounds with similar chemical properties). A coulometric borane monitor was developed by Braman et al (6), which was based on the oxidation of boranes with electrolytically generated iodine. This instrument, with a sensitivity of approximately ±0.02 ppm (40 μg./m.^{3}) pentaborane in air, unfortunately was susceptible to overloading when excessive concentrations of electroactive materials present in the air desensitized it.

A boron hydride monitor that converted hydrides to boric acid by pyrolysis and then determined the presence of boric acid by colorimetry was described by Fristom, Bennet, and Berl (7). However, this monitor was not automated. Kuhns, Forsythe, and Masi (8) have outlined a hand-held device for automatic detection of boron hydrides in air, based on the reduction of boron hydrides with triphenyltetrazoleum chloride. These devices provided sensitivity in the 0.1 ppm range, but were not continuous monitors and were subject to interference from all oxidizing or reducing material in air.

Hill and Johnston (9) reported on ultraviolet spectrophotometric detection of decaborane, but this method proved unsuitable for use with other boranes. Braman and Gordon (10) and Braman (11) reported on a direct-reading, automatic, temperature-insensitive instrument capable of detection of pentaborane down to concentrations of 0.05 ppm (100 μg./m.^{3}).

This device used the flame-emission principle of detection: in the presence of boranes, a small hydrogen flame takes on a greenish cast, the intensity of which is proportional to the borane concentration. This device can be operated continuously over an 8-hour period using a self-contained hydrogen supply and storage batteries. Development of an atmospheric monitoring device to detect and record several toxic components, including pentaborane, is reported by the Mine Safety Appliances Company (12). This device also features pyrolysis to form boric oxide from borane, followed by detection of the boron by using n-amylamine reagent. The company claims a detectability of 0.01 ppm (20 μg./m.^{3}), with rapid response and continous operation.

In a comprehensive 1964 review of analytical methods for determining the presence of boron, Nemodruk and Karalova (13) listed the following methods. Photometric methods are widely used, and are especially suitable for determining micro amounts, at concentrations of 0.05 to 0.20 μg., these methods produce 90% or better accuracies. Coulometric titration is said to provide determination of amounts as low as 0.2 μg. Spectral analysis is one of the most widely used methods in metallurgy, capable of determining 5×10^{-5}% concentrations of boron. Fluorimetric analysis is also used, with sensitivity to 0.1 μg. of boron. And radioactive methods are also available, based on the ability of boron's nuclei to absorb thermal neutrons, this method is considered simple, and highly efficient, is capable of providing 100 determinations in 8 hours, and has sensitivity to 0.1 μg.

Measurement in Water

Boron may be monitored in water by the following instrumental techniques (14), according to *LBL-1,* Vol. 2: direct reading emission spectrometry; potentiometric. Boron may be measured in water by the atomic absorption technique down to a detection limit of 2.6 ppm as shown earlier in Table 11.

An analysis developed by P. Cohen and W.D. Fletcher (15) permits determining the boron concentration in an aqueous solution of boron-containing material. The analyzer has means for adding a polyhydroxyl compound to increase the electrical conductivity of the solution and means for controlling the solution temperature, cell means for continuously measuring electrical conductivity of the solution and for comparing the conductivity of the solution with that of an established standard.

In the past analyses usually have been based upon manual titrimetric or colorimetric methods. Such methods, however, have not been suitable because of time delays of the order of at least 15 to 20 minutes. Presently available commercial analyzers are generally unsuitable either because of their insensitivity to small changes in boron concentration, or to excessive delays inherent in their mode of operation. Moreover, commercial boron analyzers of the automatic type have been very expensive and require excessive attendance and maintenance. Thus, there has been a need for accurate and fast automatic methods for analyzing the boric acid content of water.

It has been found that a reliable automatic boron analysis may be provided which is simple, relatively low in cost of manufacture, rapid, and sensitive to small changes in boron concentration. The analysis is based upon the increase in solution electrical conductivity which occurs when a polyhydroxyl compound is added to a solution containing boric acid or a borate salt.

References

(1) Durocher, N.C., "Air Pollution Aspects of Boron and Its Compounds," *Report PB 188,085,* Springfield, Va., Nat. Tech. Info. Service (Sept. 1969).
(2) Davis, W.E. and Assoc., *Report PB 210,677,* Springfield, Va., Nat. Tech. Info. Service (May 1972).
(3) Patterson, J.W. and Minear, R.A., "Wastewater Treatment Technology," 2nd Ed., *Report PB 216,162,* Springfield, Va., Nat. Tech. Info. Service (Feb. 1973).
(4) A. Waggott, "An Investigation of the Potential Problem of Increasing Boron Concentrations in Rivers and Water Courses," *Water Research,* 3:749-765, 1969.

(5) W. Lowe, "The Origin and Characteristics of Toxic Wastes, with Particular Reference to the Metal Industries," *Water Poll. Control* (London) :270-280, 1970.
(6) Braman, R.S., D.D. DeFord, T.N. Johnson, and L.J. Kuhn, "A Coulometric Borane Monitor," *Anal. Chem.* 32 (Sept. 1960).
(7) Fristom, G.R., L. Bennett, and W.G. Berl, "Integrating Monitor for Detecting Low Concentrations of Gaseous Boron Hydrides in Air," *Anal. Chem.* 31 (October 1959).
(8) Kuhns, L.J., R.H. Forsythe, and J.F. Masi, "Boron Hydride Monitoring Devices," *Anal. Chem.* 28 (Nov. 1956).
(9) Hill, W.H., and M.S. Johnston, *Anal. Chem.* 27, (1955).
(10) Braman, R.S., and E.S. Gordon, "The Design of a Portable Instrument for the Detection of Atmospheric Pentaborane and Other Toxic Gases, Institute of Electrical and Electronic Engineers," *Trans. Instrumentation and Measurement,* volume IM-14 (1965).
(11) Braman, R.S., "Research and Development of an Automatic Beryllium and Boron Monitor," Armour Research Foundation of Illinois, *Institute of Technology Report ARF3202-3* (1962).
(12) Mine Safety Appliances Company, Final Report, "Development of an Atmospheric Monitoring System," (1961).
(13) Nemodruk, A.A., and Karalova, Z.K. *Analytical Chemistry of Boron,* Academy of Sciences of the U.S.S.R. (Jerusalem: Silvan Press, 1964).
(14) Lawrence Berkeley Laboratory, "Instrumentation for Environmental Monitoring – Water," *Publ. No. LBL-1,* Vol. 2, Berkeley, Univ. of Calif. (Feb 1, 1973).
(15) Cohen, P. and Fletcher, W.D., U.S. Patent 3,468,764; Sept. 23, 1969; assigned to Westinghouse Electric Corp.

BROMIDES

Measurement in Water

Bromides may be monitored in water by molecular absorption spectroscopy (colorimetry) (1).

References

(1) Lawrence Berkeley Laboratory, "Instrumentation for Environmental Monitoring – Water," *Publ. No. LBL-1,* Vol 2, Berkeley, Univ. of Calif. (Feb. 1, 1973).

CADMIUM

Health problems from cadmium were suddenly brought into focus when high cadmium intake was identified as the probable cause of the terrible Itai Itai disease, which has been prevalent in the Minimata area of Northern Japan since World War II, as pointed out by C.R. Parker and D.P. Sandoz (1). This disease, really poisoning, resulted in gross deformities, pain as bones snapped under weight of the body and in over fifty cases, death. Since that time, investigations have revealed numerous other illnesses which appear to be related to intake of lesser amounts of cadmium. The tragic deaths and deformities experienced since World War II by more than 200 people in Northern Japan have been attributed to cadmium from mining wastes which polluted the water supply, resulting in contamination of rice growing areas. The damage caused by cadmium in this case was increased by lower than normal amounts of calcium and protein in the diet, which appear to increase the proportion of cadmium retained in the body.

Cadmium is found in nature, most generally in low concentrations associated with a similar metal, zinc. It is a soft silver-blue metal, familiar to us as the anticorrosion coating often used on outdoor fittings such as gate latches, roofing nails and screws. It is present as an impurity in the more common galvanized zinc coatings (0.03%) and is also used in many other everyday items such as paint, some pottery pigments and plastics. Automobile tires also contain cadmium which is introduced as an impurity in the zinc used in the production of tires.

Cadmium and cadmium compounds are toxic substances by all means of administration, producing acute or chronic symptoms varying in intensity from irritations to extensive disturbances resulting in death, according to Athanassiadis (2). However, despite increasing use of this metal and increasing attention to its toxic nature, the exact manner in which it affects human or animal organisms is not yet known. Cadmium is toxic to practically all systems and functions of the body, and is absorbed without regard to the levels of cadmium already present, thereby indicating the lack of a natural homeostatic mechanism for the control of organic concentrations of cadmium.

Inhalation of cadmium fumes, oxides, and salts often produces emphysema, which may be followed by bronchitis. Prolonged exposures to airborne cadmium frequently cause kidney damage resulting in proteinuria. Cadmium also affects the heart and liver. Statistical studies of people living in 28 U.S. cities have shown a positive correlation between heart diseases and the concentration of cadmium in the urban air. Cadmium may also be a carcinogen. While there is little evidence to support this conclusion from studies of industrial workers, animal experiments have shown cadmium may be carcinogenic.

No data were found that indicated deleterious effects produced by airborne cadmium on commercial or domestic animals. However, experiments with laboratory animals have shown that cadmium affects the kidneys, lungs, heart, liver, and gastrointestinal organs, and the nervous and reproductive systems. No data were found on the effects of cadmium air pollution on plants or materials.

Many workers regularly exposed to cadmium dust and fumes show symptoms of lung diseases similar to emphysema, and as many as 25% in one survey showed symptoms of kidney stones. In cases of inhalation of large quantities of cadmium, severe and irreparable lung damage has resulted. The metal tends to accumulate in several organs of the body, notably the kidneys and the liver. Excessive concentrations (200 μg./g.) causes severe kidney damage, resulting in breakdown of kidney functions and excessive excretion of protein. This can result from an intake of only 2 to 3 times the average intake of a U.S. adult, which is well within the bounds of possibility in areas of contamination or for a heavy smoker.

There is also concern since one statistical survey of cadmium in the air in U.S. cities showed a correlation between cadmium levels and increased arteriosclerotic heart disease, high blood pressure and decreased life expectancy. Much more work is required to firmly establish the connection, but there are now enough indicators to cause concern. Statistical data is reinforced by knowledge that experimental animals have shown marked biochemical changes, including liver damage and decreased life expectancy, while their organs contain lower cadmium concentrations than those normally found in adult humans.

Thus, the health effects of cadmium, both proven and probable are increased blood pressure, increased incidence of arteriosclerotic disease and reduced life expectancy. As intake increases kidney damage, protein excretion, anemia and emphysema may result. Chronic high exposures can result in liver and lung damage, degenerative bone disease and death (1).

The metals industry is the major source of emissions of cadmium into the atmosphere (2). Cadmium dusts and fumes are produced in the extraction, refining, and processing of metallic cadmium. Since cadmium is generally produced as a by-product in the refining of of other metals, such as zinc, lead, and copper, plants refining these materials are sources of cadmium emissions as well as of the basic metal. Also, because cadmium is present in small quantities in the ores of these metals, cadmium emissions may occur inadvertently in the refining of the basic metal.

Common sources of cadmium air pollution occur during the use of cadmium. Electroplating, alloying, and use of cadmium in pigments can produce local contaminations of the atmosphere. Also, since cadmium is added to pesticides and fertilizers, the use of these materials can cause local air pollution. In 1964, the average concentration of cadmium in the ambient air was 0.002 $\mu g./m.^3$, and the maximum concentration was 0.350 $\mu g./m.^3$.

It is difficult to get a realistic idea of "natural" levels of cadmium in the air, since so much has been added by man. However, it is generally agreed that natural levels are 1 $ng./m.^3$ or less, such as are found in country areas far removed from pollution and in a few clean cities (1). In contrast, yearly averages as high as 65 $ng./m.^3$ have been measured for Chicago while concentrations above 1,000 $ng./m.^3$ have been measured near smelters and other sources of cadmium emissions.

Cadmium taken through the lungs is about 60 times more poisonous than when taken through the digestive tract. Yet a large proportion of the population continues to expose itself to an additional source of airborne cadmium from burning tobacco. It is not widely known that tobacco contains between 1 and 2 micrograms of cadmium per gram, that is, there is 1 to 2 micrograms of cadmium in every cigarette.

Refining cadmium bearing ores results in the release to the atmosphere of more than two million pounds per year in the United States. Cadmium is almost always found with zinc, but in much lower concentrations. We use enormous quantities of zinc, often for corrosion protection as on galvanized water pipes, downpipe and other steel products. During scrap metal recovery processes in the U.S. an estimated 2.3 million pounds of cadmium went "up the chimney" in 1968.

The total usage of cadmium is around 12 to 13 million pounds per year in the U.S., and 31 million pounds in the world. Apart from the corrosion protection, the metal has a host of uses such as in alloys and nickel-cadmium batteries. In plastics it is a valuable stabilizer, assisting in the manufacturing process and extending the lifetime of the product. It is also used as a pigment in some paints. During the manufacture of these goods some cadmium escapes to the atmosphere. Many such products end their useful lifetime in incinerators, releasing much of the metal to the air. The total cadmium emitted to the atmosphere was estimated to be 4.6 million pounds in the U.S. in 1968. Most of this eventually settled onto the soil to be taken into plants by roots then into food products. Some settled in water catchment areas and was swept into our water supplies, and some we breathed directly.

Many industrial users exert strict controls over their effluents, but others do not, either from lack of suitable recovery equipment or from incorrect operating procedures. It is known that large quantities of cadmium find their way into our water supplies from electroplating and coating installations and from other industries using cadmium. The exact amounts are not known, but are probably large. We can get an idea of the extent of this problem from the knowledge that silt dredged from the bottom of Lake Erie contained 130 parts per million of cadmium. This is 3 times more than soils immediately adjacent to factories which are known to emit large quantities of cadmium, and is many times larger than the normal soil levels of several parts per million.

Air pollution control procedures are employed at some metal refinery plants in order to recover the valuable cadmium that would otherwise escape into the atmosphere. Electrostatic precipitators, baghouses, and cyclones are effectively used for abatement. However, little information has been found on the specific application of these procedures for the purpose of controlling cadmium air pollution. The procedures for recovering cadmium from exhaust in a copper extraction plant collected significant quantities of valuable cadmium, at the same time reducing local air pollution levels.

No information has been found on the economic costs of cadmium air pollution or on the costs of its abatement according to Athanassiadis (2). British emission standards for cadmium (as the element) are 39 mg./cu. m. at STP and Australian standards are 23 mg. per cu. m. at STP, according to Stern (2a).

Cadmium, usually considered a toxic metal, and under certain circumstances a nephrotoxic one, has been found in the kidneys of all adult human beings from many cities, of the world, according to Cothern (3). It is virtually absent in newborns and accumulates with age. There is no known biological function for cadmium in man (4). A marked correlation (r = 0.76) exists between the concentration of cadmium in the air and death rates from hypertension and arteriosclerotic heart diseases (5). Zinc showed a similar correlation with r = 0.56.

Feeding the environmental toxicant cadmium (75 mg./kg.) to young Japanese quail for 4 weeks produced growth retardation, severe anemia, low concentrations of iron in the liver and high concentrations of cadmium in the liver. Dietary ascorbic acid supplements (0.5 and 1.0% by weight) almost completely prevented the anemia and improved the growth rate but did not markedly alter concentrations of iron or cadmium in the liver (6). It has been found that 0.1 ppm of cadmium is toxic to some aquatic insects and 0.01 to 10 ppm is toxic to fish (7).

Cadmium in concentrations of 5 ppm were given via their drinking water to mice. This level showed an innate toxicity in terms of survival of male mice (8). Hypertension was induced in rats by adding 5 ppm cadmium to their drinking water. Females were more susceptible at all ages. Cadmium was found in the livers and kidneys of all rats. Along with the hypertension in older rats went salt (sodium chloride) hunger (9)(10). Rats fed 10 ppm of cadmium in water showed in 2 to 4 months chronic arterial hypertension (11). Thus, cadmium is a serious health hazard in water supplies in the ppm range.

Cadmium is extremely toxic to fish. Concentrations between 0.008 and 0.01 are lethal to 50% of a test batch of trout (12). Concentrations of $Cd(OAc)_2$ between 25 and 50ppm produced marked osteoporosis (13). In concentrations between 0.01 and 10 ppm the fish all overturned in a few days except in the lowest concentrations. In the upper range hyperplasia and necrosis of the epithelium of secondary gill lamellae were evident. Death probably resulted from interference in respiration. Gill damage was less pronounced in the low concentrations. There were morphological and histological changes in the internal organs especially the spleen, which was nodular in appearance, congested and structurally changed. The liver was also affected. The heart, skeletal muscle and brain had degenerated (14).

The permissible concentration of cadmium in domestic water supplies is 0.01 ppm as shown earlier in Table 3. The permissible cadmium concentration in irrigation water on a continuous use basis is 0.005 ppm and is 0.05 ppm on a short term use basis in fine-textured soil as shown earlier in Table 5.

Effluents to storm sewers or streams should contain less than 0.05 ppm of cadmium according to earlier standards. Now, however, new toxic pollutant effluent standards for cadmium are being promulgated by the EPA as outlined in the Federal Register for Dec. 27, 1973. The new proposed standards say that for streams having a flow of less than 10 cu. ft./sec. or lakes having an area of less than 500 acres, there shall be no cadmium discharged. In faster flowing streams or larger lakes, the limit is 40 µg./l. (approximately 0.04 ppm) except that the allowable limit is only 10% of this amount if the low flow of the stream is less than 10 times the waste stream flow. The proposed allowable discharge into salt water bodies is 320 µg./l. (approximately 0.32 ppm).

We ingest cadmium from a number of sources, the air, our food, and tobacco. Studies indicate that the average American adult ingests a total of 50 to 60 micrograms/day (compared with 15 to 20 in Sweden and 80 µg. in Japan). The typical adult cadmium intake in the U.S.A., is estimated to be 50 µg. from food; 1 to 10 µg. from air, depending on local air content; 1 to 10 µg. from water; and 2 to 4 µg. from tobacco for heavy smokers. Fortunately, most of this is passed through the body with only about 2 µg./day retained although this amount may rise to 3 to 4 µg./day for a person in a polluted environment or for a heavy smoker. However, even 2 µg./day stored in the body daily adds up to an appreciable amount over a period of years (1).

Potential industrial sources of cadmium wastewaters reported by McKee and Wolf (15) and Santaniello (16) are:

Metallurgical alloying
Ceramics
Electroplating
Photography
Pigment works
Textile printing
Chemical industries
Lead mine drainage

Of the total industrial cadmium use, 90% is reported to be utilized in electroplating, pigments, plastics stabilizers, alloying and batteries. Most of the remaining 10% is used for television tube phosphors, golf course fungicides, rubber curing agents and nuclear reactor shields and rods (17).

Considering the industries listed above, only the electroplating industry is discussed to any extent in the literature with regard to cadmium bearing wastes and the treatment thereof. High concentrations of cadmium in lead mine drainage have been reported, however (15). Cadmium wastewater concentrations reported in the literature are summarized in Table 21. A report on groundwater contamination from plating wastes gives a concentration of 1.2 mg./l. cadmium in a groundwater recharge basin receiving plant wastes. Private wells in the area reported 0.3 to 0.6 mg./l., while one test well contained 3.2 mg./l. of cadmium (18).

TABLE 21: CADMIUM CONCENTRATIONS REPORTED FOR INDUSTRIAL WASTEWATERS

Process	Cadmium Concentration	Reference
Automobile Heating		
Control Manufacturing	14 - 22*	(19)
Plating Rinse Waters		
Automatic Barrel Zn & Cd Plant	10 - 15	(20)
Mixed Manual Barrel and Rack	7 - 12	(20)
Plating Rinse Waters		
(large installations)	15 ave. - 50 max.	(21)
Plating Rinse Waters		
0.5 gph dragout	48	(22)
2.5 gph dragout	240	(22)
Plating Bath	23,000	(22)
Bright Dip and Passivation Baths	2,000 - 5,000	(20)
Lead Mine Acid Drainage	1,000	(15)

*24-hour composite on days when plating baths were dumped. Cadmium levels much reduced due to dilution when solution dragout was the only contributing source.

Source: Report PB 216,162

Measurement in Air

Sampling Methods: Dust and fumes of cadmium compounds may be collected by any method suitable for collection of other dusts and fumes; the impinger, electrostatic precipitator, and filter are commonly used. The National Air Sampling Network uses a high volume filtration sampler (23).

Quantitative Methods: Emission spectroscopy has been used by the National Air Pollution Control Administration for cadmium analysis of samples from the National Air Sampling Network (23)(24). The samples are ashed and extracted to eliminate interfering elements.

The minimum detectable cadmium concentration by emission spectroscopy is 0.011 μg./m.^{3} for urban samples and 0.004 μg./m.^{3} for nonurban samples. The different sensitivities result from the different extraction procedures required for urban samples (25).

Thompson et al (23) have reported that the National Air Pollution Control Administration uses atomic absorption to supplement analyses obtained by emission spectroscopy. The method has a minimum detectable limit of 0.0002 μg./m.^{3} based on a 2,000 m.^{3} air sample. Emission spectroscopy has also been used to determine cadmium and other trace materials in mammalian tissues (26). A resolution of 9,350 μg./100 grams dry tissue was reported.

Polarographic methods (27)(28) have also been used to determine cadmium concentrations in air. Ashing of the sample is followed by neutralization and precipitation of cadmium sulfate. The sample is recorded using a polarographic radiometer. This procedure has also been used to determine the presence of cadmium in urine, providing a measure of concentrations between 10,000 to 70,000 μg./liter. A colorimetric method (29)(30) was used to determine the concentration of cadmium in the air in a cadmium stearate plant. Dithizone was the coloring agent. The method has a sensitivity of 50 μg. for cadmium if less than 10,000 μg. of interfering metals are present.

Cadmium may be monitored in air by the following instrumental techniques (31): colorimetric, atomic absorption, emission spectroscopy. Environmental cadmium analysis by flameless atomic absorption has been reviewed by C.R. Parker and D.P. Sandoz (1). Atomic absorption can be applied to 67 metals, many of which concern the environmentalist. The atomic absorption spectrophotometer can be coupled with the carbon rod atomizer to provide an extremely sensitive method of analysis which exhibits a significant (130-fold) improvement in cadmium sensitivity with the carbon rod atomizer compared to conventional flame atomic absorption. It enables one to analyze water, air, food and body tissues as well as other materials for metals such as cadmium, at levels anywhere between "clean" natural waters and uncontaminated foods to grossly contaminated situations, and even higher levels.

Measurement in Water

The USPHS has set an upper limit of 10 μg./liter for cadmium in drinking water. Federal regulations now require monitoring of both supplies and effluents to minimize air intake from this source and to minimize the spread of cadmium. Cadmium may be monitored in water by the following instrumental techniques (32): atomic fluorescence, electrochemical.

The measurement of cadmium in water by colorimetric determination with dithizone has been described by C.C. Ruchhoft et al (33). As shown earlier in Table 11, cadmium may be measured by atomic absorption using a flame down to 1.0 ppb levels and using flameless atomic absorption down to 0.02 ppb detection levels.

References

(1) Parker, C.R. and Sandoz, D.P., *VIA (Varian Instrument Applications)* 7 (2). 18-21. (1973).
(2) Athanassiadis, Y.C., "Air Pollution Aspects of Cadmium and Its Compounds," *Report PB 188,086;* Springfield, Va., Nat. Tech. Info. Service (Sept 1969).
(2a) Stern, A.C., in *Industrial Pollution Control Handbook,* H.F. Lund, Ed., New York, McGraw-Hill Book Co. (1971).
(3) Cothern, C.R., "Determination of Trace Metal Pollutants in Water Resources and Stream Sediments," *Report PB 213,369,* Springfield, Va., Nat. Tech. Info. Service (1972).
(4) Schroeder, H.A. and Balassa, J.J., *J. Chron. Dis.* 14, 236 (1961).
(5) Carroll, R.E., *Jour. Am. Med. Assoc.* 198, 177 (1966).
(6) Fox, M.R.S. and Fry, B.E., *Science* 169, 989 (1970).
(7) Warnick, S.L. and Hall, H.L., *J. Water Poll. Control Fed.* 41, 280 (1969).
(8) Schroeder, H.A., Virton, W.H. and Balassa, J.J., *J. Nutrition* 80, 39 (1963).
(9) Schroeder, H.A., *Amer. J. Physiol.* 207, 62 (1964).
(10) Schroeder, H.A., Nason, A.P., Prior, R.E., Reed, J.B. and Haessler, W.T., *Amer. J. Physiol,* 214, 469 (1968).

(11) Schroeder, H.A., Baker, J.T., Hansen, N.M., Jr., and Wise, R.A., *Arch. Envir. Health* 21, 609 (1970).
(12) Her Majesty's Stationers Office, *Water Pollution Research* (1967).
(13) Maehara, T., *Nippon Seikeigakagakkai Zasshi* (1967).
(14) Her Majesty's Stationers Office, *Water Pollution Research* (1969).
(15) J.E. McKee and H.W. Wolf, *Water Quality Criteria,* 2nd ed., California State Water Quality Control Board, Publication No. 3-A, 1963.
(16) R.M. Santaniello, "Air and Water Pollution Quality Standards, Part 2: Water Quality Criteria and Standards for Industrial Effluents," *Industrial Pollution Control Handbook,* Herbert F. Lund, ed., McGraw-Hill Book Co., New York, 1971.
(17) Anonymous, "Metals Focus Shifts to Cadmium," *Env. Sci. Technol.,* 5:754-755, 1971.
(18) M. Lieber and W.F. Welsch, "Contamination of Ground Water by Cadmium," *Jour. Amer. Wat. Works Assoc.,* 46:541-547, 1954.
(19) S.M. Gard, C.A. Snavely and D.J. Lemon, "Design and Operation of a Metal Wastes Treatment Plant," *Sew. Ind. Wastes,* 23:1429-1438, 1951.
(20) W. Lowe, "The Origin and Characteristics of Toxic Wastes with Particular Reference to the Metals Industries," *Water Poll. Control* (London), 1970: 270-280.
(21) H.L. Pinkerton, "Waste Disposal. Inorganic Wastes," *Electroplating Engineering Handbook,* 2nd ed., A. Kenneth Graham, ed., Reinhold Publishing Co., New York, 1962.
(22) N.L. Nemerow, *Theories and Practices of Industrial Waste Treatment,* Addison Wesley, Reading, Mass., 1963.
(23) Thompson, R.J., G. B. Morgan, and L.J. Purdue, "Analyses of Selected Elements in Atmospheric Particulate Matter by Atomic Absorption," Preprint, presented at the Instrument Society of America Symposium, New Orleans, La. (May 5-7, 1969).
(24) "Air Pollution Measurements of the National Air Sampling Network – Analyses of Suspended Particulates, 1957-1961," U.S. Dept. of Health, Education, and Welfare, *Public Health Service Publication No. 978,* U.S. Government Printing Office, Washington, D.C. (1962).
(25) "Air Quality Data from the National Air Sampling Networks and Contributing State and Local Networks," 1966 ed., U.S. Dept. of Health, Education, and Welfare, *National Air Pollution Control Administration Publication No. APTD 68-69,* U.S. Government Printing Office, Washington, D.C. (1968).
(26) Butt, E.M., "Trace Metals in Health and Disease," Air Pollution Medical Research Conference, San Francisco, Calif. (1960).
(27) Ferret, D.J. et al, "A Comparative Study of Three Recently Developed Polarographs," *Analyst* 81:506 (1956).
(28) Levine, L., "Polarographic Determination of Toxic Metal Fumes in Air," *J. Ind. Hyg. Toxicol.* 27:171 (1945).
(29) Saltzman, B.E., "Colorimetric Micro-Determination of Cadmium with Dithizone," *Anal. Chem.* 25:493 (1953).
(30) Setterlind, A.N., et al, "Microanalysis of Cadmium by the Diphenylthiocarbazone (Dithizone) Method," Department of Public Health, State of Illinois (1943).
(31) Lawrence Berkeley Laboratory, "Instrumentation for Environmental Monitoring – Air," *Publ. LBL-1,* Vol. 1, Berkeley, Univ. of Calif. (Feb. 1, 1973).
(32) Lawrence Berkeley Laboratory, "Instrumentation for Environmental Monitoring – Water," *Publ. LBL-1,* Vol. 2, Berkeley, Univ. of Calif. (Feb. 1, 1973).
(33) Ruchhoft, C.C., Moore, W.A., Terhoeven, G.E., Middleton, F.W. and Krieger, H.L., *Report PB 215,059,* Springfield, Va., Nat. Tech. Info. Service (Dec. 1, 1949).

CALCIUM

Measurement in Water

Calcium may be monitored in water by the following instrumental techniques (1): atomic (flame) emission, spark source mass spectrometry, x-ray fluorescence. Calcium can be measured by atomic absorption down to a 0.04 ppb detection limit as shown earlier in Table 11.

References

(1) Lawrence Berkeley Laboratory, "Instrumentation for Environmental Monitoring – Water," *Publ. No. LBL-1,* Vol. 2, Berkeley, Univ. of Calif. (Feb. 1, 1973).

CARBOHYDRATES

Measurement in Water

A process for detecting traces of soluble carbohydrates in a return flow of water in an industrial process has been developed by J.A. Tjebbes and F.J. Deak (1). In the method a flow stream to be analyzed is led through the apparatus comprising a combination of means for deaeration, filtration, ion exchange or electrodialysis, catalytic reaction, measuring electric conductivity or hydrogen ion activity, and recording the values measured. During the passage of the branch flow through the means for the catalytic reaction any soluble carbohydrates present in the branch flow are oxidized with oxygen to dissociated uronic acids, the conductivity or hydrogen ion activity of which is determined as a measure of the content of soluble carbohydrates in the return flow.

References

(1) Tjebbes, J.A. and Deak, F.J., U.S. Patent 3,492,094; Jan. 27, 1970; assigned to Svenska Scokerfabrik AB.

CARBON

By carbon, one does not mean sooty particles but carbonaceous matter, particularly in the sense of organic water pollutants having appreciable oxygen demand.

Measurement in Water

Increasing concern with the problem of water pollution and waste treatment has brought about a need for a rapid and precise method for determining carbonaceous matter in aqueous systems. The advantages of rapid analytical methods permitting immediate evaluation of corrective treatments at water pollution sources are manifest.

A process developed by J.L. Teal, C.E. Van Hall, V.A. Stenger and J.W. Safranko (1) enables the determination of total carbon content of an aqueous sample by passage through a combustion zone followed by analysis of the evolved gases for CO_2, preferably by a nondispersive infrared analyzer.

A process developed by I.A. Capuano (2) involves the steps of heating a flowing stream of an aqueous solution of interest in the presence of oxygen to oxidize the organic carbon constituents contained therein to carbon dioxide, reacting the carbon dioxide with an electrolytic solution and then conductometrically analyzing the reaction product of the carbon dioxide and the electrolytic solution, the electrolytic solution alone being first passed through a reference measuring element to effect a comparison base, and using the conductometrically produced differential signal as an indication of the concentration of organic carbon constituents present at any time in the aqueous solution being analyzed.

From previous other types of analyses it was known that samples encountered in continuous operation could be expected to contain those impurities for which certain specific removal steps were incorporated into the design. It was expected for instance that the stream being monitored might contain inorganic carbon constituents in the form of dissolved carbon dioxide or soluble carbonates such as sodium carbonate and on this expectation provision was included in the design for precipitation and removal of such constituents by the barium hydroxide method and the filtering step.

$$Ba(OH)_2 + CO_2 \longrightarrow BaCO_3 + H_2O$$

$$Ba(OH)_2 + Na_2CO_3 \longrightarrow BaCO_3 + 2NaOH$$

Similarly, a manganese dioxide chamber was included to remove any oxides of nitrogen and sulfur formed in the heating step, the occurrence of trace quantities of nitrogen and sulfur in the sampled stream being also expected. A powdered antimony chamber was included in the flow arrangement to remove any chlorine produced incidentally by the reduction of chlorine compound impurities which were also expected in the sample stream.

A device developed by H.H. Seward (3) can be used to measure total organic carbon in water by colorimetry, by measuring the ultraviolet absorbance. It has been found that generally there is a good correlation between the absorption of near ultraviolet light in water and its organic carbon content. By near ultraviolet is meant the range of 1800 to 4000 A. including the mercury lines at 1849, 1942, 2536 and 3650 A. especially the region around the 2536 A. line. It has been found that even in distilled water, absorption at these wavelengths is predominantly by the small residue of dissolved organic carbon.

A simple quartz test tube functions as the sample container and the necessary lens system concentrating light from an ultraviolet "ozone lamp" through the sample to an ultraviolet detector. Rectified line current provides power for the ultraviolet source resulting in pulses at 60 herz frequency. Pulses derived respectively from the ultraviolet detector and from a reference detector, are compared to yield an electrical current proportional to the absorbance of the sample.

A process developed by H.K. Staffin and R.J. Ricci (4) provides a method for continuously analyzing organic carbon contained in a fluid medium, which is automatic and functions continuously without requiring manual pretreatment of a sample. The apparatus, sometimes referred to for brevity as a TOC monitor, i.e., total organic carbon monitor, comprises three sections:

1. A sample pretreatment system which efficiently removes inorganic carbon present in the effluent stream;
2. An oxidation system which converts the organic carbon into CO_2;
3. An infrared analyzer to measure quantitatively CO_2 which is proportional to the TOC in the system.

Broadly regarded, the system comprises a reservoir holding a quantity of effluent at a constant head. An outlet from the reservoir feeds the sample to a tank whereat it is treated with hydrochloric acid to shift the equilibria of the carbonates prior to feeding the acidified sample to a nitrogen sparge column. This latter is essentially a countercurrent, multistage gas stripper, wherein CO_2 is removed from the sample to result in an effluent virtually free of inorganic carbon and containing no dissolved CO_2.

The carbonate-free sample is delivered at some predetermined nominal rate, by means of a positive displacement metering pump, into a fluidized bed reactor. The reactor consists of a bed of aluminum oxide or equivalent particles which is heated and caused to behave analogously to a fluid by forcing a gas therethrough. A reactor of this type is disclosed in U.S. Patent 3,350,915. Oxygen is supplied to the reactor at some multiple, e.g., 10 times, the stoichiometric requirement to oxidize the maximum amount of TOC which may be present. The gas which serves to fluidize the bed comprises the input of oxygen and the products of vaporization and combustion of the sample.

The effluent of the reactor is a stream of water, CO_2 and O_2, plus oxidation products of other noncarbonaceous compounds present in the water sample. The effluent from the reactor is cooled and the major part of the water removed in a condenser prior to entering a nondispersive infrared analyzer for CO_2 analysis. CO_2 as thus measured, is proportional to the TOC in the sample stream.

A method developed by Y. Takahashi, R.T. Moore and T.M. Stephens (5) permits directly selectively analyzing organic carbon or other predetermined compounds in a sample-containing fluid such as wastewater. The organic carbon in the wastewater sample is either oxidatively pyrolyzed to carbon dioxide or reductively pyrolyzed to methane, which products are then analyzed. Prior to pyrolysis, the inorganic carbonate and bicarbonate salts present in the water are removed by acidification to carbonic acid and vaporization of the carbon dioxide component in a volatilization zone. The vapors, including certain volatile organic carbon compounds, are directed in a carrier gas through a packed detention column for sorption of the organic carbon compounds on the packing while the carbon dioxide is vented. Thereafter, the carrier gas flow is reversed and the sorbed compounds are desorbed and conveyed to a pyrolysis zone wherein they are pyrolyzed along with the unvaporized sample and converted to a gaseous pyrolysis product which is quantitatively analyzed. The wastewater may be supplied in a boat to the volatilization zone and the boat moved into the pyrolysis zone during pyrolysis.

A process developed by D.C. Freeman, Jr., and L.J. Rogers (6) involves the determination of total inorganic carbon (TIC). In determining the total organic carbon (TOC) in a fluid, an indirect procedure may be used whereby total carbon (TCA) is measured by a part of the system employed, and total inorganic carbon (TIC) by another. The total organic carbon content (TOC) is then determined by subtraction. In the determination of total carbon several methods can be used and instruments embodying such methods are commercially available (e.g., Beckman Instruments Inc., Carbonaceous Analyzer, and Union Carbide Corp. Total Carbon Analyzers Model 1212 and 1202).

The aforementioned devices function by periodically injecting a measured quantity of sample water onto a hot catalyst or into a rapid combustion chamber wherein all carbon bearing materials are converted to CO_2. This newly formed gaseous component is then swept by a carrier gas into a sensing system (usually a nondispersive infrared analyzer sensitized for CO_2) with the analyzer output signal being displayed by a recording or indicating device.

When a second channel is provided which is sensitized to inorganic compounds only, a further set of information is obtained. This "total inorganic" channel (TIC) usually involves a lower temperature reaction chamber which employs a catalyst such as zinc chloride. Again, a measured quantity of water is periodically injected onto the heated catalyst and vaporized; the inorganic components generating the product CO_2 in direct proportion to the number of carbon atoms they represent. Organic components are not affected. The resulting CO_2 gas is then swept into a detecting device, such as a nondispersive infrared analyzer sensitized for CO_2 and the signal output displayed as before noted.

The two readings of generated CO_2 obtained are usually scaled by conventional calibration procedures to read parts per million (ppm) by weight directly. The difference between the two readings then yield the quantity of ppm of total organic carbon (TOC) i.e.:

$$TOC = TCA - TIC$$

The present state of the art has made possible the manufacture of reliable, automatic instruments for measuring total carbon (TCA). These instruments provide long periods of accurate operation without need for repair or other attention. However, systems for measuring the total inorganic carbon (TIC) suffer from a fundamental problem in that the acids or salts employed as the catalyst are water-soluble and tend to be washed away fairly rapidly by the water sample being analyzed and by the application, if used, of rinsing fluid. The operating lifetimes of such devices range from several days to 2 or 3 weeks. It is therefore an object of the process to provide a method for replenishing the catalyst used in instruments for total inorganic carbon measurements to thereby increase the life of such instruments indefinitely.

References

(1) Teal, J.L., Van Hall, C.E., Stenger, V.A. and Safranko, J.W., U.S. Patent 3,296,435; Jan. 3, 1967; assigned to Dow Chemical Co.
(2) Capuano, I.A., U.S. Patent 3,322,504; May 30, 1967; assigned to Union Carbide Corp.
(3) Seward, H.H., U.S. Patent 3,535,044; Oct. 20, 1970.

(4) Staffin, H.K. and Ricci, R.J., U.S. Patent 3,607,071; Sept. 21, 1971; assigned to Procedyne Corp.
(5) Takahashi, Y., Moore, R.T. and Stephens, T.M., U.S. Patent 3,703,355; Nov. 21, 1972; assigned to Dohrmann Div., Envirotech Corp.
(6) Freeman, D.C., Jr. and Rogers, L.J., U.S. Patent 3,672,841; June 27, 1972; assigned to Ionics, Inc.

CARBON DIOXIDE

Measurement in Air

Carbon dioxide may be monitored in air by the following instrumental techniques (1): infrared spectroscopy, gas chromatography.

A carbon dioxide sensing system has been developed by Guenther (2). It is a low cost system for sensing carbon dioxide content of ambient air. It comprises a thin layer of a chemical reversibly absorbent for carbon dioxide, the pH of which changes in relation with the amount of absorbed carbon dioxide at any time; an indicator present in the film which changes color in accordance with the pH, and a photoelectric system to produce an electric signal proportionate to the color change. In this system, substantial independence of humidity is gained by the use as principal solvent in the thin layer of an ionizing solvent having a vapor pressure in the range of 0 to 10 mm. at temperatures up to 150°F., and compatible with the other components of the system.

Measurement in Water

A device has been developed by O.K. McFarland, G.M. Illich, Jr., and W.C. Ling (3) for continuously measuring the partial pressure of CO_2 in a stream of liquid. This device includes an ion sensitive electrode having a bicarbonate buffer solution. The electrode is positioned in a sampling cavity which is maintained continuously full of the liquid which is to be monitored and the buffer solution is separated from the cavity by a semipermeable Teflon membrane. A sampling stream of the liquid is directed into the body of liquid in the cavity and against the membrane and at least a portion of the CO_2 in the stream passes through the membrane to vary the pH of the buffer solution as a function of the partial pressure of the CO_2 in the sample stream.

The change in pH is measured to provide an indication of the partial pressure in the sample. Calibration of the monitor is effected by draining the chamber and exposing the membrane to a nebulized gaseous mixture of the liquid and a standard CO_2 of a known partial pressure.

References

(1) Lawrence Berkeley Laboratory, "Instrumentation for Environmental Monitoring – Air," *Publ. No. LBL-1,* Vol. 1, Berkeley, Univ. of Calif. (Feb. 1, 1973).
(2) Guenther, K. R., U.S. Patent 3,754,867; Aug. 28, 2973; assigned to Bjorksten Research Laboratories.
(3) McFarland, O.K., Illich, G.M., Jr. and Ling, W.C., U.S. Patent 3,689,222; Sept. 5, 1972; assigned to Abbott Laboratories.

CARBON MONOXIDE

Scientists at Argonne National Laboratory, Argonne, Illinois, confirmed in 1972 the growing belief that the common air pollutant, carbon monoxide, plays a much more complex role in nature than previously had been suspected (1). C.M. Stevens and his colleagues found that nature produces at least 10 times as much carbon monoxide as do man's combustion processes. They found that a worldwide buildup of carbon monoxide does not appear to be the problem it once was feared to be. The Argonne scientists' conclusions stem from a three-year study of the ratios of carbon and oxygen isotopes in carbon monoxide in samples of air. The ratios of carbon-13 to carbon-12 and oxygen-18 to oxygen-16 vary significantly in carbon monoxide from different sources. Stevens' group identified at least four types of carbon monoxide, one of which is man-made or "technological" carbon monoxide. Each type shows quite different isotopic ratios. The analysis indicates that nature produces some 3.5 billion tons of carbon monoxide annually, while man produces about 270 million tons. The major natural source (85%) is oxidation of methane emitted by decaying organic matter.

The average global concentration of carbon monoxide is about 0.1 part per million by volume and appears not to be increasing. Other research indicates that carbon monoxide is removed from the air in several ways: photochemical reactions involving hydroxyl radical (OH) in the troposphere, extending up to 10 miles from the earth's surface; the same types of reactions in the stratosphere, starting immediately above the troposphere; and consumption by bacteria in soil and by green plants. Stevens calculates at any rate that a given parcel of carbon monoxide that enters the air is removed by natural processes in about one month.

The Argonne results do not diminish the need to abate carbon monoxide emissions from man-made sources. Such sources produce 95% or more of the toxic gas in urban areas, where it can build to harmful levels, far above the average global concentration. Stevens and his colleagues plan to do a more comprehensive study of carbon monoxide. They plan to set up two sampling stations in the Northern Hemisphere and two in the Southern Hemisphere. At each station the air would be sampled twice weekly, year-round. Stevens plans also to extend the isotopic analyses to include carbon-14. The technique to be used is new, however, and the group is not certain how well it will work.

Carbon monoxide air quality standards have been set by the states of California, New York and Pennsylvania as pointed out by A.C. Stern (1a). Pennsylvania has a standard of 25 ppm over a 24-hour averaging time. New York specifies 60 ppm for a 1-hour averaging time and California, 120 ppm for that same period. Limits of 15 to 30 ppm for 8-hour averaging times are specified by California and New York.

Measurement in Air

A review has been published by J.N. Driscoll and A.W. Berger of Walden Research Corp. of Cambridge, Mass. (2) covering the state of the art of chemical methods for sampling and analysis of carbon monoxide from stationary fossil fuel combustion sources in the concentration range of 10 to 1,000 ppm. A bibliography of 267 references is appended to the report. Methods proposed by an intersociety committee have been published (2a) covering:

1. Preparation of carbon monoxide standard mixtures,
2. Manual colorimetric method for CO in the atmosphere,
3. Infrared absorption method for CO in the atmosphere,
4. Nondispersive infrared method for CO in the atmosphere, and
5. Hopcalite method for CO in the atmosphere.

Carbon monoxide has been qualitatively and quantitatively determined by passing the gas to be tested through an elongated bed of granular colorimetric indicator contained in a transparent tube, thereby developing a color change or stain that progresses along the length of the bed, the length of stain depending on the concentration of CO in the gas and the volume of gas passed through the tube. Colorimetric indicators used for such analysis consist of an adsorbant carrier impregnated with a reagent that changes color on reaction with CO, such as alkali metal palladosulfites disclosed in U.S. Patent 2,569,895 and palladium sulfate-ammonium molybdate reagent disclosed in U.S. Patent 2,487,077. Commonly, the elongate bed of indicator is contained in a sealed transparent tube between layers of adsorbant, such as silica gel or alumina gel, that adsorb and prevent interfering gases, such as water vapor, hydrocarbons or organic vapors, from reaching the indicator when in use. When using these indicator tubes, the ends are opened or broken off and the tube is connected to a pump or aspirator that forces a predetermined volume of sampled gas through the tube.

A process developed by P.W. McConnaughey (3) provides a protective adsorbent composition to adsorb any CO sealed in the tube during manufacture which could develop a false color indication in storage and prior to use.

An evaluation study of length-of-stain gas indicator tubes for measuring CO concentration of from 0.1 to 10.0% in air has been published by E.C. Klaubert and J.C. Sturm of the U.S. Department of Transportation (4).

A device developed by E.S. Mayo, Jr. (5) provides an improved palladium-containing agent together with a photoelectric detector for continuously sensing color changes in the reagent. The improved reagent consists essentially of palladium chloride and a self-regenerating amount of hydrochloric acid with respect to the palladium chloride, both adsorbed onto an inert porous matrix or solid carrier such as silica gel. A practical regenerating amount of the hydrochloric acid is when the weight ratio thereof to $PdCl_2$ is from about 0.65 to about 2.0.

The carbon monoxide detection or indication by the above described reagent representatively occurs as follows:

$$PdCl_2 + CO + H_2O \longrightarrow Pd^\circ + 2HCl + CO_2$$

The palladium chloride ($PdCl_2$), which is light in color, is reduced by the carbon monoxide (CO) present in the surrounding atmosphere and coming into contact therewith to dark palladium metal (Pd°) which is evidenced by a darkening of the reagent. The silica gel merely serves as a support for the $PdCl_2$ and the HCl, and the darkening that takes place is due to the formation of metallic palladium. Inert inorganic carriers such as silica or alumina gel are preferred. The degree of darkening occurring depends upon the amount of Pd° formed and thus upon the amount of CO present in the atmosphere contacting the reagent.

Dangerous concentrations of CO could be indicated solely by this darkening effect; however, it is preferred that an alarm or annunciator apparatus be coupled with the reagent so as to give notice of the presence of CO rather than depend upon intermittent visual observation of the reagent and judgment of the relationship of the degree of color change to gas concentration.

An important aspect of the present reagent is the fact that it is self-regenerative, i.e., the above reaction that proceeds to the right when CO is present is driven to the left during the absence of significant amounts of the toxic gas. This driving force is internally supplied by the HCl originally adsorbed into silica gel along with the $PdCl_2$. The presence of this acid causes the conversion of the palladium metal to the chloride thereof and thereby returns the gel to its original light color. The reagent is dried before use, but is preferably used in the presence of a relative humidity of at least 20%.

Another type of carbon monoxide detector relies on the reaction of CO with a catalyst which causes a temperature change which in turn will affect some type of temperature sensitive element. This element in turn is connected to a control system for providing some type of alarm or indication readings. Most such gas detecting devices suffer from the same deficiency in that they are incapable of compensating for a wide range of variation in ambient temperature. It is possible in these devices for the ambient temperature to reach a temperature such that the heat of the atmosphere will be sufficient to actuate the circuitry and provide a false reading as to the presence of the gas.

A device developed by J.M. Ward (6) is designed to overcome this ambient temperature problem. The present carbon monoxide detector comprises two parallel mounted cylinders of metal having a high coefficient of linear expansion and spaced apart from one another on a common base. A chemical catalyst of the type capable of generating heat when exposed to a minimal concentration of carbon monoxide is mounted within a porous tube placed within one of the cylinders.

Free air is allowed to enter the catalyst-containing cylinder from its base and circulate about the catalyst. The heat generated by the catalytic reaction of carbon monoxide in air will cause the linear expansion of the cylinder thereby closing the microswitching mechanism mounted partly at the top of each cylinder. By using two cylinders having a high coefficient of linear expansion, changes in ambient temperature may be compensated for.

Carbon monoxide may be monitored in air by the following instrumental techniques (7): infrared absorption (reference method), gas chromatography – flame ionization detection.

The EPA-designated reference method for CO in air is nondispersive infrared spectroscopy. This method has been described in detail by Smith and Nelson (8). The essentials of the system are shown in the diagram in Figure 25.

FIGURE 25: CARBON MONOXIDE MONITORING SYSTEM USING NONDISPERSIVE INFRARED SPECTROSCOPY

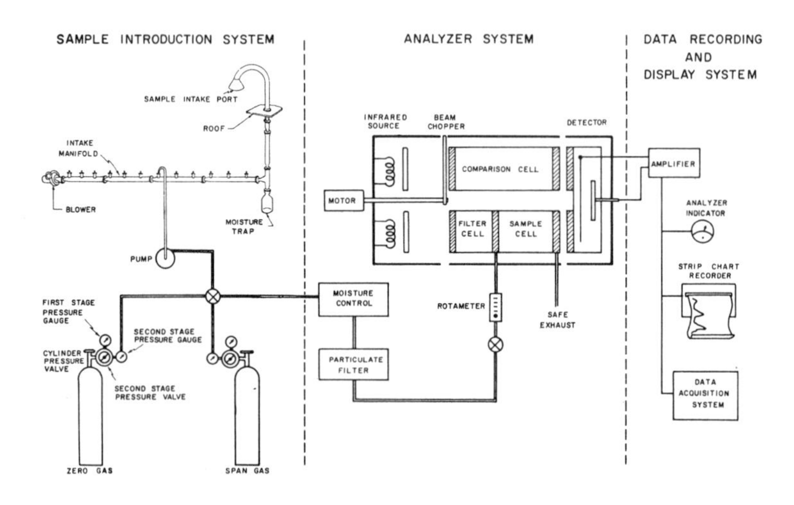

Source: Report PB 222,512

A continuous monitoring process for CO measurement in off-gases from processes such as the argon degassing of steel has been developed by R. Frumerman (9). According to the process, off-gases from the degassing vessel are divided and a known small portion thereof passed through the sample system. Into the effluent gases there is injected a tracer gas such as nitrous oxide (N_2O) at a known flow rate and the mixed off-gases and tracer gas are passed through two gas analyzers in a series at a controlled analyzer inlet pressure. One analyzer continuously delivers an output signal representative of the volume of mol fraction of CO in the sample and the other continuously gives an output signal representative of the volume or mol fraction of tracer gas in the sample. A signal indicative of the tracer gas input flow rate is continuously divided by the tracer gas analyzer output and the result continuously multiplied by the output signal from the CO analyzer to produce the instantaneous rate of CO evolution.

Measurement of the CO concentration in the atmosphere is difficult because there are no suitable instruments capable of measuring the very low concentrations found in the cleaner parts of the troposphere such as the North Pacific Maritime air having a CO concentration of 0.03 ppm. The only continuous CO meter described in the literature that has a sensitivity to less than 0.1 ppm utilizes the reduction of red mercuric oxide wherein the amount of mercury vapor released by the reduction of the oxide is measured and is proportional to the CO concentration. Such an instrument was described by Robbins, Borg and Robinson in 1967 in a paper entitled, "Carbon Monoxide in the Atmosphere" presented at the 60th Annual Meeting of the Air Pollution Control Association in Cleveland, Ohio – Paper No. 67–28. While this instrument is continuous and adequately sensitive for the investigation of tropospheric CO concentrations, it still possesses many problems of operation and calibration which prevents its use in areas of polluted air where possible interfering gases are present.

A number of detecting and measuring instruments have been developed and sold in the past for both sensing and/or measuring minute quantities of trace gas with a condensation nuclei meter which has a projected sensitivity on the order of 1 part in 10^{14}. Condensation nuclei is a generic name given to small airborne particles which are characterized by the fact that they serve as the nuclei on which water will condense as in a fog or cloud. Such condensation nuclei may have particle sizes lying in a size range extending from slightly above molecular size on the order 1×10^{-8} cm. in radius to 1×10^{-4} cm. in radius, although the most significant portion of condensation nuclei numerically lie in the size range between 5×10^{-5} and 5×10^{-7} cm. in radius.

The method by which known condensation nuclei meters operate relies on the property of condensation nuclei particles to act as the nucleus of a water droplet. If a sample of air or other gaseous atmosphere-containing condensation nuclei is drawn into a chamber and its relative humidity brought up to 100%, adiabatic expansion of the gaseous sample will cause the relative humidity to rise instantaneously to a value greater than 100%. The moisture of the air will then tend to condense about the condensation nuclei particles as centers.

These particles then grow, due to the condensation of the water about them from the original submicroscopic size to the size of fine fog droplets which then may be measured by known light-scattering and light absorption techniques. It is this growth of water about the particle that gives the necessary magnification to obtain a detectable sensitivity on the order of 1 part by weight in 10^{14} parts by weight of air for small particles. One such condensation nuclei meter has been described in U.S. Patent 2,684,008, issued July 20, 1954, to Bernard Vonnegut. U.S. Patent 2,897,059, issued July 28, 1959, for "Process and Apparatus for Gas Detection," Frank W. Van Luik, describes a method and apparatus for detecting gaseous carbon compounds such as carbon monoxide and carbon dioxide (CO_2).

This apparatus has an extremely high order of sensitivity and employs a condensation nuclei meter in conjunction with a conversion apparatus that converts gaseous carbon compounds to metallic carbonyls that then are sensed and measured by the condensation nuclei meter. The process and apparatus described in U.S. Patent 2,897,059, is quite satisfactory for many purposes; however, this instrument is capable of detecting only the total concentration of CO and CO_2 and has no ability to discriminate between the two gases.

A similar device for CO determination by nickel carbonyl formation and decomposition has been developed by K. Slater (11). This device involves introducing the gas mixture to be tested into a closed reaction chamber, exposing the gas mixture to nickel in the presence of heat for conversion of any carbon monoxide present to gaseous nickel carbonyl, passing the resultant product through a temperature environment sufficiently high to cause decomposition of the nickel carbonyl and deposition of nickel on a filter and measuring by an electrical sensor the rate of change in electrical conductivity of the filter as nickel is deposited thereon as an indication of the concentration of carbon monoxide in the gas mixture.

What is desired at this stage of development in the art for the reasons enumerated above, is a method and apparatus which detects only CO and will not respond to CO_2 or other interfering gases even in rather high concentration on the order of 1 part in a hundred (1%).

In a unique device developed by E.G. Walther (10) such a method and apparatus for detecting carbon monoxide in a sample gas is provided. This method and apparatus employs filtering a specimen portion of a sample gas to free it of naturally occurring condensation nuclei. The filtered, carbon monoxide-bearing gaseous specimen is then reacted with a reactant polymeric material to induce thermoparticulate production of condensation nuclei from the polymer. Many materials such as plastics when they are heated sufficiently will emit numerous particles due to thermoagitation which can serve as condensation nuclei.

Such emission has been termed "thermoparticulate production" by early investigators, C.B. Murphy and C.D. Coyle in a paper entitled, "Thermoparticulate Analysis" presented in 1966 at the Applied Polymer Symposium No. 2 and appears in the report of that symposium on pages 77-83. The condensation nuclei particles induced by thermoparticulate production in the presence of the CO-bearing gas specimens are then detected and measured by a condensation meter in a conventional manner. The number of condensation nuclei produced and counted provides a measure of the quantity of CO gas present in the specimen being analyzed. The reactant polymer is heated to a temperature just below that required for appreciable thermoparticulate production due to thermal agitation alone.

It is preferred that the temperature to which the reactant polymer is heated, be controlled in accordance with the concentration of CO present in the sample gas. For this purpose, a temperature-controlling means and scale range changing means of the indicator of the condensation nuclei meter are interconnected whereby scale range change is achieved automatically with changes in temperature of the reactant polymeric material in a manner such that for increasing CO concentrations the reactant temperature is lowered and the scale of the indicator means is adjusted to read higher CO concentrations.

The reactant polymeric material preferably is selected from the group consisting of styrene-divinyl benzene copolymer, polyethylene, chlorinated polystyrene and copolymer of styrene. It is also preferred that the polymeric material be in the form of finely divided globular beads and the filtered CO-bearing gas specimen to be reacted is passed over the surface of the beads in a manner to maximize surface exposure of the polymeric material to the gas specimen.

As mentioned above, one advanced type of CO monitor suitable for monitoring concentrations of less than 0.1 ppm utilizes the reduction of mercuric oxide by the CO-containing stream followed by measurement of the mercury vapor evolved as an index of CO concentration.

An improved device of this type is claimed by R.C. Robbins (12). Figure 26a shows the essential component of this instrument. Air, which is drawn into the apparatus and which is to be analyzed first passes through a dehydrating column **10**. Thereafter, it can have one of two paths. One of these, when an air sample valve **12** is opened and a reference air valve **14** is closed, is through an air flow meter **16**. The other of these paths, when the air sample valve **12** is closed and the reference air valve **14** is opened, is through a cell **18** containing an oxidizer, such as silver oxide and thereafter through the air flow meter **16**.

FIGURE 26: CARBON MONOXIDE ANALYZER USING MEASUREMENT OF EVOLVED MERCURY VAPOR FROM CO REDUCTION OF MERCURIC OXIDE

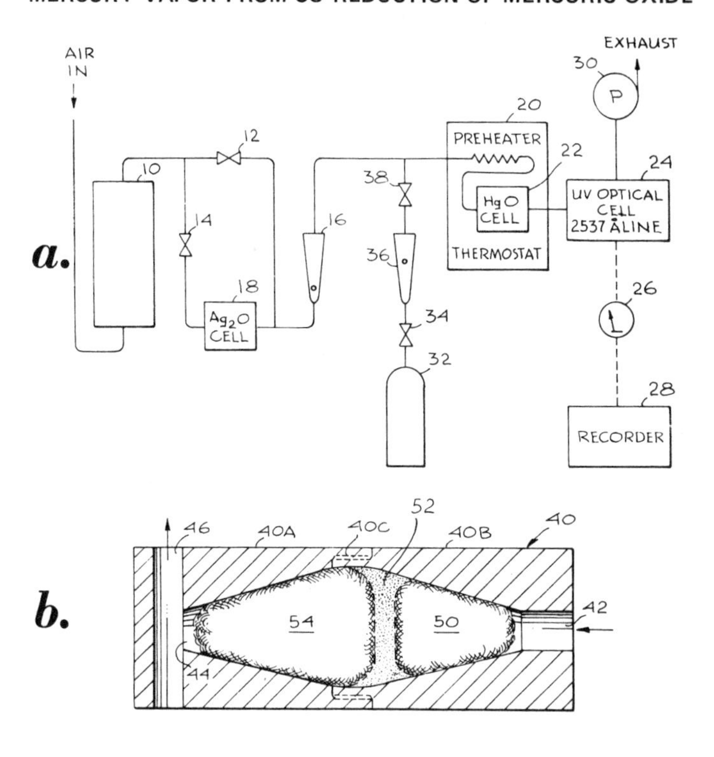

Source: R.C. Robbins; U.S. Patent 3,420,636; Jan. 7, 1969

The air passing through the air flow meter then passes through a preheater **20** which heats it to a temperature on the order of 210°C., and then through the mercury oxide cell **22**, also operated at this temperature. Thereafter, the gas passes through an ultraviolet photometer **24** which detects the amount of mercury vapor present, which may be indicated on a meter **26** and/or on a recorder **28**. The exhaust pump **30** serves to pull the gas through the entire apparatus. Also connected to the system before the preheater **20** is a container **32** of calibration gas which for carbon monoxide can contain 100 ppm CO in high purity N_2. The container **32** is connected by a valve **34** to a flow meter **36**. The flow meter is then connected through a calibration valve **38** to the tubing leading to the preheater **20**.

The mercury oxide cell shown in Figure 26b comprises a cross-sectional view of the cell. The cell consists of a stainless steel cylinder **40** which is made of two pieces respectively, **40A, 40B,** which threadably engage one another as by the threaded region **40C**. The center of the cylinder **40** is hollow and expands from the inlet opening **42** to a maximum diameter substantially at the threaded region **40C** and thereafter reduces in diameter to an exit region **44**. The end of the opening **44** abuts a right angle passage **46** through the cylinder **40**, and exits into the photometer through this passage **46**. The cell contains two wads of a heat resistant porous material **50**, such as Pyrex wool, which fill both conical sections of the cell between openings **42** and **44** except for the volume occupied by the mercuric oxide.

The key feature of the present apparatus is that the HgO cell **22** is maintained under such conditions that a very low mercury vapor background is present whereby the sensitivity of the apparatus is greatly increased. The cell is made of a stainless steel container within which a disc of red mercuric oxide powder is supported between two pads of glass wool. The cell is maintained at a temperature of 210° ± 15°C., which would produce about 2% by volume of mercury vapor at thermal equilibrium. Yet, because of the unique design and operating conditions of this sytem the pseudo-steady state concentration of mercury vapor in the air stream from thermal dissociation is less than 1 part per million.

In summarization, the recommended method of operation of the system shown in Figure 26 is as follows. First, the reference baseline is established. This is done by passing sample air through the dryer **10** and through the Ag_2O cell **18**, through the flow meter **16**, and into the preheater **20** and the hot mercuric oxide cell **22**. Next, the system is calibrated. This is done by metering, using the calibration flow meter **36**, a known concentration of the CO gas in the container **32** into a reference sample stream of air which has passed through dryer **10** and through the Ag_2O cell **18** before the calibration gas is added.

The sample rate of air is measured by the air flow meter **16**. Two or three known concentrations of the CO are metered into the reference air in this manner to provide calibration over a range of CO concentration levels. Thereafter, the reference air valve **14** is closed as is the calibration valve **38** and valve **34** and the air valve sample **12** is opened to permit the intake of a known quantity of the air the CO contents of which is to be measured. The air flow meter **16** is used for this purpose.

Measurement in Water

A device developed by V.J. Linnenbom and J.W. Swinnerton (13) permits measuring trace amounts of carbon monoxide dissolved in water along with light hydrocarbons including methane. The carbon monoxide is stripped from the water sample with a stream of helium, the effluent gas stream dried and passed through a pair of cold traps with joint adsorption of the methane and carbon monoxide in the second cold trap. The methane and carbon monoxide are collected from the second cold trap in a stream of helium and separated in a molecular sieve chromatograph from which the gas stream passes through a reduction zone in which the carbon monoxide is quantitatively reduced to methane and then to a hydrogen flame-ionization detector in which the carbon monoxide is detected as methane.

References

(1) American Chemical Society, *Chemistry in 1972,* Wash., D.C., Am. Chem. Soc. (1973).

(1a) Stern, A.C., in *Industrial Pollution Control Handbook,* H.F. Lund, Ed., New York, McGraw-Hill Book Co. (1971).

(2) Driscoll, J.N. and Berger, A.W., *Report PB 209,269,* Springfield, Va., Nat. Tech. Info. Service, (July 1971).

(2a) American Public Health Assoc, *Methods of Air Sampling and Analysis,* Wash. D.C., (1972).

(3) McConnaughey, P.W., U.S. Patent 3,312,527; Apr. 4, 1967; assigned to Mine Safety Appliances Co.

(4) Klaubert, E.C. and Sturm, J.C., *Report No. PB 211,003,* Springfield, Va., Nat. Tech. Info. Service, (Nov. 1971).

(5) Mayo, E.S., Jr., U.S. Patent 3,245,917; April 12, 1966; assigned to E.I. du Pont de Nemours & Co.

(6) Ward, J.M., U.S. Patent 3,577,322; May 4, 1971; assigned to Armstrong Cork Co.

(7) Lawrence Berkeley Laboratory, "Instrumentation for Environmental Monitoring – Air," *Publ. No. LBL-1,* Vol. 1, Berkeley, Univ. of Calif. (Feb 1, 1973).

(8) Smith, F. and Nelson, A.C., Jr., "Reference Method for the Continuous Measurement of Carbon Monoxide in the Atmosphere," *Report PB 222,512,* Springfield, Va., Nat. Tech. Info. Service, (June 1973).

(9) Frumerman, R., U.S. Patent 3,520,657; July 14, 1970; assigned to Dravo Corp.

(10) Walther, E.G., U.S. Patent 3,640,688; Feb. 8, 1972; assigned to Environment/One Corp.

(11) Slater, K., U.S. Patent 3,518,058; June 30, 1970.

(12) Robbins, R.C., U.S. Patent 3,420,636; Jan. 7, 1969; assigned to Stanford Research Institute.

(13) Linnenbom, V.J. and Swinnerton, J.W., U.S. Patent 3,545,929; Dec. 8, 1970; assigned to the Secretary of the Navy.

CHLORIDES

Chlorides are usually encountered in the air in the form of hydrogen chloride. As noted in the section of this book dealing with hydrogen chloride, HCl is emitted from the combustion of chlorine-containing coals and from the incineration of chlorine-containing plastics (1).

Chlorides are present in practically all waters. They may be of natural origin, or derived (a) from sea-water contamination of underground supplies; (b) from salts spread on fields for agricultural purposes; (c) from human or animal sewage; or (d) from industrial effluents, such as those from food processing, paper works, galvanizing plants, water softening plants, oil wells, and petroleum refineries (2). Herbert and Berger (3) have reported waste chloride concentrations from a Kraft process paper mill of 350 to 1,760 mg./l.

An industrial waste study on blast furnaces and steel mills (4) reveals that average size steel mills using advanced production technology produce from their pickling line 9,700 pounds per day of chloride as the ferrous salt, and 4,500 lbs./day as hydrochloric acid. The waste stream from the pickling line contains a total chloride content of 7,000 mg./l. This is diluted to as little as 15 mg./l. chloride upon mixing of the pickling line waste stream (approximately 0.24 MGD) with the waste flow from other steel processes (approximately 113 MGD).

Oil refineries and petrochemical processes are major producers of high chloride wastes. Crude oil desalting is a significant source of chloride in petroleum refinery wastes. Berger (5) describes a refinery effluent containing total dissolved solids of 1,163 mg./l., primarily as sodium chloride dissolved out of crude oil by desalting. Petrochemical processes, including chlorination, oxidation, polymerization and alkylation produced chloride wastes, usually associated with calcium or aluminum (6).

The Solvay process, widely used in making soda ash, yields waste waters high in chloride. Soda ash is a major raw material of many industries, including glass, chemicals, pulp and paper, soap and detergent, aluminum and water treatment. The Solvay process uses sodium chloride brine solution, and yields a substantial amount of waste calcium chloride (7). Jones has reported that the chloride content of a typical soda ash manufacturing waste is 90,000 mg./l. (8). While some calcium chloride is recovered by distillation, this practice is not sufficiently prevalent to prevent the Solvay process having a major waste disposal problem (7).

Food processing industry wastewaters may also be high in chloride content. Burr and Byrne (9) have described a 400,000 gal./day waste flow from an olive processing facility in California, which contained 3,500 to 6,000 mg./l. of chloride. The chloride represents 35 to 40% of the total dissolved solids concentration of that waste. Another report (10) has cited chloride levels of 36 to 4,280 mg./l. of chloride from olive processing, and 24,220 mg./l. of chloride from cucumber pickling.

Frequently, chloride ion is added to a wastewater as a result of waste treatment required to remove some less desirable constituent. Sodium chloride and hydrochloric acid are often used to regenerate spent ion exchange resins used in water purification. Chloride ions on the regenerated resins then exchange with undesirable anions as the waste water passes through the ion exchange bed. Another example of chloride addition is chlorination treatment of cyanide waste streams.

The permissable concentration of chlorides in domestic water supplies is 250 ppm and the desirable concentration is less than 25 ppm as shown earlier in Table 3. Effluents to storm sewers or streams should contain a maximum of 150 ppm of chloride ion.

Measurement in Air

A manual colorimetric method for determination of the chloride content of the atmosphere by mercuric nitrate titration in the presence of a mixed diphenylcarbazone-bromphenol blue indicator has been described in the literature (10a).

A process for the measurement of chloride ions in air has been developed by R.H. Jones (11). The sample gas stream is dissolved in substantially pure water to dissolve the hydrogen chloride therein, thus producing chloride ions. Then the stream containing the chloride ions is delivered to a silver chloride saturation chamber and thereafter to a silver ion sensing measuring cell. The change in silver ion content is a measure of the chloride ion introduced to the system.

Figure 27 shows the arrangement of equipment used in this process. The gas sample is delivered into the apparatus through an inlet line **34**, the flow of the gas being controlled by a flow regulator **36** in the line. The water used in the process enters the apparatus through a second inlet line **38** having therein a second flow regulator **40**. The gas and water enter a contractor **42** through a line **43** where the constituent in the gas being measured dissolves completely in the water thereby forming chloride ions and the insoluble component of the gas stream separates from the water and exits from the contactor **42** via a vent **44**.

The effluent from the contactor **42** passes through a line **45** to the saturation chamber **20** containing silver chloride and the effluent from the chamber **20** passes to the silver ion sensing electrode **26**. It is desirable to provide a temperature controller in this apparatus. The temperature controller may be located between the contactor **42** and saturation chamber **20** in the line **45**; although, it could also be positioned in the line **43**.

In utilizing the apparatus shown, the sample gas stream and substantially pure water are delivered to the lines **34** and **38**, respectively, at a predetermined and constant flow rate so that when the silver ion concentration of the effluent from the chamber **20** is measured, the chloride ion concentration and, therefore, the amount of hydrogen chloride in the sample gas, may be indirectly determined.

FIGURE 27: MONITOR FOR MEASURING CHLORIDE ION CONTENT OF GASEOUS STREAM

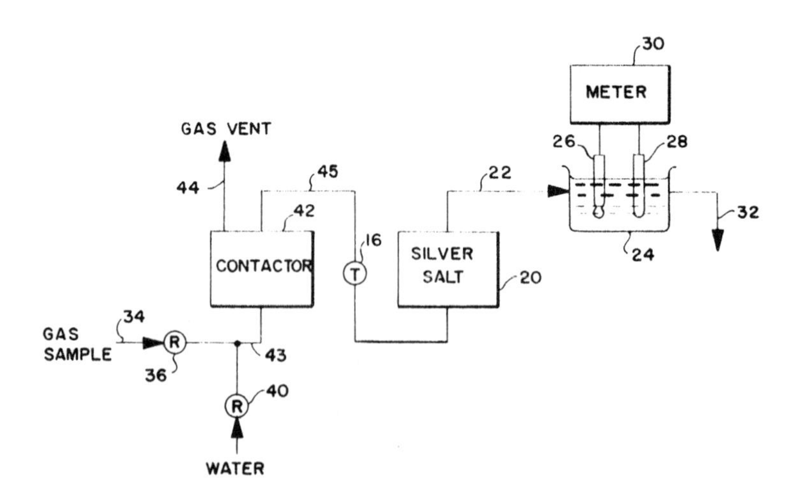

Source: R.H. Jones; U.S. Patent 3,457,145; July 22, 1969

Measurement in Water

Chlorides may be monitored in water by the following instrumental techniques (12):

Ion Selective Electrode
Molecular Absorption (Colorimetric)
Potentiometric (Ag-AgCl Electrode)

A colorimetric test paper method for determining chloride ion in solution has been developed by C.H. Hach (13). The test paper is specially impregnated with a continuous zone of linearly increasing concentration corresponding to a particular concentration range for the analysis being carried out, e.g., as 0 to 25,000 parts per million for the chloride determination, and further impregnated with an indicator to signal the particular concentration of the test source.

When the analysis is for chloride ion, the titrant can be silver nitrate and is present in an amount corresponding to a range from about 0 to 100 parts per million free chlorine; the indicator can be silver chromate and includes a buffer capable of maintaining the pH of the liquid test source at a predetermined range.

A device for the measurement of chloride ions in water has been developed by R.H. Jones (11). It involves determining the chloride ion content of an aqueous stream by saturating the stream with sufficient silver chloride so that the chloride ions in the stream and silver ions and chloride ions from the dissolved silver chloride reach equilibrium with the solid silver chloride, and thus maintain the solubility product relationship. By measuring the aqueous stream with a silver ion sensitive glass electrode system, the original chloride ion content may be calculated.

Figure 28 shows a suitable form of apparatus for the conduct of this process. The apparatus includes an inlet line **10** for the aqueous stream which is connected to a chamber **12** containing a cation exchange resin. A flow regulator **14** is provided in the line so that the sample stream will pass at a relatively low rate to ensure that the analysis will not be disturbed because of incomplete saturation by the silver salt as might occur if too high a flow were provided. However, it is not necessary that the flow rate be highly constant in this method. A temperature controller **16** is also provided in the line **10** for maintaining the temperature of the incoming aqueous stream substantially constant.

FIGURE 28: MONITOR FOR MEASURING CHLORIDE ION CONTENT OF LIQUID STREAM

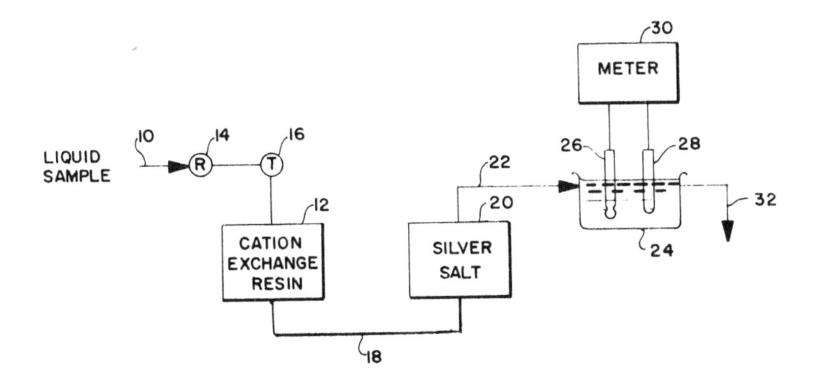

Source: R.H. Jones; U.S. Patent 3,457,145; July 22, 1969

The amount of solid silver chloride dissolved in the stream will change appreciably with temperature when the chloride ion content of the incoming sample stream is low because the solubility of silver is not suppressed as would be the case if the chloride ion content of the sample stream were high. Therefore, the temperature controller is of utmost importance when the chloride ion content of the stream is low. However, if the chloride ion content of the stream is relatively high, the temperature controller could be eliminated and relatively accurate determinations of the chloride ion content still could be made.

The cation exchange resin in chamber **12** is provided to remove any sodium ions which might exist in the stream and exchange those ions with other cations which will not interfere with the measurement of silver ions subsequently in the analysis. Examples of suitable cations exchange resins are Rohm and Haas IR. 120 and Dowex 50.

The effluent from the chamber **12** is delivered through a line **18** to a saturation chamber **20** which contains silver chloride when the chloride ion concentration of the sample stream is being determined. The silver chloride is provided in solid form and in a sufficient amount to allow the aqueous stream to become completely saturated with silver chloride so that the chloride ions in the sample stream and silver and chloride ions from the slightly soluble silver chloride will reach chemical equilibrium with the solid silver chloride.

Therefore, the effluent from the chamber **20** passing through the line **22** to the container **24** will have a silver ion content which is a function of the chloride ion concentration of the incoming stream as explained previously. A glass electrode **26** having a silver ion sensitive glass bulb and a reference electrode **28** are immersed in the solution in the container **12** and are connected to a conventional high impedance voltmeter **30** for measuring the silver ion content of the stream. The stream leaves the container **24** through an outlet **32**. It is to be understood that the meter **30** may read in terms of silver ion content or may be calibrated directly in chloride ion content if the temperature is maintained constant as by the controller **16**.

It is to be further understood that the cation exchange resin in the chamber **12** may be eliminated if the aqueous stream contains no or few sodium ions which would interfere in the silver ion measurement of the effluent from the chamber **20**.

References

(1) Patterson, J.W. and Minear, R.A., "Wastewater Treatment Technology," 2nd Ed., *Report PB 216,162;* Springfield, Va., Nat. Tech. Info. Service (Feb. 1973).
(2) J.E. McKee and H.W. Wolf, "Water Quality Criteria," California State Water Quality Control Board, *Publication No. 3-A,* 1963.
(3) A.J. Herbert and H.F. Berger, "A Kraft Bleach Waste Color Reduction Process Integrated with the Recovery System," *Proc. 15th Purdue Industrial Waste Conf.,* 15:49-57, 1962.
(4) Anonymous, "The Cost of Clean Water," Vol. III, *Industrial Waste Profiles, No. I - Blast Furnaces and Steel Mills,* U.S. Dept. Interior, Washington, D.C., 1967.
(5) M. Berger, "The Disposal of Liquid and Solid Effluents from Oil Refineries," *Proc. 21st Purdue Industrial Waste Conf.,* 21:759-767, 1966.
(6) E.F. Gloyna and D.L. Ford, "The Characteristics and Pollutional Problems Associated with Petrochemical Wastes - Summary", U.S. Dept. Interior, Washington, D.C., 1970.
(7) Anonymous, "The Economics of Clean Water," Vol. III, *Inorganic Chemistry Industry Profile,* U.S. Dept. Interior, Washington, D.C., 1970.
(8) M.W. Jones, "Construction and Operation of a Chloride Holding Basin," *Proc. 16th Purdue Industrial Waste Conf.,* 16:186-192, 1961.
(9) D. Burr and P.J. Byrne, "City and Industry Cooperate to Solve Brine and Waste Problems," *Public Works,* 1971:46-48 January.
(10) Anonymous, "Reduction of Salt Content of Food Processing Liquid Waste Effluent," *U.S. EPA Report 12060* DXL 01/71, 1971.
(10a) American Public Health Assoc., "Methods of Air Sampling and Analysis," Washington, D.C., 1972.
(11) Jones, R.H., U.S. Patent 3,457,145; July 22, 1969; assigned to Beckman Instruments, Inc.
(12) Lawrence Berkeley Laboratory, "Instrumentation for Environmental Monitoring - Water," *Publ. No. LBL-1,* Vol. 2, Berkeley, Univ. of Calif., Feb. 1973.
(13) Hach, C.H., U.S. Patent 3,510,263; May 5, 1970; assigned to Hach Chemical Co.

CHLORINATED HYDROCARBONS

U.S.S.R. air quality standards for some chlorinated hydrocarbons are as follows:

	Single Exposure, ppm	24-hr. Average, ppm
Carbon Tetrachloride	0.7	–
Chlorobenzene	0.02	0.02
Dichloroethane	0.75	0.25
Epichlorohydrin	0.05	0.05

These may be compared with German engineering society standards as follows:

	Max. Once in 4 Hours, ppm	Long-term, ppm
Carbon Tetrachloride	1.5	0.5
Chlorobenzene	3.0	1.0
Chloroform	6.0	2.0
Trichloroethane	15.0	5.0

These are quoted from various sources by A.C. Stern (1). Some water quality standards for chlorinated hydrocarbon pesticides have been presented earlier in Table 3.

Polychlorinated biphenyls (PCB's) have been singled out for particular attention by the EPA and included in their list of Proposed Toxic Pollutant Effluent Standards published in the Federal Register for December 27, 1973. The concern is based on the fact that they are among the most persistent organic chemicals in the environment, degrading very slowly, plus their fairly wide use as hydraulic fluids, heat-exchange fluids, plasticizers, etc.

The proposed standard specifies no PCB discharge into streams having a flow less than 10 cu. ft./sec. or lakes having an area less than 500 acres. Faster flowing streams or larger lakes are permitted to receive 280 μg./l. (salt water bodies 10 μg./l.). These regulations are to apply unless the receiving stream flow is less than 10 times the waste stream flow in which case the limits are to be reduced by a factor of ten.

Measurement in Air

The determination of chlorinated hydrocarbons in the air by the method of microcombustion has been reported in a Survey of U.S.S.R. Literature on Air Pollution and Related Occupational Diseases by B.S. Levine (2).

The method described is based on the quantitative oxidation of chlorinated hydrocarbon vapor in a combustion chamber equipped with a platinum coil heated to redness. The combustion products are then passed through an absorber solution and the ionic chlorine determined nephelometrically. With an appropriately prepared standard scale accurate determinations can be made in solutions containing 0.001 mg. of chlorine in 1 ml. Control tests were made with ethylene chloride, chloroform, carbon tetrachloride and trichlorethylene.

A portable apparatus for the determination of chlorinated hydrocarbons in the air by the microcombustion method was constructed. The microcombustion method described proved to be accurate for the determination of thousandths of a milligram of chlorine within 30 to 40 minutes. A new microabsorber is described which assures complete absorption of products of hydrocarbon combustion. Air samples are aspirated into gas pipettes filled with a saturated solution of sodium sulfate or into vacuum gas pipettes.

Measurement in Water

Chlorinated hydrocarbons can be monitored in water by the following instrumental techniques (3):

Gas Chromatography
Gas Chromatography/Mass Spectrometry/Computer

References

(1) Stern, A.C., in *Industrial Pollution Control Handbook,* H.F. Lund, Ed., New York, McGraw-Hill Book Co. (1971).
(2) Levine, B.S., Nat. Bur. of Stds., Inst. for Applied Tech. 3, 23-27 (May 1960).
(3) Lawrence Berkeley Laboratory, "Instrumentation of Environmental Monitoring - Water", *Publ. No. LBL-1, Vol. 2,* Berkeley, Univ. of Calif. (Feb. 1973).

CHLORINE

Low concentrations of chlorine gas cause irritation of the eyes, nose, and throat of humans (approximately 3,000 $\mu g./m.^3$ or 1 ppm is the threshold value). Excessively prolonged exposures to low concentrations, or exposure to higher chlorine concentrations, may lead to lung diseases such as pulmonary edema, pneumonitis, emphysema, or bronchitis. Recent studies indicate that other residual effects may occur, such as a decrease in the diffusing capacity of the lungs. Furthermore, there is evidence which suggests that continuous exposure to low concentrations may cause premature ageing and increased susceptibility to lung diseases, according to Stahl (1).

Chlorine is also a phytotoxicant which is stronger than sulfur dioxide but not as strong as hydrogen fluoride. Several episodes have been reported in which chlorine emissions from accidental leaks or spillages have resulted in injury and death to humans, animals, and plants. Material damage from chlorine is possible, since chlorine has strong corrosive properties.

The major production processes involve the electrolysis of alkali chloride solution. Possible sources of chlorine pollution from these processes include the liquefaction process, the filling of containers or transfer of liquid chlorine, and the emission of residual gas from the liquefication process.

No information is currently available on the concentrations of chlorine gas in ambient air. However, chlorine has been classified and is presently analyzed as one of the "oxidants" of the air. Effective methods are available for controlling chlorine emissions. In some cases the chlorine or its by-products can be recovered for further use. No information has been found on the economic costs of chlorine air pollution or on the costs of its abatement.

The U.S.S.R. air quality standard for chlorine in air is 0.033 ppm on a single exposure basis or 0.01 ppm on a 24-hour averaging basis. The Federal Republic of Germany specifies 0.2 ppm on a basis not to be exceeded more than once in any 8-hour period and 0.1 ppm on a long-term exposure basis. Canadian standards call for a 0.1 ppm maximum on a single exposure basis. Emission standards are 77 ppm for chlorine in the U.K. and in Australia as quoted by A.C. Stern (1a).

Measurement in Air

There appears to be no specific method for the analysis of chlorine according to Stahl (1). Most of the methods used rely on the oxidizing property of chlorine, and thus the presence of other oxidizing agents, such as ozone, bromine, nitrogen oxides, sulfur oxides, etc., can sometimes seriously interfere with the analysis.

Sampling Methods: Chlorine gas samples are collected either in impingers containing a reactive liquid solution or on a solid which is impregnated with a reactive substance. The common liquid solution used in the United States is dilute sodium hydroxide solution (approximately 0.01 to 0.1 N), (2) (3). This basic solution converts the chlorine gas to equal amounts of chloride ions and hypochlorite ions. When the solution is made acidic, the reaction is reversed and the chlorine gas is regenerated. However, the hypochlorite ion can further slowly decompose (under basic conditions) to give more chloride ions and oxygen. The amount of hypochlorite ion that decomposes will result in a corresponding loss in chlorine gas when the solution is acidified.

Other common absorber solutions react with chlorine gas to directly or indirectly produce a color change in the solution, which is taken as an indication of the amount of chlorine in the sampled air or gas. A variety of solutions are used, which are discussed in the follow sections.

The chlorine gas can also be absorbed on solids containing reactive substances similar to those used with the absorber solutions. A color change results which indicates the amount of chlorine that is present. The solid support is usually either paper or silica gel. The former is generally used for both qualitative and quantitative measurements, while the latter, found in the commercially available "detector tubes", is usually reserved for quantitative measurement. Some of the reactive substances are discussed in the following section.

Qualitative and Semiquantitative Methods: Many types of chlorine-indicator papers have been described in the literature. Among the most common are the starch-iodide papers, (4)(5)(6)(7). The basic reaction of chlorine gas with the potassium iodide of the paper yields free iodine, which then reacts with the starch to produce a blue color. (However, other oxidants can also turn the paper blue.) The limit of detection is approximately 2 to 6 ppm. A starch-iodide paper also coated with glycerin and sulfurous acid is reported to turn brown or black and to have a sensitivity of 0.25 to 12 ppm, (8). A buffered cadmium-iodide-starch paper is reported to be free of interference from nitric oxide (9).

Other common papers used include bromide-fluorescein papers, (4)(10)(11)(12) which change from yellow to red or rose, with a sensitivity of approximately 10 ppm; and o-tolidine papers, (4)(9)(11)(13) which turn yellow or bluish green (depending on other reagents present), with a sensitivity of approximately 2 ppm. Other indicator papers that have been reported include dimethylaniline (7), aniline (4)(14)(15)(16), o-phenetidine (17), and benzidine (18).

Rapid semiquantitative determination of chlorine gas can be made with commercially available gas-detecting tubes (19)(20)(21). These tubes contain a solid-coated reactive material which changes colorimetrically when exposed to a specific gas or to certain types of gases. A given volume of the gas sample is passed over the absorbent, and the amount of absorbent that changes color (measured by the length that is affected) is used to determine the amount of particular gas being tested (in this case, chlorine gas). Reactive substances that have been used as indicators for chlorine are o-tolidine (22), bromide-fluorescein (23) (24) and tetraphenylbenzidine (25).

Continuous sampling instruments for detection of chlorine gas have been developed, based on the reflectance from a chlorine-sensitive coated paper (26)(27)(28). An alarm system can be used with the instruments to warn when a certain limit has been reached. Sensitivity ranges from 300 to 9,000 $\mu g./m.^3$ (0.1 to 3 ppm) of chlorine.

Quantitative Methods: Most quantitative methods of analysis for chlorine gas are based on colorimetric reactions. A sensitive reagent commonly used is o-tolidine (3)(29)(30)(31) (32)(33). The air sample can either be passed directly into an acid solution of o-tolidine or collected in dilute sodium hydroxide, which can later be acidified and the o-tolidine then added. The latter method has the advantage that the time allowed for the development of color can be controlled and maximized. Acidification yields a more stable yellow-to-orange color. Color comparisons can be made with standardized color solutions by visual methods or in a spectrophotometer at 435 and 490 mμ. Approximately 3 liters of air must be sampled to detect chlorine concentrations of 3,000 μg./m.^{3}. This method is reported to be better than 99% efficient. However, the presence of other oxidants, such as chlorine dioxide, ozone, ferric and manganic compounds, and nitrates, may interfere with this method.

In a method used in England (34), the reagent 3,3'-dimethylnaphthidine turns mauve in the presence of chloride gas. This reagent is about eight times more sensitive than o-tolidine. Other oxidizing agents, such as bromine, chlorine dioxide, nitrogen dioxide, etc., interfere with the test.

One reagent is reported to be specific for chlorine gas, with a sensitivity of 1,000 μg./m.^{3}, even in the presence of such oxidants as bromine, ozone, and nitrogen oxide (23). The method is based on the oxidation of arsenious anhydride in an alkaline solution in the presence of potassium iodide and starch. The solution turns blue in the presence of chlorine.

Other reagents used for determination of chlorine gas that have been reported in the literature include: dimethyl p-phenylenediamine (35)(36), methyl orange (5)(37)(38), iodide-starch (23), iodide-starch-arsenic oxide (39)(40), benzidine acetate (23), bromide-fluorescein (23), resorufin (41), 4,4'-tetramethyldiaminodiphenylmethane (42), and rosaniline hydrochloride (43). An air analyzer for chlorine gas was reported, based on polarographic measurements with a special galvanic cell (44).

Chlorine may be monitored in air by manual wet chemical means (45). The methyl orange method for the determination of the free chlorine content of the atmosphere has been described in detail as the result of the deliberations of an Intersociety Committee (45a).

A thin-layer electrochemical monitoring device has been developed by W.D. Shults and J.R. Kuempel (46) which is applicable to the monitoring of various air pollutants, including chlorine.

A thin layer of electrolyte solution is exposed on one side to the atmosphere and on the other side to a mercury pool electrode. Electrochemical reactions which take place at the mercury electrode surface due to the presence of certain pollutants alter the current flow through the mercury pool at a given voltage, thereby providing an indication of the amount of the particular pollutant under study.

A process developed by Lerner (47) for detecting chlorine gas entails mixing copper particles with a silver activated blue phosphor compound and exposing this mixture to an environment suspected of containing chlorine gas to convert a portion of the copper to copper chloride. The exposed mixture is reacted by heating it to a temperature sufficient to cause the copper chloride to accelerate diffusion of copper atoms into the crystal lattice network of the phosphor. The reacted mixture is excited with a source of radiant energy and the spectral emission of the reacted mixture is then detected and compared with the spectral emission of the blue phosphor compound to determine if a shift in spectral emission, which is indicative of the presence of chlorine, has occurred.

An apparatus developed by O.K.C. Mang (48) is one in which an electronegative gas, such as chlorine, is detected in a gaseous mixture by passing an aqueous electrolyte across the surfaces of first and second spaced electrodes in the direction from the first electrode to the second electrode. The electrodes are externally connected by a current detection circuit. The electronegative gas component dissolving in the electrolyte with the formation of negative ions causes a current to flow through the current detection circuit.

Measurement in Water

Chlorine in dissolved form can be monitored in water by the following instrumental techniques (49):

Colorimetric
Electrochemical

A method has been developed by J.J. Morrow, Jr. (50) for determining free chlorine in solution in an aqueous sample which involves the addition of a soluble bromide to the sample followed by introduction of the sample as an electrolyte into a cell subject to cathodic polarization and measurement of the depolarizing action of the sample in the cell.

References

(1) Stahl, Q.R., "Air Pollution Aspects of Chlorine Gas," *Report PB 188,087,* Springfield, Va., Nat. Tech. Info. Service (Sept. 1969).

(1a) Stern, A.C. in *Industrial Pollution Control Handbook,* H.F. Lund, Ed., New York, McGraw-Hill Book Co. (1971).

(2) Devorkin, H., et al, *Air Pollution Source Testing Manual,* Air Pollution Control District, Los Angeles County, Calif. (1965).

(3) Stern, A.C., Ed., *Air Pollution,* Vol. II, New York: Academic Press, pp. 99,325 (1968).

(4) Guatelli, M.A., "Analytical Procedures for the Detection of Gases," *Rev. Asoc. Bioquim. Arq. 18:*3 (1954).

(5) Krivoruchko, F.D., "Determination of Chlorine in the Atmosphere," *Gigiena i Sanit.* 3:53 (1953).

(6) Permissible Immission Concentration of Chlorine Gas, *Verein Deutscher Ingineure, Kommission Reinhaltung der Luft,* Duesseldorf, Germany, VDI 2106 (1960).

(7) Weber, H.H., "Simple Methods for the Detection and Determination of Poisonous Gases, Vapors, Smokes, and Dusts in Factory Air," *Zentr. Gewerbehyg. Unvallverhuet 23:*177 (1936).

(8) Silverman, L., "Sampling and Analyzing Air for Contaminants," *Air Conditioning, Heating Ventilating,* p. 88 (1955).

(9) Studinger, J., "Short Scheme of Analysis for the Detection of Poison Gases," *Chem. Ind.* (London) *15:*225 (1937).

(10) Leclerc, E., and R. Haux, "Modern Methods of Detection and Determination of Industrial Gases in the Atmosphere," *Rev. Universelle Mines 12:*293 (1936).

(11) Leroux, L., "The Detection of Toxic Gases and Vapors," *Rev. Hyg. Med. Prevent. 57:*81 (1935).

(12) Zhitkova, A.S., *Some Methods for the Detection and Estimation of Poisonous Gases and Vapors in the Air, A Practical Manual for the Industrial Hygienist,* Service to Industry, West Hartford, Conn. (1936).

(13) Cook, D.A., "Test Strips for Detecting Low-Chlorine Concentrations in the Air," U.S. Patent 2,606,102 (Aug. 2, 1952).

(14) Cox, H.E., "Tests Available for the Identification of Small Quantities of the War Gases," *Analyst 64:*807 (1939).

(15) Liberalli, M.R., "Identification of War Gases," *Rev. Quim. Farm. (Rio de Janeiro) 4:*49 (1939).

(16) Zais, A.M., "Identification of Gas Warfare Agents," *J. Chem. Educ. 21:*489 (1944).

(17) Moureau, H., et al, "Photochemical Transformation of Chloropicrin to Phosgene. I. New Reagents Sensitive and Specific for These Two Substances," *Arch. Maladies Prof. Med. Travail Securite Sociale 11:*445 (1950).

(18) Russkikh, A.A., "Rapid Method for the Determination of Chlorine in the Atmosphere," *Tr. Khim. i Khim, Tekhnol. 1*(1):157 (1960).

(19) Ferris, B.G., Jr., W.A. Burgess, and J. Worcester, "Prevalence of Chronic Respiratory Disease in a Pulp Mill and a Paper Mill in the United States," *Brit. J. Ind. Med.* (London) *24*(1):26 (1967).

(20) Kusnetz, H.L., et al, "Calibration and Evaluation of Gas Detecting Tubes," *Am. Ind. Hyg. Assoc. J. 21:*361 (1960).

(21) "M.S.A. Chlorine-in-Air Detector," Mine Safety Appliances Co., Pittsburgh, Pa., *Bull. 0805-2* (1959).

(22) Kitagawa T., and Y. Kobayashi, "Chlorine-Gas Concentration in Some Plants Manufacturing Electrolytic Sodium Hydroxide," *Japan Analyst 3:*42 (1954).

(23) Demidov, A.V., and L.A. Mokhov, "Rapid Methods for the Determination of Harmful Gases and Vapors in the Air," Translated by B.S. Levine, *U.S.S.R. Literature on Air Pollution and Related Occupational Diseases 10:*114 (1962).

(24) Fenton, P.F., "Detection of War Gases," *J. Chem. Educ. 21:*488 (1944).

(25) McConnaughey, P.W., "Field Type Colorimetric Testers for Gases and Particulate Matter," presented at the annual meeting of the American Industrial Hygiene Association, Atlantic City, N.J. (1958).

(26) Allen, E.J., and N.R. Angvik, "Watchdog System Detects Chlorine Leaks at Seattle," *Water Works Eng. 114:*614 (1961).
(27) Kinnear, A.M., "An Alarm for the Detection of Toxic Industrial Gases," *Chem. Ind.* (London) 1959:361 (1959).
(28) Sundstrom, C., "Apparatus for Detecting Chlorine Gas Leaks from Containers," U.S. Patent 2,606,101 (1952).
(29) Fukuyama, T., et al, "Estimation of Toxic Gases in Air. VII. Chlorine Determination," *Bull. Inst. Public Health* (Tokyo) *4:*10 (1955).
(30) "Methods for the Detection of Toxic Gases in Industry - Chlorine," Dept. of Science and Industrial Research, London, Eng. *Leaflet No. 10* (1939, reprinted 1947).
(31) Pendergrass, J.A., "An Air Monitoring Program in a Chlorine Plant," *Am. Ind. Hyg. Assoc. J. 25*(5):492 (1964).
(32) Porter, L.E., "Free Chlorine in Air. A Colorimetric Method for its Estimation," *Ind. Eng. Chem.* 18:730 (1926).
(33) Wallach, A., and W.A. McQuary, "Sampling and Determination of Chlorine in Air," *Am. Ind. Hyg. Assoc. Quart. 9:*63 (1948).
(34) "Methods for the Detection of Toxic Substances in Air - Chlorine," *Booklet No. 10,* Ministry of Labor, Her Majesty's Factory Inspectorate, London (1966).
(35) Alfthan, K., and A.C. Jarvis, "New Indicator for Chlorine," *J. Am. Water Works Assoc. 20:*407 (1928).
(36) Polezhaev, N.G., "Determination of Active Chlorine in the Air," *U.S.S.R. Literature on Air Pollution and Related Occupational Diseases 3:*29 (1960).
(37) Ryazanov, V.A., "Limits of Allowable Concentrations of Atmospheric Pollutants," Translated by B.S. Levine, U.S. Department of Commerce, Office of Technical Services, Washington, D.C. Book 2 (1955).
(38) Takhirov, M.T., "Determination of Limits of Allowable Concentrations of Chlorine in Atmospheric Air," *USSR Literature on Air Pollution and Related Occupational Diseases 3:*119 (1960).
(39) Mal'chevskii, A.N., "A Method for the Rapid and Accurate Determination of Small Concentrations of Chlorine in Air," *Hig. i Sanitariya 11-12:*37 (1938).
(40) Popa, I., and L. Armasescu, "Determination of Gaseous Chlorine in an Industrial Atmosphere," *Farmacia* (Bucharest) *7:*499 (1959).
(41) Eichler, H., "Detection of Chlorine and Bromine in Air, Gas Mixtures and Solutions by the Formation of Iris Blue," *Anal. Chem. 99:*272 (1934).
(42) Masterman, A.T., "Color Tests for Chlorine, Ozone, and Hypochlorites with Methane Base," *Analyst 64:*492 (1939).
(43) Litvinova, N.S., and N. Ya Khlopin, "Rapid Determination of Active Chlorine in the Presence of Hydrogen Chloride in the Atmosphere of Industrial Establishments," *Gigiena i Sanit. 4:*45 (1953).
(44) Ersepke, Z., and J. Baranek, "Analyzer for Chlorine Content of Air," *Chemicky Pramysl* (Prague) *16*(8):496 (1966).
(45) Lawrence Berkeley Laboratory, "Instrumentation for Environmental Monitoring - Air," *Report LBL-1, Vol. 1,* Berkeley, Univ. of Calif. (Feb. 1973).
(45a) Amer. Public Health Assoc., *Methods of Air Sampling and Analysis,* Washington, D.C. (1972).
(46) Shults, W.D. and Kuempel, J.R.; U.S. Patent 3,713,994; January 30, 1973; assigned to U.S. Atomic Energy Commission.
(47) Lerner, M.L.; U.S. Patent 3,748,097; July 24, 1973; assigned to Zenith Radio Corp.
(48) Mang, O.K.C.; U.S. Patent 3,761,377; September 25, 1973.
(49) Lawrence Berkeley Laboratory, "Instrumentation for Environmental Monitoring - Water," *Report LBL-1,* Vol. 2, Berkeley, Univ. of Calif. (Feb. 1973).
(50) Morrow, J.J., Jr; U.S. Patent 3,413,199; November 26, 1968; assigned to Fischer and Porter Co.

CHROMIUM

The exposure of industrial workers to airborne chromium compounds and chromic acid mists, particularly the hexavalent chromates, has been observed to produce irritation of the skin and respiratory tract, dermatitis, perforation of the nasal septum, ulcers, and cancer of the respiratory tract. Chromium metal is thought to be nontoxic. Hexavalent compounds appear to be much more harmful than trivalent compounds, with the toxic effects depending on solubility. Two effects that appear to be particularly important in relation to air pollution are hypersensitivity to chromium compounds and induction of cancers in the respiratory tract. Exposure of industrial workers in the chromate-producing industry has shown an incidence of deaths from cancer of the respiratory tract which is over 28 times greater than expected. Time concentration relationships for induction of cancer are not known, according to Sullivan (1).

No evidence of damage by airborne chromium to animals or plants has been found. Chromic acid mists have discolored paints and building materials. In 1964, atmospheric concentrations of total chromium averaged 0.015 $\mu g./m.^3$ and ranged as high as 0.350 $\mu g./m.^3$. Although the exact sources of chromium air pollution are not known some possible sources are the metallurgical, refractory, and chemical industries that consume chromite ore; chemicals and paints containing chromium; and cement and asbestos dust. Particulate control methods should be adequate for chromium-containing particles. No information has been found on the economic costs of chromium air pollution or on the costs of its abatement.

Chromium occurs in nature as both the trivalent (Cr^{+3}) and the hexavalent (Cr^{+6}) ion. Hexavalent chromium present in industrial wastes is primarily in the form of chromate ($CrO_4^{=}$) and dichromate ($Cr_2O_7^{=}$). Chromium compounds are added to cooling water to inhibit corrosion. They are employed in manufacture of paint pigments, in chrome tanning, aluminum anodizing and other metal cleaning, plating and electroplating operations. In the metal plating industry, automobile parts manufacturers are one of the largest producers of chromium-plated metal parts. Frequently, the major source of waste chromium is the chromic acid used in such metal plating operations (2). Table 22 is a compilaton of typical sources and concentrations of chromate wastewaters.

TABLE 22: Cr^{+6} WASTEWATER SOURCES AND TYPICAL CONCENTRATIONS

	Chromium (VI) Concentration, mg./l.		
Industrial Source	Average	Range	Reference
Leather Tanning	40		(3)
Wood Preserving		0.23 - 1.5	(4)
Cooling Tower Blowdown	31.4		(5)
Cooling Tower Blowdown		8 - 10.7	(6)
Cooling Tower Blowdown		10 - 60	(3)
Bright Dip Rinse		1 - 6	(7)
Bright Dip Bath		10,000 - 50,000	(3)
Bright Dip Bath		20,000 - 75,000	(8)
Bright Dip Bath		200 - 600	(8)
Anodizing Bath	173		(9)
Anodizing Bath		15,000 - 52,000	(8)
Anodizing Rinse	49		(9)
Anodizing Rinse		30 - 100	(8)
Plating	1,300		(10)
Plating	600		(3)
Plating		100,000 - 270,000	(8)
Plating		60 - 80	(11)
Electroplating	140		(12)
Electroplating	41	15 - 70	(13)

Source: Report PB 216,162.

Chromium in industrial wastes occurs predominantly as the hexavalent (Cr^{+6}) form, in chromate ($CrO_4^{=}$) and dichromate ($Cr_2O_7^{=}$) ions. Hexavalent chromium treatment frequently involves reduction to the trivalent (Cr^{+3}) form prior to removing the chromium from the industrial waste. Thus trivalent chromium in industrial waste may result from one step of the waste treatment process itself, that of chemical reduction of hexavalent chromium. Industries which employ trivalent chromium directly in manufacturing processes include glass, ceramics, photography and textile dyeing.

In addition to the predominant hexavalent form, some few mg./l. of trivalent chromium may be encountered in plating wastes, even prior to chemical reduction of Cr^{+6}. For example, Anderson and Iobst (14) report for one process waste a hexavalent chromium concentration ranging from 0.0 to 18.0 mg./l. and a total chromium concentration of 2.9 to 31.8 mg./l. By difference, their data indicate a trivalent chromium concentration in this one waste stream of 2.9 to 13.9 mg./l. These same authors report for an acid bath waste stream a hexavalent chromium content of 122 to 270 mg./l. and trivalent levels of 37 to 282 mg./l. Thus, according to these authors, trivalent chromium represents a major component of this acid waste, even before reduction of hexavalent chromium. After reduction, the trivalent content of this latter waste stream would be approximately 160 to 550 mg./l. Evaluation of similar plating rinse data presented by Germain, et al (15), indicates trivalent chromium levels of 2.5 to 15 mg./l. prior to treatment.

Hexavalent chromium has been considered a toxic metal for many years. Trivalent chromium is less toxic although it raises the glucose tolerance of rats and rabbits (16). However, it is not clear whether either is an essential element for man. Schroeder et al (16) feels that chromium has not been definitely proven to be a carcinogen. The toxicity of hexavalent chromium to fish is fairly high in that 1.2 to 5.0 ppm is fatal to many species (16).

Considerable evidence does exist to show that chromium in the atmosphere inhaled by man can be a carcinogen. Hill and Warden (17) suggest that chromyl chloride may play an important role in the formation of lung cancer in chromate workers. Hueper (18) pointed out that the existence of a serious lung cancer hazard for producers of chromates from chromite ore and tar handlers of certain chromium pigments (zinc chromate, barium chromate and lead chromate) is well documented among workers in these plants.

Hueper also reported the development of squamous cell carcinomas of the lung and sarcomas of the lung and of the soft tissues of the thigh of rats which had received intrapleural or intramuscular implants of powdered chromite ore roast suspended in sheep fat. Shelley (19) has pointed out that chromium in welding fumes was the cause of eczematous eruptions of the palms of the workers.

Calcium chromate, sintered calcium chromate and sintered chromium trioxide when introduced in pellet form into the pleural cavity of rats produce cancers, mainly sarcomas, in the majority of animals at the site of implantation within 14 months (18). This effect is related to the solubility of these compounds in the watery medium. It has been reported that as little as 0.7 ppm of chromium in the water can be toxic to some aquatic insects. Thus it can be concluded that chromium may not be an essential element to man, that it may be a carcinogen and that hexavalent chromium is toxic to man at fairly low levels.

The acute minimal lethal dose in dogs for chromic acid is stated as 330 mg./kg. for intravenous sodium chromate, corresponding to 75 mg. of the element. The latter salt causes preterminal hypotension, hypocholesteremia, and hyperglycemia. Chronic toxicity was observed in several species with Cr^{+6} in concentrations of more than 5 ppm in the drinking water. Subacute toxicity and accumulation of chromium has been found (20) in the tissues of rabbits that were given 5 ppm (as chromate) in the drinking water. The element accumulation also occurred in rats when given at this level, but caused no changes in the growth rates, food intake, or blood analysis. Even 25 ppm in the drinking water failed to produce changes in these parameters or in the histological appearance of the tissues after 6 months. Dogs tolerated up to 11.2 ppm of Cr^{+6} in the water for 4 years without any ill effects. Growing chickens showed no detrimental symptoms when they were fed 100 parts per million in the diet.

Toxicity of trivalent chromium appears to be restricted to parenteral administration. No reports of oral toxicity of trivalent chromium are known. All these results point out the very low toxicity of chromium except for poorly soluble chromates directly implanted in tissues in very high amounts or deposited in the respiratory tract. The therapeutic:toxic ratio for intravenously injected Cr^{+6} is approximately 1:10,000.

Chromium is poorly absorbed in the gastrointestinal tract, the hexavalent better than the trivalent form. Affinity for testes, bones, liver, and spleen is high, that for muscle and brain, low. Although poorly soluble chromate dusts lead to increased incidence of respiratory diseases, the toxicity of soluble hexavalent and trivalent compounds is very low.

The permissible concentration of hexavalent chromium in domestic water supplies is 0.05 ppm as shown earlier in Table 3. The amount of chromium which can be tolerated in irrigation waters is 5.0 ppm for continuous use and 20.0 ppm for short term use in fine textured soil as shown earlier in Table 5. The maximum allowable concentration of chromium in an aqueous effluent discharged to a storm sewer or stream should be 0.5 ppm for hexavalent chromium and 1.0 ppm for trivalent chromium.

Measurement in Air

Sampling Methods: Dusts and fumes of chromium compounds may be collected by any method suitable for collection of other dusts and fumes; the impinger, electrostatic precipitator, and filters are commonly used. The National Air Sampling Network uses a high-volume filtration sampler (21). Chromic acid mists may be collected in an impinger using water or caustic solutions (22).

Quantitative Methods: Emission spectroscopy has been used by the National Air Pollution Control Administration for chromium analysis of samples from the National Air Sampling Network (21)(23). The samples are ashed and extracted to eliminate interfering elements. The minimum detectable chromium concentration by emission spectroscopy is 0.0064 $\mu g./m.^3$ for urban samples and 0.002 $\mu g./m.^3$ for nonurban samples. The different sensitivities result from different extraction procedures required for urban samples (24).

Thompson et al (21) have reported that the National Air Pollution Control Administration uses atomic absorption to supplement results obtained by emission spectroscopy. The method has a minimum detectable limit of 0.002 $\mu g./m.^3$ based on a 2,000 $m.^3$ air sample.

West (22) indicates that the ring oven technique for the determination of chromium is sensitive to 0.15 $\mu g.$ and has a normal range of 0.3 to 1.0 $\mu g.$ The chromium is oxidized to the hexavalent state, and the method, therefore, measures total chromium. Diphenylcarbazide is used as the indicator.

A multichannel flame spectrometer has been used by Iida and Fuwa (25)(26) to measure magnesium, calcium, copper, manganese, and chromium simultaneously in biological materials. Their initial work indicated a sensitivity of 1.0 $\mu g./ml$; however, phosphates seriously interfered with the application of the analysis (25). In subsequent studies (26), they added 8-hydroxyquinoline and perchloric acid, which not only eliminated the interference, but increased the sensitivity of the test to two of the elements, calcium and manganese (0.1 $\mu g./ml.$).

The concentration of chromic acid mists in air can be estimated by a direct field method described by Ege and Silverman (27)(28)(29). This is a spot-test method using phthalic anhydride and sym-diphenylcarbazide. A coulometric method for determining chromium in biological materials is described by Feldman et al (30). The method is sensitive to 0.2 $\mu g.$ of chromium.

Pierce and Cholak (31) have described an atomic absorption method for the determination of chromium in biological systems. The sample must be ashed. The method is sensitive to 0.1 $\mu g.$ of chromium or 0.05 $\mu g./g.$

The Mine Safety Appliances Co. (32)(33) has a lightweight sampler for estimating quickly the chromic acid mist concentrations in the atmosphere. The operating principle of this instrument is based on the method described above by Ege and Silverman (27)(28)(29).

Measurement in Water

Hexavalent chromium may be measured by a Technicon CSM-6 in a normal range of 0 to 5 ppm and down to a detection limit of 0.05 ppm as shown earlier in Table 12. Conventional flame atomic absorption may be used to measure chromium down to a detection limit of 7 ppb as shown earlier in Table 11.

A colorimetric analysis technique for chromates and dichromates has been described by V.M. Marcy and S.L. Halstead (34). According to these investigators, the best known and most widely used indicator for chromate is 1,5-diphenylcarbohydrazide (hereinafter called DPCH), which forms a pink or reddish-violet color in the presence of chromate ion. The intensity of color formed in a test solution treated by a known amount of DPCH is proportional to the concentration of chromate ion in it. No more useful indicator for chromate has been found. A great disadvantage to the use of DPCH as an indicator, however, has been its instability in virtually all media heretofore used as a solvent for it.

In research leading to this process, it was found that the use of glycerine or related polyhydric alcohol as the solvent for DPCH enables the preparation of an indicator solution for chromate having hitherto unrealized stability. An excellent indicator solution can be prepared, for example, by dissolving 2.5 g. of DPCH in 750 ml. of glycerine heated to about 110°C. Fifty ml. of 85% orthophosphoric acid is then added and the mixture diluted to one liter with distilled water.

A process developed by C.A. Noll, E.C. Feddern and L.J. Stefanelli (35) permits determining the chromium concentration in an aqueous medium containing either or both trivalent chromium or hexavalent chromium. The method generally entails acidifying the aqueous medium, oxidizing any trivalent chromium present in the aqueous medium to hexavalent chromium, and adding to the medium a composition comprising a water-soluble cobaltous salt and a chelating agent of the acetic acid derivative type. The color intensity of the resulting medium is then measured and compared to the intensity of samples of aqueous medium containing known quantities of chromate in its hexavalent state which have been treated in the same manner. Chromium may be monitored in water by the following instrumental techniques (36):

Chemiluminescence
Electrochemical
Spark Source Mass Spectrometry
X-ray Fluorescence

References

(1) Sullivan, R.J., "Air Pollution Aspects of Chromium and Its Compounds," *Report PB 188,075,* Springfield, Va., Nat. Tech. Info. Service (Sept. 1969).
(2) Patterson, J.W. and Minear, R.A., "Wastewater Treatment Technology," 2nd Ed., *Report PB 216,162,* Springfield, Va., Nat. Tech. Info. Service (Feb. 1973).
(3) P.N. Cheremisinoff and Y.H. Habib, "Cadmium, Chromium, Lead, Mercury: A Plenary Account for Water Pollution, Part I - Occurrence, Toxicity and Detection," *Water and Sewage Works,* 119:(July) 73-85, 1972.
(4) E.H. Teer and L.V. Russell, "Heavy Metal Removal from Wood Preserving Wastewater," presented at 27th Indust. Waste Conf., Purdue Univ., 1972.
(5) J.A. Landy, "Chromate Removal at a Saudi Arabian Fertilizer Complex," *Jour. Wat. Poll. Control Fed.,* 43, 1971.
(6) E.W. Richardson, E.D. Stobbe, and S. Bernstein, "Ion Exchange Traps Chromates for Reuse," *Envir. Sci. Tech.,* 2:1006-1016, 1968.
(7) E.J. Donovan, Jr., "Treatment of Wastewater for Steel Cold Finishing Mills," *Water Wastes Engng.,* (November) F22-F25, 1970.
(8) W. Lowe, "The Origin and Characteristic of Toxic Wastes, with Particular Reference to the Metal Industry," *Wat. Poll. Control* (London), 1970:270-280.
(9) J. Germain, C. Vath, and C. Griffin, "Solving Complex Waste Disposal Problems in the Metal Finishing Industry," presented at Georgia Wat. Poll. Central Assoc. Meeting, Sept., 1968.

(10) N.H. Hansen and W. Zabben, "Design and Operation Problems of a Continuous Automatic Plating Waste Treatment Plant at the Data Processing Division, IBM, Rochester, Minnesota," Proc. 14th Purdue Industrial Waste Conf., pp. 227-249, 1959.
(11) V. Crowle, "Effluent Problems as They Affect the Zinc Die-Casting and Plating-on-Plastics Industries," *Metal Finish. Jour.* 17:194:51-54, 1971.
(12) P.I. Avrutskii, "Control of Chromium (VI) Concentration in Waste Waters," *Chem. Abst.* 70:206-207, 1969.
(13) Anonymous, "An Investigation of Techniques for Removal of Chromium from Electroplating Wastes," *U.S. EPA Publication 12010 EIE* 13/71, 1971.
(14) J.S. Anderson and E.H. Iobst, Jr., "Case History of Wastewater Treatment in a General Electric Appliance Plant," *Jour. Wat. Poll. Control Fed.,* 10:1786-1795, 1968.
(15) J.E. Germain, C.A. Vath, and C.F. Griffin, "Solving Complex Waste Disposal Problems in the Metal Finishing Industry," presented at Georgia Wat. Poll. Control Assoc. Meeting, Sept., 1968.
(16) Schroeder, H.A., Balassa, J.J. and Tipton, I.H., *J. Chron. Dis.* 15, 941 (1961).
(17) Hill, W.H. and Worden, F.X., *Am. Ind. Hyg. Assoc. J. 23,* 186 (1962).
(18) Hueper, W.C. et al, *Am. Ind. Hyg. Assoc. J. 21,* 530 (1960).
(19) Shelley, W.B., *J. Am. Med. Assoc. 189,* 170 (1964).
(20) Cothern, C.R., "Determination of Trace Metal Pollutants in Water Resources and Stream Sediments," *Report PB 213,369;* Springfield, Va., Nat. Tech. Info. Service (1972).
(21) Thompson, R.J., G.B. Morgan, and L.J. Purdue, "Analyses of Selected Elements in Atmospheric Particulate Matter by Atomic Absorption," Preprint. Presented at the Instrument Society of America Symposium, New Orleans, La. (May 5-7, 1969).
(22) West, P.W., "Chemical Analysis of Inorganic Pollutants," Chapter 19 in *Air Pollution,* Vol. II, 2nd ed., A.C. Stern, Ed. (New York: Academic Press, 1968).
(23) "Air Pollution Measurements of the National Air Sampling Network - Analyses of Suspended Particulates," 1957-1961, *U.S. Public Health Serv. Publ. 978* (1962).
(24) "Air Quality Data from the National Air Sampling Networks and Contributing State and Local Networks," 1966 ed., U.S. Dept. of Health, Education, and Welfare, *National Air Pollution Control Administration Publication No. APTD 68-9,* U.S. Govt. Printing Office, Washington, D.C. (1968).
(25) Iida, C., and K. Fuwa, "Studies with a Multichannel Flame Spectrometer, Part I," *Anal. Biochem. 21:*1 (1967).
(26) Iida, C., and K. Fuwa, "Studies with a Multichannel Flame Spectrometer, Part II," *Anal. Biochem. 21:*9 (1967).
(27) Ege, J.F., Jr., and L. Silverman, "Stable Colorimetric Reagent for Chromium," *Anal. Chem. 19:*693 (1947).
(28) Silverman, L., and J.F. Ege, Jr., "A Rapid Method for Determination of Chromic Acid Mist in Air," *J. Ind. Hyg. Toxicol. 29:*136 (1947).
(29) Silverman, L., and J.F. Ege, Jr., "Chromium Compounds in Gaseous Atmospheres," U.S. Patent 2,483,108 (1949); *CA-Bull. Cancer Progr. 44:*490 (1950).
(30) Feldman, F.J., G.D. Christian, and W.C. Purdy, "Coulometric Determination of Chromium in Biologic Materials," *Am. J. Clin. Pathol. 49*(6):826 (1968).
(31) Pierce, J.O., and J. Cholak, "Lead, Chromium, and Molybdenum by Atomic Absorption," *Arch. Environ. Health 13*(2):208 (1966).
(32) "MSA Chromic Acid Mist Detector," *Mine Safety Appliances Co. Bull. 0811-8* (1959).
(33) "MSA Samplair," in *Catalog of Industrial Safety Equipment,* Mine Safety Appliances Co. Cat. 7-B, Sect. 3, p. 38, Pittsburgh, Pa. (1957).
(34) Marcy, V.M. and Halstead, S.L., U.S. Patent 3,216,950; November 9, 1965; assigned to Calgon Corp.
(35) Noll, C.A., Feddern, E.C. and Stefanelli, L.J., U.S. Patent 3,656,908; April 18, 1972; assigned to Betz Laboratories, Inc.
(36) Lawrence Berkeley Laboratory, "Instrumentation for Environmental Monitoring - Water," *Publ. No. LBL-1,* Vol. 2, Berkeley, Univ. of Calif. (Feb. 1973).

COBALT

Inhalation of metallic cobalt dust may cause pulmonary symptoms. Contact with the powder may cause dermatitis. Ingestion of the soluble salts may produce nausea and vomiting by local irritation. As shown earlier in Table 5, the allowable cobalt concentration in irrigation water is 0.2 ppm on a continuous use basis and 10.0 ppm on a short term use basis in fine textured soil.

Measurement in Air

A rapid, selective procedure for the determination of 1 to 20 μg. of cobalt in airborne particulates has been described by W.F. Davis and J.W. Graab of NASA's Lewis Research Center (1).

The method utilizes the combined techniques of low-temperature ashing and atomic absorption spectroscopy. The airborne particulates are collected on analytical filter paper. The filter papers are ashed, and the residues are dissolved in hydrochloric acid. Nickel, chromium, and cobalt are determined directly with good precision and accuracy by means of atomic absorption. The effects of flame type, burner height, slit width, and lamp current on the atomic absorption measurements are reported.

Measurement in Water

In the conventional analytic determination of cobalt ions, a blue colored solution is obtained with a salt of thiocyanic acid. This conventional indicator, however, has several drawbacks. In the first place, the test is only qualitative and is impaired by the presence of copper ions therein which yield a green or brown color with the test paper.

Furthermore, high zinc ion concentrations, for example, 1,000 mg./l., prevent the detection of smaller amounts, e.g., 10 to 25 mg./l., of cobalt (II) ions. In addition, cobalt ions cannot be detected if permanganate ions are also present therein. Further, when an absorbent carrier is impregnated with a solution containing salts of thiocyanic acid, a concentration gradient is produced, thus making a semiquantitative determination of the cobalt ions impossible. Consequently, the field of application of this conventional indicator is very limited.

There thus remains a need for an indicator composition and a process for effectively and efficiently performing both qualitative and semiquantitative tests on a cobalt containing solution which can be carried out in a very short period of time and which do not exhibit the aforementioned disadvantages of conventional indicators.

According to D. Schmitt, A. Stein and W. Baumer (2), these prior art problems can be overcome by an improved indicator composition. This improved indicator for the colorimetric detection of cobalt ions is formed by impregnating an absorbent carrier with a composition comprising a water-soluble rhodanide salt, an alkali-metal or ammonium thiosulfate, an alkali metal fluoride and one or more of emulsifier and/or wetting agent, a tertiary amine and a quaternary ammonium salt.

This indicator ensures identical readings at the same cobalt concentration, because identical colorings of the indicator are obtained. Thus, a comparison with a color scale can easily be made, thereby permitting semiquantitative tests on a solution containing cobalt ions. Cobalt can also be monitored in water by atomic absorption techniques down to a detection limit of 16 ppb as shown earlier in Table 11.

References

(1) Davis, W.F. and Graab, J.W., *Report NASA-TM-X-2672* (Nov. 1972).
(2) Schmitt, D., Stein, A. and Baumer, W., U.S. Patent 3,697,225; October 10, 1972; assigned to Merck Patent GmbH, Germany.

COPPER

Copper and copper compounds are emitted to the atmosphere, primarily as a result of the metallurgical processing of primary copper and to a lesser extent from the production of iron and steel. The only other significant emission source is the combustion of coal. Quantitative estimates of these various emissions have been made by W.E. Davis and Associates of Leawood, Kansas (1).

Primary sources of copper in industrial wastewater streams are metal process pickling baths and plating baths. Brass and copper metal working requires periodic oxide removal by immersing the metal in strong acid baths. Solution adhering to the metal surface, referred to as "dragout," is rinsed from the metal and contaminates the waste rinse water. Similarly, metal parts undergoing copper or brass plating drag out some of the concentrated plating solution. Copper concentrations in the plating bath depend upon the bath type, and may range from 3,000 to 50,000 mg./l. (2).

For a given bath, the rinse water concentration will be a function of many factors including drainage time over the bath, shape of the parts and their total surface area and the rate of rinse water flow. One author (3) has suggested a typical rinse water level would be 1% of the process bath concentration. Untreated process wastewater concentrations of copper typical of plating and metal processing operations are summarized in Table 23.

Jewelry manufacturers employ copper plating either directly or as a base metal for silver and other precious metal surfaces. Copper is also employed in the alkaline Bemberg rayon process, as cupro-ammonium salts. Copper bearing acid mine drainage also contributes significant quantities of dissolved copper to waste streams. Hill (18) has reported copper concentrations of 3.2 and 3.9 mg./l. for two mines, while another discharge contained only 0.12 mg./l. Wixson, et al (19), have cited mine waters, mine tailings ponds and acid mine drainage (primarily with respect to lead mining) as possible sources of copper.

Teer and Russel (20) have reported wood preserving to generate copper containing aqueous wastes. Dean, et al (21), indicated additional potential sources of copper bearing wastes, which included pulp and paper mills, paper and paper board mills, fertilizer manufacturing, petroleum refining, basic steel works and foundaries, nonferrous metal-works and foundaries, and motor vehicle and aircraft plating and finishing.

TABLE 23: CONCENTRATIONS OF COPPER IN INDUSTRIAL WASTEWATERS

Process	Copper Concentration, mg./l.	Reference
Plating Wash	20 - 120	(4)
Plating Wash	0 - 7.9	(5)
Brass Dip	2 - 6	(5)
Brass Mill Rinse	4.4 - 8.5	(6)
Copper Mill Rinse	19 - 74	(6)
Metal Processing	204 - 370	
Brass Mill Wash		(7)
Tube Mill	74	
Rod and Wire Mill	888	
Brass Mill Bichromate Pickle		(7)
Tube Mill	13.1	
Rod and Wire Mill	27.4	
Rolling Mill	12.2	
Copper Rinse	13 - 74	
Brass Mill Rinse	4.5	
Appliance Manufacturing		(8)
Spent Acids	0.6 - 11.0	
Alkaline Wastes	0 - 1.0	
Typical Large Plater		(9)
Rinse Waters	up to 100 (20 avg.)	
Plating Operations	6.4 - 88	(10)
Automobile Heater Production	24 - 33 (28 avg.)	(11)

(continued)

TABLE 23: (continued)

Process	Copper Concentration, mg./l.	Reference
Silver Plating		(12)
Silver Bearing	3 - 900 (12 avg.)	
Acid Wastes	30 - 590 (135 avg.)	
Alkaline Wastes	3.2 - 19 (6.1 avg.)	
Brass Industry		(12)
Pickling Bath Wastes	4.0 - 23	
Bright Dip Wastes	7.0 - 44	
Business Machine Corp.		(12)
Plating Wastes	2.8 - 7.8 (4.5 avg.)	
Pickling Wastes	0.4 - 2.2 (1.0 avg.)	
Copper Plating Rinse Water	5.2 - 41	(13)
Copper Tube Mill Waste	70 (avg.)	(14)
Copper Wire Mill Waste	800 (avg.)	(14)
Plating Plants	2.0 - 36.0	(15)
Brass and Copper Wire Mill	75 - 124	(16)
Copper and Brass Pickle	60 - 9	(17)
Copper and Brass Bright Dip	20 - 35	(17)
Plating	20 - 30	(17)
Plating	10 - 15	(17)
Plating	3 - 8	(17)
Large Plater	11.4	(8)

Source: Report PB 216,162.

Forage with too much copper can cause fatal hemolytic anemia and hepatic necrosis in cattle and sheep. Copper toxicity is a fundamental cause of Wilson's disease (22). It has been recognized that this disorder is inherited as an autosomal recessive trait and is the only example of significant copper toxicity in man. This has made clear that man has had to develop a mechanism, transmitted from one generation to the next which prevents copper poisoning in all the rest of us.

The geographic location has influence on the biological priorities related to copper intake. In Florida cattle with 60 ppm Cu on dry weight basis were often showing severe diarrhea, achromotrichia, bone change, and anemia and severe reproductive disturbances. Yet in the Netherlands and in Oregon values of 25 ppm Cu are found in the liver with the animals apparently in normal health (23).

Although not much information appears to be available concerning the effects of copper in water supplies, it is known that 0.01 to 1.7 ppm Cu is the toxic range for fish and as little as 0.027 ppm Cu can be toxic to aquatic insects (24).

Copper in ionic form is very toxic to the photosynthesis of the green alga, *Chlorella pyrenoidosa* and the diatom, *Nitzschiz palea,* in concentrations of copper normally found in natural waters, indicating that copper is not ordinarily present in ionic form but is complexed to organic matter such as polypeptides (25).

Copper concentrations of 0.04 ppm were acutely toxic to chinook salmon fry and concentrations of 0.02 ppm increased mortality and inhibited growth. The maximum acceptable concentration for the flathead minnow is between 0.03 and 0.08 ppm (26). The survival of gammarus, an invertebrate, was markedly decreased between 12.9 and 6.2 ppb. Newly hatched amphipods grew to the adult stage only in concentrations less than or equal to 4.6 ppb (27).

Concentrations between 0.56 and 3.2 ppm resulted in the fatty metamorphosis of the liver, necrosis in the kidney and gross changes in gill architecture in the winter flounder.

A concentration of 0.18 ppm produced and extracted appearance in the gill lamellae. The epithelial layer appeared vacuolated, the basilamellar region was reduced in thickness and the lamellar mucus cells were few, while chloride cells appeared in their stead. Under an electron microscope, vacuolation in epithelial vesicles, myelin like figures, various membrane bound vesicles and apical homogeneous layers of reduced thickness were evident. Increased amounts of particulate matter adhering to the external surface of the epithelial cells were also apparent (28).

The permissible concentration of copper in domestic water supplies is 1.0 ppm; it is desirable, however, that it be virtually absent as noted earlier in Table 3. The allowable copper concentration in irrigation water is 0.2 ppm on a continuous use basis and is 5.0 ppm on a short term use basis in fine textured soil as noted earlier in Table 5. The maximum allowable concentration of copper to be discharged to a storm sewer or stream is less than 0.5 ppm.

Measurement in Air

Copper may be monitored in air by the following instrumental techniques (29):

Emission Spectroscopy
Atomic Absorption

Measurement in Water

Copper may be monitored in water by the following instrumental techniques (30):

Atomic Fluorescence
Electrochemical
Spark Source Mass Spectrometry
X-ray Fluorescence

Copper may be measured using the Technicon CSM-6 in the 0 to 10 ppm range down to a detection limit of 0.1 ppm as shown earlier in Table 12. Copper may be measured down to a detection limit in the 1 to 5 ppb range by a variety of techniques as indicated earlier in Table 11. Those techniques include:

Flame Atomic Absorption
Flameless Atomic Absorption
Neutron Activation
Polarography

The determination of copper by a spectrofluorometric method has also been reported by K.H. Mancy (31). The use of tetrachlorotetraiodofluorescein-o-phenanthroline as an indicator gave a sensitivity of 0.001 μg./ml. with only cyanide ion interfering.

References

(1) Davis, W.E. and Assoc., *Report PB 210,678,* Springfield, Va., Nat. Tech. Info. Service (May 1972).
(2) Patterson, J.W. and Minear, R.A., "Wastewater Treatment Technology," *Report PB 216,162,* Springfield, Va., Nat. Tech. Info. Service (Feb. 1973).
(3) A. Golomb, "Application of Reverse Osmosis to Electroplating Waste Water Treatment. Part II. The Potential of Reverse Osmosis in the Treatment of Some Plating Wastes," *Plating,* 59:316-319, 1972.
(4) W.A. Parsons and W. Rudolfs, "Lime Treatment of Copper Pyrophosphate Plating Wastes," *Sew. Ind. Wastes Eng.,* 22:313-315, 1951.
(5) R.W. Simpson and K. Thompson, "Chlorine Treatment of Cyanide Wastes," *Sew. Ind. Wastes Eng.,* 21:302-304, 1950.
(6) F.X. McGarvey, "The Application of Ion Exchange Resins to Metallurgical Waste Problems," *Proc. 7th Purdue Industrial Waste Conf.,* 7:289-304, 1952.

(7) F.X. McGarvey, R.E. Tenhoor, and R.P. Nevers, "Brass and Copper Industry: Cation Exchangers for Metal Concentration from Pickle Rinse Waters," *Ind. Eng. Chem.,* 44:534-541, 1952.
(8) J.S. Anderson and E.H. Iobst, Jr., "Case History of Wastewater Treatment in a General Electric Appliance Plant," *Jour. Wat. Poll. Control Fed.,* 40:1786-1795, 1968.
(9) H.L. Pinkerton, "Waste Disposal," in *Electroplating Engineering Handbook,* 2nd ed., A. Kenneth Graham, ed.-in-chief, Reinhold Pub. Co., New York, 1962.
(10) W.S. Wise, "The Industrial Waste Problem, IV. Brass and Copper Electroplating and Textile Wastes," *Sew. Ind. Wastes,* 20:96-102, 1948.
(11) C.M. Gard, C.A. Snavely, and D.J. Lemon, "Design and Operation of a Metal Wastes Treatment Plant," *Sew. Ind. Wastes,* 23:1429-1438, 1951.
(12) N.L. Nemerow, *Theories and Practices of Industrial Waste Treatment,* Addison Wesley, Reading, Mass. (1963).
(13) G.E. Barnes, "Disposal and Recovery of Electroplating Wastes," *Jour. Wat. Poll. Control Fed.,* 40:1459-1470, 1968.
(14) J.A. Tallmange, "Nonferrous Metals," in *Chemical Technology, Vol. 2, Industrial Waste Water Control,* C. Fred Gurnham, ed., Academic Press, New York, 1965.
(15) Battelle Columbus Laboratories, "An Investigation of Techniques for Removal of Cyanide from Electroplating Wastes," *U.S. EPA Report 12010 EIE 11/71,* 1971.
(16) Volco Brass and Copper Co., "Brass Wire Mill Process Changes and Waste Abatement, Recovery and Reuse," *U.S. EPA Report 12010 DPF 11/71,* 1971.
(17) W. Lowe, "The Origin and Characteristics of Toxic Wastes with Particular Reference to the Metals Industries," *Wat. Poll. Control* (London), 1970:270-280.
(18) R.D. Hill, "Control and Prevention of Mine Drainage," presented at the Environmental Resources Conference on Cycling and Control of Metals, Battelle Columbus Laboratories, 1972.
(19) B.G. Wixson, E. Bolter, N.L. Gale, J.C. Jennett, and K. Purushothaman, "The Lead Industry as a Source of Trace Metals in the Environment," presented at the Environmental Resources Conference on Cycling and Control of Metals, Battelle Columbus Laboratories, 1972.
(20) E.H. Teer and L.V. Russell, "Heavy Metals Removal from Wood Preserving Wastewater," presented at the 27th Indust. Waste Conf. Purdue Univ., 1972.
(21) J.G. Dean, F.L. Bosqui and K.H. Lanouette, "Removing Heavy Metals from Waste Water," *Env. Sci. Technol.,* 6:518-522, 1972.
(22) Scheinberg, Z.H., *Proc. Third Annual Conf. on Trace Substances in Envir. Health,* p. 79, Columbia, Mo. (1969).
(23) Davis, G.K., *Proc. Third Annual Conf. on Trace Substances in Envir. Health,* p. 135, Columbia, Mo. (1969).
(24) Warnick, S.L. and Bell, H.L., *J. Water Pollution Control Fed.* 41, 280 (1969).
(25) Nielsen, E.S., *Marine Biology* 6, 2 (1970).
(26) Mount, R. and Stephan, C., *J. Fishery Reserve Board of Canada* 26 (1969).
(27) Arthur, J. and Leonard, E., *J. Fishery Reserve Board of Canada* 27 (1970).
(28) Baker, J.J., *J. Fishery Reserve Board of Canada* 26 (1969).
(29) Lawrence Berkeley Laboratory, "Instrumentation for Environmental Monitoring - Air," *Publ. LBL-1,* Vol. 1, Berkeley, Univ. of Calif. (Feb. 1973).
(30) Lawrence Berkeley Laboratory, "Instrumentation for Environmental Monitoring - Water," *Publ. LBL-1,* Vol. 2, Berkeley, Univ. of Calif. (Feb. 1973).
(31) Mancy, K.H., *Instrumental Analysis for Water Pollution Control,* Ann Arbor, Mich., Ann Arbor Science Publishers (1971).

CYANIDES

Cyanides, such as hydrogen cyanide, may be encountered as air pollutants from industrial operations. For example, as noted by M. Sittig (1), the stripping of waste waters from coke oven plants can produce gaseous effluents containing HCN. Cyanide, as sodium cyanide (NaCN) or hydrocyanic acid (HCN) is a widely used industrial material. Cyanide waste streams result from ore extracting and mining, photographic processing, coke furnaces, synthetics manufacturing, case hardening and pickling of steel, and industrial gas scrubbing. A major source of waste cyanide is the electroplating industry. Electroplaters use cyanide baths to hold metallic ions such as zinc and cadmium in solution. "Dragover" of the plating solution, containing cyanide ions and metal-cyanide complexes, contaminates rinsing baths. Table 24 presents typical cyanide levels reported for plating wastewaters.

TABLE 24: CONCENTRATIONS OF CYANIDE IN PLATING WASTEWATERS

Process	Average, mg./l.	Range, mg./l.	References
Plating Rinse	2	0.3 - 4	(2)
Plating Rinse	700		(3)
Plating Rinse		10 - 25	(4)
Plating Rinse	32.5		(5)
Plating Rinse	25		(6)
Plating Rinse		60 - 80	(7)
Plating Rinse		30 - 50	(8)
Plating Rinse	55.6	1.4 - 256	(2)
Bright Dip		15 - 20	(8)
General (Separate Cyanide)	72	9 - 115	(9)
General (Combined Stream)	28	1 - 103	(9)
Alkaline Cleaning Bath		4,000 - 8,000	(8)
Plating Bath		45,000 - 100,000	(10)
Plating Bath			(8)
Brass		16,000 - 48,000	
Bronze		40,000 - 50,000	
Cadmium		20,000 - 67,000	
Copper		15,000 - 52,000	
Silver		12,000 - 60,000	
Tin - Zinc		40,000 - 50,000	
Zinc		4,000 - 64,000	

Source: Report PB 216,162.

The permissible concentration of cyanides in public waters is 0.20 mg./l. and it is desirable that cyanide be completely absent. The maximum allowable concentration of cyanides in effluents to storm sewers or streams was 0.05 ppm as of 1971. As of Dec. 29, 1973, the EPA has issued new toxic pollutant effluent standards covering cyanides. They say that into streams having a flow of less than 10 cu. ft./sec. or lakes having an area of less than 500 acres, there shall be no cyanides discharged. Bodies exceeding these flow and area limits (or salt water bodies) may receive 100 μg./l. of cyanides (approximately 0.1 ppm). However, when stream flow is less than ten times effluent flow, those allowable limits are to be reduced by a factor of ten.

Measurement in Air

A number of methods for quantitative determination of hydrogen cyanide in air have been pinpointed by W.E. Ruch in "Quantitative Analyses of Gaseous Pollutants," Ann Arbor, Mich., Ann Arbor Science Publishers, Inc. (1970).

For example, cyanides in the ppm range may be determined by reacting the sample with chloramine-T to produce cyanogen chloride. The cyanogen chloride in turn may be reacted with nicotinamide to give a fluorescent blue solution whose color can be compared with similar solutions made from known standards.

Measurement in Water

It is well known in the art that numerous procedures are available for the estimation of cyanide. These methods may be classified as follows.

(1) Noncolorimetric Methods:
- (a) Titrimetric methods involving visual end point detection
- (b) Titrimetric methods involving instrumental end point determination
- (c) Polarographic methods
- (d) Gas chromatography

(2) Colorimetric Methods:
   (a) Methods involving formation of a metal complex
   (b) Colorimetric methods based on the Konig reaction.

The above cited methods are not truly specific for the cyanide ion. The Konig synthesis (the reaction of cyanogen bromide or chloride with pyridine and an aromatic amine to form a dye) permits the determination of cyanide directly on the original sample and is generally considered the best for small amounts of cyanide.

A simpler, direct determination process for the determination of submicrogram quantities of cyanide has however been developed by G.G. Guilbault and D.N. Kramer (12). It involves:

(a) adding an unknown solution contained in a phosphate buffer and having a pH of 6.5 to 7.5 and a concentration up to 0.1 M to a solution of a reagent selected from the group consisting of p-benzoquinone, N-chloro-p-benzoquinoneimine, 2,5-dichloro-4-benzoquinone and a substituted quinone monoxime represented by the formula:

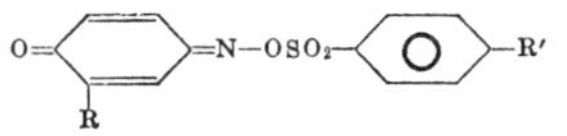

O=⟨ring, R⟩=N—OSO₂—⟨O⟩—R′

R is selected from the group consisting of H or CH_3, and
R' is selected from the group consisting of CH_3, H, NO_2, OCH_3 and bromine
and a solvent taken from the group consisting of dimethylformamide and dimethylsulfoxide.

(b) registering the developed green fluorescent solution at excitation and emission wavelengths of 400 to 500 mμ, respectively,
(c) recording the calibration plots of fluorescence readings versus cyanide concentration whereby the unknown concentration of cyanide may be determined.

Another colorimetric technique for cyanide determination has been developed by J.A. Platte (13) which is suitable for monitoring a continuous sample. This method essentially comprises fractionally distilling the sample to be tested, and adding to the distillate a predetermined amount of a prepared solution including a metal ion and a metal complexing agent capable of exhibiting alterations of color or color density in the presence of the metal ion. The color or absorbence of the distillate so treated may then be compared to a standard. Within the limits explained below, the absorbence is reduced in direct proportion to the amount of cyanide present in the solution.

Distillation of the sample is utilized in order to separate the cyanide, or at least a known portion thereof, from substances in the sample medium known sometimes as "interfering substances." These are substances which interfere with the reaction of analytical reagents with the tested substance. Methods of distillation of batch samples are known in the art and in fact are recommended as part of standard procedures. In the case of flowing samples which are to be continuously monitored, an excellent method and apparatus for continuous distillation which assures a constant fractionation is described in detail in U.S. Patent 3,147,082.

However, any method of preparing a sample free of interfering substances will be satisfactory. A base, usually sodium hydroxide, is added to the distillate to prevent the evolution of hydrogen cyanide gas.

The indicator solution is preferably a solution of copper and sodium diethyldithiocarbamate, or copper and bis-cyclohexanone oxaldihydrazone. Copper and copper-complexing agents are preferred because the specific color indicators for copper are quite sensitive to minute changes in the copper indicator solution. Zinc may also be used. In this case, the preferred complexing agents are dithizone, 2-carboxy-2'-hydroxy-5'-sulfoformazylbenzene (Zincon), and 1-(1-hydroxy-2-naphthylazo)-5-nitro-2-naphthol-4-sulfonate (Chrome Black-T).

Theoretically, about 1.6 milligrams of cyanide is required to complex one milligram of zinc. Similarly, 1.24 mg. of cyanide is required to complex 1 mg. copper. These calculations are based upon the formula for the metal complex, $Zn(CN)_4$, for example, and may be calculated in the same manner for any metal. Of course, calibrations can be made in terms of parts per million of cyanide in the same relation only to the absorbence of the sample without regard to the number of milligrams of metal actually complexed.

A continuous analysis may be run on a colorimetric analyzer by continuously adding to a flowing sample at a known rate a reagent solution of known strength. Such an apparatus is capable of recording changes in the absorbence of the sample treated with the reagent. A written record of cyanide concentration in terms of parts per million may thus be produced.

A device developed by R. Hoffmann, H.-J. Schuster, W. Gruhl, H. Michel and K. Noske (14) permits automatically measuring the cyanide content and/or the pH value of industrial waste waters.

Figure 29 shows bottom and sectional elevation views of a suitable type of apparatus for the conduct of this process. As shown there, a measuring vessel **2** is equipped with an electrochemical analyzing device **1**. Measuring vessel **1** suitably communicates with the reaction chamber containing the waste waters (not shown) by means of an inlet pipe **3** and is connected to a vessel or line containing a calibration solution (not shown) through an inlet pipe **4**.

FIGURE 29: APPARATUS FOR MEASUREMENT OF CYANIDE CONCENTRATION IN WASTEWATERS

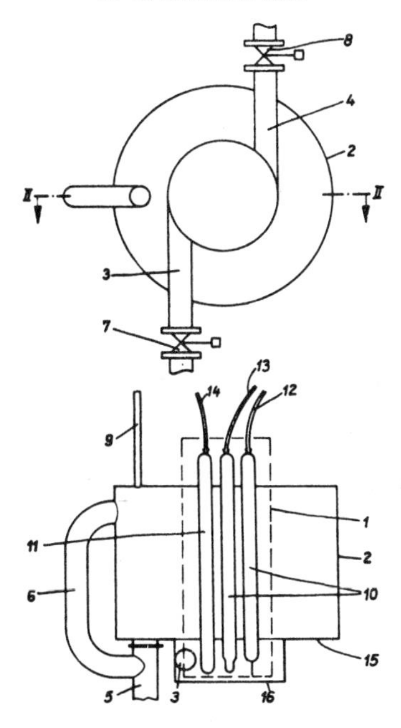

Source: R. Hoffmann, H.-J. Schuster, W. Gruhl, H. Michel and K. Noske; U.S. Patent 3,394,080; July 23, 1968

Measuring vessel **2** is provided with an outlet pipe **5** into which there discharges an overflow tube **6**, tube **6** communicating with vessel **2** near the upper edge. Inlet pipes **3** and **4** are provided with valves **7** and **8** respectively, the opening and closing of these valves being adapted to be controlled by automatic program control means (not shown). A tube **9** which communicates with vessel **2** at its upper edge is utilized for ventilating vessel **2**.

The electrochemical analyzing device comprises an electrode device **10** which produces the electrical quantity which is a function of the cyanide concentration of a waste water sample and electrode **11** which produces the electrical quantity which is a function of the pH value of the waste water sample. The two electrodes comprising electrode device **10** preferably are a silver electrode and a saturated calomel (mercurous calomel) electrode respectively and the electrode comprising electrode device **11** is suitably a pH single rod electrode.

The electrodes are adapted to be connected by means of leads **12**, **13** and **14** through electrical devices (not shown) such as, for example, amplifiers, to the automatic program control means. The latter means suitably contains therein valve **7** of inlet pipe **3** and valve **8** of inlet pipe **4** in addition to electrical stages which control the quantities of the decontaminating and alkalizing agents added to the waste water in the reaction chamber.

At its bottom, measuring vessel **2** is provided with a recess **16** into which the lower ends of electrodes **10** and **11** extend. Such construction and arrangement ensures that the electrodes are always maintained moist and, thereby, are prevented from drying out. Inlet pipes **3** and **4** preferably communicate with measuring vessel **2** in the region of recess **16**. Thus, in response to the openings of valves **7** and **8** respectively, the electrodes are flooded with the appropriate liquid. It is found to be advantageous for inlet pipes **3** and **4** to be attached to measuring vessel **2** in a manner such that the samples of waste water as well as the pure water or calibration solution are caused to flow into the measuring vessel in a path tangential thereto.

In the operation of the measuring device a command from the automatic program control means applied to a device contained therein for controlling the movement of valve **7** in inlet pipe **3**, causes valve **7** to be opened to permit the introduction into measuring vessel **2** of a sample amount of waste water contained in the waste water containing reaction chamber. After the period of time necessary to produce the electrical quantities representative of the pH and the cyanide concentration respectively of the introduced sample, such two produced electrical quantities are scanned by the automatic program control means.

Valve **7** is automatically caused to be closed by the program control means and the sample of the waste water which fills in ensuring vessel **2** is drained therefrom, through pipes, pipes being automatically controlled by means not shown. Next, in response to a command from the automatic program control means to a device which controls the opening and closing of valve **8** in inlet pipe **4**, valve **8** is caused to be opened to permit vessel **2** to be filled with the calibration or pure water solution. With measuring vessel **2** filled with the calibration solution, a chosen period of time is provided which is necessary to enable the electrochemical analyzing device to respond to the calibration solution and thereby to be reset at its calibrated or comparison standard setting in preparation for the receipt of the next forthcoming sample of waste water to be analyzed for pH and cyanide concentration.

At the end of this last mentioned period in the program control means, there is initiated an examination of the operational capability or capacity of the electrodes. Thereafter, valve **8** is caused to be closed in a response to a command from the automatic program control means and the calibration solution is automatically caused to be drained from measuring vessel **2** through pipe **5**, whereby the electrochemical analyzing device is now ready to be employed to analyze the next waste water sample.

The electrical quantities produced upon the introduction of the waste water sample into measuring vessel **2** and scanned by the automatic program control means are employed to influence the operation of the devices which control the addition of decontaminating and alkalizing agents into the reaction chamber. The program in the automatic program control means is preferably designed such that the alkalizing agents are added to the waste water containing reaction chamber prior to the adding of the decontaminating agents, the latter addition not being permitted to occur until the most favorable pH value required for the decontamination reaction has been attained.

It has been found advantageous to employ a calibration solution rather than relatively pure water if it is desired to enable continuous examination of measuring device **2** as to its operational safety. Thus, if some wear or even damage should possibly occur at one or more electrodes of the electrochemical analyzing device during its operation, then the electrical magnitudes produced by the chemical analyzing device when filled with the calibration solution would have values that would deviate from the standard values or norms that should be produced at this juncture. Such deviations may be employed to automatically actuate an annunciator or optical device, and/or to automatically interrupt the operation of the measuring device **2** and the decontamination process.

Cyanides may be monitored in water by the following instrumental techniques (15):

Ion Selective Electrode
Molecular Absorption (Colorimetric)

As shown earlier in Table 12, the Technicon CSM-6 can be used to monitor cyanides in the 0 to 3 ppm range with a detection limit of 0.03 ppm.

References

(1) Sittig, M., *Pollutant Removal Handbook,* Park Ridge, N.J., Noyes Data Corp. (1973).
(2) F.E. Bernardin, "Detoxification of Cyanide by Adsorption and Catalytic Oxidation on Granular Activated Carbon," presented at 44th Annual Conf., Wat. Poll. Cont. Fed., 1971.
(3) N.H. Hansen, "Design and Operation Problems of a Continuous Automatic Plating Waste Treatment Plant at the Data Processing Division, IBM, Rochester, Minnesota," *Proc. 14th Purdue Industrial Waste Conf.,* 227-249, 1959.
(4) L.T. Palla and R.G. Spicher, "Cyanide Treatment in Profit and Cure," presented at 26th Indust. Waste Conf., Purdue Univ., 1971.
(5) K.S. Watson, "Treatment of Complex Metal-Finishing Wastes," *Sew. Indust. Wastes,* 26:182-194, 1954.
(6) Anonymous, "Ozone Counters Waste Cyanide's Lethal Punch," *Chem. Engr.,* 24:63-64, 1958.
(7) V. Crowle, "Effluent Problems as They Affect the Zinc Die-Casting and Plating-on-Plastics Industries," *Metal Finishing Jour.,* 17:51-54, 1971.
(8) W. Lowe, "The Origin and Characteristics of Toxic Wastes, with Particular Reference to the Metal Industries," *Water Poll. Control* (London), 1970:270-280.
(9) Anonymous, "A State of the Art Review of Metal Finishing Waste Treatment," *U.S. EPA Publication 12010 EIE 11/68,* 1968.
(10) J.K. Easton, "Electrolytic Decomposition of Concentrated Cyanide Plating Wastes," *Jour. Wat. Poll. Control Fed.,* 39:1621-1625, 1967.
(11) Ruch, W.E., "Analyses of Gaseous Pollutants," Ann Arbor, Mich., Ann Arbor Science Publishers, Inc. (1970).
(12) Guilbault, G.G and Kramer, D.N., U.S. Patent 3,432,269; March 11, 1969; assigned to Secretary of the Army.
(13) Platte, J.A., U.S. Patent 3,195,983; July 20, 1965; assigned to Calgon Corp.
(14) Hoffmann, R., Schuster, H.J., Gruhl, W., Michel, H. and Noske, K., U.S. Patent 3,394,080; July 23, 1968; assigned to Siemens AG, Germany.
(15) Lawrence Berkeley Laboratory, "Instrumentation for Environmental Monitoring - Water," *Publ. No. LBL-1,* Vol. 2, Berkeley, Univ. of Calif. (Feb. 1, 1973).

ETHYLENE

Ethylene does not appear to be toxic to humans or animals or cause damage to materials in the concentrations that have been found in air. In fact, 75 to 90% ethylene in oxygen has been used as an anesthetic in hospital surgery with no adverse side effects.

However, low concentrations of ethylene do have a pronounced effect on plant life according to Stahl (1). Unlike most gaseous phytotoxicants, which attack the plant tissue, ethylene disrupts the normal processes of the plant hormones or growth regulators. This disruption results in morphogenetic and physiological changes in the plant. The response to ethylene varies widely with different species of plants. Some of the commonly observed effects are growth reduction; epinasty; abscission of flowers, buds, and leaves; discoloration; and abnormal growth patterns.

Another important aspect of ethylene air pollution is the photooxidation products created by reaction of ethylene with nitrogen oxides in the atmosphere. Although ethylene is not as reactive as many other hydrocarbons, it is the most abundant (on a mol basis) of the "reactive" hydrocarbons present in the atmosphere. The major products formed from the photooxidation of ethylene are formaldehyde and carbon monoxide.

The major source of ethylene air pollution in large metropolitan areas appears to be emissions from combustion, particularly automobile exhaust. It has been estimated that in Los Angeles, ethylene emissions from automobile exhaust amount to 60 tons per day. No information was found on the contribution of industrial sources to ethylene pollution. Approximately 55% of the commercial ethylene is produced in Texas. Moreover, in one case, a polyethylene factory was shown to be a local source of ethylene pollution.

Ethylene is also produced by plant life. The importance of this source is not known in the case of open fields, but in confined areas such as greenhouses or plant storage containers, damage has been reported to certain species of plants, particularly orchids.

Only limited information is available concerning the amount of ethylene in the atmosphere. Air samples recently taken in California indicate that the average ethylene concentration for metropolitan areas ranges from 40 to 120 μg./m.^{3} (0.04 to 0.1 ppm). The maximum value observed has been as high as 800 μg./m.^{3} (0.7 ppm).

The methods used for control of volatile hydrocarbons are applicable to ethylene. These include combustion, adsorption, and vapor recovery systems. No information was found on the cost of abatement. Economic losses from plant damage due to ethylene have been reported in the literature. Greenhouse plants, particularly orchids, appear to be most affected.

A limit on ethylene concentration on an air quality basis has been set by the State of California. The limits set are 0.5 ppm on a 1-hour averaging basis and 0.1 ppm on an 8-hour averaging basis.

Measurement in Air

Sampling Methods: Ethylene samples can be collected by the grab sampling method, i.e., collection into an evacuated container (2)(3)(4)(5). Gas chromatography or infrared spectroscopy methods of analysis lend themselves to direct passing of the sample into the instrument without use of collection containers (6)(7)(8). Some authors report collecting ethylene in freeze traps (9) in mercuric solution (10) and on silica gel (11).

There are no problems in the collection of ethylene samples other than the normal difficulties in taking and transferring samples. However, in cases in which both ethylene and nitrogen oxides are present in large amounts (e.g., automobile exhaust), a reaction between the two substances may occur in the sampling vessel. This very slow reaction can be reduced to a negligible rate by diluting the sample with dry nitrogen.

Qualitative anu Semiquantitative Determination Methods: A detector tube method has been described by Kitagawa and Kobayashi (12) which has a sensitivity of about

23,000 μg./m.³ (20 ppm) with ± 5% error when a 10 cm.³ sample is used. Another adsorbent was reported (13)(14) to detect approximately 10 μg./m.³ (0.01 ppm) ethylene by using a 3,000 cm.³ sample size, which is drawn through the tube at 100 cm.³ per minute.

A portable instrument has been reported to provide a simple and reliable field method for estimating ethylene concentration in the range of 5,700 to 230,000 μg./m.³ (5 to 200 ppm) (14)(15). The air sample is drawn over mercuric oxide at 285°C. The mercury vapor released by the ethylene is passed over a strip of sensitized seleno-cyanate paper at 125°C. The length of the black coloration is used to estimate the amount of ethylene present.

Quantitative Methods: Gas chromatography methods have found widespread use for determining the presence of ethylene as well as numerous other hydrocarbons (16). Gas chromatography has gained wide acceptance because it is a simple and rapid technique. With the development of the packed and open tubular columns, automatic temperature programming, and new types of detectors, particularly the flame ionization detector, one can obtain rapid, efficient separation of many components with a sensitivity in the ppb range and even lower.

The commonly used gas chromatographic methods for detecting ethylene employ a silica gel packed column at or near room temperature, with a flame ionization detector (2)(6)(17)(18)(19)(20). Other columns that are used include alumina (3), dimethyl sulfolane (4), hexadecane (5), polypak-2 (8), multicolumn techniques (21)(22), and open tubular columns (7)(23). Gas chromatography has been used to determine ethylene in air samples (6)(18)(24), automobile exhaust (7)(18)(20)(22), blow-by (25), municipal wastes (8), agricultural wastes (3), and incinerator effluents (5). To increase the sensitivity of this method, some authors also used trapping techniques (2)(5)(26)(27).

Infrared spectroscopy has been used to determine ethylene in air samples (28), automobile and diesel exhaust (4)(29)(30)(31), and incinerators. To obtain a greater sensitivity, very long optical path lengths are needed (e.g., 300 meters) (28), generally making large samples necessary. The absorption peak at 10.5 μ (952 cm.⁻¹) is normally used, having a sensitivity of greater than 0.1 ppm.

Ethylene concentration has been determined by mass spectrometry in air samples (9) and automobile exhaust (32)(33)(34). A manometric technique has been used for determining small amounts of ethylene (10). A mercuric perchlorate solution containing the collected ethylene is acidified in a closed system, and the released ethylene is determined by the change in pressure. The sensitivity is approximately 0.2 ml. of ethylene per 2 ml. of solution.

The chemical methods developed for quantitative determination of olefins (35)(36)(37)(38)(39) are not generally applicable to ethylene because of the lack of reactivity of the ethylene and the low sensitivity of the product formed with ethylene.

References

(1) Stahl, Q.R., "Air Pollution Aspects of Ethylene," *Report PB 188,069,* Springfield, Va., Nat. Tech. Info. Serv. (Sept. 1969).

(2) Bellar, T.A., M.F. Brown and J.E. Sigsby, Jr., "Determination of Atmosphere Pollutants in the Part-Per-Billion Range by Gas Chromatography (A Simple Trapping System for Use with Flame Ionization Detectors)," *Anal. Chem. 35*:1924 (1963).

(3) Darley, E.F., et al, "Contribution of Burning of Agricultural Wastes to Photochemical Air Pollution," *J. Air Pollution Control Assoc. 16*(12):685 (1966).

(4) Linnell, R.H., and W.E. Scott, "Diesel Exhaust Analysis," *Arch. Environ. Health 5:*616 (1962).

(5) Tuttle, W.N., and M. Feldstein, "Gas Chromatographic Analysis of Incinerator Effluents," *J. Air Pollution Control Assoc. 10*(6):427 (1960).

(6) Altshuller, A.P., and T.A. Bellar, "Gas Chromatographic Analysis of Hydrocarbons in the Los Angeles Atmosphere," *J. Air Pollution Control Assoc. 13*(2):81 (1963).

(7) Jacobs, E.S., "Rapid Gas Chromatographic Determination of C_1 to C_{10} Hydrocarbons in Automobile Exhaust Gas," *Anal. Chem. 38*:43 (1966).
(8) Jerman, R.I., and L.R. Carpenter, "Gas Chromatographic Analysis of Gaseous Products from the Pyrolysis of Solid Municipal Waste," *J. Gas Chromatog. 6*(5):298 (1968).
(9) Shepherd, M., et al, "Isolation, Identification and Estimation of Gaseous Pollutants of Air," *Anal. Chem. 23:*1431 (1951).
(10) Young, R.E., H.K. Pratt, and J.B. Biale, "Manometric Determination of Low Concentrations of Ethylene," *Anal. Chem. 24:*551 (1952).
(11) Stitt, F., and Y. Tomimatsu, "Removal and Recovery of Traces of Ethylene in Air by Silica Gel," *Anal. Chem. 25:*181 (1953).
(12) Kitagawa, T., and Y. Kobayashi, "Gas Analysis by Means of Detector Tubes. VII. Rapid Method for the Determination of Ethylene," *J. Chem. Soc. Japan 56:*448 (1953).
(13) Kobayashi, Y., "Rapid Method for the Determination of Low Concentrations of Ethylene by Means of a Detecting Tube," *Yuki Gosei Kagaku Kyokai Shi 14:*137 (1957).
(14) Stitt, F., A.H. Tjensvold and Y. Tomimatsu, "Rapid Estimation of Small Amounts of Ethylene in Air. Portable Instrument," *Anal. Chem.* 23:1138 (1951).
(15) Stitt, F., Y. Tomimatsu, and A.H. Tjensvold, "Determination of Ethylene in Gases," U.S. Patent 2,648,598 (August 11, 1953).
(16) Stern, A.C. (Ed.), *Air Pollution, II,* 2nd ed. (New York: Academic Press, p. 116, 1968).
(17) Bellar, T.A., M.F. Brown, and J.E. Sigsby, Jr., "Evaluation of Various Silica Gels in the Gas Chromatographic Analysis of Light Hydrocarbons," *Environ. Sci. Technol.* 1(3):242 (1967).
(18) Bellar, T.A., et al, "Direct Application of Gas Chromatography to Atmospheric Pollutants," *Anal. Chem.* 34(7):763 (1962).
(19) McMichael, W. F., and J.E. Sigsby, Jr., "Automotive Emissions After Hot and Cold Starts in Summer and Winter," *J. Air Pollution Control Assoc.* 16(9):474 (1966).
(20) Swartz, D.J., K.W. Wilson, and W.J. King, "Merits of Liquified Petroleum Gas Fuel for Automobile Air Pollution Abatement, *J. Air Pollution Control Assoc.* 13:154 (1963).
(21) Klosterman, D.L., and J.E. Sigsby, Jr. "Application of Subtractive Techniques to the Analysis of Automotive Exhaust," *Environ. Sci. Tech.* 1:309 (1967).
(22) McEwen, D.J., "Automobile Exhaust Hydrocarbon Analysis by Gas Chromatography," *Anal. Chem.* 38:1047 (1966).
(23) McEwen, D.J., "Temperatures Programmed Capillary Columns in Gas Chromatography," *Anal. Chem.* 35:1636 (1963).
(24) Stephens, E.R. and W.E. Scott, "Relative Reactivity of Various Hydrocarbons in Polluted Atmospheres," *Proc. Am. Petrol. Inst.* 42:665 (1962).
(25) Sigsby, J.E., Jr., and M.W. Korth, "Composition of Blow-by Emissions," preprint. Presented at the 57th Annual Meeting, Air Pollution Control Association, Houston, Tex. Paper No. 64-72 (June 21-25, 1964).
(26) Eggertson, F.T., and E.M. Nelsen, "Gas Chromatographic Analysis of Engine Exhaust and Atmosphere Determination of C_2 to C_5 Hydrocarbons," *Anal. Chem.* 30:1040 (1958).
(27) Feldstein, M., and S. Balestrieri, "The Detection and Estimation of Part Per Billion Concentrations of Hydrocarbons," *J. Air Pollution Control Assoc. 15*(4):177 (1965).
(28) Scott, W.E., et al, "Further Developments in the Chemistry of the Atmosphere," *Proc. Am. Petrol. Inst.* 37:171 (1957).
(29) Mader, P.P., et al, "Effects of Fuel Olefin Content on Composition and Smog Forming Capabilities of Engine Exhaust," Air Pollution Control District, County of Los Angeles, Calif. (1959).
(30) Schuck, E.A., H.W. Ford and E.R. Stephens, *Air Pollution Found. (Los Angeles) Rept. 26* (1958).
(31) Stephens, E.R., et al, "Auto Exhaust Composition and Photolysis Products," *J. Air Pollution Control Assoc. 8:*333 (1959).
(32) Coulson, D.M., "Hydrocarbon Compound-type Analysis of Automobile Exhaust Gases by Mass Spectrometry," *Anal. Chem.* 31:906 (1959).
(33) Rounds, F.G., P.A. Bennett, and G.J. Nebel, "Some Effects of Engine-Fuel Variables on Exhaust Gas Hydrocarbon Content," *J. Air Pollution Control Assoc.* 5:109 (1955).
(34) Walker, J.K., and C.L. O'Hara, "Analysis of Automobile Exhaust Gases," *Anal. Chem.* 26:352 (1954).
(35) Altshuller, A.P., and S.F. Sleva, "Spectrophotometric Determination of Olefins," *Anal. Chem.* 33:1413 (1961).
(36) Altshuller, A.P., and S.F. Sleva, "Vapor Phase Determination of Olefins by a Colorimetric Method," *Anal. Chem.* 34:418 (1962).
(37) Altshuller, A.P., S.F. Sleva, and A.F. Wartburg, "Spectrophotometric Determination of Olefins in Concentrated Sulfuric Acid," *Anal. Chem.* 32:946 (1960).
(38) Mader, P.P., K. Schoenemann, and M. Eye, "Detection of Nonaromatic Unsaturates in Automobile Exhaust by Spectrophotometric Titration," *Anal. Chem.* 33:733 (1961).
(39) Nicksic, S.W., and R.E. Rostenback, "Instrumentation for Olefin Analysis at Ambient Concentrations," *J. Air Pollution Control Assoc.* 11:417 (1961).

FLUORIDES

Airborne particulate and gaseous fluorides are emitted from a variety of sources in the primary aluminum, iron and steel, glass and phosphate rock processing industries as reviewed by Arthur D. Little, Inc. in *Development of Methods for Sampling and Analysis of Particulate and Gaseous Fluorides from Stationary Sources* (1). Table 25 (2) shows the fluoride concentrations found in some typical gaseous industrial effluents.

Fluoride concentrations (as HF) have been spelled out on an air quality basis by the States of Montana and New York. Montana specifies a maximum of 0.001 ppm on a 24-hour averaging basis. New York specifies 0.001 ppm for rural areas, 0.002 ppm for urban areas and 0.004 ppm for industrial areas on a 24-hour averaging basis. Pennsylvania specifies a limit of 0.007 ppm on soluble fluorides on a 24-hour basis. Canada (Ontario) has a limit of 0.001 for rural and residential areas and 0.004 ppm for commercial and industrial areas.

Emission standards for HF are 230 mg./cu. m. at STP in the United Kingdom and are 115 mg./cu. m. in Australia as reviewed by A.C. Stern (2a). Industries which discharge significant quantities of fluorides in process waste water streams are:

Glass manufacturers
Electroplating operations
Steel and aluminum producers
Pesticide and fertilizer manufacturers

Glass and plating wastes typically contain fluoride in the form of hydrogen fluoride (HF) or fluoride ion (F^-), depending upon the pH of the waste. Fluoride discharged from fertilizer manufacturing processes is typically in the form of silicon tetrafluoride (SiF_4) as a result of processing of phosphate rock (3)(4). The aluminum processing industry utilizes the fluoride compound cryolite (Na_3AlF_6) as a catalyst in bauxite ore reduction. Previously, the gaseous fluorides resulting from this process were discharged directly into the atmosphere. Wet scrubbing of the process fumes, an air pollution abatement procedure, results in transfer of the fluorides to aqueous waste streams.

Average fluoride values for aluminum reduction plants were given as 107 to 145 mg./l. in wastewater streams (5). Concentrations an order of magnitude greater have been reported for glass manufacturing, ranging from 1,000 to 3,000 mg./l. of fluoride (3).

The permissible concentration of fluorides in domestic water supplies is 0.8 to 1.7 ppm and the desirable concentration is 1.0 ppm as shown earlier in Table 3. The permissible concentration of fluorides in effluents to storm sewers or streams is 1.5 ppm.

Measurement in Air

An evaluation of the technology for the measurement of the fluoride content of process streams has been presented by J.M. Robinson, G.I. Gruber, W.D. Lusk and M.J. Santy (2). The results are discussed in the following sections under the categories of sampling, separation of fluoride from interfering ions, and analytical methods.

Sampling Procedures: Selection of a sampling technique for measuring the fluoride content of effluents from process sources is dictated by the effluent stream composition and the pollutants to be determined. For sources that emit both particulate fluorides and gaseous silicon tetrafluoride and hydrogen fluoride, chemical reactivity presents a major sampling problem. Such sources include the industrial plants manufacturing phosphate fertilizer, producing pig iron, processing iron and steel, reducing aluminum ore, and manufacturing glass and ceramics. For accurate sampling of effluent and differentiation between particulate and gaseous pollutants from such operations, the sampling technique must prevent interaction of the gaseous and particulate fluorides.

TABLE 25: CONCENTRATION OF FLUORIDE FOUND IN VARIOUS EFFLUENTS
(At standard temperature and pressure)

	Grains/ft³	w/v mg/M³	v/v ppm
NORMAL SUPERPHOSPHATE			
Den Scrubber Emissions	0.08-0.30	183-686	220-824
nominal	~0.15	343	412
Building Scrubber Emissions	~0.00035	0.80	0.96
DI-AMMONIUM PHOSPHATE			
Granulator Exhaust	0.0093	21.3	25
Dryer Duct	0.1100	250	300
Dry Screens Exhaust	0.0025	5.7	6.8
WET PROCESS PHOSPHORIC ACID			
AP-57			
Digester-Filters-Tanks	0.0011-0.0147	2.5-33.6	3-40
Scrubber Exhaust "Big Plant"	0.001-0.03	2.5-68.6	3-82.3
Scrubber Exhaust "Medium Plant"	0.0048	10	12
TRIPLE SUPERPHOSPHATE			
Scrubber Inlet	0.55	1258	1500
Scrubber Outlet	0.016	36.6	43.9
Den Scrubber Inlet	0.10	229	275
Den Scrubber Outlet	0.008	18.3	22.0
Reactor and Granulator Scrubber Exhaust	0.0021	4.8	5.8
Dryer	1.3	2970	3560
Dryer Exhaust	0.0025	5.7	6.8
Granulator Scrubber Inlet	0.48	1098	1320
Granulator Scrubber Outlet	0.030	68.6	82.3
DEFLUORINATED PHOSPHATE ROCK			
Kiln Scrubber Exhaust	0.00056	1.3	1.6
Fluosolids Scrubber Exhaust	0.0048	11.0	13.2
Prep. Feed to Kiln	0.00095	2.2	2.6
Di-Cal (from acid and limestone)	0.00020	0.5	0.6
ELEMENTAL PHOSPHOROUS			
Water Sol. F(updraft dryer)	0.0313	71.6	85.9
Emissions Particulate (updraft dryer)	0.0099	22.9	27.5
Furnace Exhaust Gas	0.0031	7.1	8.5
ALUMINUM PREBAKE ANODE			
Primary Control Process (average)	0.033	75.5	90.6
Secondary Control Process	0.00006-0.00042	0.13-0.96	0.16-1.16
ALUMINUM VERTICAL STUD SODERBERG			
Primary (average)	0.43	982	1180
Secondary Loading	0.00049	1.12	1.34
ALUMINUM HORIZONTAL STUD SODERBERG			
Primary (average)	0.01	22.9	27.5
Secondary Loading	0.00026-0.00042	0.59-0.96	0.71-1.15
IRON AND STEEL			
SINTER PLANT			
Normal Conditions, water sol. F	0.0023	5.3	6.4
Normal Conditions, particulate F	0.0011	2.5	3.0
Special Conditions, water sol. F	0.0042	9.6	11.5
Special Conditions, particulate F	0.00075	1.7	2.0
Blast Furnace Stoves	0.0027	6.1	7.3
Boiler House	0.00014	0.32	0.38
Coke Ovens	0.00068	1.6	1.9
Open Hearth, water sol. F	0.0157	35.9	43.1
Open Hearth, particulate F	0.00004	0.09	0.11

Source: Report PB 207,506

Sampling procedures for use in the measurement of fluorides in the atmosphere have been developed to prevent, to some extent, the interaction in the collection train of gaseous and particulate fluoride. Unfortunately, except for work carried out for the Office of Air Programs (formerly the National Center for Air Pollution Control) by Dorsey (6), Elfers and Decker (7) and the Manufacturing Chemists Association, no detailed methodology [other than APCO Procedure H-7, Reference (6)] is available in the open literature covering stack sampling for fluorides.

The developed techniques involve the sampling of stack effluents with a hot glass probe followed by a heated train consisting of a cyclone, filter and a Greenburg-Smith impinger containing distilled water. Particulate fluorides are collected using a high-efficiency cyclone followed by a Whatman #41 filter. Active gaseous fluorides, such as HF and F_2 react with the heated glass probe to form gaseous silicon tetrafluoride which, after passing through the heated cyclone and filter, hydrolyzes in the water of the Greenburg-Smith impinger to form fluosilicic acid and insoluble orthosilicic acid. The water-soluble particulate fluorides, total particulate fluorides, and soluble gaseous fluorides can thus be determined separately.

Some of the procedures used for sampling fluorides in ambient air may be adaptable for sampling fluoride emissions from cyclones, baghouses, electrostatic precipitators, or other so-called dry collection equipment. At least the gaseous portion of the fluoride emissions from these collectors might be adequately measured. The procedure using sampling tubes with alkaline coatings could be used if a suitable dilution technique were employed. However, the emissions from scrubbers using aqueous scrubbing liquids require other measurement methods.

Water vapor, droplets of entrained scrubber liquid, and uncaptured fluoride particulates could all be present at the scrubber exit. Considering these problems, even the sampling of particulate was viewed with concern by Lunde (8) who stated, "Adequate data are not available to evaluate the performance of the equipment installed for the collection of particulate fluorides." His comment refers to scrubbing devices using liquids to capture fluorides.

The most important constituent, the gaseous fluoride emission from the scrubber, is the constituent most difficult to separate from such a mixture. Total fluorides could be analyzed very efficiently, but the ambiguity concerning the proportion of gaseous and particulate fluoride in the emisson would remain.

Mixtures of fluorides are usually evolved by industrial processes. If there is a need to separate the particulate and gaseous fluoride components, the sample train shown in Figure 30 has frequently been used for this purpose. The particulate filter shown is a porous thimble. A variety of filters and filter holders have been used. Some portion of the sample may deposit on the inner surfaces beginning at the probe; therefore, to minimize the reaction of HF with the sample train, stainless steel parts have been used.

As particulates collect on the filter surface, the dust layer tends to become a collector for gaseous fluorides. Many dusts will absorb or adsorb HF to some degree, and two patents extol the effectiveness of aluminum oxide for retaining HF (9)(10). Limestone dust is well known for its ability to remove HF from air; however, little has been published on the reactivity of HF with dusts such as fly ash from coal burning, borates in glass making, and clays or other mineral dusts present in industrial processes. It seems clear that in a sample train such as that shown in Figure 30, the filtering section would collect a particulate sample with an indeterminate portion of the gaseous fluorides either reacted or adsorbed. The aqueous collectors would retain the remaining gaseous fluoride.

Citric acid-treated filter paper allows gaseous fluoride to pass through to be collected in a following section of the fluoride sampler. The effectiveness of this arrangement for stack sampling would be completely dependent on very small dust loading of the filter; therefore, it is not a promising method for sampling most industrial effluent gas streams

FIGURE 30: SCHEMATIC DIAGRAM OF SAMPLING TRAIN FOR DRY PARTICULATE MATTER

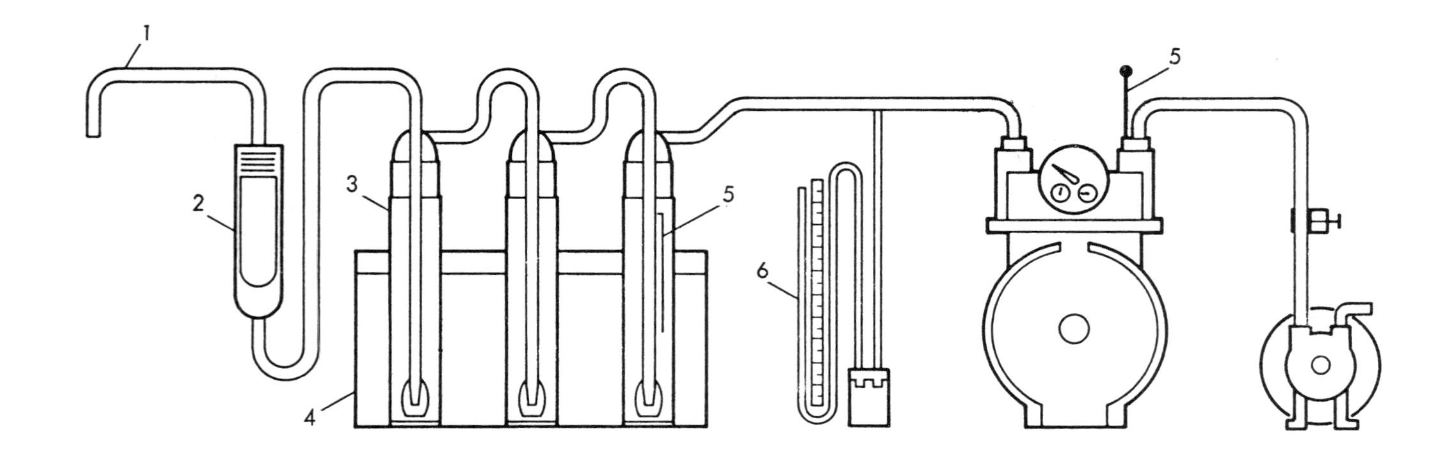

The components are: (1) sampling probe; (2) dry filter; (3) impinger (dust concentration sampler); (4) ice bath container; (5) thermometers; (6) mercury manometer; (7) Sprague dry gas meter; (8) vacuum pump; and (9) hose clamp to control gas flow rate.

Source: Report PB 207,506.

which contain appreciable amounts of dust. Other sampling trains have included insertion of a small cyclone collector ahead of the filter to trap dust larger than 25 microns, reducing the amount of dust deposited on the surface of the filter. The filter has, in some cases, been placed in the train following the aqueous collectors.

Where the quantity of particulate matter in an effluent stream is large, the separation of gaseous and particulate fluorides is difficult. However, control techniques are frequently concerned only with the determination of total fluoride content. The Greenburg-Smith impinger can be considered as the standard for collecting total fluoride though other collectors are sometimes used. As reported in a review by Farrah (11), the Greenburg-Smith impinger is fairly rugged and has collection efficiencies ranging from 90 to 98% when operated properly, at flow rates of 1.5 to 2.0 cfm. [Keenan and Fairhall (13) found that lead fume particulate collection efficiencies improved when a flow rate of 1.6 cfm was used with a standard impinger designed for use at a flow of 1 cfm.]

Pack, et al, (12) and Farrah (11) report that pure water is as good a collector as caustic solution for fluoride contaminants. The impinger collection solution is usually diluted to a constant volume and an aliquot taken from this solution for determination of the fluoride content by the separation and analytical method selected.

Using the hot glass probe sampling technique, the particulate contaminant (free of gaseous fluorides) is transferred quantitatively from the cyclone and probe by washing with acetone after which the particulate material is dried and weighed. The particulate contaminant collected on the filter paper is combined with the cyclone-collected particulate material; the filter paper is shredded; and the contents are diluted to a constant volume. If, upon acidifying, the particulate material is not dissolved, caustic fusion as described by Pack, et al (12) is required for complete recovery of fluoride. Aliquots are taken to yield the desired quantity of fluoride for the analytical method selected.

If the reactivity of gaseous fluorides could be diminished by some mechanism, difficulties in separating them from dusts could be reduced and fluoride sample collection simplified. Since SiF_4 is less reactive than HF and since HF can be converted to SiF_4 through contact with heated glass, this principle was employed in designing the sample train shown in Figure 31.

This sampling train, developed by Dorsey et al, (6) modified by Elfers and Decker (7) and described in APCO Procedure H-7, provides, with some limitations, detailed methodology for handling the total range of fluoride contaminants in most process effluent streams and for differentiation between particulate and gaseous fluorides in stack gases. Potential problems with this sampling train still await solution.

Gelatinous silica hydrate is formed by the SiF_4 hydrolyzed in the impinger solution. A similar problem with gelatinous silica was solved by using ammonium compounds (14). Whether SiF_4 would react with iron oxide dust on the surface of the filter may need to be tested; iron oxide is reported to react readily with SiF_4 (15). In some process gas streams, the heated probe could become coated with dust, carbon, or tarry materials to such an extent that the desired reaction of HF with glass could not occur.

The technique developed by Pack, et al (12) involving use of a glass fiber filter for collection of suspensoid particulate contaminants can be used instead of the cyclone and heated glass probe. The glass fiber or paper filter (which separates and collects 98% of the suspensoid particulate material) can be washed to remove soluble particulate fluorides. Filter discs treated with alkaline reagents, used instead of a complex sampling train, will also collect total fluorides very satisfactorily. However, some limits on the size of the fluoride sample collected may have to be observed to avoid exceeding the capacity for absorption of the filter while collecting a relatively large sample compared to the intrinsic fluoride content of the glass fiber. Micropore-type filters may be used to collect most, if not all, submicron-size particulate fluorides.

FIGURE 31: EPA SAMPLING TRAIN FOR FLUORIDES IN AIR

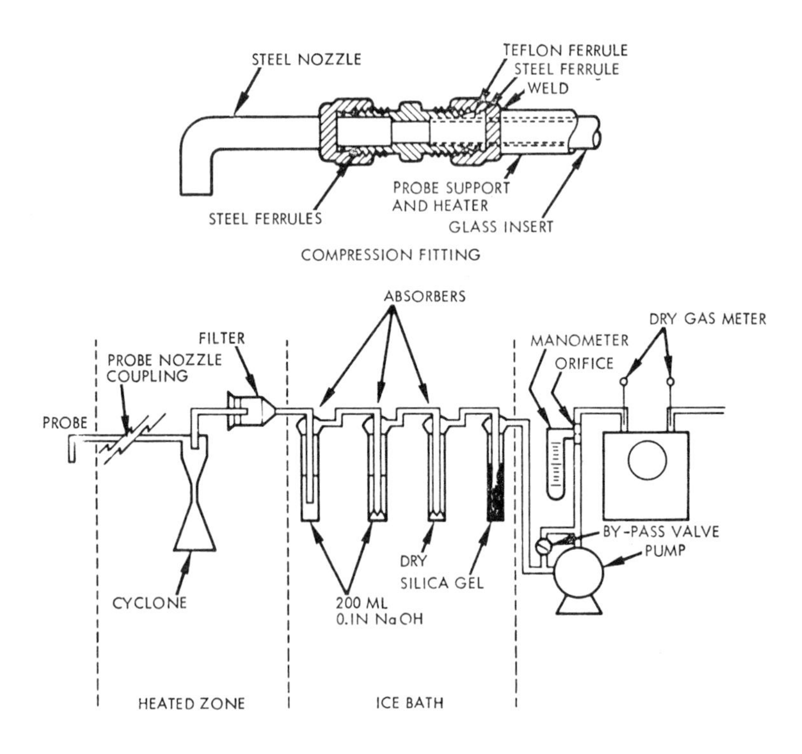

Source: Report PB 207,506.

Some process-related factors contribute to the sampling problems and are described on an industry-by-industry basis in the paragraphs which follow.

The phosphate industry uses phosphate rock as raw material which can cause problems because the rock does not have a fixed composition but varies from mine to mine and even from area to area within the same mine. Some phosphate rock behaves as though much of the fluoride was present as fluosilicate and some as fluorspar. Since the raw materials are treated differently in different processes, the form of the fluoride may be very important (16). Heat may be added as in nodulation, with SiF_4 released from the fluosilicate present; or acidulation may be used as in fertilizer manufacturing with the SiF_4 escaping but the HF derived from CaF_2 staying in the slurry to react with some of the calcium carbonate; or heat, acid, and silica may be added to the raw material as in the manufacture of defluorinated rock with nearly all of the fluoride volatilized, probably as a mixture of SiF_4 and HF.

In each instance, some pulverized phosphate rock may be entrained in the effluent gas stream along with the volatilized fluoride and water vapor released by the process reactions. Each of these mixtures of fluorides may react differently as it is drawn into and through the sampling train.

Little has been published describing fluoride effluent gas streams related to iron and steel manufacturing. However, it is reasonable to assume that: (a) fluorides added to the slag in steel furnaces may react with moisture to release HF, (b) fluorides may be converted to fluosilicates in the slag and then thermally decomposed to release SiF_4, and (c) fluorides may sublime as iron fluorides since the sublimation temperature (1800° to 2110°F.) is well below the pouring temperature of steel (15). The complexity of these reactions in the presence of dusts and moisture in the effluent gases could make sample collection very complicated and the analytical results difficult to interpret. As little as half of the fluoride added in steel furnaces is recovered in the slag from the steel processing (17).

The nature of fluoride evolution in aluminum reduction is complex with HF, cryolite, alumina, aluminum fluoride, chiolite, and possibly heavy hydrocarbons in the effluent gas from the aluminum reduction processes. The evolution mechanisms for these materials have not been completely described but fragments of the chemistry have been reported (15)(18)(19).

Aluminum fluoride dispersed as a fume in air reacts with moisture to form HF and aluminum oxide, but the rates of reaction are dependent upon vapor pressure (19) and temperature (18). Hence, as aluminum fluoride leaves an aluminum reduction cell, hydrolysis starts as soon as it encounters atmospheric moisture but diminishes rapidly as the fume cools.

Cooling may occur rapidly enough for some or much of the aluminum fluoride to remain dispersed in the effluent gas stream as an unhydrolyzed fume. Sublimed chiolite may rearrange into other solids as it condenses but in cooling probably gives rise to a fine fume that reacts slowly or not at all with moist air at ambient temperatures. Over the range of water vapor pressure and temperatures that prevail in the fluoride collection systems used in aluminum plants, there has been no really complete description of the chemical and physical states of the fluoride to be sampled at various points in emissions control systems.

Several of the devices for collecting fluorides from effluent streams have been discussed above. Some of them performed very well in sampling fluorides dispersed in ambient air and separate gaseous and particulate fluorides. For sampling industrial gas streams, too little testing has been done to demonstrate the usefulness of these devices for fluoride levels that may be far higher than those found in ambient air.

The types of sampling trains frequently used for stack sampling have been discussed in relation to industrial fluoride effluent gas streams. Since these effluent streams are usually mixtures of gaseous and solid fluorides, separation of the two phases causes problems in sample collection. Particulates deposit on the interior surfaces of probes and sampling tubes. This dust and that collected on the surface of the filter in the sampling train may absorb or adsorb significant amounts of gaseous fluorides. Reactivity of gaseous fluorides with the sampling train components may further interfere with the separation of the fluorides into gaseous and particulate samples.

Inaccuracies related to fluoride sampling and analysis of process streams are primarily caused by procedures used for collecting samples. The materials that are collected can be analyzed relatively accurately for fluoride content.

Before determining fluoride in particulate and gas fractions collected from effluents, interfering ions must be removed if any of the well-established analytical methods are to be used. Only aliquots providing the quantity of fluoride for the analytical method selected should be used in order that fluoride isolation can be performed with a minimum of work.

The separation of the fluoride ions from ions interfering in fluoride analyses such as Al^{+3} PO_4^{-3}, Cl^-, SO_4^{-2} is accomplished by (a) distillation, (b) ion exchange, or (c) diffusion. The most widely used of the separation methods is the Willard-Winter distillation (20). This method on the macro-scale is considered the standard by which newer methods are evaluated. Fluorine is separated as fluosilicic acid from interfering ions by steam distillation from solutions containing perchloric (21), sulfuric (21)(22) or phosphoric acids (23). The fluosilicic acid is swept out of the distillation flask with water vapor, the boiling point of the solution being held at a constant temperature by addition of steam or water and by regulating the heat applied to the solution. The addition of steam rather than water reduces the time required for the distillation and eliminates bumping of solution (24). When the original sample is relatively free of interfering materials and the fluoride is in a form easily liberated, a single distillation from perchloric acid is carried out at 135°C.

Samples containing appreciable amounts of aluminum, boron, or silica require a higher temperature and larger volume of distillate for separation. In this case a preliminary distillation from sulfuric acid at 165°C. is commonly used. Large amounts of chloride are separated by precipitation with silver as an intermediate step. Small amounts of chloride are held back in the second distillation from perchloric acid by addition of silver perchlorate to the distilling flask (11). The distillation method requires considerable operator time and results in a large volume of distillate for quantitative recovery (250 to 375 ml. for samples containing up to 100 mg. of fluoride).

Isolation by an ion exchange resin allows recovery of fluoride in a more concentrated form free of interfering ions. Nielsen et al (25) separated microgram quantities of fluoride on a quaternary ammonium styrene resin with recoveries approaching 95% for amounts of 20 μg. or less from mixtures including hydrofluoric acid, sodium fluoride, fluosilicic acid and calcium fluoride. The technique was used to concentrate Willard and Winter distillate and was also used directly on impinger-captured atmospheric fluorides.

Newman (26) removed interfering anions as well as cations on a single exchange resin (Di-Acidite FF). Funasaki, et al, (27) removed interfering ions PO_4^{-3}, AsO_4^{-3}, SO_4^{-2}, and CO_3^{-2} by means of Amberlite IRA-400. Elution was affected with 10% NaCl. Dowex anion exchange resin was used by Ziphin, et al, (28) to separate fluorides from PO_4^{-3} with gradient elution of the fluoride from the resin by sodium hydroxide. Nielsen (29) separated fluoride from Fe^{+3}, Al^{+3}, PO_4^{-3}, and SO_4^{-2} on the resin and removed the ions by stepwise elution with sodium acetate.

The ion exchange columns permit separations of 1 μg. to 0.1 g. of fluoride from interferences when the sample is in a few milliliters up to a liter of solution. The elution volumes usually are about 50 ml.

Diffusion methods for separating fluoride from interferences before determinations are simple and show great promise. They involve collection of fluoride in volumes ranging from a few milliliters to a liter of alkaline solution, the liberation of fluoride by treating with mineral acids, diffusion through a short distance and absorption of the fluorides in approximately 5 milliliters of alkaline solutions. These methods are generally applicable to quantities of fluorides in the 0.05 μg. to 1 mg. range. Singer and Armstrong (30) and Hall (31) suggested the use of polyethylene bottles for diffusion vessels which were sealed with stoppers.

Alcock (32) prepared a satisfactory diffusion cell of Teflon that was used at 55°C; higher temperatures released fluoride from Teflon. Taves (33) found that fluoride passes into trapping solutions in the form of methylfluorosilane if silicone grease is used for sealing the diffusion cell. In the presence of the simplest silicone, hexamethyldisiloxane, the separation of fluoride is much more rapid.

A faster diffusion method for the separation of fluoride was proposed (34); fluoride was liberated in the presence of hexamethyldisiloxane in 6N hydrochloric acid. The separation was carried out at 25°C. for 2 to 6 hours, depending on the volume of sample analyzed. Otherwise, the separation by diffusion takes place for at least 24 hours at much higher temperatures (usually 60°C.). Tusl (35) established a rapid diffusion technique using polyethylene diffusion cells to which were added a purified high vacuum silicone grease that was a homogeneous mixture of methylsilicone fluid and aerogel of silica. Following the diffusion separation, fluoride was determined by the zirconium-SPADNS colorimetric method. Stuart (36) followed the diffusion separation with fluoride determination with the fluoride specific ion electrode. He isolated 0.05 μg. to 200 μg. from a large volume of solution to a 5-milliliter solution. Because of its wide acceptance and ability to effect satisfactory separations of fluoride with a minimum of equipment, the Willard-Winter distillation technique is recommended for separating interfering ions in the wide weight range from 0.1 μg. to 1 g. of fluoride collected from plant gaseous effluents.

The distillation procedure described in Procedure H-7 appears satisfactory for most applications. Though the ion exchange isolation of 0.1 μg. to 1 mg. fluoride from samples collected from the atmosphere is useful, there appears little need for this technique for use with samples from plant effluents because of the larger quantities of fluoride in the samples. For handling a large number of samples, the diffusion separation techniques are capable of isolating fluoride from interferences and concentrating it into 5 milliliters with the possibility of labor savings.

Analytical Methods: Analytical methods are discussed in several sections as indicated below:

Spectrophotometric
Titrimetric
Instrumental
Continuous and semicontinuous

As previously noted, aliquots from the samples collected should be subjected to some process for separation of fluoride from interfering ions. The aliquots to be taken for most efficient separations should be only as large as those required by the selected analytical method. For the most accurate analyses, the aliquot should provide a mid-range fluoride concentration for that method.

Table 26 gives the applicable concentration ranges for the various analytical techniques described. The concentration ranges in the table and the following discussions are for solutions containing fluoride ions after separation from interfering ions.

Spectrophotometric Analysis — The spectrophotometric methods have been developed to the point where several are accepted as standards. After the separation of soluble fluorides from interfering ions, spectrophotometric methods can generally be used to determine fluoride with a relative precision of 5 to 10% for solutions containing 0.01 μg. to 0.2 mg. of fluoride per milliliter. The claim in some publications of better precision is largely unsubstantiated. The accuracy, except on standard solutions containing NaF, has not been established but should be about the same as the relative precision for solutions that do not contain any interfering ions. Little or no data exist concerning total system accuracy, i.e., sampling, collection, removal of interfering ions and spectrophotometric analysis.

Interaction of fluoride ion with a metal dye complex generally forms the basis for the colorimetric-type methods. The metals of the complex are from the group Th, Zr, La, Ce, Y, Bi, Fe, and Al. This group is capable of forming insoluble or slightly ionized fluorides and also insoluble phosphates which is a well-known interferent. Some of the more common dyes used for this purpose are Alizarin Red S, Eriochrome cyanine R, arsenazo, Ferron, and SPADNS. Many of these dyes function as acid base indicators and, therefore, require close control of pH in fluoride determination (11).

Many semiquantitative and qualitative techniques have been used for estimation of fluoride; while these are not spectrophotometric, they are colorimetric and a typical example is discussed here. Mavrodineauu (37) describes a color complex for fluoride ion sample on dry zirconium or thorium nitrate and a lake-forming dye (sodium alizarin) absorbed on filter paper. No interference was noted for other halogens, but sulfate and phosphate interfered. Semiquantitative results could be achieved by acid treatment and color intensity comparison.

Many color complex systems for the determination of fluoride spectrophotometrically have been described in the literature. Generally, spectrophotometric methods provide a means for measuring a 20,000 fold range of fluoride concentration directly with very good

TABLE 26: APPLICABLE CONCENTRATION RANGE OF ANALYTICAL METHODS

Technique	Applied to Aliquots of Fluoride Concentration Range	Precision	Interferences**	Comments
FLUOROMETRIC				
Morin or quercetin	0.5-20µg	±5%	Fe^{+3},$C_2O_4^{-2}$,Cl^-,Mn,NO_3	Better accuracy is claimed for these methods than for visual end-point techniques for concentration above 2mg.
Specific ion electrode	0.02µg/ml-20 mg/ml (5 ml minimum sample)	±1%	OH^- total ionic strength	Preferred method of end-point detection in most cases. Precision is better than other titrimetric procedures.
INSTRUMENTAL				
Specific ion electrode	0.03µg/ml to 30 mg/ml	5% standard deviation at low range, 2% at high range	OH^-,Al^{+3},Fe^{+3},pH adjusted	Ease of use and equal precision justifies use in most cases in place of spectrophotometric methods.
Kinetic method	0.0004 µg-0.4 mg/ml	Not established research method	SO_4^{-2},Cl^-,Al^{+3},PO_4^{-3}	Research method.
Atomic Absorption	0.005 µg/ml-4 mg/ml	~5% standard deviation	SO_4^{-2}, PO_4^{-3}	Can be used over a wide concentration range. Useful when a large number of samples are to be analyzed.
X-Ray of LaF_3	1µg-apx 10mg	~5% standard deviation	None	Research method that can be developed into a rapid method.
Radio-release of zirconium salt	10µg-100 g	~5% relative precision	PO_4^{-3},Fe^{+3},Al^{+3}	Research method.
Amperometric	0.5µg/ml-10 µg/ml	Not established	None	Research method; could be used to detect titration end-points.
Photo-activation	0.01%-5% in mg size samples (dried)	~5% relative precision	Cl,Br,S	Useful for small samples.
Mass spectrometric	0.1 mol%-100 mol% for Hf,SiF_4,CF_4,C_2F_6	~5% relative precision	None	Determination of HF difficult, useful for determining organic bound flourine.
Electrochemical				
Null point measurement of cerium (IV) to (III)	10µg/ml-1 mg/ml	None given	Al^{+3},Fe^{+3},PO_4^{-3}	Research method.
Coulometric	0.001µg/ml-100 µg/ml of F_2	None given	None	Specific method for F_2.
Gas Chromatography	1mg/cc-100% as HF or SiF_4 in gas sample	~5% relative precision	None	Could be developed into an automatic method.
SPECTROPHOTOMETRIC*				All the spectrophotometric methods suffer from interferences from ions that form more insoluble compounds with the metal of the complex than the fluoride itself. pH changes also effect most of these methods.
Lanthanum-Alizarin Complexone	0.01-0.4µg/ml	±5%	PO_4^{-3},Al,Cl^-,Fe,NO_3^-, $C_2O_4^{-2}$	Nitrate and phosphate interfere only when in excess.
Titanium-Chromatropic Acid	2µg-0.2mg/ml	±5%	PO_4^{-3},Al,Fe,$C_2O_4^{-2}$ (SO_4^{-2} in excess)	Sulfate does not interfere. K^+,Na^+,NH_4^+,Cl^- and NO_3^- do not interfere in small amounts.
Iron-Ferron	0.01-0.2 mg/ml	±5%	PO_4^{-3},Al,Fe,oxalate, SO_4^{-2}	Useful at higher fluoride concentration levels.
Thorium-Alizarin	0.01µg-0.2 mg/ml	±5%	PO_4^{-3},SO_4^{-2},Al,Fe, oxalate	Calibration not linear at higher fluoride concentrations
Zirconium-SPADNS	0.01µg-0.2 mg/ml	±5%	PO_4^{-3},Cl^-,Al,Fe,oxalate	Calibration not linear at higher fluoride concentrations
Amadac-F	0.5-4µg/ml	±5%	PO_4^{-3},Cl^-,Al,Fe,oxalate	Affected by high acid or alkali content, pH change, and total ionic strength.
Zirconium-Eriochromcyanin R	0.01µg-0.2 mg/ml	±5%	PO_4^{-3},Cl^-,Al,Fe,oxalate (SO_4^{-2} in excess)	Calibration is linear from 0.01-2µg/ml.
TITRIMETRIC				Thorium nitrate is usually preferred as titrant. Lanthanum is also used in some cases.
Visual				
Purpurin sulfonate	1µg-10 mg	>±5%	PO_4^{-3},SO_4^{-2},Al,Fe,$C_2O_4^{-2}$	
Alizarin Red S	1µg-10 mg	>±5%	PO_4^{-3},SO_4^{-2},Al,Fe,$C_2O_4^{-2}$	Listed in order of preference as indicators.
Eriochromcyanin R	1µg-10 mg	>±5%	Fe,oxalate,Cl^-,Mn,NO_3, etc.	
Photometric				
(Same metal-dye complexes can be used)	1µg-100 g	±5%	Same	Technique requires a colorimeter and is slower than visual end-point; however, operator error is reduced.

*Accuracy is not known for most methods except for standard solutions where the accuracy is the same as the precision.
**Interferences usually removed by distillation, ion exchange or diffusion.

Source: Report PB 207,506

sensitivity. Ranges for two common systems are reported as follows: Iron-Ferron, 0.01 to 0.2 mg./ml. (1 cm. cell) and 0.01 to 0.4 μg./ml. for Lanthanum-Alizarin "Complexone" reagents. Both these systems have visible spectrum absorptions. Belcher and West (38) report 200% increases in sensitivity by working in the ultraviolet region of the electromagnetic spectrum.

Decolorization of Titanium-Chromotropic acid by fluoride ion with a detection level of 2 μg./ml. (total range 2 μg. to 0.2 mg. per ml.) was proposed by Babko and Khodulina (39). No interference was observed from sulfate, but phosphate must be removed. Sensitivity to pH is a problem common to this technique. If a dye is added to the system, the resultant color change can increase sensitivity to <0.5 μg./ml. Skanavi (40) applied this method to micro quantities of fluoride with a sensitivity in the range 0.3 to 17 μg. per ml.; however, accuracy was not too good in this range. Phosphate interferes, but K^+, Na^+, NH_4^+, $SO_4^=$, Cl^-, and NO_3^- in small amounts do not cause problems.

Mal'kov et al (41) outlined a method using thorium-alizarin to form a colored complex with fluoride. Range of the method is 0.01 μg. to 0.2 mg. fluoride per milliliter. Amadac-F (42) is a mixture of alizarin complexan, lanthanum nitrate, acetic acid, partially hydrated sodium acetate and stabilizers useful for quantitative determination of fluoride in the range 15 to 50 μg. per milliliter. A color change is observed in this reaction complex which is affected by high acid or alkali content, pH change, and total ionic strength.

Green iron-Ferron complexes with fluoride to produce a color change useful for fluoride measurement in the range 0.01 to 0.2 mg./ml. Adams (43) in discussing this work proposed the use of this method for stack monitoring with removal of sulfur dioxide, an interferent, by sodium tetrachloromercurate absorber solution. A prior reference (44) utilized zirconium-Eriochrome cyanine R for the determination.

A recent spectrophotometric technique described by West et al (45) analyzes fluoride by complexing with alizarin complexan and lanthanum buffer. Determinations in the range 0.01 to 0.4 μg./ml. can be done if metals, nitrates, and phosphates are removed. That is, concentrations of <4 μg./ml. nitrate and <3 μg./ml. phosphate in 0.4 μg./ml. of fluoride are tolerable.

Because of the large volume of literature and numerous possible combinations of metal-dye-fluoride complexes, the above summary must be considered only as a few typical recognized procedures which reflect the possibilities of the spectrophotometric technique. The methods described here generally can be considered as new techniques or the later modification of older techniques. It is very difficult to make a statement of preference for any of these methods unless dynamic range of applicability of specific interference are the judgement criteria. Sensitivity and precision are nearly the same for each method.

Titrimetric Analysis – Titration methods using indicators to detect the end-point are all difficult to perform with a high degree of precision and have been superceded, to a major extent, by the use of the specific ion electrode to determine fluoride ion content either directly for relatively dilute solutions, or by the use of titrimetric methods employing a specific ion electrode for end-point determination.

The most commonly used titrants for the volumetric or titrimetric determination of fluoride in aqueous systems are thorium and lanthanum nitrate. However, because of the large variety of end-point detection procedures the classification of the titrimetric methods will be based on the detection technique utilized. End-point detection can be generally broken down into the following types: visual, photometric, fluorometric, specific ion electrode, and electrometric. While specific ion electrodes may be classed under electrometric, their relative importance dictates a separate class for this discussion. The first four types will be evaluated in this section, where titration of the total sample distillate, thus preventing dilution error, is possible.

The visual end-point detection procedures have generally been supplanted by other means for end-point detection and by spectrophotometric methods. Photometric, fluorometric, electrometric and specific ion electrode end-point detection have largely eliminated the operator perception and dilution errors present in the visual methods.

Visual — The greatest difficulty in the quantitative utilization of the color indicator end-point techniques for fluoride is that it depends on the color perception and experience of the operator. Many indicators have been suggested for improving the subtle color change; however, there still remains much to be done. Much work has been done by Willard et al (21) on these as well as other systems with the following colorimetric indicators being recommended: Purpurin sulfonate, Alizarin Red S Eriochrome cyanin R, dicyano-quinizarin, and Chrome Azurol S.

In visual procedures the sample of fluoride is titrated with thorium or lanthanum nitrate to the end-point as indicated by one of the suggested complex colorimetric (visual) dyes. Generally the methods using visual indicators to detect the end-point are used in the fluoride concentration range of 1 μg. to 10 mg. in an aliquot from 10 to 200 ml. Analysis in this range is described in ASTM Method D1606-60.

Not only do these titrations suffer from the abovementioned operator error, but other problems exist depending on the composition of the sample to be analyzed. When large amounts (above 1 mg.) of fluoride are titrated, interference may result from semicolloidal thorium nitrate, and medium-to-high concentrations of metals, nitrates, and phosphates interfere in most cases.

Allison (46) in his work, compared the determination of fluoride by both volumetric visual end-point detection and a colorimetric (spectrophotometric) method with the conclusion that the latter technique was more sensitive and should be used in the 0.5 to 50 μg./ml. range, while the volumetric was faster and more useful for concentrations above 50 μg./ml.

Photometric — The real advantage in using a photometer to determine the end-point in a fluoride determination lies in the elimination of the variable of operator perception differences. All the other parameters remain essentially the same as for the visual indicator method above.

Fluorometric — Here the dyes recommended for use are different from the visual indicator dyes because of the requirement for measuring fluorescence changes to detect the end-point. Willard et al (47) recommend two: pure sublimed morin and quercetin. The titration is again carried out using thorium nitrate, while the end-point is observed by the fluorescence change. Better accuracy is claimed for this method than for the color end-point method for fluoride concentrations greater than 2 mg. Many variables again need to be controlled, such as pH and interfering ions.

Willard et al (47) also describe a fluorometric technique for the determination of trace amounts of fluoride using aluminum-oxine or aluminum-morin systems. In these systems the fluoride complex with aluminum decreases the aluminum-oxine or morin complex. The resultant change in fluorescence of the system is measured. The range of sensitivity to fluoride is around 0.5 to 20 μg. total sample. Many variables must be controlled and standards should be run with each set of unknowns. Ions that react with aluminum or oxine or which precipitate with fluoride at pH 4.7 must be removed.

Specific Ion Electrode — The use of the fluoride specific ion electrode (lanthanum fluoride membrane electrode) for end-point detection is a recent innovation and is discussed by Lingane (48) and Frant et al (49). The conclusions reached by these and other investigators point out the usefulness of this technique. Sensitivity to fluoride over a concentration range of five orders of magnitude is easily achieved while ultimate sensitivity is down to 10^{-7} M fluoride. The electrode is very selective to fluoride, but pH and total ionic strength are very important considerations in the analysis. In typical titrations of fluoride with thorium and lanthanum nitrate, the latter yielded the best potential break with precision to ±1 mv. Far better end-point accuracy and precision were achieved using the electrode than could be achieved using color indicators for detection. The useful range of this end-point detection method is for solutions in the concentration range of 0.1 μg. to 20 mg./ml.

Schultz (50) points out that large errors can result from potentiometric titrations employing ion-selective electrodes. The error increases as the sample ion concentration decreases and as the interfering ion concentration, solubility product constant, and dilution factor increase. Of the abovementioned end-point detection methods the fluoride electrode technique is the most precise (interference removed) and generally the easiest to apply.

Instrumental Methods – Nearly every analytical instrument has been investigated for direct determination of fluoride. Many of these instruments have been previously discussed as detectors for titrimetric end-points, but in this section instrumental methods are discussed as they apply to direct determination of fluoride either as collected or after separation from interferences common to most analytical methods. The various instrumental techniques are discussed below.

Specific Ion Electrode – The accepted dynamic range for the fluoride specific electrodes is from 50 mg./ml. down to 0.10 μg. in the minimum usable volume of 5 ml. Preliminary work with this electrode has shown promise of making fluoride ion determinations virtually as simple, rapid, and precise as hydrogen ion activity measurements with the glass pH electrode. It must be remembered that fluoride activity is measured and concentration is dependent on total ionic strength as well as other factors.

Harriss et al (51) discuss the direct measurement of fluoride with the specific ion electrode and noted the speed and low cost of this analysis. Baumann (52) describes the interference from hydroxyl ion and its elimination for accurate fluoride analysis. As little as 10^{-5} M fluoride (1 μg. in 5 ml.) could be analyzed with a relative error of ≅ 10% and standard deviation of $<5\%$. He suggested that interfering ions be complexed before fluoride analysis. Electrode response time was less than one minute in these experiments. Durst et al (53) describe microchemical analysis techniques for fluoride using the electrode.

Because of the importance of the total ionic strength on the fluoride concentration measurement with the electrode and the effect of acidic or basic media on the values of fluoride, it is necessary to control or evaluate these parameters. Vanderborgh (54) used a lanthanum fluoride membrane electrode in his study of response in an acidic solution with varying ionic strength. A recent article by Bruton (55) for the useful range of the electrode points out that the known addition technique can be used for the simple and accurate analysis of fluoride. The millivolt readout for the electrode is adjusted to zero in the sample, an addition of standard fluoride is made, and the change in potential is related to fluoride concentration. The activity coefficient must remain constant for accurate measurement; where necessary this can be accomplished by the addition of a noninterfering salt.

Frant et al (49) adjusted the total ionic strength, the pH, and complexed ferric iron or aluminum (citrate used) by using a buffer in a 1/1 ratio with the samples and standards. Fluoride could be determined accurately over the entire useful electrode concentration range using a single calibration curve for a wide range of samples.

Kinetic Method – Because a kinetic method employs unusual instrumentation, the kinetic method is included under the instrumental section. Klockow, et al (56) developed a kinetic method for the determination of traces of fluoride (3.8×10^{-2} to 3.8 μg./ml.) based on strong inhibiting action. Fluorides act as a negative catalyst in the zirconium-catalyzed reaction between perborate and iodine. Kinetic measurements are accomplished by an automatic potentiostatic technique. Only small quantities of extraneous ions can be present.

Atomic Absorption – Bond et al (57) applied the depression of absorption of the magnesium line at 285.2 mμ into an atomic absorption method for fluoride in the range of

0.005 μg./ml. to 2,000 mg./l. using an air-coal gas flame. Both SO_4^{-2} and PO_4^{-3} ions must be absent. A somewhat less sensitive method (5 to 500 μg./l.) was also established based on the enhancement of zirconium absorption by fluoride in the nitrous oxide-acetylene flame. They also established an even less sensitive method (400 to 4,000 mg./l.) without interference based on the enhancement of titanium absorption. These methods offer the advantage of being rapid, and little handling of collected samples is required.

X-Ray Spectrography – An x-ray spectrographic method (58) was established for measuring fluoride collected by nearly any of the previously described sampling methods, adjusting the pH of the solution containing the fluoride, and collecting the fluoride as LaF_3 on a Millipore disc with a pore size of 2μ. The disc is washed, dried and submitted to x-ray spectrographic measurement with a tungsten target and a lithium fluoride analyzing crystal. Fluoride can be detected in the range of 1 μg. to about 10 mg. without interference.

Polarographic – MacNulty, et al (59) applied to fluoride determination the reduction of the polarographic half-wave potential at 0.3 v. versus saturated-calomel electrode (pH 4.6 in acetate buffer) for the sodium salt of 5-sulfo-2-hydroxy-α-benzene-azo-2 naphthol in the presence of aluminum. Fluoride complexes aluminum and reduces the half-wave potential. The method can detect 0.2 μg./ml. but the method precision and the maximum concentration of fluoride that could be detected was not investigated.

Radio Release – Carmichael et al (60) established a radio-release method for determination of fluoride (20 to 100 μg.). The fluoride is converted to a zirconium salt, placed in a neutron flux, and the radioactivity release measured. The relative precision is about 5%; PO_4^{-3}, Fe^{+3} and Al^{+3} interfere.

Amperometric – A British patent (61) was issued for an amperometric method for fluorides from an air sample that was collected in 0.5 M nitric acid. The fluoride is determined by a platinum wire or plate electrode and a zirconium wire electrode rotating at 300 to 1,600 rpm and maintained between -1 and +1 volt with respect to an S.C.E. The current passing between the electrodes was measured. The method can measure 0.05 μg. to 1 μg./ml. in the collecting solution. This method was not applied to detecting titration end-points, but could be considered.

Photo Activation – Kosta et al (62) demonstrated the use of photo activation for determining fluoride in the concentration range of 0.01 to 5% on as little as 1,160 μg. of sample. The method has not been applied to gas stream fluorides, but could be used to determine the fluoride content of particulates collected on a filter. Interferences from elements such as chlorine, bromine, and sulfur can be eliminated or reduced to a minimum by adjusting irradiation time, waiting period, and energy of the primary electron beam. Results obtained were in good agreement with those obtained by distillation-titration methods.

Mass Spectrometric – Mass spectrophotometric analysis of anode gases from aluminum reduction cells was accomplished by Henry et al (63) for HF, SiF_4, CF_4 and C_2F_6. The method determined the substances from 0.1 mol percent to 100 mol percent.

Electrochemical – Curran et al (64) determined fluoride by precipitating fluoride ions with lanthanum ion electrochemically generated from lanthanum hexafluoride anode. The end-point was detected with a fluoride specific ion electrode. Fluoride was determined by null point potentiometric measurement of the cerium (IV) - cerium (III) reduction potential (65) for solution containing greater than 14 μg. of fluoride per ml. of solution. The method can be applied to Willard and Winter distillates or ion exchange eluates.

A coulometric specific method was established by Kaye et al (66) for free fluorine in a gas stream. In this method gas is aspirated at constant flow rates between 100 and 300 ml./min. through 0.2M LiCl. The fluorine oxidizes the Cl^- with one mol of fluorine corresponding to two atoms of Cl^-. The quantity of fluorine is determined coulometrically using a silver anode and a platinum gauze cathode. The method determines 0.1 ppm up to about 100 ppm of fluorine.

Gas Chromatography – The analysis of various fluorine-containing compounds was investigated by Pappas et al (67). They found the high affinity of HF toward almost any surface to be a problem. By the use of Teflon columns prepared with Teflon-6 support coated with fluorocarbon oil and carrier gas spiked with HF as proposed by Knight (68) they found that HF along with other fluorine compounds in the concentration range of 0.1 mol percent to 100 mol percent could be separated and measured with a gas density balance. Air was used as the carrier gas, but greater sensitivity than the ppm level they observed could be achieved with a carrier gas such as sulfur hexafluoride. More sensitive detectors have not been investigated.

Infrared Spectrometry and Infrared Lasers – Hydrogen fluoride has an absorption band at 3961.6 $cm.^{-1}$, but this absorption band has not been used for determinations in the stack gases because of interactions with water vapor. SiF_4 has absorption bands at 1010 and 800 $cm.^{-1}$. A recent review of infrared lasers for monitoring air pollution by Hanst (69) proposed a method using a Kr laser and the infrared absorption line for HF.

Instrumental Methods Summary – With the exception of the specific ion electrode, the instrumental methods presented are useful for only special cases. Because the specific ion electrode is accurate when properly used and easy to use, the specific ion electrode is recommended for fluoride determination whenever possible.

Continuous and Semicontinuous Methods – There are no continuous or semicontinuous methods in use for the determination of the fluoride content of the various gaseous effluents from manufacturing processes and the abatement systems employed in connection with the processes. Most continuous or semicontinuous methods were developed for measuring fluoride content of the ambient atmosphere. These methods are summarized here because they can be considered as candidates for continuous monitoring of plant effluents.

The analysis of air for detecting fluorine compounds in the parts per billion concentration range is usually done by aspirating a large volume of air through distilled water or dilute alkali, concentrating the fluoride by distillation, ion exchange or diffusion (as discussed under titration and colorimetric methods), and then determining the quantities of fluorides by titrimetry or colorimetry. The distillation or ion-exchange step can be omitted only in special cases. Collection of enough fluoride for analysis may take several hours to one or two days, thus giving long term average concentrations. A method for continuous determination of fluoride content of process stream effluents is needed. The various approaches that have been developed are:

Mini-Adak Colorimetric Analyzer
Fluorescence-Quenching Method (SRI Fluoride Recorder)
Billion-Aire Ionization Detector
Current Flow Method
Specific Ion Electrode Method

Mini-Adak Analyzer – In 1956, Adams, Darra and Koppe (70) reported on a prototype photometric fluoride analyzer for use with a liquid reagent. In 1959, Adams et al (71) studied this instrument extensively and established that there was an excellent correlation between the instrument and conventional sampling and analytical procedures for total soluble ion-producing fluoride pollutants.

Basically, the instrument may be characterized as a recording flow colorimeter in which the flow forms an integral part of the air-reagent absorption system. As a fluoride analyzer, it photometrically measures and records the rate of reaction of zirconium-Eriochrome Cyanine R reagent with concentration of soluble fluorides in a sampled air stream throughout a given sampling period. High sensitivity is achieved by an unusual absorber which permits the contact of a small volume of liquid with a large volume of air (1 cu. ft. per min.). The efficiency of hydrogen fluoride absorption is reported to be 95%. The volume of liquid is kept constant by automatic addition of water to replace evaporation losses. The liquid is periodically discarded and replaced by a measured volume of fresh solution. The addition of fluoride ion to the zirconium-Eriochrome Cyanine R reagent shifts the absorption maximum to 550 mμ. The color is measured continuously by a recording colorimeter. Response of the recorder to fluoride is nearly linear at 1 scale division per μg. of fluoride ion per 15 ml. solution until 20 μg. are added. In the range from 0.75 to 35 ppb hydrogen fluoride, the standard error of estimate was about 0.8 ppb.

Fluorescence-Quenching Methods (SRI Fluoride Recorder) – Chaikin and Associates at Stanford Research Institute developed a fluoride recorder (72)(73) which was further modified (74)(75) to provide an instrument that could operate under field conditions. The method consists of drawing parallel air streams into the analyzer through warmed glass tubes; one tube coated with $NaHCO_3$ and the other clean. The $NaHCO_3$ absorbs hydrogen fluoride, but the clean tube allows it to pass. The air streams are drawn through adjacent spots on sensitized paper tape made by dipping chromatography paper in a methanol solution of 8-hydroxyquinoline and magnesium acetate.

The resulting magnesium salt of 8-hydroxyquinoline fluoresces when illuminated with ultraviolet light. The visible fluorescence is quenched by hydrogen fluoride, thus providing a quantitative measure of fluoride. The difference in emitted light from the two areas of paper type is monitored by reflecting the two beams of light onto balanced photomultiplier tubes. Differential output from the tubes is recorded on a strip chart recorder. This instrument is 50 to 100 times more sensitive than the Adak recorder and determines only hydrogen fluoride, not total fluoride. The instrument can detect hydrogen fluoride in the range 0.2 to 10 ppb and appears free of interferences by common air pollutants. However, the instrument requires additional field testing.

Billion-Aire Ionization Detector – The Billion-Aire Ionization Detector manufactured by the Mine Safety Appliance Company (76) lends itself to the detection of hydrogen fluoride and fluorine in air. When a gas is ionized in the detector between two oppositely charged electrodes, a current is conducted depending primarily on the strength of the ionizing source, the applied voltage, and the composition and pressure of the gas. With air in the detection chamber, most gaseous additives in the concentration range of several thousand parts per million will cause only small changes in ion current.

However, very small concentrations of finely divided particulate matter produce a pronounced decrease in current. The action of particles is to promote effective recombination through third body collisions and to decrease mobility through attachment. By converting a gas to particulate matter by a suitable reaction and measuring the decrease in ion current due to the presence of the particles in an ion chamber, many contaminants can be detected in the concentration of ppb. For example, HF, HCl and NO_2 can be converted to particulate aerosol by reaction with ammonia. Though the instrument is not specific for hydrogen fluoride, it provides instantaneous response for hydrogen fluoride and fluorine concentrations of 1 to 100 ppb.

Current Flow Method – Howard, et al (77) developed a portable fluoride analyzer based on the fact that the current from an aluminum-platinum internal electrolysis cell is a function of the fluoride content of an acetic acid electrolyte, after the sampled air has been scrubbed with the electrolyte. The analyzer responds to all substances which form fluoride ion in aqueous solution and is specific for fluoride in the presence of common contaminants. The method is capable of detecting from 5 to 100 ppb for a 2-liter sample.

A second electroanalytical instrument that detects fluorine but not fluorides was developed by Kaye et al (66). In this instrument, air containing fluorine oxidizes Cl^- ion in a buffered LiCl solution in a solution containing platinum and silver electrodes. The chlorine formed is reduced at the silver cathode. Insoluble AgCl is produced on the cathode so that chloride is removed from the solution. For every molecule of fluorine, two electrons flow through the coulometric circuit and two atoms of chlorine are transformed from solution to cathode. By using a pump to deliver a constant flow of air to the instrument one can determine fluorine concentrations between 5 and 1,000 ppm without interferences from other atmospheric oxidants.

Specific Ion Electrode Method – Light (78) discussed the adaptation of the fluoride ion specific electrode to the continuous monitoring of gas streams. By simply scrubbing the gaseous constituents with a suitable reagent and measuring the quantity of gas, reagent solution, and concentration of the resulting solution with the electrode automatic monitoring can be achieved. Direct application of this technique to effluent gas analysis as yet has not been reported, but recently has been applied by Mori, et al (79) to the determination of hydrogen fluoride in the atmosphere.

The hydrogen fluoride is collected by absorption on dry sodium carbonate coated glass tubes. The sodium carbonate and collected fluorides are washed to a collection container, the solution buffered, and the fluoride concentration determined with the specific ion electrode. Automatic cycling of the apparatus provides a continuous recording of the hydrogen fluoride concentration of air.

Measurement in Water

As noted in the text immediately preceding, fluoride ions in gaseous effluents are generally measured after absorption into water solution. Therefore the analytical techniques described above for measurement of fluorides in air are equally applicable to airborne fluorides which have been absorbed in aqueous scrubbing solutions and to wastewater containing fluoride contaminants and the reader is referred in either case to the preceding discussion. Fluorides may be monitored in water by the following instrumental techniques (80):

Ion Selective Electrode
Molecular Absorption (Colorimetric)

References

(1) Little, Arthur D., Inc., *Report PB 213,313,* Springfield, Va., Nat. Tech. Info. Service (Nov. 1972).

(2) Robinson, J.M., Gruber, G.I., Lusk, W.D. and Santy, M.J., "Engineering and Cost Effectiveness Study of Fluoride Emissions Control," *Report PB 207,506,* Springfield, Va., Nat. Tech. Info. Service (Jan. 1972).

(2a) Stern, A.C. in *Industrial Pollution Control Handbook,* H.F. Lund, Ed., New York, McGraw-Hill Book Co. (1971).

(3) W. Zabban and H.W. Jewett, "The Treatment of Fluoride Wastes," Proc. 22nd Purdue Industrial Waste Conf., 22:706-716, 1967.

(4) J.M. Cherry, "Fluorine Recovery in Phosphate Manufacture," *Wat. Wastes Eng.,* 7:D-5, 1970.

(5) R.O. Sylvester, R.T. Oglesby, D.A. Carlson, and R.F. Christman, "Factors Involved in the Location and Operation of an Aluminum Reduction Plant," Proc. 22nd Purdue Industrial Waste Conf., pp. 441, 1967.

(6) Dorsey, James A. et al, "Source Sampling Technique for Particulate and Gaseous Fluorides," *J. Air Pollut. Contr. Ass., 18*(1), 12-14 (1968).

(7) Elfers, Lawrence A. & Decker, Clifford E., "Determination of Fluoride in Air and Stack Gas Samples by use of an Ion-Specific Electrode," *Anal. Chem., 40*(11), 1658-61 (1968).

(8) Lunde, K.E., "Performance of Equipment for Control of Fluoride Emissions," *Ind. Eng. Chem., 50,* 293-8, (1958).

(9) Doerschuk, V.C., "Electrolytic Production of Aluminum," Canadian Patent 613,352 (Jan. 1961).

(10) Knapp, Lester L., "Treatment of Gases Evolved in the Production of Aluminum," U.S. Patent 3,503,184 (March 31, 1970).

(11) Farrah, H., "Manual Procedures for Estimation of Atmospheric Fluorides," *Air Pollution Control Assn., 17,* No. 11, 738-41 (Nov. 1967).

(12) Pack, M.R., "Taking Fluoride Samples from the Atmosphere with Glass Filters," *J. Air Pollution Control Assoc., 13,* 374-77 (1963).

(13) Keenan, R.G., *Industrial Hygiene and Toxicology, 26,* pp. 241, (1944).

(14) Barber, J.C., "Fluoride Recovery From Phosphorus Production," *Chemical Engineering Progress, 66,* No. 11, p. 56-62, (Nov. 1970).

(15) Ryss, I.G., *Chemistry of Fluorine and Its Inorganic Compounds,* State Publishing House for Scientific, Technical, and Chemical Literature, Moscow, 1956, AEC Tr 3927, Part 1 & 2 U.S.
(16) Anonymous, *Superphosphate: Its History, Chemistry and Manufacture,* U.S. Dept. of Agriculture and Tennessee Valley Authority.
(17) Great Britain *84th Annual Report Alkali Etc.,* Works by the Chief Inspectors, 1947, H.M. Stationery Office (1948) and *85th Annual Report* (1949).
(18) Semrau, Konrad T., "Emission of Fluorides from Industrial Processes - A Review," *J. Air Pollution Control Assoc., 7,* 92-108 (1957).
(19) Cochran, C.M., "Fumes in Aluminum Smelting: Chemistry of Evolution and Recovery," *Journal of Metals,* 54-57 (Sept. 1970).
(20) Willard, H.H., *Industrial Engineering Chem. Anal. 5,* p. 7 (1933).
(21) Willard, H.H. et al, *Anal. Chem. 22,* 1190 (1950).
(22) Shuey, G.A., *J. Assoc. Official Agr. Chemists, 18,* p. 156 (1935).
(23) Reynolds, D.S., *J. Assoc. Official Agr. Chemists, 18,* 108 (1935).
(24) Armstrong, W.D., *Ind. Eng. Chem. Anal. Ed., 8,* 1936, 384 (1936).
(25) Nielsen, J.P., *AMA Arch. Ind. Hyg. and Occup. Med., Vol. II,* 61 (1955).
(26) Newman, A.C.D., *Anal. Chem. Acta., 19,* 471 (1958).
(27) Funasaki, W., and Kawane, M., "Removal of Fluoride Ion with Anion-Exchange Resins," *Mem. Fac. Eng., 18,* Kyoto Univ. 44-50 (1956).
(28) Ziphin, I., et al, *Anal. Chem., 29,* 310 (1957).
(29) Nielsen, H.M., *Anal. Chem., 30,* 1004 (1958).
(30) Singer, L., et al, *Anal. Chem., 26,* 904 (1954).
(31) Hall, R.J., *Analyst, 68,* 76 (1963).
(32) Alcock, N.W., *Anal. Chem., 40,* 1397 (1968).
(33) Taves, D.R., *Anal. Chem., 40,* 204 (1968).
(34) Taves, D.R., *Talanta, 15,* 969 (1968).
(35) Tusl, J., *Anal. Chem. 41,* 352 (1969).
(36) Stuart, J.L., *Analyst, 95,* 1032 (1970).
(37) Mavrodineanu, R. "Fluoride-Ion Indicator," U.S. Patent 2,823,984, (Feb. 18, 1958).
(38) Belcher, R., *Talanta, 8,* 863-870 (1961).
(39) Babko, A.K., *Zhur. Anal. Khim., 7,* 281-284 (1952).
(40) Skanavi, M.D., *Determination of Small Amounts of Hydrogen Fluoride with the Titanium-Chromotropic Acid Reagent,* Zavodskaya Lab., *24,* 683 (1958).
(41) Mal'Kov, E.M. et al, "Novye Melody Analiza Khim," *Sostava Podzemn,* Vod., 59-64 (1967).
(42) Burdick and Jackson Lab., "Amadac-F Data-Sheets" Burdick and Jackson Lab., Muskegon, Mich.
(43) Adams, D.F., *Anal. Chem., 32,* 1312-1316 (1960).
(44) Adams, Donald F.,"An Automatic Atmospheric Fluoride Analyzer with Potential Application to Other Pollutants," *J. Air Pollution Control Assoc., 9,* 160-8 (1959).
(45) West, P.W. et al, *Env. Sci. and Technol., 4,* 487-491 (1970).
(46) Allison, R.S., "The Determination of Small Amounts of Fluorine in Glass," *J. Soc. Glass Technol. 37,* 213-18T (1953).
(47) Willard, H.H. et al, *Anal. Chem., 24,* 862-865 (1952).
(48) Lingane, J.J., *Anal. Chem., 39,* 881 (1967).
(49) Frant, M.S., et al, *Scil., 154,* 1553 (1966).
(50) Schultz, F.A., *Anal. Chem.,* Vol. 43, 1971, p. 502.
(51) Harriss, R.C. et al, "Specific Ion Electrode Measurements on Bromine, Chlorine and Fluorine in Atmospheric Precipitation," *J. Appl. Meteorol., 8*(2), 299-301 (1969).
(52) Baumann, E.W., *Anal. Chim. Acta., 42,* 127-132 (1968).
(53) Durst, R.A., et al, *Anal. Chem., 39,* 1483-1485 (1967).
(54) Vanderborgh, N.E., *Talanta, 15,* 1009-1013 (1968).
(55) Bruton, L.G., *Anal. Chem., 43,* 579 (1971).
(56) Klockow, D., *Anal. Chem., 42,* 1682 (1970).
(57) Bond, A.M., "Determination of Fluoride by Atomic Absorption Spectrometry," *Anal. Chem., 40*(3) :560-563 (March 1968).
(58) Reva, B., *Anal. Chem. Acta., 43,* 245 (1968).
(59) MacNulty, B.J., et al, "Polarographic Determination of Fluoride," *Nature, 1952,* 169, 888-9.
(60) Carmichael, I., et al, *Analyst, 94,* 737 (1969).
(61) Anonymous, British Patent 1,156,915, (July 10, 1965).
(62) Kosta, L., et al, *Anal. Chem., 42,* 831 (1970).
(63) Henry, Jack L. et al, "Mass Spectrometric Examination of Anode Gases from Aluminum Reduction Cells," *J. Metals 9,* Aime Trans. 209, 1384-5 (1957).
(64) Curran, D.J., *Anal. Chem., 40,* 267 (1968).
(65) O'Donnell, T.A., *Anal. Chem.,* 33, 337 (1961).

(66) Kaye, Samuel et al, "Electroanalytical Determination of Molecular Fluorine in the Atmosphere," *Anal. Chem. 40*(14), 2217-18 (1968).
(67) Pappas, W.S. et al, *Anal. Chem., 40,* 2178 (1968).
(68) Knight, H.S., *Anal. Chem., 30,* 2030 (1958).
(69) Hanst, P.L., *Applied Spectroscopy, 24,* 161 (1970).
(70) Adams, D.F. et al, Instrument Society of America, New York, (Sept. 1956).
(71) Adams, Donald F. et al, "Automatic Atmospheric Fluoride Pollutant Analyzer," *Anal. Chem. 31,* 1249-54 (1959).
(72) Chaikin, S.W. and Parks, T.D., "Hydrogen Fluoride Detector," U.S. Patent 2,741,544, (April 10, 1956).
(73) Clauss, J.K., "Study of Fluorescent Tape Sensitive to Hydrogen Fluoride," *D.R.C. Report,* Stanford Research Institute, (May 1957).
(74) Ivie, J.O., "Atmospheric Fluorometric Fluoride Analyzer," *J. Air Pollution Control Assn., 15,* No. 5, 195-7 (May 1965).
(75) Thomas, M.D., *An Atmospheric Fluoride Recorder,* Am. Soc. Testing Materials, Spec. Tech. Publ. No. 250, 49-57 (1959).
(76) Strange, J.P., "Continuous Parts Per Billion Recorder for Air Contaminants," *J. Air Pollution Control Assoc., 10,* 423-6 (1960).
(77) Howard, O.H., "A Portable Continuous Analyzer for Gaseous Fluorides in Industrial Environments," *AMA Arch. Ind. Health, 19,* 355-64 (1959).
(78) Light, T.S., "Industrial Analysis and Control with Ion Selective Electrode," NBS Sp-314.
(79) Mori, M. et al, presented at the 64th Annual Meeting of the Air Pollution Control Association, Atlantic City, New Jersey, June 22-July 2, 1971.
(80) Lawrence Berkeley Laboratory, "Instrumentation for Environmental Monitoring - Water," *Publ. No. LBL-1,* Vol. 2, Berkeley, Univ. of Calif. (Feb. 1, 1973).

HALOGENS

The reader is also referred to the sections of this book on "Chlorine" and "Fluorides" for additional information.

Measurement in Air

A device developed by P.T. Gilbert, Jr., (1) provides an inexpensive and reliable means for the continuous monitoring of a test sample for the presence of a halogen or halogen compound. The sample, in vapor or gas form, is passed over a heated indium surface where the sample reacts with the indium to form a volatile indium halide. These indium halides are burned in a flame which imparts such energy to the indium halide as to cause them to photometrically radiate. Such radiation is spectrally unique to indium halides offering a distinctive means of detection both quantitatively and qualitatively with ability to discriminate against interfering compounds heretofore not possible.

Various methods have been utilized for the determination of halogen and halogen compounds in air, e.g., flame emission, electron capture, electrochemical detection and the liberation of ions from hot surfaces. One prior art method is the Beilstein test which provided a qualitative determination of volatile halogen compounds by a flame photometric method in which a flame, burning in contact with metallic copper, is colored green in the presence of chlorine, bromine, iodine and certain other substances.

A second method and apparatus for producing the characteristically colored Beilstein flame consists of mixing the test gas (the gas to be tested for the presence of the halogen or halogen containing compound) with a fuel gas, passing the mixture over heated copper and subsequently igniting the gases. In this system the copper was not in direct contact with the flame but was indirectly heated by preheating the gas mixture. This preheating was accomplished by either heating the conduit through which the gas mixture passed or by burning the gas mixture in an atmosphere lean in combustion support gases such that not all of the fuel gases are consumed.

In the latter system the hot gases from the first flame are passed over the copper and subsequently burned in a second flame which constitutes the Beilstein flame. This second method has the advantage over the standard Beilstein test in that it provides a degree of quantitative precision.

All of these methods suffer from one or more disadvantages, such as insufficient sensitivity, limited concentration range, nonlinearity, instability of the zero reading, poor precision or reproducibility, slow response, dependence of response on the nature of the halogen compound, the inability to continuously monitor, or the complexity of the equipment.

The original Beilstein test in which the copper is brought into direct contact with the flame affords a degree of specificity to chlorine, bromine and iodine, but has a very low sensitivity. The second method mentioned above where dual flames are utilized although having a much higher sensitivity affords no specificity.

The essential features of this photometric indium halide detector are shown in Figure 32. Referring to the drawing, there is illustrated a first flame supporting structure comprising tube **11** adapted to be connected to a source of fuel gas such, for example, as hydrogen. A second flame supporting structure comprising tube **12** encloses the tip of tube **11** and has an inlet **13** adapted to be connected to a source of flame supporting gas such as air. The sample to be tested for the presence of a halogen or halogen compound may be entrained into the flame **14** by any suitable means such as by adding the sample to either the fuel gas or the support gas.

FIGURE 32: FLAME PHOTOMETRIC DEVICE FOR HALOGEN DETECTION

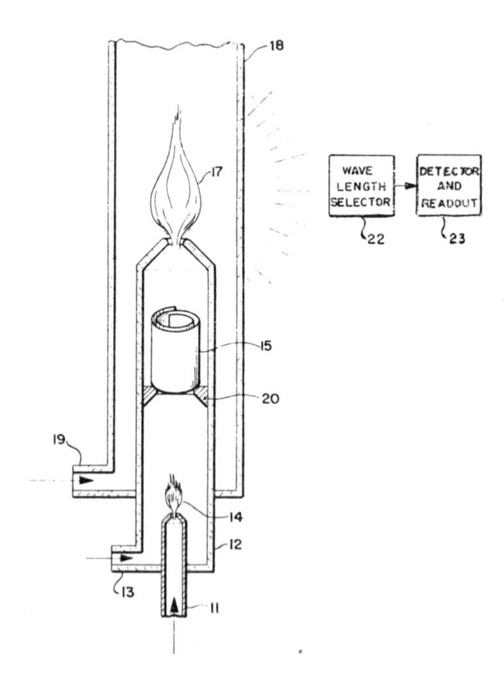

Source: P.T. Gilbert, Jr., U.S. Patent 3,504,976; April 7, 1970

Alternatively, the sample may be brought alongside the flame by a jet, not shown, and entrained into the burning flame. In some instances, it is merely desired to test for the presence of the halogens in air. In this instance, no sample atomizer is necessary and the burning of hydrogen in the presence of air is sufficient. Liquids may be atomized into the fuel or support gas in any suitable manner well-known in the art and solids may be provided by the method described by Gilbert, "Direct Flame Photometric Analysis of Powdered Materials," *Analytical Chemistry* 34, 1025 (1962).

Located within the upper portion of tube **12** is an indium insert **15** over which the hot gases from flame **14** pass. The quantity of combustion supporting gas admitted to tube **12** is insufficient to allow complete combustion of the fuel gases in flame **14**. The remaining fuel gases are burnt in a second flame **17** as they escape from tube **12**. If desired, the tube **12** may be surrounded by a chimney **18** having an air or other flame supporting gas inlet **19**.

The metallic indium insert **15** may be provided in any number of forms other than the loosely coiled spiral illustrated in the drawing. The insert may be formed as a screen or as a tube. Since the operating temperatures are believed to be above the melting temperature of indium a coating of indium on any suitable substrate is most suitable. Beryllium copper forms one such suitable substrate. The indium insert may be supported in any suitable manner such as by shelf **20**. In one test, an indium coated copper tube was suspended within tube **12** by a palladium wire.

The first flame supporting tube **11** may be of any suitable material. It is generally preferred, however, that tube **12**, supporting the second flame, be of nonmetallic construction and may be composed of Vycor or Pyrex. The chimney **18** is preferably of Pyrex to allow radiant energy to pass therethrough. It is obvious however that chimney **18** may be omitted. Where a halogen or halogen compound is present, the intensity of the indium line at 4511 A. is greatly increased over the background signal in the absence of the halogen or halogen compound. The stability of the device is good and the instrument will respond well to concentrations as high as 4500 μg./l. (Micrograms of a halogen per liter of support gas). The background reading in the absence of a halogen is equivalent to approximately 10 μg./l.

Referring again to the drawing, radiant energy emitted from flame **17** passes any suitable wavelength selector **22** to a detector connected to readout **23**. Wavelength selection may be performed by any suitable means such as filters or a monochromator. The wavelength selector may be set or selected to pass any line or wavelength characteristic of the particular indium halide if the system is to be made selective to a particular halogen or may be made sensitive to the indium line at 4511 A. if the presence of any halogen is to be detected.

The method and apparatus disclosed affords an inexpensive, reliable means for the continuous monitoring of a test sample for the presence of a halogen or halogen compound. The system may be made sensitive to the halogens as a class or made specific to a particular halogen by the selection of an appropriate wavelength.

Although the burner has been illustrated and described as the dual flame type, the heating of the indium may be accomplished by indirectly heating the fuel and support gases by heating the conduit as is known in the art. Further, the indium insert may be indirectly heated by any suitable means, e.g., electrically. In these instances, the flame **14** and its supporting structure may be omitted.

A flame ionization device developed by H.-G. Zimmerman (2) for the detection of halogens or halogen compounds includes an enclosed body forming a reservoir for an alkali material which includes a portion formed of a substance permeable to vapor of the alkali. A heater is provided to vaporize and diffuse the alkali through the permeable material. First and second electrodes near the permeable materials establish an electric field and indicate the amplitude of current flow in the presence of a halogen vapor.

References

(1) Gilbert, P.T., Jr., U.S. Patent 3,504,976; April 7, 1970; assigned to Beckman Instruments, Inc.
(2) Zimmerman, H.G., U.S. Patent 3,535,088; October 20, 1970; assigned to Bodenseewerk Perkin-Elmer & Co., GmbH, Germany

HALOGEN COMPOUNDS

One category of halogen compounds which has come up for particular attention recently is that of polychlorinated biphenyls (PCB's). Indeed, these are one of nine categories of materials singled out by the Environmental Protection Agency for inclusion in a new set of toxic pollutant effluent standards published in the Federal Register for December 27, 1973.

According to the new standards, streams having a flow of less than 10 cu. ft./sec. or lakes having an area of less than 500 acres shall receive no PCB discharges. Faster flowing streams or larger lakes may receive a maximum of 280 μg./liter (salt water bodies only 10 μg./l.) unless the low flow of the stream is less than ten times the waste flow in which case the 280 is reduced to 28 μg./liter.

Another category of halogen compounds of particular environmental concern is pesticides. The reader is referred to the section on "Pesticides" for details regarding such materials as aldrin, dieldrin, DDT and toxaphene.

Measurement in Air

The reader is referred to the section immediately preceding this one on "Halogens" since the two patented methods described there are applicable to halogen compounds as well as to halogens. The reader is also referred to the earlier section on "Chlorinated Hydrocarbons."

A device applicable to the detection of halogen compounds which makes use of the discovery that certain metallic materials are volatilized or vaporized at an increased rate when heated in the presence of a halogen has been described by A. Karmen, (1). The rate of vaporization of the metallic material is measured by a flame ionization detector or flame photometry device. The metal could be an alkali metal, barium, strontium, calcium and mixtures of the same, and could be in the form of a pure metallic material, a metallic amalgam, a metallic salt and mixtures of the same.

The gaseous material undergoing investigation may be subjected to two flame ionization procedures, the first functioning to simultaneously combust the gaseous materials while detecting its organic constituents, and the combusted gaseous material from the first flame impinging upon a probe including the metallic material to cause increased rates of vaporization of the same under the influence of element-containing compounds in the gaseous material, this increased rate of vaporization being detected by a second flame ionization means which is not sensitive to the organic compounds which were in the original gaseous material since these compounds have been subjected to combustion producing various non-detected carbon, hydrogen and oxygen containing vapors.

Consistent with the preferred embodiment separate monitoring of the two flames will provide a more detailed picture of the make-up of the gaseous material. This double flame ionization technique can preferably be utilized in conjunction with gas chromatography by feeding the effluent from the gas column to the first flame.

Additionally, the double flame detector described above can be improved by interposing an electrostatic shield between the two flames which defines the volume of gas of which the electrical conductivity is to be measured, this shield preferably including the metallic material to thereby simultaneously serve as the probe. It is to be understood that the use of the term "probe" herein is intended to include any conventional form of such members as well as the preferred foraminous element such as a screen which serves in the dual capacity as a metallic vapor source and an electrostatic shield.

Quantitative results can be readily realized by comparing the increased rate of vaporization of the metallic material caused by a gaseous mixture having an unknown quantity of a compound including the particular element defined above with increased rates of vaporization of the metallic material caused by various gaseous materials having known quantities of compounds including such elements.

Figure 33 is a drawing showing a suitable form of apparatus for the conduct of this process. As shown in the figure, the device includes a base or support **10** provided with an air or oxygen-containing material inlet **12**, a hydrogen inlet **14** (it being understood that the hydrogen gas may be replaced by other operable materials such as carbon monoxide) and a further inlet **16** connected to a source of gaseous material to be investigated, preferably the effluent from a column of a conventional gas chromatograph apparatus (not shown).

Inlets **14** and **16** each communicate with a common passageway **18** which terminates in a burner device schematically shown at **20** having a flame **22**. It will be readily understood that the burner device **20** may be supported in any desired manner. In the particular construction shown in the drawings the flame **22** is confined within a lower chamber **24** defined by the lower portion **26** of a continuously extending housing **28** which is grounded in any conventional manner (not shown).

FIGURE 33: FLAME IONIZATION DEVICE FOR HALOGEN DETECTION

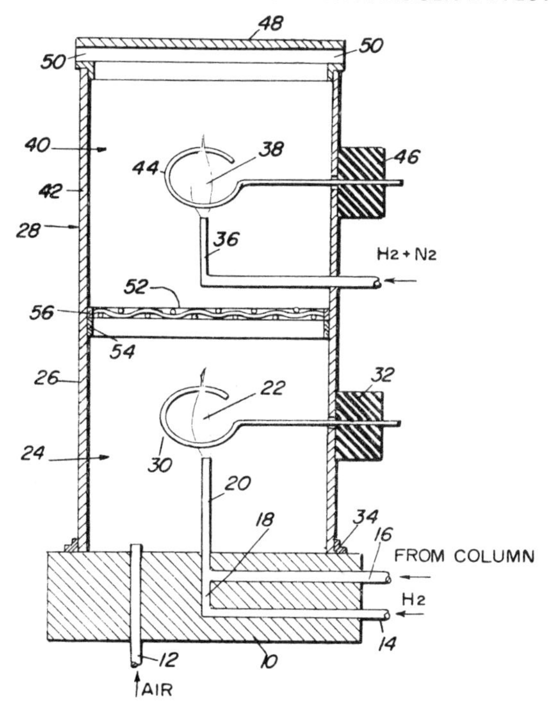

Source: A. Karmen; U.S. Patent 3,425,806; February 4, 1969

The electrical conductivity of the flame **22** may be readily monitored by any of a variety of well-known means, a ring electrode **30** circumscribing the flame **22** being shown as illustrative. This ring electrode may be supported in any desired manner, a block **32** secured to the housing **28** being shown in the drawings.

Of course, the ring electrode **30** must be insulated from the other electrode which is defined by the grounded housing **28**. Conductors (not shown) may connect the electrodes to a conventional monitoring means (not shown) such as an electrometer or the like in a well-known manner. A flanged angular bracket **34** is shown as supporting the housing **28** on the base 10.

An upper burner device **36** having a flame **38** is illustrated as confined in an upper chamber **40** defined by the upper portion **42** of the housing **28**, this burner similarly being fed by a source of hydrogen and nitrogen (which is a conventional carrier gas in a chromatograph column) and being supported as desired above the lower burner device **20**. Similar means for monitoring the electrical conductivity of the flame **38** are provided in the form of a ring electrode **44** carried by a block **46** and insulated from the housing **28**. A removable cover member **48** is carried by the upper portion **42** of the housing **28**, vents **50** in the cover member **48** being provided to permit escape of combustion products and other gaseous materials rising from the interior of the housing **28**.

Although, as pointed out hereinabove, the probe may take any conventional form, according to the preferred embodiment shown in the drawings, it is comprised by a foraminous electrically conductive member **52** in the form of a screen or the like removably supported on an angular bracket **54** to permit replacement or retreating of the same, as necessary.

The screen **52** may be constructed by spot welding stainless steel mesh (50 by 50 Surgaloy mesh, Davis and Geck Co.) or platinum wire gauze 52 mesh (Fisher Scientific Co.) around the circumference of a flat stainless steel ring **56**. The screen **52** thus acts as an electrostatic shield between the flames **22** and **38** to permit accurate and separate monitoring of the electrical conductivities of each flame. The metallic material which volatilizes or vaporizes at an increased rate when heated in the presence of a halogen or phosphorus may either be an integral part of the wires forming the screen **52** or the screen may be dipped into a solution of a salt of the metallic material as described hereinabove.

The gaseous material being studied is fed to the lower burner device **20** through the inlet **16** and intermixed with hydrogen or the like from the inlet 14, combustion being supported by the air entering through passageway 12 to form the flame **22**. In effect, the lower burner device **20** functions to combust the gaseous material fed through the inlet **16** while simultaneously acting as a flame ionization means to provide a general record of the constituents of the gaseous material.

The combusted gaseous materials rising from the flame **22** impinge upon the probe or screen **52** concomitantly heating the same and causing an increased rate of vaporization of the metallic material in or on the screen due to the presence of the halogen in the original gaseous material. The vaporized metallic material diffuses upwardly from the screen **52** to be received by the flame **38** of the upper burner device **36** wherein a second flame ionization technique is performed by monitoring the electrical conductivity of the same thereby detecting the presence of the halogen containing compound indirectly.

The curve resulting from the upper flame ionization procedure will be specific to the halogen containing material since it is nonresponsive to the combusted products of the other organic materials in the original gaseous material which form carbon dioxide, water or other carbon, oxygen and hydrogen containing compounds. Additionally, no response will be received from various other materials such as sulfur or nitrogen containing compounds in the gaseous mixture.

A device developed by J. Waclawik and S. Waszak, (2) is one in which a gas contaminated with a small concentration of gaseous halogenated organic compounds or hydrogen halides or both is continuously passed through a furnace at 800°C. filled with a palladium catalyst whereby the contaminants are burned to form a combustion product which is a hydrogen halide, a halogen or both. Thereafter, the combustion product is passed into contact with a bromide-bromate mixture or an iodide-iodate mixture whereby the combustion products are quantitatively converted into the free halogen, which can be determined by known continuous methods, preferably in the galvanic cell of a gas analyzer.

This method may be used for continuously determining gaseous halogenated organic compounds, such as trichloroethylene, tetrachloroethylene, chloroform, methyl bromide, or phosgene, and hydrogen halides, such as hydrogen chloride, hydrogen bromide, or hydrogen fluoride, which are borne in air or in other gases as contaminants present in very small amounts on the order of 0.02 ppm.

The operation of the method will be described with reference to the apparatus diagram in Figure 34. Air contaminated with gaseous trichloroethylene in amounts up to 3.0 ppm was passed at a constant rate through a furnace **F** at a temperature of 800°C. filled with a palladium catalyst. In the furnace, trichloroethylene was quantitatively burned to yield carbon dioxide and, in part, chlorine and hydrogen chloride.

Then the combustion products were passed through a tube **T** filled with a bromide-bromate mixture to release an equivalent amount of bromine, which was subsequently determined by the method and in the apparatus described in Polish Patent 50,961. Briefly, this was effected by forming an electrolyte containing the mixture, adding the electrolyte to a galvanic cell comprising a housing **1** made of glass or another material which is resistant to both the action of the gas to be examined and the electrolyte employed, a gold electrode **2** having a large surface area in the form of a foil, and a silver electrode **3**. The cell is filled with a solution of potassium chloride in hydrochloric acid.

The cell is connected with the spiral **4** and container **6** by means of separator **5**. The determination of chlorine consists in passing air of constant speed through the spiral **4** in which the gas comes into contact with the electrolyte. In the separator **5**, the gas is separated from liquid, the gas flows outwardly and the electrolyte, together with the chlorine dissolved therein, flows to the galvanic cell, in which, on gold electrode **2** the reduction of chlorine to chloride ions takes place, while the silver electrode oxidizes to silver ions according to the reaction:

$$Cl_2^\circ + 2e \longrightarrow 2\,Cl^- \quad \text{on the gold electrode}$$
$$2\,Ag^\circ - 2e \longrightarrow 2\,Ag^+ \quad \text{on the silver electrode}$$

- -

$$2\,Ag + Cl_2 \longrightarrow 2\,AgCl$$

After passing through the galvanic cell, the electrolyte flows to the container **6** from which it thereafter flows to the spiral **4**. Thus the electrolyte circulates in a closed cycle. To the outer circuit of the cell a microammeter is connected, measuring the intensity of the current flowing during the work of the cell. The concentration of chlorine in the gas is proportional to the intensity of this current.

Using an automatic registration to a Fe-Konstantan thermocell (not shown) it is possible to determine from 0.00 to 3.00 ppm of chlorine with an accuracy up to 0.3 ppm, and with the use of a compensator (scaled from 0 to 2 mv.) (also not shown) it is possible to determine from 0.00 to 1.5 ppm with an accuracy up to 0.02 ppm. After a certain working time of the cell, the silver electrode **3** is subjected to regeneration, i.e., to the reduction of formed silver chloride to silver.

Measurement in Water

A device developed by H.D. Law (3) permits the detection of certain organic polyhalogen

FIGURE 34: DEVICE FOR HALOGEN COMPOUND DETECTION BY COMBUSTION FOLLOWED BY GALVANIC HALOGEN DETECTION

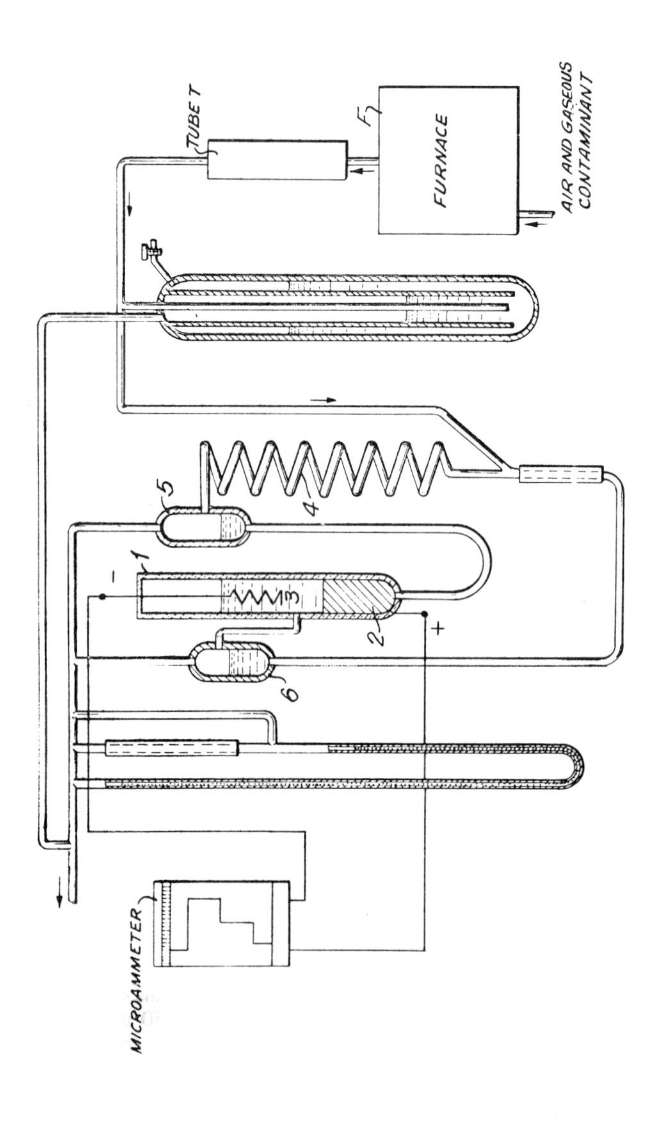

Source: J. Waclawik and S. Waszak; U.S. Patent 3,546,079, December 8, 1970

compounds in aqueous fluids, the system comprising a solid acid addition salt of pyridine such as pyridine citrate or pyridine tartrate and a solid caustic material such as caustic soda or caustic potash.

It is known that a red color is formed when halogenated organic compounds such as chloroform, bromoform, or iodoform are heated with pyridine and caustic alkali in aqueous medium, and this color reaction has been used for the detection of polyhalogen compounds including trichloroacetic acid. The test is applicable to a wide variety of organic compounds containing two or more halogen atoms joined to a single carbon atom and which in the presence of alkali are degraded to a reactive carbene species, e.g., as follows:

$$Cl_3CH + OH \rightleftharpoons Cl_3C^- + H_2O$$

$$Cl_3C^- \rightarrow :CCl_2 + Cl^-$$

While this test is effective, it has heretofore always required the use of liquid reagents and the application of external heat.

A simple method has been discovered, using only solid reagents and for which no external source of heat is necessary, by which it is possible to detect the presence of halogen-containing compounds of the aforesaid type in aqueous liquids containing them.

This method for detecting the presence, in an aqueous liquid, of an organic halogen-containing compound comprises adding, to a predetermined quantity of the liquid, a solid, preferably nonhygroscopic, acid addition salt of a pyridine base unsubstituted in the 2- and 4-positions, and either solid caustic soda or solid caustic potash, the weights and weight ratio of the alkali and acid addition salt being such that on admixture with the aforesaid predetermined quantity of aqueous liquid, the mixture heats spontaneously to at least about 80°C. and a liquid pyridine base separates as a supernatant layer.

It is apparent that the color produced is proportional to the amount of polyhalogen present and by the expediency of proper color standards, a quantitative as well as qualitative test method may be achieved.

References

(1) Karmen, A., U.S. Patent 3,425,806; February 4, 1969; assigned to Secretary of Health, Education and Welfare.
(2) Waclawik, J., and Waszak, S., U.S. Patent 3,546,079; December 8, 1970.
(3) Law, H.D., U.S. Patent 3,472,626; October 14, 1969; assigned to Miles Laboratories, Inc.

HYDROCARBONS

Air quality standards for gasoline in air have been set in West Germany at 60 ppm on a basis not to be exceeded more than once in any 4 hours and at 20 ppm on a long-term exposure basis. Emission standards for hydrocarbons from incinerators have been set by the San Francisco Bay Area Air Pollution Control District at 50 ppm maximum. Federal emission standards for motor vehicles specify hydrocarbon concentrates of 275 to 410 ppm depending on engine size.

Measurement in Air

Air quality criteria have been developed by EPA for only one group of volatile organic substances, nonmethane hydrocarbons (1). The lack of additional differentiation among volatile organic substances does not result so much from lack of laboratory analytical methods as from lack of routine monitoring techniques. Gas chromatographs with flame ionization detectors have been utilized in mobile laboratories to analyze atmospheric samples for identification and quantitation of 30 to 60 different aliphatic and aromatic hydrocarbons (2)(3). The chromatographs have been used both for periodic monitoring and for grab sampling. The number of components analyzed has depended on the substrates used and the atmospheric concentration levels. Process-type gas chromatographs for nonmethane hydrocarbons are in development.

The gas chromatographic technique for carbon monoxide and methane utilized in former years in laboratory photochemical studies (4)(6) was developed into a convenient monitoring instrument (7)(8). In combination with a capability for measuring total hydrocarbons, such a gas chromatographic analyzer provides a highly specific and sensitive means of analyzing carbon monoxide and nonmethane hydrocarbons over a wide range of atmospheric concentrations.

As a carbon monoxide analyzer, the gas chromatograph is much more sensitive and specific than current nondispersive carbon monoxide analyzers. This approach can also be extended to include monitoring for other gaseous hydrocarbons. The direct analysis for methane is much more desirable than the earlier attempt to use substrates to remove hydrocarbons other than methane (9)(10).

This type of technique has given erratic results in routine monitoring activities because of the care necessary to maintain the characteristics of the substrate utilized to provide the specificity required. Another technique developed for methane and other hydrocarbons involves selective combustion, with subsequent detection by a water sorption sensor (11).

Analysis of nonmethane hydrocarbons has several fundamental limitations: (1) the measurements cannot be utilized to determine the effectiveness of control, over a given period, of any particular source of organics; (2) response characteristics of the flame ionization detector differ for various hydrocarbons; and (3) the flame ionization detector does not respond to formaldehyde and shows reduced response to other aldehydes.

Even a detailed gas chromatographic analysis for hydrocarbons will not make it possible to follow control of even the major sources of hydrocarbon. This difficulty results from a lack of specific hydrocarbons that can be used as unique tracers of contributions from individual emission sources. However, acetylene and ethylene can be used as indicators of control of combustion sources. In downtown high-traffic-density areas, these hydrocarbons also should serve as indicators of the level of hydrocarbons from vehicular exhaust. Ethylene measurements would be of direct interest in some areas because of its plant-damage characteristics.

Some steps must be taken to optimize flame ionization detector response to minimize deficiencies in response to individual hydrocarbons in atmospheric analysis. Calibration with a single saturated hydrocarbon is the accepted procedure. Such a calibration limits the accuracy of comparative measurements of atmospheres in which the hydrocarbon composition varies.

Hydrocarbons may be monitored in air by the following instrumental techniques (12):

Gas Chromatography - Flame Ionization Detection
(reference method)
Flame Ionization Detection
Mass Spectrometry
Ultraviolet Absorption

The EPA designated reference method for hydrocarbons in air is flame ionization detection using gas chromatography.

As described by L.A. Elfers (13), a hydrogen flame ionization type monitor may be used for hydrocarbon detection in air. Such a device was shown earlier in this volume as Figure 7.

A device developed by L.E. Hakka, G.L. Bata and J.E. Hazell (14) is a device for detecting components such as hydrocarbons in a gas mixture such as air. More particularly, it is a detection device consisting of a thin flexible strip comprising two layers of different material laminated together, one of such layers being a better solvent for the component being detected than the other. The swelling of the solvent layer in the presence of the component will cause the strip to deflect in proportion to the concentration of the component, thereby giving a quantitative determination of the presence of the component.

Various devices for detecting components in gas mixtures are known, particularly for detecting leaks in hydrocarbon gas pipelines. These previously known devices are generally more complex and expensive than the device.

A process for determining the amount of reactive hydrocarbon constituents in a gas sample has been described by W.B. Innes (15). In this process, air is mixed with a sample and passed through a suitable oxidation catalyst for oxidizing the constituent. The temperature rise in the catalyst is proportional to the concentration of the constituent in the sample and is measured immediately adjacent the inlet to the bed. The temperature of the air and sample is maintained above the dew point thereof to prevent condensation of water and absorption of reactable components therein. Provisions are made to prevent sorption of reactable components in the inlet line upstream of the reactive bed.

A process for the continuous determination of combustible gases and vapors such as hydrocarbons has been described by M. Siano and G.G. Guidelli (16). In this process the ambient air flows through a reaction chamber in contact with an oxidation catalyst in the presence of a plurality of detecting means sensitive to the varying radiant energy emitted along the catalyst in the direction of flow of the ambient air. Means for receiving and comparing these different detected values are provided.

A process developed by E.C. Betz (17) provides a system for determining the hydrocarbon impurity concentration in a gaseous atmosphere where a known impurity gaseous stream and an unknown impurity gaseous stream are passed through parallel conversion zones to produce separate conversion product streams after catalytic oxidation in each case. These separate product streams are then passed over suitable detection devices for the generation of an output signal which is then quantitatively correlated with the impurity content of the unknown stream. In a typical operation, the sample gaseous atmosphere comprises air contaminated with hydrocarbons.

Another process for hydrocarbon determinations on the basis of a combustion technique is that of H.-P. Neubert (18). This is a process for the measurement of the concentration of combustible gases, such as methane, in which the gas mixture is catalytically burned. More particularly it utilizes an apparatus that does not give a false indication at high combustible gas concentration, comprising a detector filament and a substantially identical compensator filament each adapted to be brought into contact simultaneously with a sample of a gas mixture to be tested, the detector filament alone being adapted to cause oxidation of combustible constituents therein.

A device developed by A.A. Poli, Jr. and S.D. Delaune (19) is a device for simultaneously measuring the amount of carbon monoxide and total hydrocarbons in a gas sample, comprising first and second means for measuring hydrocarbons, a conduit for delivering part of the sample to the first means, a second conduit for simultaneously delivering the rest of the sample to the second means, and a reactor in the second conduit for converting carbon monoxide to methane, whereby the portion of the sample delivered by the second conduit will contain a greater amount of hydrocarbons than the part delivered by the first conduit.

The measuring device used is a hydrogen flame analyzer in which the air being sampled is delivered to a flame of pure hydrogen. Any hydrocarbons carried into the flame result in the ionization of carbon atoms. An electrical potential across the flame, and an ion collector electrode supported above the flame, produce an electric current proportional to the hydrocarbon count. This is measured by an electrometer circuit, the output signal from which will indicate the total hydrocarbons in the sample.

A process developed by P.W. Van Luik, Jr. (20) involves detecting and measuring certain organic solvents in an atmosphere or flowing gas stream by converting the same into airborne particulates and determining the concentration of these particulates as a measure of the gas or vapor concentration.

Recent investigations have shown that extremely low concentrations, on the order of several parts per million, of certain types of organic gases and vapors, such as the unsaturated hydrocarbon, cyclohexene, for example, may be detected by converting the gas or vapor into small airborne particulates. The particulates are nucleogenic in that they are capable of functioning as condensation nuclei and are detected by known highly sensitive condensation nuclei measuring techniques.

One mechanism for converting these unsaturated hydrocarbon gases and vapors into the desired particulate form is photochemical in nature; i.e., conversion takes place by subjecting the hydrocarbon vapors and gases to ultraviolet radiation in the presence of a minute quantity of mercury vapor. A method and apparatus for the photochemical conversion of certain hydrocarbon vapors is described and claimed in U.S. Patent 3,102,192.

In addition to detecting unsaturated hydrocarbons, certain classes of organic gases and vapors may also be detected by a similar though different technique. It has been found that organic solvents, such as the aromatic hydrocarbon solvents as exemplified by toluene, naphtha, and benzene, alcohol solvents as exemplified by ethyl-alcohol, methyl-alcohol and acetone, and aliphatic hydrocarbon solvents as exemplified by the aliphatic solvent marketed under the names Solvesso 100 and Solvesso 150, may be detected and measured by their ability to inhibit the formation of particulates of the condensation nuclei type in other gases or vapors.

Certain vapors, such as mercury, for example, when exposed to ultraviolet radiation, are easily converted by a photochemical process to airborne particulates which act as condensation nuclei. However, the conversion of mercury (Hg) vapor to particulates is inhibited by the presence of organic solvents to a degree which is a function of the solvent concentration. It will be appreciated, therefore, that this inhibiting property of the solvents makes it possible to detect many of these organic solvents.

An analytical apparatus whose function is based on the second technique described above is shown in Figure 35. The apparatus shown there is one for detecting an organic solvent liberated during the curing cycle of a polymeric material, such as the one designated by the name Alkanex, which is disclosed in U.S. Patent 2,936,296, and which, for example, may be a cross-linked polyester of ethylene glycol, glycerin and a lower dialkyl ester of isophthalic acid or terephthalic acid or mixtures.

To this end, a particle free stream of air or gas is passed through a charcoal filter **1** where various unwanted gases or vapors are removed by absorption on charcoal filter materials **2** contained therein. The stream flows into an electric oven, illustrated at **3**, which contains a small beaker or container **4** housing the uncured polymeric material which evolves the organic solvent as it goes through its curing cycle.

One physical example of a polymeric material which emits an organic solvent in the course of its curing solvent is the Alkanex resin mentioned above and which usually contains varying percentages of an aliphatic hydrocarbon solvent known as Solvesso 100 or Solvesso 150 (Socony Mobil Corp.). The organic solvent liberated from the polymeric material in the container **4** is entrained in the stream passing through oven **3** and passes through a copper finned heat exchanger **5** to reduce the stream temperature. The cooled stream flows through a glass wool filter **6** to remove any particulates in the air. During calibration of the device, valve **6'** is opened and the ambient air stream bypasses chamber **3** containing the organic solvent vapor.

FIGURE 35: APPARATUS FOR HYDROCARBON MEASUREMENT BY CONVERSION TO PARTICULATES

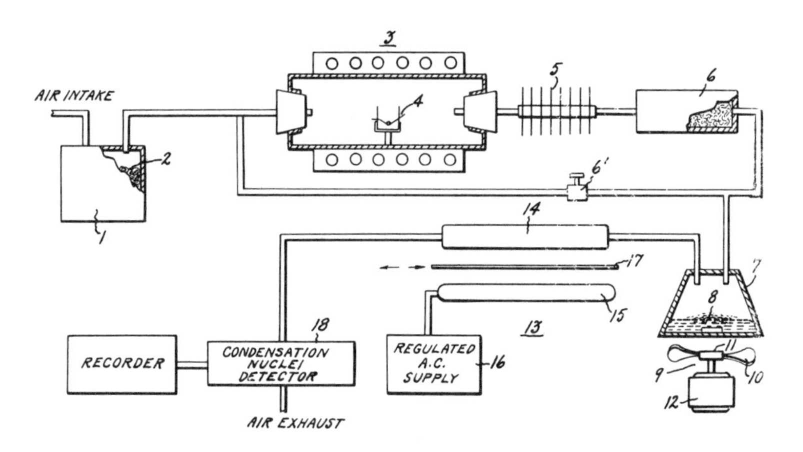

Source: F.W. Van Luik, Jr.; U.S. Patent 3,206,449; September 14, 1965

During operation valve **6'** is closed, and the stream passes through chamber **4** and flows directly through a mercury vapor source which consists of glass flask **7** containing a mercury pool **8**. Mercury pool **8** is continually agitated by means of a magnetic stirring and coupling mechanism **9** which comprises a fan blade **10** having a magnetic disc **11** mounted thereon. Fan blade **10** is driven by a driving source, such as the motor **12**, rotating the magnetic disc. Since mercury is a conductive liquid, rotation of the magnetic disc produces eddy currents therein, stirring the mercury and continuously exposing a fresh mercury surface to the air stream passing through source **7**.

It has been found that in the absence of such stirring, surface oxidation of the mercury takes place and the device may not always function with the desired effectiveness. The air stream containing the organic solvent vapor absorbs mercury vapor from pool **8** and passes through a converter, shown generally at **13**, where the organic solvents and mercury vapors are exposed to ultraviolet radiation.

The converter comprises a quartz tube **14**, through which the gas stream passes from mercury source **7**. The quartz tube which is pervious to ultraviolet, is positioned in a field of ultraviolet radiation produced by a low pressure mercury vapor lamp **15** which is energized from a source of regulated AC power **16**. The low pressure mercury vapor lamp produces ultraviolet radiation concentrated in the 2537 Angstrom (A.) band.

A shutter, shown generally at **17**, is positioned between the mercury lamp and the quartz tube and may be selectively moved in a lateral direction to control the amount and degree of the ultraviolet radiation which impinges on the quartz tube and irradiates the gas stream. In the absence of organic vapors in the air stream, the ultraviolet radiation from the mercury lamp converts the mercury to particulates by a photochemical reaction to establish an ambient level of nuclei concentration in the chamber **14** which is measured in a condensation nuclei measuring device **18** connected to the chamber.

The presence of organic solvents in the air stream inhibits formation of nuclei by the photochemical conversion of the mercury and reduces the nuclei concentration to provide an indication of the presence of the solvents. The reaction by which these organic solvents inhibit the photochemical conversion of Hg is not completely understood. At this point, it is sufficient to point out that it has been found that the presence of minute concentrations of organic solvents, on the order of 1 to 2 ppm, is sufficient to reduce the particle concentration substantially.

The condensation nuclei measuring device shown at **18** is of the type which contains a humidifying device to bring the nuclei or airborne particles in the air stream to 100% relative humidity. The humidified sample is then subjected to an adiabatic expansion, by means of which the gaseous sample is subjected to a fixed pressure differential. The adiabatic expansion cools the gas stream, and by virtue of the cooling action a condition of supersaturation comes into being so that the gas sample which had previously been at 100% relative humidity is now at some higher percentile.

Since the supersaturated condition is an unstable one, excess vapor begins to condense about the suspended airborne particles and forms droplets which grow rapidly in size and abstract sufficient water vapor from the sample to reduce the saturation level to 100% at the new temperature. The droplets formed by condensation of water about the airborne particles are measured by means of an electrooptical system, which includes a light scattering arrangement, by means of which the droplet concentration is determined and expressed in terms of a current flowing in an electrical output circuit, which current may be directly calibrated either in particles per cubic centimeter in parts per million of the gas or the vapor.

A condensation nuclei measuring device such as this is described in detail in U.S. Patent 2,684,008, and may be utilized with this system. A similar automatically operated condensation nuclei measuring device is described in an article, entitled "Cloud Chamber for Counting Nuclei in Aerosols," by Bernard G. Saunders in *Review of Scientific Instruments,* vol. 27, No. 5, May 1956, pages 273-277. In the condensation nuclei measuring device described in the Saunders' article, periodically actuated solenoid valves control the admission of the humidified samples into the device and their subsequent expansion to form droplets, which are measured either photographically or electrooptically to provide an output indication which is a measure of the particle concentration.

In additon to the above described nuclei measuring devices, it will be understood by those skilled in the art that many other and different types of measuring devices may be used with equal facility in detecting and measuring the particles formed in quartz tube **14**.

In operation, control valve **6'** is first opened to prevent any organic solvent from passing into the rest of the instrumentality. Ultraviolet lamp **15** is energized and shutter **17** is adjusted to establish a predetermined particle conversion level which constitutes the reference or ambient level prior to the introduction of the sample solvents. To achieve adequate sensitivity and accuracy with this instrumentality, the shutter position was adjusted to produce an irradiation intensity sufficient to establish a particle concentration in excess of 30,000 particles per cubic centimeter.

After the ambient or reference nuclei concentration level is stabilized, the polymeric solvent mixture is introduced into the oven and the valve **6'** is closed to permit any organic vapor liberated during the curing process to pass through mercury vapor source **7** and into the quartz tube. The liberation of any such organic solvent is manifested by a reduction in the number of particles produced in the quartz tube so that the output reading or indication from the condensation nuclei measuring device **18** is reduced proportionately. Any such drop in the output indication is a sign that an organic vapor is present in the air stream and, hence, that the polymeric material is undergoing its curing cycle by emitting the organic vapor. When the curing process is terminated, the polymeric material or organic solvent is no longer emitted, terminating its inhibiting effect on the conversion of the mercury vapor and the ambient reference nuclei concentration level rises to its ambient value.

A device developed by T. Hirschfeld and N. Matsu (21) permits detecting and quantitatively measuring the presence of unsaturated and aromatic hydrocarbons, particularly vapors of partially burned and residual unburned hydrocarbons in the exhaust emissions of gasoline internal combustion engines. The system includes a partially shielded membrane adapted to react with aromatic and unsaturated hydrocarbons to produce derivatives having strong absorption characteristics in the ultraviolet region. There is also provided an ultraviolet fluorescence photometer including an ultraviolet source and a visible fluorescent screen between which the exposed membrane is introduced. The photometer includes a gray scale or step wedge for visually comparing the ultraviolet absorption in the unshielded section of the membrane against a calibrated standard including the shielded section of the membrane and the gray scale.

A device developed by T. Hirschfeld (22) involves detecting the vapors of aromatic and unsaturated hydrocarbon by reacting the vapor with a film of polytetrafluoroethylene sulfonic acid, and examining the ultraviolet transmission characteristics of the reacted film.

A device developed by R. Villalobos (23) permits the measurement of pollutants in ambient air and specifically the measurement of methane, ethylene, acetylene, CO, CO_2 and total hydrocarbons in ambient air. The apparatus employs a hydrogen flame ionization detector and the gases employed in the analysis instrument are only those gases required for combustion in the flame detector. Air is employed as the carrier gas in a total hydrocarbon analysis line to the detector.

The carrier air stream is passed through an oxidizer catalyst to completely oxidize the hydrocarbons in the carrier air to provide a true "zero-hydrocarbon" reference level in the ionization detector against which the total hydrocarbon content of the ambient sample is compared. Hydrogen is employed as the carrier gas in a gas chromatograph analyzer for separating the hydrocarbons and carbon monoxide in an air sample.

The hydrogen carrier gas simultaneously provides the hydrogen requirements of a methanator in the chromatograph line, which converts carbon monoxide and carbon dioxide into methane so that a relative measurement may be made in the ionization detector for these constituents, as well as the hydrogen gas requirements of the detector.

A device developed by J.L. Dennis (24) is one in which the hydrocarbon gas in a sample is measured by a double beam detector responding to the transmission of absorption properties of the gas in response to applied radiation from a laser source. Monochromatic light from the laser source is split into two radiation beams, a sample path beam and a reference path beam, and directed to a radiation detector which sequentially responds to the reference energy and the sample energy. A difference in the absorption by the sample path energy with respect to the reference path energy produces two, different level, output signals from the detector.

The sample energy signal and reference energy signal are respectively directed to a sample channel circuit and a reference channel circuit. Outputs from the individual circuits are amplified logarithmically and applied to inputs of a differential amplifier where the ratio of the sample signal to the reference signal produces a voltage for actuating a meter indicator. For a hydrocarbon gas content in a sample in excess of a predetermined limit, an alarm circuit actuates to give an audible warning of an excessive gas content.

A device for measuring olefins in gasoline samples provides an indirect air pollution detector as described by R.S. Silas (25). Research in the area of air pollution from gasolines has revealed that a large portion of the pollution caused by combustion of gasoline in automobiles results from light olefins, particularly C_5 and lighter olefins, that exist in normal gasolines traveling through the engine uncombusted and exiting to the atmosphere along with the combustion products. Upon entering the atmosphere in a gaseous form lighter olefins exhibit a tendency to combine very readily with various materials in the atmosphere, such as sulfur, oxygen and halogens and produce air pollutants commonly known as smog.

Generally, air pollution becomes acute when a gasoline contains in excess of 5% by weight olefins of C_5 or lighter based on the weight of the gasoline. Conventional methods for determination of olefins in a stream take a considerable amount of time, and must be conducted by highly skilled personnel in permanent laboratory installations. Examples of such analytical methods are fluorescent indicator absorption, mass spectrometry, chemical methods (ASTM Method D-875) and conventional chromatography.

In accordance with the process, a rapid and accurate determination of the quantity of olefins having up to a predetermined number of carbon atoms, particularly C_5 and lighter hydrocarbons, in a stream, and particularly in gasoline, can be conducted by relatively nonskilled personnel in field locations by use of the process portable device. Broadly, the process comprises chromatographically separating in a chromatographic column olefins having up to one more carbon atom than a predetermined number of carbon atoms from the remainder of the olefins in a stream that contains both olefins and nonolefins.

After the chromatographic separation, then the olefins having up to the predetermined number of carbon atoms are time separated from the olefins having one more carbon atom than the predetermined number in the stream that contains nonolefins and chromatographically separated olefins. This is accomplished by terminating the flow of the chromatographic eluate at a time after the olefins having up to a predetermined number of carbon atoms, but before the olefins having one more carbon atom than the predetermined number pass a given point.

The stream containing both olefins having up to the predetermined number of carbon atoms and nonolefins is then tested to determine the quantity of olefins present therein, all of which are olefins having up to the predetermined number of carbon atoms. In another embodiment, a restrictor column is operated in series with the chromatographic column to reduce the time necessary for backflushing the chromatographic columns after a test has been run by placing a greater backflushing pressure on the chromatographic column than the pressure used for chromatographic separation.

Measurement in Water

A process developed by C.A. Boyd and H. Kartluke (26) permits detecting the presence and measuring the content of aliphatic hydrocarbons in a body of water. The body of water is separated into two samples; ultrasonic cavitation is induced in one of the samples to form hydrogen and hydroxyl ions and a compound between the aliphatic hydrocarbon and the hydroxyl ions. Such compound is then reacted with a reagent to form a reaction product. The amount of hydrocarbon in the body of the water is then determined by comparing the reaction product of one sample with the other, or detecting the amount of hydrogen ions.

References

(1) Environmental Protection Agency, "National Primary & Secondary Air Ambient Quality Standards", *Federal Register* 36 (84), 8186 (Apr. 30, 1971).
(2) Scott Research Laboratories, "1970 Atmospheric Reaction Studies in the New York City Area", Contract CPA 70-178, Durham, N.C., Air Pollution Control Office (1971).
(3) Altshuller, A.P., Lonneman, W.A., Sutterfield, F.D., Kopczynski, S.L., *Envir. Sci. & Technol.* 6 (1971).
(4) Altshuller, A.P., & Cohen, I.R., *Intern. J. Air & Water Poll.* 8, 611 (1964).
(5) Altshuller, A.P., Cohen, I.R. and Purcell, T.C., *Can. J. Chem.* 44, 2973 (1966).
(6) Altshuller, A.P., Kopczynski, S.L., Lonneman, W.A., Becker, T.L. and Slater, R., *Envir. Sci. & Technol.* 1, 899 (1967).
(7) Stevens, R.K. and O'Keefe, A.E., *Anal. Chem.* 42, 143A (1970).
(8) Stevens, R.K., O'Keefe, A.E. and Ortman G.C., "A Gas Chromatographic Approach to the Semi-Continuous Monitoring of Atmospheric Carbon Monoxide & Methane", *Proc. of the 11th Conference on Methods on Air Pollution,* Berkeley, Calif. (Mar. 30–Apr. 1, 1970).
(9) Ortman, G.C., *Anal. Chem.* 38, 644 (1966).
(10) Altshuller, A.P., Ortman, G.C., Saltzman, B.E. and Neligan, R.E., *J. Air Pollution Control Assoc.* 16, 87 (1966).
(11) King, W.H., *Envir. Sci. & Technol.* 7, 1136 (1970).
(12) Lawrence Berkeley Laboratory, "Instrumentation for Environmental Monitoring - Air", *Publ. LBL-1,* Vol. 1, Berkeley, Univ. of Calif. (Feb. 1, 1973).
(13) Elfers, L.A., "Field Operations Guide for Automatic Air Monitoring Equipment", *Report PB 204,650,* (originally issued as *Report PB 202,249*), Springfield, Va., Nat. Tech. Info. Service (July 1971).

(14) Hakka, L.E., Bata, G.L. and Hazell, J.E.; U.S. Patent 3,674,439; July 4, 1972; assigned to Union Carbide Canada Ltd.
(15) Innes, W.B.; U.S. Patent 3,725,005; April 3, 1973.
(16) Siano, M. and Guidelli, G.G.; U.S. Patent 3,553,461; January 5, 1971.
(17) Betz, E.C.; U.S. Patent 3,567,394; March 2, 1971; assigned to Universal Oil Products Co.
(18) Neubert, H.P.; U.S. Patent 3,497,423; February 24, 1970; assigned to Auer GmbH, Germany.
(19) Poli, A.A., Jr. and Delaune, S.D.; U.S. Patent 3,692,492; September 19, 1972; assigned to Mine Appliances Co.
(20) Van Luik, P.W., Jr.; U.S. Patent 3,206,449; September 14, 1965; assigned to General Electric Co.
(21) Hirschfeld, T. and Matsu, N.; U.S. Patent 3,697,226; October 10, 1972; assigned to Block Engineering, Inc.
(22) Hirschfeld, T.; U.S. Patent 3,695,847; October 3, 1972.
(23) Villalobos, R.; U.S. Patent 3,762,878; October 2, 1973; assigned to Beckman Instruments, Inc.
(24) Dennis, J.L.; U.S. Patent 3,761,724; September 25, 1973; assigned to Resalab, Inc.
(25) Silas, R.S.; U.S. Patent 3,653,840; April 4, 1972; assigned to Phillips Petroleum Co.
(26) Boyd, C.A. and Kartluke, H.; U.S. Patent 3,436,188; April 1, 1969; assigned to Aeroprojects, Inc.

HYDROGEN CHLORIDE

Hydrogen chloride reacts rapidly with the moisture in the air and is generally found in the ambient atmosphere as hydrochloric acid. The acid at low concentrations, 15,000 to 75,000 $\mu g./m.^3$ (10 to 50 ppm), irritates primarily the mucous membranes of the eyes and upper respiratory tract in both humans and animals. Prolonged exposures to low concentrations of hydrochloric acid can also erode the teeth.

Work is intolerable after more than 60 minutes in atmospheres containing approximately 75,000 to 150,000 $\mu g./m.^3$ (50 to 100 ppm) of hydrochloric acid. Higher concentrations (approximately 1,500,000 $\mu g./m.^3$ or 1,000 ppm) can attack the mucous membranes, causing inflammation of the upper respiratory system and resulting in pulmonary edema or spasm of the larynx, which can be fatal. A wide variety of plant life is susceptible to the toxic effects of hydrogen chloride or hydrochloric acid. Several examples of plant damage due to hydrochloric acid emissions have been reported in the literature.

The primary effect on plants is a discoloration or bleaching of the leaves. The threshold for visible damage was originally reported as 75,000 to 150,000 $\mu g./m.^3$ (50 to 100 ppm) of hydrogen chloride. However, recent studies indicate that the threshold for many plants is less than 10 ppm for 4-hour exposures. The strong acidic properties of hydrochloric acid make it extremely corrosive to most metals.

Some countries have established ambient air quality standards for hydrochloric acid, including Russia (approximately 15 $\mu g./m.^3$ or 0.009 ppm as a 24-hour average), as noted by Stahl (1). Air quality standards have been set for hydrogen chloride in various other countries. In West Germany, the limit not to be exceeded more than once in any 2 hours is 1.0 ppm; the limit on a long-term exposure basis is 0.5 ppm. Canada (Ontario) has a more stringent standard of 0.04 ppm on a single exposure basis (30-minute averaging time).

Hydrogen chloride is produced by three main processes in the United States: acid-salt, direct synthesis, and the by-product process from chlorination of organic compounds. The latter process, the major production source, has shown a steady increase. Many organic chlorinating processes and other organic processes involving chlorinated compounds may produce hydrogen chloride as a by-product, but it may not be economically feasible to recover the product for other uses. Hydrochloric acid is produced by absorbing the hydrogen chloride gas into water.

Hydrochloric acid (or hydrogen chloride) is primarily used to manufacture inorganic and organic chemicals. Other major uses include metal production, oil-well acidizing, metal and industrial cleaning, and food processing.

Another potential source of emissions of hydrochloric acid is the heating or burning of chlorinated materials. The burning of coal appears to be a possible major source of hydrochloric acid pollution, since the chloride in coal is converted to over 90% hydrogen chloride during the burning process. Similarly, the burning of chlorinated plastics and paper can be an emission source, and possibly the burning of fuel oil and gasoline as well.

No information has been found on the concentrations of hydrochloric acid in the atmosphere. Hydrogen chloride emissions from commercial processes can be effectively controlled by the equipment that is available today. Systems now in use include packed water scrubbing towers and cooled absorption towers. Emissions from coal burning may be decreased by the addition of basic salts to the coal before burning.

No information has been found on costs for the control of hydrochloric acid pollution or for costs incurred by damage to humans, animals, plants and materials as a result of hydrochloric acid emissions.

Measurement in Air

All of the common methods of analysis for hydrochloric acid in air depend upon (a) measurement of acidity, (b) measurement of chloride ions, or (c) combination of the two measurements. Therefore, other strong acids (e.g., sulfuric acid, nitric acid) or other chloride salts (e.g., sodium chloride) can cause serious interference.

Sampling Methods: In most methods, hydrochloric acid is collected in impingers containing water (2)(3)(4). The high solubility of hydrochloric acid in water yields excellent absorption efficiency (5). The results of one investigation indicate that certain phosphate salts, particularly silver phosphate, have a high collection efficiency for hydrochloric acid but do not absorb sulfur dioxide (6). However, hydrobromic acid is also absorbed.

Preliminary studies indicate that liquid crystals may make effective and selective collection materials for hydrochloric acid and other gaseous pollutants (7). Acid mists have also been collected on paper (8), gelatinous film (9)(10)(11), thin metal films (12), and metal-coated glass slides (13)(14).

Qualitative Methods: The presence of hydrochloric acid or any other strong acid in air can be determined by passing the air over moist pH indicators such as methyl orange, congo red, or blue litmus papers. Hydrochloric acid or other chlorides can also be detected by bubbling the air through water and adding silver nitrate. Depending on the concentration of chloride, the solution may show a slight turbidity, or yield a white to grayish precipitate which will not change on addition of nitric acid but will dissolve or disappear upon addition of ammonia.

Quantitative Methods: In the absence of other strong acids, hydrochloric acid samples which have been collected in water may be determined quantitatively by the usual direct titration methods with standard bases (3). A spectrophotometric method for determination of strong acids has been reported (15)(16). After the air sample is passed through water, methyl red is added to the acidic solution and the optical density at 530 mμ. is determined. Normal concentrations of carbon dioxide and 1,000 μg. of sulfur dioxide per $m.^3$ of air do not interfere.

The size and quantity of acid mists have been determined by the use of both gelatinous films (9)(10)(11), and metal-coated glass slides (13)(14). The latter method is not affected by relative humidity. Through use of an electron microscope, it is possible to detect acid droplets with diameters of less than 0.1μ (14).

In the absence of soluble chloride salts, hydrochloric acid samples dissolve in water and may be determined by the standard methods for chloride determination (17) including the Mohr method (17)(18), absorption indicator method (19)(20), and modified Volhard method (21). These methods have a sensitivity of approximately 18,000 μg. of chloride per liter.

For very small amounts of chloride in air, the determination can be made turbidimetrically or nephelometrically (17)(22)(23). The hydrochloric acid solution is treated with silver nitrate and the turbidity determined spectrophotometrically; sensitivity is in the microgram range. Recently, a method has been developed for determining chloride colorimetrically (24)(25). This method is based on the yellow color produced by the iron (III) chloro complex in perchloric acid with a sensitivity of approximately 1,000 μg./m.^{3} of chloride. This method suffers from interference from mercury and sulfate ions and is affected by the relative humidity of the air. There is a neutron activation technique for determining chloride in particulates with a sensitivity of approximately 0.06 μg./m.^{3} (26).

An automated analysis method for chloride is used by the National Air Sampling Network (27). This method is based on the reaction of chloride with mercury thiocyanate to yield thiocyanate ions. Thiocyanate ions react with ferric ions to form the stable red complex, hexacyanatoferrate ion, which can be determined spectrophotometrically at 460 mμ. The relative standard deviation for chloride is 1.0 ± 0.1 μg./ml.

Since air samples usually contain other acids as well as chlorides, these methods are not generally applicable. There are no methods available that are free from interference by all other possible pollutants. However, an air sample containing a mixture of sulfur oxides, hydrochloric acid, and other chlorides can be analyzed by combining the methods, as reported in two papers (2)(4). The air sample is passed through an aqueous solution of hydrogen peroxide. The aqueous solution is then divided into three parts for three separate analyses: sulfate, chloride and hydrogen ion. From these results the amount of each component can be determined.

References

(1) Stahl, Q.R., "Air Pollution Aspects of Hydrochloric Acid", *Report PB 188,067,* Springfield, Va., Nat. Tech. Info. Service (Sept. 1969).

(2) Alekseyeva, M.V. and Elfimova, E.V., "Fractional Determination of Hydrochloric Acid Aerosol and of Chlorides in Atmospheric Air", *Gigiena i Sanit.* 23 (8):71 (1858).

(3) Devorkin, H., et al., *Air Pollution Source Testing Manual,* Air Pollution Control District, Los Angeles County, Calif. (1965).

(4) Stern, A.C. (Ed.), *Air Pollution, II,* 2nd ed. (New York:Academic Press, pp. 101, 325; 1968).

(5) Leithe, W., and Petschl, G., "Comparative Absorption Tests for Determination of Gaseous Air Contaminants in Wash Bottles," *Fresenius' Z. Anal. Chem.,* 226 (4):352 (1967).

(6) Chaigneau, M. and Santarromano, M., "Dosage de l'acide chlorhydrique en presence d'anhydride sulfureux a l'aide de reactifs solides", *Mikrochim. Acta 1965,* 976 (1965).

(7) Fergason, J.L., et al., "Detection of Liquid Crystal Gases (Reactive Materials)", Westinghouse Electric Corp., Pittsburgh Research Laboratories (Tech. Rept. RADC-TR-64-569, 1965).

(8) Comins, B.T., "Determination of Particulate Acid in Town Air", *Analyst* 88:364 (1963).

(9) Gerhard, E.R., and Johnstone, H.G., "Microdetermination of Sulfuric Acid Aerosol", *Anal. Chem.,* 27:702 (1955).

(10) Lodge, J.P., Ferguson, J. and Havlik, B.R., "Analysis of Micron-Sized Particles", *Anal. Chem.* 32:1206 (1960).

(11) Waller, R.E., "Acid Droplets in Town Air", *Intern. J. Air Water Pollution* 7:733 (1963).

(12) Lodge, J.P. and Havlik, B.R., "Evaporated Metal Films as Indicators of Atmospheric Pollution", *Intern. J. Air Water Pollution* 3:249 (1960).

(13) Hayashi, H., Koshi, S. and Sakabe, H., "Determination of Mist Size by Metal Coated Glass Slide", *Bull. Natl. Inst. Ind. Health* 6:35 (1961).

(14) Horstman, S.W., Jr. and Wagman, J., "Size Analysis of Acid Aerosols by a Metal Film Technique", *Amer. Ind. Hyg. Assoc. J.* 28:523 (1967).

(15) Goldberg, E.K., "Photometric Determination of Small Amounts of Volatile Mineral Acids (Hydrochloric and Nitric) in the Atmosphere", *Hyg. Sanitation* 31 (9):440 (1966).

(16) Manita, M.D. and Melekhina, V.P., "A Spectrophotometric Method for the Determination of Nitric and Hydrochloric Acids in the Atmospheric Air in the Presence of Nitrates and Chlorides", *Hyg. Sanitation* 29 (3):62 (1964).

(17) *Standard Methods for the Examination of Water and Wastewater,* 11th ed. (New York: American Public Health Association, 1960).

(18) "Standard Methods of Test for Ammonia in Industrial Water and Industrial Waste Water", *ASTM Designation D1426-58,* American Society for Testing Materials, Philadelphia, Pa. (1958).

(19) Fajans, K. and Wolff, H., "The Titration of Silver and Halogen Ions with Organic Dyestuff Indicators", *Z. Anorg. Allgem. Chem.* 137:221 (1924).
(20) Kolthoff, I.M. and Stenger, V.A., *Volumetric Analysis,* 2 (New York:Interscience, 1947).
(21) Caldwell, J.R. and Moyer, H.V., "Determination of Chloride", *Ind. Eng. Chem.* 7:38 (1935).
(22) Demidov, A.V. and Mokhov, L.A., "Rapid Methods for the Determination of Harmful Gases and Vapors in the Air", Translated by B.S. Levine, *USSR Literature on Air Pollution and Related Occupational Diseases* 10:32 (1962).
(23) Pimenova, Z., "Absorption of Hydrogen Chloride in Air Analysis", *Gigiena i Sanit.* 13(11):31 (1948).
(24) West, P.W. and Coll, H., "Direct Spectrophotometric Determination of Small Amounts of Chloride", *Anal. Chem.* 28:1834 (1956).
(25) West, P.W. and Coll, H., "Spectrophotometric Determination of Chloride in Air", *Proceedings of the Symposium on Atmospheric Chemistry of Chlorine and Sulfur Compounds, Cincinnati, Ohio 1957* (1959).
(26) Keans, J.R. and Fisher, E.M.R., "Analysis of Trace Elements in Air-Borne Particulates, by Neutron Activation and Gamma-Ray Spectrometry", *Atmospheric Environ.* 2:603 (1968).
(27) Thompson, Dr. R.J., Acting Chief, Laboratory Services Section, Division of Air Quality and Emission Data, National Air Pollution Control Administration, Cincinnati, Ohio, as reported in Reference (1).

HYDROGEN SULFIDE

Hydrogen sulfide is highly toxic to humans and at concentrations over 1,000,000 μg./m.^{3} quickly causes death by paralysis of the respiratory system. At lower concentrations, hydrogen sulfide may cause conjunctivitis with reddening and lachrymal secretion, respiratory tract irritation, psychic changes, pulmonary edema, damaged heart muscle, disturbed equilibrium, nerve paralysis, spasms, unconsciousness, and circulatory collapse, according to Miner (1).

The odor threshold for hydrogen sulfide lies between 1 and 45 μg./m.^{3}. Above this threshold value, the gas gives off an obnoxious odor of rotten eggs, which acts as a sensitive indicator of its presence. At these concentrations, no serious health effects are known to occur. At 500 μg./m.^{3}, the odor is distinct; at 30,000 to 50,000 μg./m.^{3} the odor is strong, but not intolerable; at 320,000 μg./m.^{3}, the odor loses some of its pungency, probably due to paralysis of the olfactory nerves. At concentrations over 1,120,000 μg./m.^{3}, there is little sensation of odor and death can occur rapidly. Therefore, this dulling of the sense of smell constitutes a major danger to persons exposed to high concentrations of hydrogen sulfide. Hydrogen sulfide produces the same health effects on domestic animals as on man, and at approximately the same concentrations.

An episode occurred at Poza Rica, Mexico, where the accidental release of hydrogen sulfide from a natural gas plant killed 22 persons, hospitalized 320 people and killed 50% of the commercial and domestic animals and all the canaries in the area. No measurements were made of the environmental hydrogen sulfide concentrations at the time of the episode. In Terre Haute, Ind., hydrogen sulfide emanations from an industrial waste lagoon caused foul odor, public complaints, and discomfort. However, very few people sought medical attention. Hydrogen sulfide concentration in the atmosphere during the episode ranged between 34 and 450 μg./m.^{3}.

There is little evidence that significant injury to field crops occurs at hydrogen sulfide concentrations below 60,000 μg./m.^{3}. At higher concentrations, the hydrogen sulfide injures the younger plant leaves first, then middle-aged or older ones.

Hydrogen sulfide combines with heavy metals in paints to form a precipitate which darkens or discolors paint surfaces. Air concentrations as low as 75 μg./m.^{3} have darkened paint after a few hours' exposure. White-lead paints often fade in the absence of hydrogen sulfide due to oxidation of the sulfite to sulfate. Paint darkening has occurred in Jacksonville, Fla., New York City, South Brunswick, N.J., Terre Haute, Ind., and in the areas near

Lewiston, Idaho and Clarkston, Wash. Hydrogen sulfide will also tarnish silver and copper. The sulfide coating formed on copper and silver electrical contacts can increase contact resistance and even weld the contacts shut.

Hydrogen sulfide is produced naturally by biological decay of protein material, mainly in stagnant or insufficiently aerated water such as swamps and polluted water. The background air concentration due to this source is estimated to be between 0.15 and 0.46 $\mu g./m.^3$. Industrial emitters of hydrogen sulfide are refineries, kraft paper mills, coke-oven plants, natural gas plants, chemical plants manufacturing sulfur-containing chemicals, viscose rayon plants, food processing plants, and tanneries. The emission of hydrogen sulfide and other organic sulfides, the cause of the foul odor in the vicinity of kraft paper mills, is the major kraft paper mill air pollution problem.

The average concentrations of hydrogen sulfide in the urban atmosphere vary from undetectable amounts to 92 $\mu g./m.^3$ (based on limited data). However, measurements as high as 1,400 $\mu g./m.^3$ have been recorded, but these have generally been in the vicinity of high hydrogen sulfide emissions.

The largest reduction in hydrogen sulfide emission from kraft mills was achieved by the addition of the black liquor oxidation process to the chemical recovery system. In addition, hydrogen sulfide emissions can be minimized if 2 to 4% excess oxygen is maintained in recovery-furnace flue gas, if the furnace is not operated above design conditions, and if noncondensible gases from the digesters and multiple-effect evaporators are scrubbed in a wet scrubber using weak caustic or chlorine water or are incinerated.

In refineries and natural gas plants, the hydrogen sulfide associated with sour gas stream is generally extracted in an absorption tower using absorbents such as aqueous amines. The hydrogen sulfide is recovered from the absorbent and in most cases further processed to produce valuable products such as sulfur or sulfuric acid.

The hydrogen sulfide associated with coke-oven gases is often removed by iron oxide-impregnated wood shavings. Liquid scrubbers using ammonium carbonate and sodium arsenate solutions can also be used. In chemical plants, liquid scrubbers and incinerators are used to minimize hydrogen sulfide emissions. Proper design, cleaning and aeration of sewers can minimize emissions of hydrogen sulfides. The most effective method of eliminating emissions of this material from sewage plants is to enclose the process and vent the gases formed to an incinerator.

Hydrogen sulfide air quality standards have been set by several states in the United States. Thus, California specifies a 0.1 ppm level on a 1-hour averaging basis as do New York and Pennsylvania. Missouri and Montana specify standards of 0.03 to 0.05 ppm on a 30-minute averaging basis.

Soviet standards are 0.005 ppm on both a single-exposure and 24-hour average basis. East Germany specifies 0.2 ppm on a not-to-be-exceeded in 8-hour basis and 0.1 ppm on a long-term exposure basis. Emission standards for H_2S in the United Kingdom are 5.0 ppm; the same is true for Australia.

Measurement in Air

The methods used in air pollution studies for hydrogen sulfide analysis are mainly based on iodometric methods such as titrating with iodine, methylene blue methods, molybdenum blue methods, and various modifications of the lead acetate paper and tile methods. The methylene blue method is based on precipitating cadmium sulfide from alkaline suspension of cadmium hydroxide by hydrogen sulfide in a known air sample. The alkaline suspension of cadmium hydroxide is contained in a standard impinger (0–1 CFM), through which is drawn the sample of air to be analyzed. The sulfide ion is then reacted with a mixture of p-aminodimethylaniline, ferric ion, and chloride ion (ferric chloride) to yield methylene blue. The concentration of hydrogen sulfide is then determined optically by a colorimeter or spectrophotometer (2)(3)(4). This method is good for hydrogen sulfide determinations down to the μg./m.^{3} range. Lahmann and Prescher (5) found that the cadmium sulfide suspensions are unstable at low concentrations and are decomposed by light. From tests run on samples of air containing from 7 to 170 μg./m.^{3}, they concluded that the method of sampling (i.e., amount of light which penetrates the sample) will affect the analytical results.

Recently Bamberger and Adams (6) claimed to have minimized these problems by adding arabinogalactan, avoiding exposure to light, and analyzing the samples as soon as possible after collection. By this procedure they claim a sensitivity of a few ppb of hydrogen sulfide in a 240 liter sample taken over a 2-hour period. This improvement gives about an 80% recovery of hydrogen sulfide.

Variations on the methylene blue method include the absorption of the hydrogen sulfide in zinc acetate instead of the cadmium salts. However, this variation is not as accurate as the cadmium salt method because zinc acetate loses hydrogen sulfide during sampling by air stripping and aging if allowed to stand for more than 2 hours. In addition, the collection efficiency for cadmium hydroxide is reported to be higher than for zinc acetate (7).

The molybdenum blue method is based on absorbing the hydrogen sulfide from the air sample in an acid solution of ammonium molybdate. The color developed in the ammonium molybdate by the hydrogen sulfide is determined optically by a colorimeter (8)(9).

The cadmium sulfide method is an example of the iodometric methods for determining hydrogen sulfide concentration in air. A known quantity of air is passed through two bubblers in series containing ammoniacal cadmium chloride solution. The collected samples are then stripped of any trapped sulfur dioxide, and the cadmium sulfide precipitate is dissolved in concentrated hydrochloric acid. The solution is then titrated with iodine using starch as the indicator. The hydrogen sulfide concentration in the air can be calculated from the amount of iodine added. Other cadmium solutions, cadmium acetate, for example, can be used as the absorbing solution (2). This method is accurate to 700 μg./m.^{3} for a 30-liter air sample.

The spot method using paper or tiles impregnated with lead acetate has been widely used to measure low concentrations of hydrogen sulfide in the atmosphere. The tiles are preferred in air pollution work. The unglazed tiles are impregnated with lead acetate and exposed in a place where they will be protected from rain. After exposure, the shade of the tiles is compared with known standards to estimate the concentration of hydrogen sulfide. In general, this method does not give accurate quantitative results, but rather, an indication of relative exposures of various localities to hydrogen sulfide (2)(4).

Gilardi and Manganelli (10) did experimental studies on the light absorbence of lead acetate-impregnated tile surfaces after exposure to various concentrations of hydrogen sulfide to develop an accurate quantitative measurement technique. From their experiments they concluded the following:

(1) Exposure unit $(\frac{\text{mg.-hr.}}{\text{m.}^3})$ is a useful parameter in representing hydrogen sulfide exposure.
(2) Average concentrations of hydrogen sulfide between 150 and 1,500 μg./m.^{3} can be determined by the measurement of surface absorbency of lead acetate.
(3) Fading of darkened tiles is accelerated by air turbulence and light.

Chiarenzelli and Joba (11) also found that the tiles faded on standing and that oxidation products formed. From these facts they concluded that the lead acetate tile method is unsatisfactory for periods greater than a day or two.

Automatic tape samplers based on lead acetate-impregnated filter paper have been developed for field air pollution application which continuously measure the hydrogen sulfide content of the atmosphere. The AISI or Hemeon tape sampler draws a known quantity of air through lead acetate-impregnated filter paper. If hydrogen sulfide is present in the atmosphere, a dark spot is formed which is measured by determining the optical density of the spots as compared to a standard (12). Sanderson, Thomas and Katz (13) reported that field experience has shown that large measurement errors can occur due to fading of the color of the precipitated lead sulfide spots by action of light, sulfur dioxide, ozone, or any other substance capable of oxidizing lead sulfide. This fading may even occur during the sampling period.

The fading can occur in a short time and a negative result is therefore not indicative of the absence of hydrogen sulfide (13). Other factors that affect the accuracy of the AISI tape sampler are relative humidity of the air and the consistency of absorbence of the blank paper (13). On the positive side, High and Horstman (7) reported obtaining results with the AISI tape sampler that were in reasonably good agreement with results obtained by the methylene blue method. They stated that the lead sulfide stains produced on the lead acetate filter paper did not fade significantly during an 8-week storage period when stored in vapor- and moistureproof bags.

Pare (14) suggested that mercuric chloride-impregnated filter paper be used in tape sampler paper as an improvement over the lead acetate-impregnated paper. He reported that the mercuric chloride paper tape is sensitive and reliable for determination of hydrogen sulfide in air and the spots are stable even in the presence of high levels of ozone, nitrogen oxides, and sulfur dioxide. He stated that it provided an adequate sensitivity on the order of 700 μg./m.^{3}. Dubois and Monkman (12) reported that although the spots formed by hydrogen sulfide on the mercuric chloride tape are resistant to fading effects, sulfur dioxide in the air results in a substantial change in hydrogen sulfide threshold of the tape.

Falgout and Harding (15) reported a method based on drawing air through a silver membrane filter. The hydrogen sulfide reacts to form silver sulfide, which results in a decrease in the reflectance of the silver surface. The reflectance of the membrane is measured before and after exposure, and the decrease in reflectance is proportional to the hydrogen sulfide exposure. This method is also sensitive to mercaptans in the air. Other methods have been used whereby silver coupons or coupons coated with lead-base paint are exposed to air.

The sulfide formed is removed and analyzed by the methylene blue method. Silver tarnishing of coupons as measured by light reflectance has also been used as a tool for measuring relative concentrations of hydrogen sulfide at various locations (7). Detector tubes containing inert particles coated with silver cyanide or lead acetate have been developed for testing for hydrogen sulfide (16). The sensitivity of this method is 0.04 μg. with a detection limit of 140 μg./m.^{3}.

Gas chromatographs with minimum detection threshold for hydrogen sulfide of 150 μg./m.^{3} have been used in air pollution and industrial work (17)(18). Adams and Koppe (19) determined that with a gas chromatograph, the bromine microcoulometric titration cell had the greatest potential for sulfur-specific analysis at a sensitivity required for direct analysis of small-volume samples. Hydrogen sulfide concentrations in the range of 15 μg./m.^{3} to 1,200,000 μg./m.^{3} can be measured by an electrolytic titrator which is preceded by a gas scrubber train to remove interfering sulfur compounds (20). Lahmann (21) reported that a sensitive instrument based on galvanic measuring cells has been developed in Germany for measuring the hydrogen sulfide content of air.

Hydrogen sulfide may be monitored in air by the following instrumental techniques (22): tape samplers, coulometric, colorimetric and gas chromatography — flame photometric detection.

An interesting detector for hydrogen sulfide has been developed by G.A. Brennan (23). It consists of an easily seen, standardized surface, such as that of a mast, mounted in the effluent discharging from a potential source of air pollution, such as a chimney, to provide a visible surface on which can be accumulated an informative record of the condition of the effluent, since the device was installed.

A major problem in the enforcement of air pollution control ordinances, is that sources of potential pollution may operate at infrequent intervals, or pollute only intermittently, or only at night. Where the pollutant is in the form of gases or vapors which are visible, the problem becomes even more difficult. Making door to door inspections of potential sources of pollution has been the most effective solution. This may require a large policing force, and can be a costly procedure.

Another alternative may be a sophisticated and expensive piece of detecting and monitoring equipment which provides a continuous instrumental record of effluent emissions. The object of the development is to provide a simple, inexpensive, visual recording device, which can be installed in the effluent of established exhaust outlets, to record the presence of both particulate and certain gaseous air pollutants, by either an accumulation on, or eroding of the surface of the indicator.

On industrial installations, or wherever noxious gases may be present in the effluent, it is desirable to have the surface of the collector coated with a chemical which will undergo a visible change in the presence of the gas or gases concerned. Lead carbonate is one such chemical which will turn from white to a dark grey or black in the presence of hydrogen sulfide. The chemical may be in the form of a pigment, in a paint or lacquer vehicle. To reduce the sensitivity of the chemical to a gas, one or more coats of a protective, transparent coating may be applied over the chemical surface. This should have a limited porosity to the gas, so that the underlying chemical will react, but more slowly than in the uprotected state.

A process developed by J.E. Lovelock (24) is a process for the detection of trace contaminants such as H_2S in air. A gas sample suspected of containing a contaminant is introduced into a reactor containing a reagent which converts the contaminant into an electron absorber. The presence of the contaminant is detected by passing the effluent from the reactor through an electron capture detector. If the gas sample contains oxygen then it is desirable to separate the oxygen from the stream entering the detector.

Thus, H_2S by reaction with a fluorinating agent in the reactor can be converted to SF_6. SF_6 is a gas which is more readily separated by chromatographic means from other constituents of the atmosphere than H_2S and is a strong electron absorber. Concentrations of H_2S in the atmosphere as low as parts per 10^{10} can be detected.

A process developed by J.B. Risk and F.E. Murray (25) permits continuously determining amounts of hydrogen sulfide as low as 7.5 ppm in a sample gas stream containing a mixture of hydrogen sulfide and sulfur dioxide. The process is based on the fact that sulfur dioxide gas strongly absorbs ultraviolet radiation in the wavelength region of 2850 A. whereas the other components of the gas stream, including hydrogen sulfide, are very weak absorbers in this wavelength region.

In accordance with the process, a stream of sample gas is diluted with air and the resultant mixture is split into two streams, the first stream being passed through a furnace in which the hydrogen sulfide present is catalytically oxidized to sulfur dioxide and then passed through the sample cell of a double beam ultraviolet analyzer while the second stream proceeds directly and without change to the reference cell of the ultraviolet analyzer. The difference in the sulfur dioxide content of the two streams is thus measured and since it is directly proportional to the hydrogen sulfide present in the original gas sample, the apparatus can be calibrated to read hydrogen sulfide directly.

The ultraviolet analyzer has as a source of radiation a hydrogen lamp which is fitted with filters so as to transmit a narrow band of radiation having a maximum intensity at 2850 A. The analyzer is a commercial item available from Analytic Systems Co., Pasadena, Calif.

Measurement in Water

Hydrogen sulfide in dissolved form can be monitored in water by the following instrumental techniques (26): colorimetric and iodometric.

References

(1) Miner, S., "Air Pollution Aspects of Hydrogen Sulfide", *Report PB 188,068,* Springfield, Va., Nat. Tech. Info. Service (Sept. 1969).

(2) Jacobs, M.B., "Techniques for Measuring Hydrogen Sulfide and Sulfur Oxides", *Geophysical Monograph Series No. 3* (1959).

(3) Jacobs, M.B., "Recommended Standard Method for Continuing Air Monitoring for Hydrogen Sulfide — Ultramicrodetermination of Sulfides in Air", *J. Air Pollution Control Assoc.* 15(7):314 (1965).

(4) Stern, A.C., *Air Pollution,* Vol. II (New York: Academic Press, 1968).

(5) Lahmann, E. and Prescher, K.E., "Intermittent Determination of Hydrogen Sulfide in the Atmosphere", *Staub* 25(12):3 (1965).

(6) Bamberger, W.L. and Adams, D.F., "Improvements in the Collection of Hydrogen Sulfide in Cadmium Hydroxide Suspension", *Environ. Sci. Tech.* 3(3):258 (1969).

(7) High, M.D. and Horstman, S.W., "Field Experience in Measuring Hydrogen Sulfide", *American Ind. Hyg. Assoc. J.* 26:366 (1965).

(8) Buch, M. and Stratmann, "The Determination of Hydrogen Sulfide in the Atmosphere", *Staub* 24:241 (1964).

(9) Sakurai, H. and Toyama, T., "An Improved Method of Colorimetric Determination of Hydrogen Sulfide in Air", *Japan. J. Ind. Health* 5(11):689 (1963).

(10) Gilardi, E.F. and Manganelli, R.M., "A Laboratory Study of a Lead Acetate Method for the Quantitative Measurement of Low Concentrations of Hydrogen Sulfide", *J. Air Pollution Control Assoc.* 13(7):305 (1963).

(11) Chiarenzelli, R.V. and Joba, E.L. "The Effects of Air Pollution on Electrical Contact Materials: A Field Study", *J. Air Pollution Control Assoc.* 16(3) (1966).

(12) Dubois, L. and Monkman, J.L. "The Analysis of Airborne Pollutants", Background Papers Prepared for the National Conference on Pollution and Our Environment, Canadian, Council of Resource Ministers, Montreal, Canada (Oct.-Nov. 1966).

(13) Sanderson, H.P., Thomas, R. and Katz, M. "Limitations of the Lead Acetate Impregnated Paper Tape Method for Hydrogen Sulfide", *J. Air Pollution Control Assoc.,* 16(6):328 (1966).

(14) Pare, J.P., "A New Tape Reagent for the Determination of Hydrogen Sulfide in Air", *J. Air Pollution Control Assoc.* 16(6):325 (1966).

(15) Falgout, D.A. and Harding, C.I., "Determination of H_2S Exposure by Dynamic Sampling with Metallic Silver Filters", *J. Air Pollution Control Assoc.* 18:15 (1968).

(16) Ruch, W.E., *Chemical Detection of Gaseous Pollutants* (Ann Arbor, Michigan: Ann Arbor Science Publishers, 1968).

(17) Applebury, T.E. and Shaer, M.J., "Analysis of Kraft Pulp Mill Gases by Process Gas Chromatography", Preprint. Montana State University, Department of Chemical Engineering (1968).

(18) "A Manual for Direct Gas Chromatographic Analysis of Sulfur Gases in Process Streams", *Atmospheric Pollution Technical Bulletin* 30 (New York: National Council for Stream Improvement, Inc., 1966).

(19) Adams, D.F. and Koppe, R.K., "Direct GLC Coulometric Analysis of Kraft Mill Gases", *J. Air Pollution Control Assoc.* 17(3) (1967).

(20) Thoen, G.N., Defass, G.G. and Austin, R.R., "Instrumentation for Quantitative Measurement of Sulfur Compounds in Kraft Gases", *Tappi* 51(6) (1968).

(21) Lahmann, E., "Methods for Measuring Gaseous Air Pollutants", *Staub* 25(9) (1965).

(22) Lawrence Berkeley Laboratory, "Instrumentation for Environmental Monitoring – Air", *Publ. LBL-1,* Vol. 1, Berkeley, Univ. of Calif. (Feb. 1, 1973).

(23) Brennan, G.A.; U.S. Patent 3,698,871; October 17, 1972.

(24) Lovelock, J.E.; U.S. Patent 3,725,009; April 3, 1973.

(25) Risk, J.B. and Murray, F.E.; U.S. Patent 3,300,282; January 24, 1967; assigned to British Columbia Research Council.

(26) Lawrence Berkeley Laboratory, "Instrumentation for Environmental Monitoring – Water", *Publ. LBL-1,* Vol. 2, Berkeley, Univ. of Calif. (Feb. 1, 1973).

IODIDES

Measurement in Water

Iodides may be monitored in water by the following instrumental techniques (1): ion selective electrode and molecular absorption (colorimetric).

Reference

(1) Lawrence Berkeley Laboratory, "Instrumentation for Environmental Monitoring — Water", *Publ. LBL-1,* Vol. 2, Berkeley, Univ. of Calif. (Feb. 1, 1973).

IRON

Inhalation of iron and iron oxides is known to produce a benign siderosis (or pneumoconiosis). However, in addition to the benign condition, there may be very serious synergistic effects as well as other undesirable effects, such as chronic bronchitis. In the laboratory, iron oxide has been shown to act as a vehicle to transport the carcinogens in high local concentrations to the target tissue. Similarly, sulfur dioxide is transported in high local concentrations deep into the lung by iron oxide particles. The relationships between dose and time and these conditions have not been determined. No evidence of animal or plant damage was found in a survey by Sullivan (1).

Soiling of materials by airborne iron or its compounds may produce economic losses. For example, iron particles have been observed to produce stains on automobiles, requiring them to be repainted. Iron oxide particulates may also reduce visibility.

The results from the National Air Sampling Network showed that iron concentrations ranged up to 22 μg./m.^{3}, with an average of 1.6 μg./m.^{3} in 1964. The most likely sources of iron pollution are from the iron and steel industry. The validity of this conclusion has been demonstrated by the decrease in iron concentration during steel strikes as well as by analysis of iron in the stack emissions. The iron pollution may be controlled by particulate removal equipment, such as electrostatic precipitators, venturi scrubbers, and filters.

Air pollution control cost the steel industry approximately $102 million in 1968. Fume control equipment costs for basic oxygen furnaces range between $3 and $7.5 million. This represents 14 to 19% of the total plant cost. Operating costs average $0.15 to $0.25 per ton of steel. Potential industrial sources of dissolved iron species are in wastewaters are reported (2)(3)(4) to be:

Mining operations	Textile mills
Ore milling	Food canneries
Chemical industries (organic, inorganic, petrochemical)	Tanneries
Dye industries	Titanium dioxide production
Metal processing industries	Petroleum refining
	Fertilizers

Iron exists in the ferric or ferrous form, depending upon conditons of pH and dissolved oxygen concentration. At neutral pH and in the presence of oxygen, soluble ferrous iron (Fe^{+2}) is oxidized to ferric iron (Fe^{+3}), which readily hydrolyzes to form the insoluble precipitate, ferric hydroxide, $Fe(OH)_3$. Therefore, acidic and/or anaerobic conditions are necessary in order for appreciable concentrations of soluble iron to exist.

In addition, at high pH values ferric hydroxide will solubilize due to the formation of the $Fe(OH)_4^-$ anion. Ferrous and ferric iron may also be solubilized in the presence of cyanide, due to the formation of ferro- and ferricyanide complexes. Such species present considerable problems for both iron and cyanide treatment. Perhaps the most significant industrial source of soluble iron waste is spent pickling solution. Pickling baths are employed to remove oxides from iron and steel during processing, or prior to plating of other metals on the surface. These baths contain strong acid solutions; usually sulfuric acid of 5 to 20% concentration by weight, although more recently hydrochloric acid has come into wider use. The baths accumulate soluble ferrous and ferric iron until their concentration interferes with product quality. At that time the bath must be replaced. According to Bramer (5), few industrial wastes have had the quantity of research and process development effort directed toward their treatment as has spent pickling solution.

Acid mine drainage waters are also high in soluble iron and represent a formidable treatment problem. It has been reported that 75% of the mines producing these wastes are inactive and abandoned (6). In one area, the flow from abandoned mines represents 90% of the total acid mine drainage of the region (7). Iron concentrations reported for some industrial wastewaters and mine waters are given in Table 27.

TABLE 27: IRON CONCENTRATIONS REPORTED FOR INDUSTRIAL WASTEWATERS

Process	Soluble Iron Concentration, mg./l.	Reference
Mine Waters		
Mine drainage	360 (36)*	(8)
Mine drainage	10 - 200, 93 avg.	(6)
Mine adit	17 - 218	(6)
Tailings pond	3,200	(6)
Acid mine water	501	(9)
Mine waste	67 - 70 (64 - 67)*	(10)
4 coal mines	122 - 330 (8 - 313)*	(11)
3 coal mines	40 - 150 (21 - 150)*	(12)
Motor Vehicle Assembly		
Body assembly	4	(13)
Vehicle assembly	3	(13)
Steel Processing		
Waste pickle liquor	96,800	(14)
Waste pickle liquor	70,000	(15)
Pickle bath rinse	200 - 5,000	(16)
Pickle bath rinse	60 - 1,300, 210 avg.	(17)
Steel cold finishing mills	60 - 150	(18)
Metal Processing		
Appliance manufacturer	0.09 - 1.9	(19)
Automobile heating controls	1.5 - 31	(20)
Appliances		
Mixed wastes	0.2 - 20	(21)
Spent acids	25 - 60	
Chrome plating	40	(15)
Plating wastes	2 - 4	(22)

*Values in parentheses are for ferrous iron levels.

Source: Report PB 216,162

As regards pollution by insoluble iron, McKee and Wolf (2) have pointed out that ferric oxides are used as pigments in certain paints. In addition, the authors indicate iron oxide is used as a polishing powder for glass, metal and ceramic materials. Such operations very likely represent a source of iron pollution, due to the colloidal nature of polishing compounds and the fact that water is usually involved as a lubricant and rinsing agent. In the processing of phosphoric acid, iron phosphates are formed and collected as a sludge, representing a potential waste source in the absence of efficient solids removal (3).

Iron and steel production represents a major source of small iron particles from the furnaces (5)(23), by virtue of air pollution control measures involving flue gas scrubbing. Steel mill wash waters may contain from 1,000 to 2,000 mg./l. suspended solids, with their metal composition roughly reflecting the furnace charge. 50% of blast furnace suspended solids may be less than 50 microns in diameter, while open hearth steel furnace solids may be even smaller (23). Ferromanganese steel processing produces the highest percentage of semicolloidal particles, with 50% being less than 10 microns in diameter (5).

Milling of iron or steel products yields solids in the quench tank in the form of rust scale. The total iron loss during milling is roughly 2.5% of the finished product weight (5). Much of this scale is of large, readily settleable size, but as metal product thickness decreases so does the scale size, with finishing mill scale at the micron level. Quench water (23) and casting equipment cleaning water (24) in iron foundries represent another source of finely divided iron and iron oxide particles. Dean, et al (4), have listed the following industries as potential sources of iron which may in part be particulate, or converted to particulate form during treatment:

Organic chemicals	Inorganic chemicals
Petrochemicals	Fertilizers
Alkalis	Petroleum refining
Chlorine	

Iron has long been known as an essential element because of its importance in the hemoglobin of blood. Iron is not considered as a pulmonary irritant (25)(26) although Bonser et al (27) claim industrial proof of iron causing lung cancer. It almost seems that at some time, someone has shown everything to be a carcinogen – iron seems to be the least likely carcinogen among the trace metals, however.

The (filterable) iron content of domestic water supplies should be below 0.3 ppm and it is desirable that it should be virtually absent as shown earlier in Table 3. The iron content of aqueous effluents discharged to a storm sewer or stream should be below 2.0 ppm.

Measurement in Air

Sampling Methods: Dusts and fumes of iron compounds may be collected by any method suitable for collection of other dusts and fumes; the impinger, electrostatic precipitator, and filter are commonly used. The National Air Sampling Network uses a high volume filtration sampler (28).

Quantitative Methods: Emission spectroscopy has been used by the National Air Pollution Control Administration for iron analysis of samples from the National Air Sampling Network (28)(29). The samples are ashed and extracted to eliminate interfering elements. The minimum detectable iron concentration by emission spectroscopy is 0.084 $\mu g./m.^3$ for urban samples and 0.006 $\mu g./m.^3$ for nonurban samples. The different sensitivities result from the different extraction procedures required for urban samples (30).

Thompson et al (28) have reported that the National Air Pollution Control Administration uses atomic absorption to supplement analyses obtained by emission spectroscopy. The method has a minimum detectable limit of 0.01 $\mu g./m.^3$ based on a 2,000 $m.^3$ air sample. Atomic absorption spectroscopy has been applied to the analysis of iron in air by Sachdev, Robinson and West (31). The sensitivity is 50 $\mu g./m.^3$ of solution.

A method for determining iron in the ambient air has been described by West et al (32). The dust sample is collected on filter paper with a high volume sampler and analyzed by the ring oven technique, using ferrocyanide. The limit of detection is 0.015 $\mu g.$, or approximately 2 $\mu g./m.^3$.

A flame emission spectrophotometry method for determining mass concentration of Fe_2O_3 aerosol in exposure chambers is reported by Crider, Strong, and Barkley (33). This is a continuous monitoring system sensitive to 3 $\mu g.$ of $Fe_2O_3/m.^3$. At the 100 to 1,000 $\mu g./m.^3$ level, the precision was $\pm$12%.

A procedure for determining the iron concentration in plant tissue and soil extracts was set forth by Paul (34). In this procedure the ferrous ion forms a complex compound with 1,10-phenanthroline, which is then determined colorimetrically.

Strackee (35) has measured the iron content of air by electron spin resonance spectrometry. Butt et al (36) has recorded an emission spectrographic method for determining the trace metal content (including iron) in the human liver, kidney, lung, brain, spleen, and heart.

Brief et al (37) has described a method for collecting and determining the air concentration of iron pentacarbonyl. The method has a sensitivity of 1μ of iron or 71 μg./m.^{3}. Bulba and Silverman (38) have developed a method of producing aerosols of iron oxides. A stream of nitrogen is passed through iron pentacarbonyl, after which it is mixed with an oxygen stream. This mixture is passed through a furnace which causes the oxidation of iron carbonyl to iron oxide.

Mallik and Buddhadev (39) have reported two spot-test methods for the determination of iron. Both methods use phenyl-2-pyridylketoxine with color development with (a) sodium carbonate and (b) ammonia. Interfering ions are copper, cobalt and cyanide. The limit of identification is about 0.05 μg. Iron may be monitored in air by x-ray fluorescence (40).

Measurement in Water

Iron may be monitored in water by the following instrumental techniques (41): atomic fluorescence, chemiluminescence and spark source mass spectrometry. As shown earlier in Table 11, iron may be measured by flame atomic absorption down to a 3 to 5 ppb detection limit and by flameless atomic absorption down to a 0.6 ppb detection limit.

As shown earlier in Table 12, the Technicon CSM-6 may be used to measure iron in the 0 to 10 ppm range with a detection limit of 0.10 ppm. According to K.H. Mancy (42), the analysis of iron with the Auto Analyzer makes use of intensely colored chelates.

A method developed by R.I. Fryer and B.Z. Senkowski (43) is a colorimetric method for quantitatively determining the iron content of aqueous solutions where the iron is present in its water-soluble ferrous state. It involves treating the ferrous ions with 5-(2-pyridyl)-2H-1,4-benzodiazepines and water-soluble salts thereof to produce a brilliant purple colored solution which can be measured quantitatively by standard colorimetric means.

Techniques have been developed for quantitatively determining the iron content of various aqueous solutions by means of reacting the iron in either its ferric or ferrous state with various color forming reagents such as ammonium thiocyanate, bathophenanthroline, dipyridyl, etc. These techniques have been of importance in determining the iron content of various iron-containing materials such as ores, foods, beverages, such as wines, etc.

However, this process has suffered from several disadvantages due to the fact that known color forming reagents are too sensitive to extraneous sources. In many cases, interfering bodies or contaminants do not produce with these color forming reagents, a sufficient color differentiation between the blank and the aqueous sample to be tested. This makes quantitative determination of the iron content in this sample difficult to carry out by standard colorimetric instruments such as a Beckman Spectrophotometer.

In contrast to these problems encountered in the practice of methods in the prior art, this method provides an extremely important diagnostic and analytical tool. The described method can be used to determine the iron in various materials rapidly and accurately. In addition to being a rapid and accurate method for making the determination, the results obtained by the test method are characterized by a high degree of reproducibility.

References

(1) Sullivan, R.J., "Air Pollution Aspect of Iron & Its Compounds", *Report PB 188,088,* Springfield,

Va., Nat. Tech. Info. Service (Sept. 1969).
(2) McKee, J.E. and Wolf, H.W., *Water Quality Criteria,* 2nd ed., California State Water Quality Board Publication, No. 3-A, 1963.
(3) Anonymous, *The Economics of Clean Water, Vol. III, Inorganic Chemicals Industry Profile,* U.S. Dept. Interior, Washington, D.C., 1970.
(4) Dean, J.C., Bosqui, F.L. and Lanouette, K.H., "Removing Heavy Metals from Waste Water", *Env. Sci. Technol.,* 6:518-522, 1972.
(5) Bramer, H.C., "Iron and Steel", in *Chemical Technology, Vol. 2, Industrial Wastewater Control,* C. Fred Gurnham, ed., Academic Press, New York, 1965.
(6) Hill, R.D., "Control and Prevention of Mine Drainage", presented at the Environmental Resources Conference on Cycling and Control of Metals, Battelle Columbus Laboratories, Columbus, Ohio, 1972.
(7) Wilmoth, R.C. and Hill, R.D., "Neutralization of High Ferric Iron Acid Mine Drainage", *U.S. EPA Report 14010 ETV* 8/70, 1970.
(8) Mihok, E.A., Duel, M., Chamberlain, C.E. and Selmeczi, J.G., "Mine Water Research. Limestone Neutralization Process", U.S. Bur. Mines, Rep. Invest. No. 7191; *Chem. Abstr.* 70: No. 6443, p. 1969.
(9) Kunin, R. and Downing, D.G., "Ion Exchange System Boasts More Pulling Power", *Chem. Eng.,* 79:67-69, June 28, 1971.
(10) Rex Chainbelt, Inc., "Reverse Osmosis Demineralization of Acid Mine Drainage," *U.S. EPA Report 14010 FQR,* 03/72, 1972.
(11) Bituminous Coal Research, Inc., "Optimization and Development of Improved Chemical Techniques for the Treatment of Coal Mine Drainage", *U.S. EPA Report 14010 EIZ,* 01/70, 1970.
(12) Bituminous Coal Research, Inc., "Studies of Limestone Treatment of Acid Mine Drainage. Part II", *U.S. EPA Report EIZ* 12/71, 1971.
(13) Anonymous, *The Cost of Clean Water, Vol. III, Industrial Waste Profiles, No. 2, Motor Vehicles and Parts,* U.S. Dept. Interior, Washington, D.C., 1967.
(14) Southern Research Institute, "An Electromembrane Process for Regenerating Acid from Spent Pickle Liquor", *U.S. EPA Report 12010 EQF* 3/72, 1972.
(15) Parsons, W.A., *Chemical Treatment of Sewage and Industrial Wastes,* National Lime Association, Washington, D.C., 1965.
(16) Nemerow, N.L., *Theories and Practice of Industrial Waste Treatment,* Addison Wesley Publishing Co., Reading, Mass., 1963.
(17) Armco Steel Corporation, "Limestone Treatment of Rinse Waters from Hydrochloric Acid Pickling of Steel", *U.S. EPA Report 12010 DUL* 02/71, 1971.
(18) Donovan, E.J., Jr., "Treatment of Wastewater for Steel Cold Finishing Mills", *Wat. Wastes Eng.,* 7:F22-F25, November, 1970.
(19) Watson, K.S., "Treatment of Metal Finishing Wastes", *Sew. Ind. Wastes,* 26:182-194, 1954.
(20) Gard, C.M., Snavley, C.A. and Lemon, D.J., "Design and Operation of a Metal Works Waste Treatment Plant", *Sew. Ind. Wastes,* 23:1429-1438, 1951.
(21) Anderson, J.S. and Iobst, E.H., Jr., "Case History of Wastewater Treatment in a General Electric Appliance Plant", *Jour. Wat. Poll. Control Fed.,* 40:1786-1795, 1968.
(22) Battelle Columbus Laboratories, "An Investigation of Techniques for Removal of Cyanide from Electroplating Wastes", *U.S. EPA Report 12010 EIE* 11/71, 1971.
(23) Kemmer, F.N., "Pollution Control in the Steel Industry", in *Industrial Pollution Control Handbook,* Herbert F. Lund, ed., McGraw-Hill Book Co., New York, 1971.
(24) Barzler, R.P. and Giffels, D.J., "Pollution Control in Foundry Operations", in *Industrial Pollution Control Handbook,* Herbert F. Lund, ed., McGraw-Hill Book Co., New York, 1971.
(25) Kleinfeld, M., Messite, S., Shapiro, J., Kooyman, O. and Levin, E., *Arch. Environ. Health* 16, 392 (1968).
(26) Amdur, M.O. and Underhill, D., *Arch. Envir. Health* 16, 460 (1968).
(27) Bonser, G.M., Faulds, J.S. and Stewart, M.S., *Am. J. Clin. Pathol.* 25, 126 (1955).
(28) Thompson, R.J., Morgan, G.B. and Purdue, L.J., "Analyses of Selected Elements in Atmospheric Particulate Matter by Atomic Absorption," preprint, presented at the Instrument Society of America Symposium, New Orleans, La. (May 5-7, 1969).
(29) "Air Pollution Measurements of the National Air Sampling Network – Analyses of Suspended Particulates", *1957-1961, U.S. Dept. of Health, Education, and Welfare, Public Health Service Publication No. 978,* U.S. Government Printing Office, Washington, D.C. (1962).
(30) "Air Quality Data from the National Air Sampling Networks and Contributing State and Local Networks," 1966 ed., *U.S. Dept. of Health, Education and Welfare, National Air Pollution Control Administration Publication No. APTD 68-9,* U.S. Government Printing Office, Washington, D.C. (1968).
(31) Sachdev, H.L., Robinson, J.W. and West, P.W., "Determination of Manganese, Iron and Nickel in Air and Water by Atomic Absorption Spectroscopy", *Anal. Chim. Acta* 38:499 (1967).

(32) West, P.W., et al, "Transfer, Concentration and Analysis of Collected Airborne Particulates Based on Ring Oven Techniques", *Anal. Chem.* 32:943 (1960).
(33) Crider, W.L., Strong, A.A. and Barkley, N.P., "Flame Emission Instrument for Selectively Monitoring Fe_2O_3 Aerosol in Animal Exposure Chambers", Preprint, Public Health Service, Cincinnati, Ohio, National Center for Air Pollution Control (1967).
(34) Paul, J., "Simultaneous Determination of Iron and Aluminum in Plant Tissues and Soil Extracts", *Mikrochim. Acta,* 6:1075 (1966).
(35) Strackee, L., "Electron Spin Resonance of Ferromagnetic Particles in Air-borne Dust", *Nature* 218:497 (1968).
(36) Butt, E.M., et al, "Use of Emission Spectrograph for Study of Inorganic Elements in Human Tissues", *Am. J. Clin. Pathol.* 24:385 (1954).
(37) Brief, R.S., Ajemian, R.S. and Confer, R.G., "Iron Pentacarbonyl: Its Toxicity, Detection, and Potential for Formation", *Am. Ind. Hyg. Assoc. J.* 28(1):21 (1967).
(38) Bulba, E. and Silverman, L.A., "Recirculating Aerosol Tunnel", read before the 59th annual meeting, Air Pollution Control Association, San Francisco, Calif. (June 23, 1966).
(39) Mallik, M.L. and Buddhadev, S., "Specific Spot Test for Iron", *Anal. Chim. Acta* 23(3):225 (1960).
(40) Lawrence Berkeley Laboratory, "Instrumentation for Environmental Monitoring – Air", *Publ. LBL-1,* Vol. 1, Berkeley, Univ. of Calif. (Feb. 1, 1973).
(41) Lawrence Berkeley Laboratory, "Instrumentation for Environmental Monitoring – Water", *Publ. LBL-1,* Vol. 2, Berkeley, Univ. of Calif. (Feb. 1, 1973).
(42) Mancy, K.M., "Instrumental Analysis for Water Pollution Control", Ann Arbor, Mich. Ann Arbor Science Publishers (1971).
(43) Fryer, R.I. and Senkowski, B.Z.; U.S. Patent 3,506,403; April 14, 1970; assigned to Hoffman-La Roche Inc.

LEAD

As has been pointed out by S.K. Hall (1), lead is widely used in industry and in everyday life but does pose serious health hazards. Indeed, Hall ascribes the decline of the Roman empire to extensive lead poisoning in the ruling class.

Air quality standards for lead have been set by the states of Montana and Pennsylvania at 0.005 mg./m.^{3} on a 30-day averaging time basis. The Soviet air quality standard for lead is 0.0007 mg./m.^{3} on a 24-hour averaging time basis. The Canadian (Ontario) standard is 0.02 mg./m.^{3} on a 30-minute averaging time basis. Emission standards for lead have been set in the United Kingdom as follows:

115 mg./m.^{3} for up to 3,000 cfm of exhaust
115 mg./m.^{3} for up to 3,000 to 10,000 cfm of exhaust
23 mg./m.^{3} for up to 10,000 to 140,000 cfm of exhaust
12 mg./m.^{3} for up to over 140,000 cfm of exhaust

Lead is used as an industrial raw material for storage battery manufacture, printing, pigments, fuels, photographic materials, and matches and explosives manufacturing. It is one of the most widely used nonferrous metals in industry. Despite its wide use, and the industry-associated lead bearing wastewaters which must result from its use, there is relatively little information in the technical literature on the treatability of lead-bearing wastes, or on associated waste treatment costs.

The storage battery industry is the largest consumer of lead, followed by the petroleum industry in producing gasoline additives (1). In a survey of eight battery manufacturing plants, it was reported that lead losses per battery manufactured ranged from 4.54 to 6,810 mg. and that water usage ranged from 11 to 77 gallons per battery (2). Highest lead concentrations originated from the plate forming area. This water has a low pH, which leads to a high concentration of dissolved lead. The particulate lead is very fine, and is reported to require long detention periods (above 24 hours) for effective settling (2).

The use of lead (as sodium plumbite) in petroleum refining produces a lead sludge, from which lead is normally recovered. This lead recovery process has been briefly discussed by Hill (3). Lead is also a common constituent of plating wastes, although not so frequently encountered as copper, zinc, cadmium and chromium. Fales (4) has reported the lead content of wastewater from one engine parts plating plant as ranging from 2.0 to 140.0 mg./l.

The higher concentrations resulted from dumping of spent plating bath solution. The author reported that the quantity of lead discharged per day might amount to 112 lbs. (50,848 grams) from a single lead-plating bath. Pinkerton (5) has reported lead content of plating bath rinse waters which, as expected, contain lower lead concentrations than presented by Fales for spent plating bath solutions. Rinse waters contained 0 to 30 mg./l. of lead (5).

A lead solder-glass frit mixture is used to fuse the front glass panel to the funnel of television tubes. Quality control of color picture tubes may require the salvage and reuse of the glass envelopes. Separation of the two components is accomplished by dissolving the frit in dilute nitric acid, which yields an acidic wastewater containing lead and fluoride. Lead wastewater levels of 400 mg./l. have been reported from such a salvage operation (6).

Pollution from lead mines, mining, and smelting have also been reported in the literature (7)(8)(9). Wixson, et al discussed the possibility of stream pollution from a new lead mining region in southeastern Missouri, but provided only general information (7). Ettinger (10) has mentioned the occurrence of lead in streams at concentrations up to 0.04 mg./l., particularly in waters receiving metal processing industrial wastes.

Analysis of lead mine wastewaters by Bolter, et al (11), are presented in Table 28. Initial lead levels shown represent wastewater from the milling process, and final values are those found following long-term detention of the wastewater in holding ponds. Pond detention times were not reported. Initial levels ranging from 0.018 to 0.098 mg./l. and effluent levels of 0.022 to 0.055 mg./l. were found (11). With one exception, lead reductions of approximately 45% were observed. This is likely due primarily to sedimentation of the particulate lead contained in the wastewater.

TABLE 28: LEAD MINE MILLING PROCESS WASTEWATER

Lead Concentration, mg./l.		
Initial	Final*	Percent Removal
0.040	0.022	45
0.018 - 0.036	0.038	+
0.051	0.030	42
0.098	0.055	44

*Effluent from holding pond

Source: E. Bolter, J.C. Jennett and B.G. Wixson; 27th Indust. Waste Conf., Purdue Univ., 1972

TABLE 29: LEAD CONTENT OF MINE DRAINAGE

Source	pH	Lead, mg./l.
Tailing pond	2.0	0.67
Mine	2.6	0.3
Mine	3.2	2.5
Mine	-	0.02

Source: R.D. Mill; Conf. on Recycling and Control of Metals, Battelle-Columbus, 1972

Municipal water pumped from an abandoned mine shaft in Australia, which was used as a municipal water supply infiltration gallery, was reported to contain 0.31 mg./l. of lead (9). Mill (12) has reported significant levels of lead in acid mine drainage, as shown in Table 29. At the reported pH values, the lead would be predominately in the soluble form. Table 30 summarizes reported lead levels in industrial wastewaters.

The permissible concentration of lead in domestic water supplies is below 0.05 ppm and it is desirable that it be totally absent as shown earlier in Table 3. According to Table 5, the permissible lead concentration in irrigation water for continuous use is 5.0 mg./l. and for short term use in fine textured soil is 20.0 mg./l. The maximum allowable concentration of lead in effluents being discharged to a storm sewer or stream is 0.1 ppm.

TABLE 30: REPORTED LEAD LEVELS IN INDUSTRIAL WASTEWATERS

Industry	Lead, mg./l.	Reference
Battery manufacture		(2)
Particulate	5 - 48	
Soluble	0.5 - 25	
Plating	2 - 140	(4)
Plating	0 - 30	(5)
Television tube manufacture	400	(6)
Mine drainage	0.02 - 2.5	(12)
Mining process water	0.018 - 0.098	(11)
Tetraethyl lead manufacture		(13)
Organic	126.7 - 144.8	
Inorganic	66.1 - 84.9	
Tetraethyl lead manufacture	45	(14)

Source: Report PB 216,162

Measurement in Air

Lead may be monitored in air by the following instrumental techniques (15): colorimetric, emission spectroscopy, x-ray fluorescence and neutron activation. The ASTM standard method for the determination of lead in air is the dithizone colorimetric which bears ASTM test # D3112.

Measurement in Water

Lead may be monitored in water by the following instrumental techniques (16): electrochemical, spark source mass spectrometry and x-ray fluorescence. As noted earlier in Table 11, lead may be determined by flame atomic absorption down to a 10 to 20 ppb detection limit and by flameless atomic absorption down to a 1.0 ppb detection limit. The determination of lead in water by atomic fluorescence spectroscopy has been discussed by R.F. Browner, R.M. Dagnall and T.S. West (17). The atomic fluorescence characteristics of lead are described in air-acetylene, nitrous oxide-hydrogen, and argon-oxygen-hydrogen flames. An electrodeless discharge tube is used as the source of excitation. A detection limit of 0.01 microgram per milliliter of lead in aqueous solution is obtained by measurement of the direct line fluorescence at 405.8 nm. in the argon-oxygen-hydrogen flame. The effect of 100-fold excesses of 30 cations and anions is examined; only aluminum interfered significantly. Effects of multipass optics and signal collection mirrors are examined and their effect on signal-noise ratios is discussed. The study confirmed that lead can be determined more sensitively by atomic fluorescence than by atomic absorption or flame photometry.

A spectropolarimetric tritrimetric method for the determination of various heavy metals, including lead, has been reported by R.J. Palma, Sr. and K.H. Pearson of the Department of Chemistry of Texas A & M University (18). Spectropolarimetric methods of analyses of Cd (II), Pb (II) and Bi (III) were developed, based on the stereospecific ligand, D-(-)-1,2-propylenediaminetetraacetic acid [D(-)PDTA]. The optical rotation of the solution was monitored with a photoelectric polarimeter to determine the end point of the titration. The effects of pH, wavelength, and dilution were discussed and the optimum conditions for the determination of each metal were established. The range of accurate analyses at 365 nm. was 0.1 minus 0.001 M for Cd (II), and 0.1 minus 0.001 M for Pb (II), Hg (II) and Bi (III). The range of the average deviations from visual colorimetry was 0.05 minus 0.29%.

A technique developed by R. Bloch, R. Bauer and B.F. Phillips (19) is one in which polyvalent metal ions such as lead ions can be determined colorimetrically by using a semipermeable polymeric membrane containing a chelating agent and a solvent therefor, which chelating agent changes color upon contact with the metal ions being determined. Various methods are known for the determination of such metal ions in solution. The most widely used of these methods are color reactions which are more or less specific for particular metal ions. The use of most color reactions, however, involves the preparation of reagents and other laboratory procedures which may be cumbersome and difficult, especially for laboratories staffed with relatively unskilled personnel or for use in the field.

It has been found that immobilizing a chelating agent and using it in the solid phase provides a convenient means for the determination of metal ions in solution. To immobilize the chelating agent it is dissolved in a suitable solvent and incorporated into a semipermeable polymeric matrix. The polymeric matrix containing the dissolved chelating agent is then cast in the form of a membrane. The membrane may be plasticized by means of a plasticizer added to the polymeric matrix or the solvent for the chelating agent may serve as a plasticizer for the membrane. The resulting membrane containing the chelating agent and solvent and/or plasticizer can be used to give a colorimetric response to various polyvalent metal ions in solution including iron, cobalt, nickel, copper, lead and mercury.

The following is one specific example of the operation of the technique. To 60 ml. of freshly distilled cyclohexanone were added 5 grams of polyvinyl chloride. The mixture was rapidly slurried and heat was applied until all particles were dissolved. The resulting solution was then cooled to 25°C. and brought to a volume of 100 ml. by adding cyclohexanone. A 0.75% solution of dithizone in dipentylphthalate (1 ml.) and 2 ml. of dipentylphthalate were than added to 20 ml. of the above solution of polyvinyl chloride in cyclohexanone. The composite solution resulting was then poured upon glass microscope slides and dried at 60°C. for 40 minutes to yield a green colored membrane. The green membrane when dipped into a 10^{-6} M solution of lead nitrate changed to a red color. Higher concentrations of lead salts caused the color to become darker.

References

(1) Hall, S.K., "Lead Pollution and Poisoning", *Env. Sci. Tech.* 6:30-35, 1972.
(2) Anonymous, "Evaluation of Lead Plant Wastewater Treatment Methods", *IIT Research Institute Report No. IITRI-C8213-2,* 1972.
(3) Hill, J.B., "Waste Problems in the Petroleum Industry", *Ind. Eng. Chem.,* 31 (11):1361-1363, 1939.
(4) Fales, A.L., "A Plating Waste Disposal Problem", *Sew. Wks. Jour.,* 20:857-860, 1948.
(5) Pinkerton, H.L., "Waste Disposal", in *Electroplating Engineering Handbook, 2nd ed.,* A. Kenneth Graham, editor-in-chief, Reinhold Publ. Co., New York, 1962.
(6) Rohrer, K.L., "An Integrated Facility for the Treatment of Lead and Fluoride Wastes", *Indust. Water,* 1971:36-39 (Sept./Oct.), 1971.
(7) Wixson, B.G., Bolter, E.A., Tibbs, N.H. and Handler, A.R., "Pollution from Mines in the New Lead Belt of Southeastern Missouri", *Proc. 24th Purdue Indust. Waste Conf.,* pp. 632-643, 1969.
(8) Jennett, J.C., and Wixson, B.G., "Treatment and Control of Lead Mining Wastes in S.E. Missouri", *Proc. 26th Purdue Indust. Waste Conf.,* (in press), 1971.
(9) Anonymous, "Annual Report Western Australian Govt. Chem. Labs.", Dept. of Mines, Australia, 1962.
(10) Ettinger, M.B., "Lead in Drinking Water", *Water Waste Eng.,* 4 (3):82-84, 1967.
(11) Bolter, E., Jennett, J.C. and Wixson, B.G., "Geochemical Impact of Lead-Mining Wastewaters on Streams in Southwestern Missouri", presented at 27th Ind. Waste Conf., Purdue Univ., 1972.
(12) Mill, R.D., "Control and Prevention of Mine Drainage", presented at Conf. on Recycling and Control of Metals, Battelle-Columbus, 1972.
(13) Nozaki, M. and Hatotani, H., "Treatment of TEL Mfg. Wastes", *Water Res.,* 1(2):167-177, 1967.
(14) Robb, L.A., "Waste Water Treating Facilities at Ethyl Corporation of Canada, Ltd.", *Proc. 8th Ontario Industrial Waste Conf.,* pp. 90-96, 1961.
(15) Lawrence Berkeley Laboratory, "Instrumentation for Environmental Monitoring – Air", *Publ. LBL-1, Vol. 1,* Berkeley, Univ. of Calif. (Feb. 1, 1973).
(16) Lawrence Berkeley Laboratory, "Instrumentation for Environmental Monitoring – Water", *Publ. LBL-1, Vol. 2,* Berkeley, Univ. of Calif. (Feb. 1, 1973).
(17) Browner, R.F., Dagnall, R.M. and West, T.S., *Anal Chim. Acta* 50, 375-81 (June 1970).
(18) Palma, R.J., Sr. and Pearson, K.H., *Anal. Chim. Acta* 49, 497-504 (Mar. 1970).
(19) Bloch, R., Bauer, R. and Phillips, B.F.; U.S. Patent 3,635,679; January 18, 1972; assigned to Miles Laboratories, Inc.

LEAD ALKYLS

Measurement in Air

A process has been developed by A.O. Walker (1) for determining the presence and quantities of volatile organic lead compounds in air. Air suspected of containing such compounds is passed over iodine crystals to sublime at least one-half thereof. Lead iodine compounds are formed and cooled on a porous substrate. Free iodine is removed from the substrate; thereby developing a yellow color on the substrate if lead was present in the air.

Several methods have been proposed for determining the presence and concentration of volatile lead compounds in the atmosphere. See, for example, U.S. Patent 3,071,446 and the article by Snyder et al appearing in *Analytical Chemistry*, volume 20, pages 772 to 776 (1948). The methods described in the above references require the performance of several procedural steps before the actual determination of lead in an air sample can be made. These steps are complex and require a trained chemical operator to perform them.

The method of this process, on the other hand, makes use of a relatively simple apparatus which can be operated to determine the presence and quantity of volatile lead impurities in the atmosphere which can be operated by a person not skilled in the field of chemistry. After a few hours of training, any person of reasonable intelligence can be taught to operate the apparatus to make determinations of the quantity of volatile lead compounds contained in the atmosphere.

Figure 36 shows a suitable form of apparatus for use in this process. The apparatus comprises a sampling tube **10** which has an air inlet **12** and an air outlet **14**. Near the air inlet is a porous support **16** upon which is placed iodine crystals **18**. Positioned above the iodine crystals is a porous substrate **20** which is capable of retaining lead iodine compounds and free iodine.

Both the porous support and the porous substrate are preferably constructed of a chemically inert material such as glass wool or fiber glass matting. The bottom **22** of the porous substrate is preferably hemispherical which increases its surface area. Connected to the air outlet of the sample tube by means of a ground joint **24** is a flow meter **26**. At the outlet **28** thereof, a T **30** is fitted thereto by means of rubber stopper **32**. The side arm **34** of the T is connected to a suitable air moving or vacuum device such as aspirator **36**. The outlet **38** of the T is fitted with a rubber or plastic tube **40** to which is affixed a closing means such as screw clamp **42**.

An optional feature resides in the utilization of air heating means **44** which is a hollow tube fitted to inlet **12** of the sampling tube. Affixed to the heater is a heating coil **46** which is connected to a suitable power source **48**.

In operation, air to be sampled for the presence and quantity of volatile organic lead compounds is drawn into the sampling tube through inlet **12** by the suction applied to the system by operation of the aspirator **36**. The particular quantity of air passing through the sample tube is regulated by the opening or closure of the screw clamp. The exact amount may be regulated carefully by adjusting the screw clamp and observing the operation of the flow meter. As the air passes into the sample tube, the iodine is sublimed and reacts with the volatile organic lead compounds to form lead iodine compounds. Both the lead iodine compounds and the free iodine are captured by the porous substrate.

The amount of iodine used should be an amount which will be at least half sublimed at the end of a particular test period. In most cases one hour is adequate. If a larger amount of iodine is used the volatile lead iodine compound is not completely released and will not be observed on the test plug. Obviously, enough iodine must be present at the end of the test to still react with any lead compounds present in the air being tested. In preparing the sample tube, a typical form would use a glass tube approximately 5 inches long having a 7 mm. interior diameter. It would then be fitted with a 2 cm. plug of

FIGURE 36: APPARATUS FOR MEASURING LEAD ALKYLS IN AIR

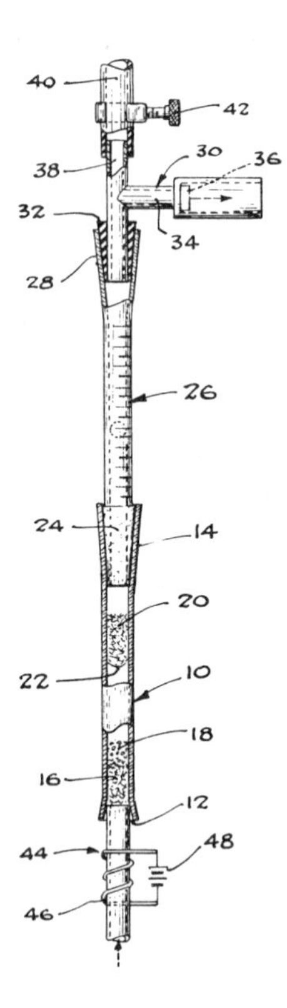

Source: A.O. Walker; U.S. Patent 3,453,081; July 1, 1969; assigned to Nalco Chemical Company

glass wool so arranged that the hemispherical surface previously mentioned is toward the upstream direction. This plug is used to filter the lead iodine. About ½ cm. of iodine crystal would then be added to the tube and another glass wool plug inserted leaving about 1 inch space between the two plugs. The flow rate of air through the crystals should be sufficient to cause them to tumble. In the apparatus described, a flow rate of 5 to 6 cu. ft./hr. is satisfactory.

It will be understood that the above is only illustrative of a typical sample tube and that it may be constructed of different size tubing and that the amount of iodine crystals used will be capable of variation depending upon the temperature of the incoming air, the flow rate of the air and the surface area of the iodine crystals exposed to the air. These calculations may be readily made utilizing well-known information with respect to the vapor pressure of iodine. See, for example, page 44 of *General Chemistry* by Linus Pauling, W.H. Freeman & Co. (1958).

After the test has been run for a sufficient period of time, the sample tube is removed from the apparatus and placed on the outlet **38** of the T **34**. The air current is then allowed to pass through the sample tube in a reverse direction which effectuates removal of free iodine from the glass wool from the porous substrate. This allows the development of a yellow color in the porous substrate, thereby indicating the presence of volatile organic lead compounds in the air sample. Intensity of color in the substrate is then compared with standardized substrates to determine the concentration of lead in micrograms per cubic foot. The standards may be prepared by inserting the end of a sample plug into a quart bottle to which has been added a standard lead containing solution. This permits the quantity of lead drawn through the sample tube to be determined by visual comparison with the thus prepared standard plugs.

The yellow color developed on the porous substrate is believed to be caused by the formation of lead iodide. It is well-known that when alkyl lead compounds are reacted with halogens at about room temperature, a complex series of reactions occur with one of the final products being a lead iodide. When the airflow is reversed through the substrate to remove free iodine, the yellow color becomes clearly visible due to the removal of the brown color caused by the free iodine.

When it is desired to determine only the presence of a minimum amount of volatile organic lead compounds in the atmosphere, the described apparatus may be slightly modified to achieve this result. When it is desired to determine the presence of at least 4 micrograms per cubic foot, the sampling tube is made of a transparent material and is packed with large amounts of iodine and air to be sampled is passed therethrough for a longer period of time.

If an amount in excess of 4 micrograms per cubic foot of volatile organic lead compound is present, a yellow color will become visible on the walls of the sampling tube in about 24 hours. To develop this yellow color during the 24 hour period, a lower flow rate should be employed. In a typical form, a flow rate of 2 to 3 cubic feet per hour would be used in an 8 mm. diameter sampling tube which contains 2 to 3 inches of iodine crystals. For this simpler qualitative type test, it is desirable that the apparatus be run for a minimum of 12 hours.

References

(1) Walker, A.O.; U.S. Patent 3,453,081; July 1, 1969; assigned to Nalco Chemical Company.

MAGNESIUM

Measurement in Water

Magnesium may be monitored in water by spark source mass spectrometry (1). The use of atomic absorption analyses for determination of a variety of metals, including magnesium has been described by M.J. Fishman and D.E. Erdmann (2).

An atomic absorption spectrophotometer was automated with a sampler, proportioning pump manifold and strip chart recorder. Copper, lithium, manganese, potassium, sodium, iron and zinc were determined by direct aspiration from the sampler. The proportioning pump manifold was added to the system for the determination of calcium, magnesium and strontium. It introduces, automatically, an interference suppressant to the samples prior to aspiration.

References

(1) Lawrence Berkeley Laboratory, "Instrumentation for Environmental Monitoring - Water," *Publication LBL-1,* Volume 2, University of California, Berkeley, Calif. (February 1, 1973).
(2) Fishman, M.J. and Erdmann, D.E., *Atomic Absorption Newsletter* 9, No. 4, 88 to 89 (July and Aug. 1970).

MANGANESE

Inhalation of manganese oxides may cause chronic manganese poisoning or manganic pneumonia. Chronic manganese poisoning is a disease affecting the central nervous system, resulting in total or partial disability if corrective action is not taken. Some people are more susceptible to manganese poisoning than others. Manganic pneumonia is a croupous pneumonia often resulting in death. The effect of long exposure to low concentrations of manganese compounds has not been determined, according to Sullivan (1). Manganese compounds are known to catalyze the oxidation of other pollutants, such as sulfur dioxide, to more undesirable pollutants, for example, sulfur trioxide. Manganese compounds may also soil materials.

The most likely sources of manganese air pollution are the iron and steel industries producing ferromanganese. Two studies, one in Norway and one in Italy, have shown that the emissions from ferromanganese plants can significantly affect the health of the population of a community. Other possible sources of manganese air pollution are manganese fuel additives, emissions from welding rods, and incineration of manganese-containing products, particularly dry cell batteries. Manganese may be controlled along with the particulates from these sources. Air quality data in the United States showed that the manganese concentration averaged 0.10 μg./m.^{3} and ranged as high as 10 μg./m.^{3} in 1964.

U.S.S.R. air quality standards dictate a manganese concentration of 0.03 mg./m.^{3} on a single exposure basis or 0.01 mg./m.^{3} on a 24 hour average basis. Czechoslovakian air quality standards are also 0.01 mg./m.^{3} on a 24 hour basis. In Czechoslovakia, manganese emissions (expressed as MnO_2) above 0.1 kg./hr. require reporting to the government.

McKee and Wolf (2) have reported that manganese and its salts are used in the following industries and appear in their wastewaters: steel alloy; dry cell battery; glass and ceramics; paint and varnish; ink and dye; and match and fireworks. Among the many forms and compounds of manganese, only the manganous salts and the highly oxidized permanganate anion are appreciably soluble. The latter is a strong oxidant which is reduced under normal circumstances to insoluble manganese dioxide. The highly soluble manganous chloride is used in dyeing operations, linseed oil driers and electric batteries, while the equally soluble manganous sulfate is used in porcelain glazing, varnishes and production of specialized fertilizers. (2).

Hill (3) reports manganese concentrations of 1.2 to 8.0 mg./l. for mine waters. Kunin and Downing (4) cite a value as high as 50 mg./l. for an acid mine water. Since the chemistry of manganese is similar to that of iron (5) it would be expected that any pickling operation involving manganese steel alloy results in dissolution of manganous ion and the presence of this ion in pickling and rinse solutions. In contrast to iron, the divalent (manganous) form is not readily oxidized to the insoluble manganic form, other than at elevated pH (5).

The permissible concentration of (filterable) manganese in domestic water supplies is below 0.05 mg./l. and it should preferably be absent as shown earlier in Table 3. The tolerance for manganese in irrigation water on a continuous use basis is 2.0 mg./ml. and on a short-term use basis in fine textured soil is 20 mg./l. as shown in Table 5. The maximum allowable concentration of manganese in an effluent discharged to a storm sewer or stream is 1.0 ppm.

Measurement in Air

Sampling Methods – Dusts and fumes of manganese compounds may be collected by any methods suitable for collection of other dusts and fumes; the impinger, electrostatic precipitator and filters are commonly used (6)(7).

Quantitative Methods – Many methods of analysis have been used for the determination of manganese. In several methods, such as atomic absorption, emission spectrography, and neutron activation, the chemical species does not matter. The essential requirement is a representative sample in the activation site. With atomic absorption, all of the manganese analyzed must be in solution or suspension (8)(9)(10)(11)(12). With chemical methods, care must be taken to insure that the manganese is not only in solution, but also in a single oxidation state. The most common method of determining manganese in air samples is a colorimetric method. In this procedure, the manganese is oxidized to permanganate by periodate ions. The color of the permanganate ion is very intense and follows Beer's law over a large range. A 0.5 ft.^{3} sample collected in a midget impinger is sufficient to determine manganese at a 5,000 μg./m.^{3} concentration (11). The following sensitivities are listed for the different analytical methods.

Method	Sensitivity (μg./g.)
Colorimetric permanganate	0.1
Emission spectrography*	10 - 1,000
Neutron activation*	0.001 - 1
Atomic absorption*	0.01 - 20
Flame photometry*	0.1 - 500

*Sensitivity depends on interfering elements and these sensitivities were not determined on air samples.

The emission spectrographic method is used to analyze samples collected by the National Air Sampling Network. Working standards are made by diluting a solution containing 20 μg./0.05 ml. by ½, ¼, ⅛ ... $1/1{,}024$. The minimum concentration detectable by the emission spectrograph is 0.011 for urban air and 0.0060 for nonurban air. The difference in sensitivity is due to different extraction methods used in preparing the sample for emission spectrography (6)(13). Manganese may be monitored in air by the following instrumental techniques (14): Colorimetric and Emission Spectroscopy.

Measurement in Water

Manganese may be monitored in water by the following instrumental techniques (15): Direct Reading Emission Spectrometry and Spark Source Mass Spectrometry.

As described by W. Baumer et al (16), an indicator for the detection of manganese (II) ions comprises an absorbent support carrier impregnated with a substantially insoluble sulfate salt of an aromatic amine oxidizable by MnO_2 to a colored oxidation product. Atomic absorption and neutron activation may be used for manganese determination down to a detection limit of 0.1 ppb as shown earlier in Table 11.

References

(1) Sullivan, R.J. "Air Pollution Aspects of Manganese and Its Compounds," *Report PB 188,079;* Springfield, Virginia, National Technical Information Service (September 1969).

(2) McKee, J.E. and Wolf, H.W., "Water Quality Criteria," California State Water Quality Control Board *Publication No. 3-A,* (1963).

(3) Hill, R.D., "Control and Prevention of Mine Drainage," presented at the Environmental Resources Conference on Cycling and Control of Metals, Battelle Columbus Laboratories, Columbus, Ohio, (1972).

(4) Kunin, R. and Downing, D.G., "Ion Exchange System Boasts More Pulling Power," *Chemical Engineering,* 78, 67 to 69, (June 28, 1971).

(5) Cotton, F.A. and Wilkinson, G., *Advanced Inorganic Chemistry,* Interscience Publishers, New York, (1962).

(6) "Air Pollution Measurements of the National Air Sampling Network, Analyses of Suspended Particulates, 1957 to 1961", U.S. Department of Health, Education and Welfare, Public Health Service *Publication No. 978,* U.S. Government Printing Office, Washington, D.C. (1962).

(7) Elkins, H.B., *The Chemistry of Industrial Toxicology,* New York, Wiley, (1959).

(8) Gordon, C.M. and Larson, R.E., "Activation Analysis of Aerosols," *NRL Quarterly on Nuclear Science and Technology,* Naval Research Lab., Washington, D.C. pages 17 to 22, (Jan. 1, 1964).

(9) Horwitz, W. (Ed.), *Official Methods of Analysis of the Association of Official Agricultural Chemists,* 10th Ed., Association of Official Agricultural Chemists, Washington, D.C. (1965).

(10) Kolthoff, T.M., Elving, P.J. and Sandell, E.B., *Treatise on Analytical Chemistry, Part 2–Analytical Chemistry of the Elements,* volume 7, S, Se-Te, F Halogens, Mn, Re New York, Interscience (1961).

(11) *Manual of Analytical Methods,* "Determination of Manganese in Air," American Conference of Governmental Industrial Hygienists, Cincinnati, Ohio (1958).
(12) Willard, H.H., Marritt, Jr., L.L. and Dean, J.A. *Instrumental Methods of Analysis,* 4th Edition, Princeton, New Jersey, Van Nostrand (1965).
(13) "Air Quality Data from the National Air Sampling Networks and Contributing State and Local Networks," 1966 Edition, U.S. Department of Health, Education and Welfare, National Air Pollution Control Administration *Publication No. APTD 68-9,* U.S. Government Printing Office, Washington, D.C. (1968).
(14) Lawrence Berkeley Laboratory, "Instrumentation for Environmental Monitoring - Air," *Publication LBL-1,* Volume 1, Berkeley, University of California (February 1, 1973).
(15) Lawrence Berkeley Laboratory, "Instrumentation for Environmental Monitoring - Water," *Publication LBL-1,* Volume 2, Berkeley, University of California (February 1, 1973).
(16) Baumer, W., Schmitt, D. and Stein, A.; U.S. Patent 3,770,379; November 6, 1973; assigned to Merck Patent GmbH.

MERCAPTANS

Measurement in Air

Mercaptans may be monitored in air by gas chromatography combined with flame photometric detection (1). In a method developed by an Intersociety Committee (2), mercaptans may be collected by absorption in a mercuric acetate-acetic acid solution. The collected mercaptans are subsequently determined by the spectrophotometric measurement of the red complex produced by reaction between mercaptans and a strongly acid solution of N,N-dimethyl-p-phenylenediamine and ferric chloride. This method permits measuring mercaptan contents below 200 $\mu g./m.^3$ (100 ppb).

Measurement in Water

A technique has been developed for the quantitative determination of thiols in water by D.R. Grassetti (3). It is useful in determining the presence of thiols in various samples by selective visualization in thin layer or paper chromatography, or electrophoresis, but also adapts itself, when working with organic solutions, to the use of spectrophotometographic or simple colorimetric devices.

The process is one where the reagent chemical 2,2'-dithiobis-(5-nitropyridine) reacts rapidly and irreversibly with the thiol compound to form a thione derivative of the reagent chemical and a disulfide of the thiol compound. The resulting thione manifests a characterizing spectrum having absorption at useful wavelengths, from an analytical standpoint, in both the ultraviolet and the visual ranges.

References

(1) Lawrence Berkeley Laboratory, "Instrumentation for Environmental Monitoring - Air," *Publication LBL-1,* Volume 1, Berkeley, University of California (February 1, 1973).
(2) American Public Health Association, *Methods of Air Sampling and Analysis,* Washington, D.C. (1972).
(3) Grassetti, D.R.; U.S. Patent 3,597,160; August 3, 1971; assigned to Arequipa Foundation.

MERCURY

Since 1969, much attention has been directed to mercury in the environment, and consequently, much has been learned. The public recognition of mercury as a potential health hazard followed reports of numerous incidents of human and wildlife illnesses and deaths attributed to mercury poisoning. As a result of this awareness, there arose several excellent reviews of mercury in the environment [see references (2) through (16)] and a valuable bibliography (17). Reviews dealing specifically with mercury in air appeared as it became evident that mercury exists abundantly in air as well as water and soil (18)(19)(20). This brief introduction is intended to acquaint the reader with mercury in the atmospheric environment. The nature of environmental mercury is frequently misunderstood. Authorities differ widely in their attitudes from firm complacency to frenzied alarm. Yet, certain facts remain, as noted by the Lawrence Berkeley Laboratory (21).

Firstly, even background (natural) concentrations exceed standards in some cases. Tuna and swordfish caught in unpolluted waters near Malaya and Africa had mercury concentrations exceeding the Food and Drug Administration tentative standard of 0.5 ppm (11). The air above ore deposits may contain as much as 20 μg./m.^{3} of mercury (6), higher than the Environmental Protection Agency's recent suggestion of 1 μg./m.^{3} maximum for public, long-term average exposures (19).

Secondly, as a result of man's activities, mercury concentrations are elevated above background levels, natural hazards are enhanced, unsafe conditions are created and human sickness and deaths have occurred. In industrial and research work areas where mercury is used, air in poorly ventilated spaces can easily contain over 50 μg./m.^{3} of organic mercury, the maximum safe concentration in occupational air as recommended by the American Conference of Government Industrial Hygienists (22).

Uncontrolled plant emissions into air and water can cause the surrounding ambient air to contain average mercury concentrations over 1 μg./m.^{3} (19). Similarly, liquid wastes can induce high levels of mercury in local waters and wildlife. Perch caught upstream from a pulp and paper mill outfall contained from 0.18 to 0.70 ppm mercury; those caught downstream from it contained from 1.90 to 3.02 ppm mercury (2).

Thirdly, because of the mobility of mercury, discharged mercury does not remain localized. Through natural cycles, it slowly disperses and extends itself into much broader regions to remain almost indefinitely.

Mercury is extremely mobile in the environment. Natural cycles transform and transport mercury in a wide variety of manners throughout air, water and soil. Elemental mercury is constantly outgassed from soil (14). Mercury in the bottom sediment of waterways is converted by anaerobic bacteria into methyl mercury (16), which can then evaporate into the atmosphere as dimethyl mercury (14).

In the atmosphere, many mercury compounds decompose in the presence of sunlight to form elemental mercury. Nearly all forms of airborne mercury can be adsorbed on particulate matter which can then be washed out by precipitation and returned to the surface. Because mercury is readily cycled among air, water and soil, any contribution anywhere is potentially an addition to the mercury pollution problem.

Mercury is removed from the environment when it is lost from the natural cycles. Possible routes of removal may be settling to great ocean depths or drainage to deep underground layers. Such removal does not really reduce the total amount of mercury; instead, it displaces it to regions remote from biological activity and deters its return, thus reducing the mercury found in the biosphere. The processes and rates of mercury removal, which are presently unknown are an important area of future research. Up to a century may be required for the removal of mercury from bottom sediment (2).

The occurrence of airborne mercury may be classified into three general and overlapping categories: background, occupational and public. Background concentrations in air do not seem to warrant close attention except for those near natural sources. Atmospheric concentrations ranging from 0.003 to 0.009 μg./m.^{3} have been found at 400 feet above unmineralized land (6). Concentrations over oceans may be as high as 0.002 μg./m.^{3} (23) although generally less than 0.001 μg./m.^{3} (2). At the ground surface of mercury mines, concentrations up to 20 μg./m.^{3} have been found (6). Occupational levels in many locations are significantly higher than background levels. In medical, dental and chemical laboratories, mercury vapor concentrations can range from 10 to 150 μg./m.^{3} (24). Instrument shops have been found to have average concentrations from 10 to 40 μg./m.^{3} with peak concentrations up to 120 μg./m.^{3} (25). In a mercury mine at Idria, Yugoslavia, mercury concentrations were reported to range from 1,000 to 20,000 μg./m.^{3} (6).

The occurrence of airborne mercury in residential areas is of primary concern. Scattered data showing mercury concentrations in urban air are shown in Table 31. Most of these reported values are probably low, since few of the measurement methods used were able to sample and analyze for total mercury. Other data too recent to be included in this table can be found in reference (33).

TABLE 31: MERCURY CONCENTRATIONS IN URBAN AIR

Location	Observed Range (μg/m^3)	Measured Form	Ref.
Palo Alto, CA	0.001-0.01	Vapor	(18)
Los Altos, CA	0.001-0.050	Vapor	(23)
East Chicago, IN	0.002-0.005	Particulate	(26)(27)
Niles, MI	0.002 Typical	Particulate	(26)(27)
Columbia, MO	0.0002-0.0005	Particulate	(28)
Chicago, IL	0.003-0.038	Particulate	(29)
Cincinnati, OH	0.10 Typical	(Unknown)	(2)
Charleston, WV	0.17 Typical	(Unknown)	(2)
Philadelphia, PA	0.002 Typical	Particulate	(30)
Denver, CO	0.002 Typical	Particulate	(30)
New York City, NY	0.001-0.014	Particulate	(2)(18)
Houston, TX	<0.003	Particulate	(31)
Beaumont, TX	0.05	Particulate	(31)
Toronto, CANADA	≤0.001-0.004	Particulate	(32)

The term "mercury" is conventionally applied to all chemical forms of the element mercury. Furthermore, each chemical form may exist in numerous physical forms. For example, elemental mercury is commonly known as a liquid metal but can readily evaporate into the air as mercury vapor. Once in the air, the vapor may adhere to airborne particles. Mercury in this form is referred to as particulate in Table 31. The specific chemical and physical forms of mercury in air depend upon past and present conditions and vary with location.

Regardless of form, results of mercury-in-air measurements are reported as weight of mercury per unit of volume of air, e.g., microgram per cubic meter (μg./m.^{3}). The form of the measured mercury may or may not be stated, although such information is important. Most methods based on air sampling measure only elemental mercury. Besides time and location, physical parameters such as temperature, pressure, humidity and gas velocity of the sample and parent air streams should also be stated.

In the general atmosphere, most mercury is elemental either as vapor or as adsorbed on particulates (19). Much of the atmospheric mercury probably enters already in the elemental state. In addition, there is evidence that most mercury compounds decompose in the presence of sunlight to form elemental mercury. Particulates include droplets of mercury condensate and dust particles on which mercury has adhered. Because of its great affinity for a variety of materials, an estimated 50% of elemental mercury is absorbed or adsorbed on dust (3). However, more work is needed to confirm this estimate.

Inorganic mercury compounds exist in the air as particulates. Examples are cinnabar dust generated at a mine or ore refinery, mercury salts in ocean spray and mercury oxides in pesticide aerosol. Inorganic compounds can also be formed in air by the reaction of elemental mercury or decomposition of organic mercury compounds. The inorganic compounds Hg_2Cl_2, $HgCl_2$, HgS, HgO, and other mercury salts are known to volatize to various degrees, dependent upon the relative humidity (36).

Organic mercury compounds are found both as vapor and particulates. Their vapor pressures range at least from 10^2 down to 8×10^{-7} mm. Hg at 35°C. (6) (34), dimethyl mercury being the most volatile. Methylmercuric chloride, one of the most volatile organic mercuric salts, has a vapor pressure five times that of elemental mercury; other methyl and ethyl salts have volatilities resembling that of the element (6). A large variety of organic mercury compounds, including methylmercuric chloride, dimethyl mercury, diethyl mercury, phenylmercuric chloride and phenylmercuric acetate, have been used in agriculture as plant, insect and fungus control agents (19). Particulates are generated through their use in aerosol sprays and through the adsorption or absorption of these compounds onto airborne particles.

Three million thermometers are broken in homes and hospitals across Canada each year, releasing an incredible 6.6 metric tons of mercury per year into the environment (10). But this amount is minute if compared to the hundred thousand (metric) tons which annually enter the environment.

Mercury enters the environment through a variety of routes. The high volatility of mercury readily allows it to seep up slowly through layers of earth to the surface. Such outgassing of mercury from rock and soil and the transport of mercury by natural processes cause a natural background everywhere. Higher background concentrations are found above ore deposits, in underground waters, in hot springs, and near other geothermally heated regions. The natural rate of outgassing has been estimated to be between 25,000 and 150,000 tons per year (35). Alteration of the terrestrial surface as in mining and agriculture adds to the outgassing by agitation of the land and increased surface exposure. The contribution from weathering, 230 tons per year, is relatively insignificant (36).

The world industrial production of mercury in 1968 was 8,800 tons (2). Most of this industrial mercury is eventually lost as waste into streams or the atmosphere (36). In addition, up to 20,000 tons of mercury are released annually from fossil fuel burning (37). Roasting of sulfide ores may contribute about 2,000 tons per year. Cement manufacture, 100 tons per year (35). Worldwide contribution by man, not including that from terrestrial alteration, may be up to 31,000 tons per year. It should be emphasized that there is little precise data available on the total amount of mercury which enters the environment. More work is required to make these estimates more exact.

The United States demand in 1968 was 2,590 tons, or 30% of the world industrial production for that year (2). U.S. demand rose only slightly between 1945 and 1969. In 1971, less than 2,300 tons were used in the U.S. (38), and further reductions are forecast because of the desirability of immediately reducing mercury use in paints, pesticides, textiles, and agricultural products.

The societal flow of mercury through the U.S. in 1968 is given in Figure 37 (2). Note that 49% of the industrial demand is lost or dissipated. A large portion of the balance is presumably also lost eventually to the environment, since there are few major recovery programs for mercury. Only 18% of the 1968 demand was supplied by recycled mercury. Besides the industrially produced mercury, a comparable amount is lost directly to the environment by the burning of fossil fuels, namely coal and petroleum products.

FIGURE 37: FLOW OF MERCURY IN THE UNITED STATES IN 1968

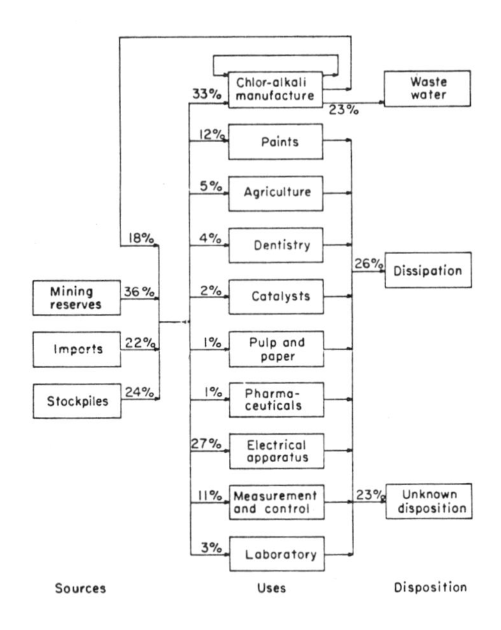

Source: R.A. Wallace, W. Fulkerson, W.D. Shults and W.S. Lyon; Report ORNL-NSF-EP-1 March 1971

Samples of 36 American coals contained from 0.07 to 33 ppm mercury; the average was 3.3 ppm (36). Samples of Soviet coals contained 0.05 to 300 ppm mercury (6). Approximately 5.0 x 10^8 tons of coal are burned each year in the U.S. (2). If the average mercury concentration of the coal burned is 3.3 ppm, 1,600 tons of mercury would be released annually from coal in the U.S. alone.

Fly ash contains only a small amount of the total mercury originally present in coal (39). This suggests that mercury in coal must be released directly into the air during combustion (2). Even if a portion were trapped in bottom ash, it would eventually reach the environment, since there is no mercury reclamation connected with fossil fuel burning (37). Data relating to the mercury content of petroleum products and crude oil are scarce. California crude oil samples have been analyzed and found to contain from 2 to 21 ppm mercury (2). Work is in progress which will give additional data on the mercury content of both coal and oil (40).

Chlor-alkali plants and primary mercury-processing plants are known to emit mercury into the atmosphere in sufficient quantities to create a public health problem. The proposed EPA regulations would limit emissions from these two sources to no more than 2.3 kg. mercury per day from each facility. Other sources are yet to be identified. However, power plants and incinerators in Illinois and Missouri are suspected of emitting up to 67 kilograms/day (2,450 kg./yr.) of elemental mercury from each installation (20). Measurements for total mercury at a 2,100 megawatt plant have indicated a daily discharge of 2.5 kg. from each of three stacks (41). Man generally absorbs 75 to 85% inhaled mercury vapor at concentrations between 50 and 350 μg./m.^{3} (19).

Experiments, however, have shown that some individuals can have nearly complete absorption at concentrations between 60 and 250 $\mu g./m.^3$ (18). Reports regarding the respiratory intake of mercury other than elemental vapor are scarce, but inorganic and organic compounds are known to be absorbed (2)(42). Elemental, and especially organic, mercury are also absorbed through the skin.

Inorganic mercury compounds, phenylmercury, and methoxyethylmercury are converted in the blood to soluble inorganic mercury salts (2)(18). Carried by the blood plasma, the mercury is widely distributed in all tissues, but accumulates primarily in the kidneys and liver prior to excretion (2)(18). Some accumulation of the mercury salts also occurs in the brain, spleen and alimentary tract (18). Nonalkyl mercury might, to a limited extent, be transformed into methylmercury within the liver (2)(12). In general, those compounds which are, or readily form, water-soluble inorganic salts are among the least toxic of the mercury compounds.

The more stable organic mercury compounds, notably alkyl mercury, are not readily converted to organic salts (2)(18). Methyl and ethyl mercury, which are lipid soluble, are carried and stored for extended periods in the blood cells (2). Because water-soluble salts are not readily formed, little alkyl mercury is found in the blood plasma and excretion does not proceed readily through the kidneys. The stability and lipid solubility of the alkyl mercury compounds allow them to penetrate into cells more easily and to be retained much longer. Accumulation in the brain and liver is more pronounced than with other mercury forms.

Methyl mercury is known to accumulate also in the fetus. Excretion appears to occur mainly via the liver into the feces; over 90%, however, can be reabsorbed as the feces passes through the gastrointestinal tract (2). Some conversion of alkyl to inorganic forms does occur in the kidneys (18), liver (18) and intestine (12). The excretion rate of methyl mercury is approximately 1% of the body burden per day (13). Its mobility, retentiveness, and preferential concentration in the brain make alkyl mercury among the most toxic of mercury forms.

Elemental mercury has transport and retention properties resembling both inorganic and alkyl mercury compounds. Converted to mercury salts, it is partially transported in the blood plasma. While able to penetrate all body tissues, it accumulates in the brain, other nervous tissue, intestinal tract and salivary glands (2). Having characteristics intermediate between inorganic and alkyl mercury, elemental mercury is somewhat more toxic than the former but far less toxic than the latter.

The toxicity of mercury is attributed to its high affinity for sulfur-containing compounds and its lesser affinity for organic ligands (18). Interference by mercury in the synthesis and function of enzymes and other proteins can result in a variety of adverse effects (18).

Acute mercury poisoning can result from inhalation of inorganic mercury at concentrations from 1,200 to 8,500 $\mu g./m.^3$ (18). The kidneys and intestinal tract are primarily involved. Symptoms are metallic taste, nausea, adbominal pain, vomiting, diarrhea, headache, salivation, and anuria (18)(42). The stomach, gums and salivary glands may become inflamed. Acute exposure to elemental mercury can also cause pulmonary irritation and neural damage. Chronic symptoms such as muscular tremor may persist in some cases. Extreme cases may lead to hemolysis, insomnia, delirium and ultimate death from exhaustion (2)(18).

Chronic mercury poisoning can result from inhalation of inorganic mercury at concentrations as low as 100 $\mu g./m.^3$. Slight anemia, hypothyroidism and increased excitability may result from occupational exposures as low as 10 to 30 $\mu g./m.^3$ elemental mercury (2). Among 642 workers in the chlor-alkali industry, low-level exposures averaging from <50 to 270 $\mu g./m.^3$ elemental mercury have been strongly correlated with loss of appetite and weight (2). Chronic poisoning by inorganic mercury affects primarily the nervous system. There may arise anxiety, insomnia, muscular tremor and other psychological distrubances. Other possible symptoms are erethism, inflammation of the gums, gastrointestinal disturbances, weakness and other symptoms similar to those in acute poisoning (2)(18)(42). The effects of phenylmercury and methoxyethylmercury are similar to those of inorganic mercury compounds, since phenylmercury and methoxyethylmercury are readily converted into inorganic forms. In one study, 26 subjects were continually exposed to 250 to 3,200 micrograms per cubic meter phenylmercuria pyrocatechin for up to 6 years without evidence of injury, suggesting that phenylmercury may be less toxic than elemental mercury (18).

Alkyl mercury poisoning has somewhat different characteristics than other mercury poisoning. Symptoms may be dormant for weeks or months after acute exposure and if brain damage occurs, effects may be irreversible. Inhalation of alkyl mercury can produce dryness and irritation in the nasopharynx and mouth. Alkyl mercury poisonings have caused permanent neurological damage resulting in impaired vision and hearing, sensory loss in limbs, ataxia and tremor. Neurological disorders such as mental retardation and convulsive cerebral palsy have occurred in infants whose mothers were exposed to methyl mercury during pregnancy (2)(18). Fetal nerve tissue may be especially sensitive to methyl mercury (12).

There are several methods for removing elemental mercury from gas streams. Although some are designed for specific applications such as in chlor-alkali plants, all may have general applicability. In one method, refrigerated cooling can be applied to condense elemental vapor into droplets that are then trapped in filters. If the process is hindered by water condensation, the gas stream should be pretreated to remove water vapor (43).

In another method, solid absorbents such as iodine-impregnated activated carbon can adsorb elemental mercury. Monsanto Enviro-Chem Systems Inc., St. Louis has developed a proprietary adsorbent with efficiencies over 99% in laboratory tests; the gas stream should be free of particulates (44). Although the Monsanto process is technically a success, the company is no longer marketing mercury vapor control systems (45).

In still another method, scrubbing solutions such as chlorinated brine can remove mercury but the solution may tend to contaminate the gas stream with other chemicals. BP Chemicals International, U.K. has proposed a scrubbing system using sodium hypochlorite and brine solution, which has achieved efficiencies of 95 to 99% and minimizes gas stream contamination (43).

In the Oak Ridge National Laboratory study, it was noted that natural gases containing mercury vapor appear to be effectively decontaminated simply by mixing with sour gases containing hydrogen sulfide. A system operating on a similar principle and incorporating a filter to remove precipitated mercuric sulfide could perhaps be developed for stack gases (2).

Major control technology for mercury emissions from stationary sources has yet to be fully developed. Some of the techniques available for emission control of other species might also be effective for mercury (41).

In enclosed facilities, occupational exposure to mercury is minimized by ventilation of mercury work areas, immediate and thorough cleanup of mercury spills and use of nonporous floors and work surfaces to facilitate cleanup (18). Mercury reagents should be properly sealed. In mines where large mercury ore surface is exposed, a spray sealant can be used to control mercury volatization (46). Cleanup of spills can be made more effective by the use of vacuum pumps and flowers of sulfur as an absorbent (47)(48).

Assuming a maximum allowable average daily intake of 30 μg. mercury per day (12)(19), and an average dietary intake of 10 μg./day (19), one finds that the average intake of mercury from air should be restricted to 20 μg. per day. If the inhalation of air is 20 cubic meters per day, the average mercury concentration should be no more than 1 μg./m.^{3} (19). Occupational health standards based on the 7 or 8 hour workday and 40 hour workweek have been adopted by the American Conference of Governmental Industrial Hygienists. These ACGIH standards suggest maximum, time weighted average values (threshold limit values) of 10 μg./m.^{3} for alkyl mercury and 50 μg./m.^{3} for other mercury (22).

The American Conference of Governmental Industrial Hygienists (ACGIH) standards are commonly adopted by the U.S. Bureau of Mines (46), U.S. Department of Labor (22)(49), and other regulatory agencies. Short discussions of health standards for airborne mercury may be found in the references (2)(12)(18). Included are standards of other nations.

Proposed National Emission Standards for mercury were published in the *Federal Register,* Volume 36, No. 235, pages 23239 to 23256 (December 7, 1971). The proposed standards, applicable to mercury ore processing facilities and mercury-cell chlor-alkali plants, would restrict emission to the atmosphere from each facility to 2.3 kg. mercury per 24 hour period.

The occurrence of aquatic mercury may be classified into three general and overlapping categories: background, discharged and public. Background concentrations in the aquatic environment do not seem to warrant close attention. The natural level of mercury in water in the northeastern United States is said to be 0.06 ppb (9). Mercury ranged between 0.05 to 0.11 ppb in water in western Canada (10). The U.S. Geological Survey analyzed the concentration of mercury in a number of rivers in the United States in 1970. Mercury was reported as 0.1 ppb in most rivers and streams (5).

Discharged sediment levels in many locations may be significantly higher than background levels. Since 1970, substantial decreases in discharges of mercury have taken place and this may be reflected by a lower level of discharged effluents. Sediment immediately downstream of a chlor-alkali plant in Saskatoon (Canada) was reported in 1971 to contain 1,800 ppm of mercury (10). Sediment in Lake Ontario near Hamilton Harbor contained 692 ppm of mercury (10). The concentration of mercury in some water samples ranged from 0.99 to 10.4 ppb (50). The occurrence of mercury in man's food chain is of primary concern. This introduction of toxic organomercury compounds into man's food chain is documented for fish (2)(17) and pigs (51).

The term "mercury" is conventionally applied to all forms of the element mercury. In an aquatic environment, mercury may exist in the elemental, +1 or +2 oxidation states. For example, it may exist as $HgCl_2$, $Hg(OH)_2$, HgClOH, Hg, or methylmercury (52)(53). Mercury compounds may also absorb onto suspended matter and sediments. Regardless of form, results of mercury analyses are reported on the basis of μg./l. (ppb). The form of the measured mercury may or may not be stated, although such information is important in terms of toxicological effects. Mercury in fish exists primarily as methylmercury (13) (52); the form so toxic to man if ingested in amounts exceeding about 1 mg./day (13).

Sources of mercury in the aquatic environment are reviewed by Caton et al (17). The sources include discharged effluents from chlor-alkali plants, wood pulp mills (13) and possible municipal wastes (10). The multiplicity of uses of mercury (over 3,000) creates many sources of contamination and thus a diverse number of routes mercury compounds may take in reaching the environment. A survey of these sources is revealing. A partial list of mercury users is indicated in Table 32.

TABLE 32: SOURCES AND USES OF MERCURY

a. Electrical industry is the largest consumer in the U.S. (26.6%).
b. Chlor-alkali industry closely follows (23.1%). However, in Canada the chlor-alkali industry is by far the largest user.
c. Industrial control equipment--mercury switches, relays, etc. (10.6%).
d. Paints--bactericide fungicide for damp and humid conditions (14.4%).
e. Agriculture--fungicides (4.6%).

(continued)

TABLE 32: (continued)

f. Dental preparations (4.1%).

g. Pulp and paper industry--fungicide and preservative (0.6%).

h. Catalysts in chemical manufacturing processes (2.5%).

i. General laboratory uses--hospitals, universities, and private labs.

Source: Interagency Committee on Environmental Mercury, "Mercury in the California Environment," July 1971

Prior to July 1970, 50 plants were dumping 130 kg. (287 lb.) of mercury per day into U.S. waterways; by September 1970, this was reduced to 18 kg. (40 lbs.) (9). Methods used to control mercury discharges into the aquatic environment have been reviewed (1)(17)(51)(54)(55). The methods include precipitation (or cementation) of mercury by a more active metal, e.g., aluminum, copper, iron, zinc. This is effective but it substitutes another metal in solution usually in considerably more than an equivalent amount.

Methods based on the precipitation of mercuric sulfide suffer from the high toxicity and offensive odor of sulfide reagents. Sorption of mercury on preformed metal sulfides, e.g., iron sulfide, cadmium sulfide, zinc sulfide, is an effective removal method. However, the mercury in solution is replaced by another ion (54).

The toxicological effects of mercury on man depend to a large extent on the form of mercury ingested. Elemental, liquid mercury "... is not a poison; a person could swallow up to a pound or more of quicksilver with no significant adverse effects ..." (5). Calomel, primarily mercurous chloride (Hg_2Cl_2), has been used medically as a cathartic, diuretic, antiseptic and antisyphlitic (56). Mercuric chloride ($HgCl_2$) is poisonous when taken orally and in a substantial dose (5)(56). Organomercury compounds including the methyl and ethyl compounds can be lethal to man (5)(51).

Elemental mercury has transport and retention properties resembling both inorganic and alkyl mercury compounds. Converted to mercury salts, it is partially transported in the blood plasma. Being somewhat lipid soluble, elemental mercury is also transported in the blood cells. While able to penetrate all body tissues, it accumulates in the brain, other nervous tissue, intestinal tract and salivary glands (2).

The toxicity of mercury is attributed to its high affinity for sulfur-containing compounds and its lesser affinity for organic ligands (18). Interference by mercury in the synthesis and function of enzymes and other proteins can result in a variety of adverse effects.

Alkyl mercury poisoning has somewhat different characteristics than other mercury poisoning. Symptoms may be dormant for weeks or months after acute exposure and if brain damage occurs, effects may be irreversible. Inhalation of alkyl mercury can produce dryness and irritation in the nasopharynx and mouth. Alkyl mercury poisonings have caused permanent neurological damage resulting in impaired vision and hearing, sensory loss in limbs, ataxia and tremor. Neurological disorders such as mental retardation and convulsive cerebral palsy have occurred in infants whose mothers were exposed to methylmercury during pregnancy (2)(18). Fetal nerve tissue may be especially sensitive to methylmercury (12).

Ingestion of 1.0 to 2.0 grams mercuric chloride is frequently fatal to human beings (57). It produces corrosion of the intestinal tract, injury to the kidney and ultimately death due to kidney failure (5). The effects of some other mercury compounds are described (56)(57). The kidney appears to be the concentrator organ for inorganic mercury (58).

The mercury content in human urine is normally zero, but may rise to 20 ppb in persons with amalgam dental fillings (58). Persons given a mercurial diuretic had urine specimens containing 20 ppb mercury or above (59). A hazardous exposure is believed to exist when urinary excretion reaches 250 ppb (58). Two children exhibited symptoms of pink disease with urinary mercury excretions of 2,000 to 5,000 ppb (59).

A limit of 0.5 ppm (500 ppb) of mercury in fish to be consumed by humans has been set by the Food and Drug Administration (9). Departments of industrial hygiene usually consider 50 ppb of mercury in urine as indicative of excessive exposure (59). A tentative standard for mercury in drinking water was set at 5 ppb by the U.S. Public Health Service in 1970 (2). The Safe Drinking Water Act of 1973 proposed a limit of 0.002 mg./l. (2 ppb). A Swedish toxicological committee recommended a maximum acceptable intake level of 0.030 mg./day of methylmercury for a 70 kg. (154 lb.) male (13).

Now, mercury has come under even more severe scrutiny as a consequence of its inclusion in the toxic pollutant effluent standards published by the EPA for comment in the *Federal Register* (December 27, 1973). Under the proposed new regulations, streams having a flow of less than 10 cu. ft./sec. or lakes having an area less than 500 acres shall receive no mercury discharges. Faster flowing streams or larger lakes can have 20 μg./l. of mercury and salt water bodies can have 100 μg./l. of mercury discharged into them. However, if the stream flow is less than 10 times the waste stream flow, the effluent limit is cut from 20 to 2.0 μg./l.

Measurement in Air

Various methods have been developed to measure mercury vapor and mercury particles in the atmosphere. Portable continuous monitoring detectors for mercury vapor are available from several companies. These detectors are based on the principle that ultraviolet light at 2537 A. is strongly absorbed by Hg vapor (60)(61). Any other vapor that absorbs light at 2537 A. or affects the accuracy of the measurement of the light intensity (such as fogs, dust and smoke) can cause interference and produce unreliable results.

Many compounds do absorb light in this range (ozone, carbon dioxide and aromatic hydrocarbons, for example). Since their sensitivity is much less, however, a high concentration is necessary to interfere with mercury vapor detection (generally a concentration about 100 to 100,000 times greater than that of mercury). The lower sensitivity of these instruments is in the range of 5 to 10 $\mu g./m.^3$ with about 2% full-scale accuracy.

Battery operated vapor detectors have been described by McMurry and Redmond (62) and also by Jacobs and Jacobs (63). Use of these detectors at concentrations of mercury above 1,000 $\mu g./m.^3$ of air requires recalibration. An apparatus was designed by Nelson et al (64) to facilitate the calibration of these detectors.

Systems have also been developed for use when other vapors which absorb at 2537 A. are present in concentrations high enough to cause serious interference. A specific system for mercury was reported by Hawkes and Williston (65) and also by James and Webb (66). Their approaches are basically similar. The air sample is split into two portions and each air stream is passed through identical absorption cells. One of the cells is preceded by a material which removes the mercury vapor from the air stream, and the difference in absorption values between the two cells is a measure of the mercury vapor.

Barringer (67) developed a method to measure mercury vapor in free-standing air which is based on absorption of light at 2537 A. By taking advantage of the pressure-broadening of the mercury emission lines, the interference due to other compounds is minimized.

Techniques have been developed which can detect lower concentrations of mercury vapor than those mentioned although they do not continuously monitor. The basic approach is to collect the vapor on absorbing materials such as gold (68), silver (68), or paper impregnated with potassium iodide (69), and then release it into an ultraviolet detector system.

A portable instrument is being developed, using the initial absorption of mercury on gold, which will be able to detect mercury vapor in the picogram range (10^{-12} grams) in the atmosphere (18).

Mercury vapor can be detected by using indicator papers (such as copper iodide paper (70), selenium sulfide paper (71)(72), or selenium paper (73), gold chloride on silica gel (74), or commercially available gas-detecting tubes (75). All of these methods are quick and simple but not very sensitive (they can generally detect about 500 to 1,000 μg./m.^{3} of mercury with approximately ± 5% accuracy). They could be useful for detecting vapor leaks, however.

A radiochemical method has been developed for the detection of mercury vapor. This method is based on isotope exchange that takes place when the vapors are passed through a solution of Hg 203-mercuric acetate (76). Several methods are available for the determination of mercury in biological materials, such as photometric (77), neutron activation (78), spectrographic (79) and chemical (80) techniques.

Numerous chemical methods have been described for the determination of mercury vapor, mercury in dust, and both organic and inorganic mercury compounds. The chemical procedures consist of collecting the mercury-containing material in impingers containing water (81)(82), alcohol (81)(82), potassium permanganate-sulfuric acid (83)(84)(85)(86), potassium permanganate-nitric acid (87), or iodine-potassium iodide solution (82). Trapping is possible with iodide-activated charcoal and mineral wool (88).

The final determination is usually made colorimetrically with dithizone (81)(86), di-beta-naphthyl-thiocarbazone (84), or Reyneke salt (87). The final determination can also be done by means of electrolysis (81) or the use of selenium sulfide paper (88). If 100 liters of air are passed through the collection media, the limit of detection is about 1 to 10 micrograms per cubic meter of mercury, with ± 5% accuracy. The final mercury determination is usually done in a laboratory.

Kudsk (83), however, has developed an on-site method of determination with a sensitivity of about 1 μg. of mercury. By careful control of the acidity of the solutions, interference from other metals is greatly minimized and causes no problems in most cases. However, problems do arise with certain organomercury compounds, especially the dialkyl derivatives (89)(90). Linch et al (89) found that a collection medium of iodine monochloride in acid gave excellent recovery of dimethyl and diethyl mercury.

Quino (90) found isopropyl alcohol to be an effective medium for collection of dibutyl mercury. The latter has also developed a simple, rapid method for the determination of dialkyl mercury compounds (by reaction with bromine followed by reaction with ditolyl mercury and dithizone) that can be used in the field with a sensitivity of about 500 micrograms per cubic meter of mercury, or if determined in the laboratory, about 2 to 12 micrograms of the mercury compound.

Hamilton and Ruthven (91) reported a technique which provides a continuous monitoring of combined mercury vapor and organomercury compounds. The technique consists of pyrolyzing the compounds to free mercury vapor, which is then detected by a spectrophotometer. By determining the mercury vapor content of the air, the amount of organic mercury can then be found by difference.

Mercury may be monitored in air by the following instrumental techniques (92): Flameless Atomic Absorption; Ultraviolet Absorption; and X-ray Fluorescence. A detailed review of mercury-in-air monitoring systems has been presented by the Lawrence Berkeley Laboratory (92). The reader is referred to that report for more detail but Table 33 shows the types and characteristics of available mercury in air monitors drawn from that report.

A device developed by H.H. Anderson, R.H. Moyer, D.J. Sibbett and D.C. Sutherland (93) is a device for detecting and sampling mercury vapor in the atmosphere, particularly for

TABLE 33: TABLE OF MERCURY-IN-AIR INSTRUMENTS

AMBIENT AIR MONITORS

Principle of Operation	Instrument Note	Measured Form	Range* (μg/m³)	Response Time	Cost	Multi-Parameter Capability	Remarks
Intermittent Analyzers							
Colorimetry							
Bendix	1	Hg vapor	0.1-2.0	minutes	$ 75	Other gases	Detector tubes
Lovibond	1	Total Hg	~25-200		$ 75	Other gases	Mechanical sampler needed
Mine Safety Appliances	1	Hg vapor	~50-1000	minutes	$ 60	None	Detector tubes
Sunshine	1	Hg vapor	~100 in 2 hour sampling		$ 225	None	Selenium sulfide paper
UV Absorption							
Bacharach	3	Hg vapor			$1,950	None	Gold collection matrix
Condensation Nuclei Formation							
Environment/One	1	Hg vapor	0.05-0.10, 1.0		$5,825	SO_2, NH_3, CO, NO_x	Silver wool collection matrix
Continuous Analyzers							
UV Absorption							
Bacharach	1	Hg vapor	10-200, 1000	seconds	$ 691	None	Completely portable, hand-held
Bacharach	2	Hg vapor	0-500	seconds	$1,617	None	Built-in alarms
Barringer	1	Hg vapor	1-10, 10,000	≥ 6 sec	~$35,000 (Canada)	None	Mobile or airborne
Beckman	1	Hg vapor	5-100, 1000	seconds	$ 895	None	Hand-held
Coleman	1	Hg vapor	≤26-260	~30 sec	$ 995	None	Readily adaptable to Hatch and Ott-type procedures for total Hg

(continued)

TABLE 33: (continued)

Principle of Operation	Instrument	Note	Measured Form	Range* ($\mu g/m^3$)	Response Time	Cost	Multiparameter Capability	Remarks
	Geomet	1	Hg vapor/ total Hg	0.001-1000	2 hrs (typ.)	$7,600	None	Silver collection matrix; cyclic operation
	Lab Data Control	1	Hg vapor	7-350, 12,000	seconds	$1,789	UV absorbers	
	Olin	1	Total Hg	0.5-50	6 min	$19,850	None	$SnCl_2$ scrubber; cyclic operation
	Solar	1	Hg vapor	0-350, 3500	seconds	$2,450	None	
	Scintrex	1	Hg vapor	0.01-0.25, 2.5	10 sec	$9,250	None	Zeeman effect
	Sunshine	2	Hg vapor	~10-300	4 sec	$ 395	UV absorbers	Hand-held
	Thermotron	1	Hg vapor	0.2-saturation	seconds	$1,134	None	Versatile flow configurations
STATIONARY SOURCE MONITORS								
Continuous Analyzers								
UV Absorption								
	DuPont	1	Hg vapor	0.1-200	30 sec	~$4,500	SO_2, NO_2	
	Geomet	1	Hg vapor/ total Hg	0.001-1000		$7,600	None	Silver collection matrix; cyclic operation
	Olin	1	Total Hg	0.5-50	6 min	$19,850	None	$SnCl_2$ scrubber; cyclic operation
	Teledyne	1	Hg vapor	800-8000	seconds	$4,875	UV absorbers	

*The minimum and maximum ranges

Source: Lawrence Berkeley Laboratory, *Publication LBL-1,* February 1, 1973

air pollution determination, utilizing sensitized absorption of the vapor on surfaces of noble metal wire grids. The wire grids operate to concrete encountered low levels of vapors. Release of mercury from the grids into a photometer for quantitation is achieved by direct passage of electrical current through the grid wire. The grids are designed to allow for ohmic heating of the absorbent wire to render possible a portable monitoring device.

The development and evaluation of an analytical method for the determination of total atmospheric mercury has been reported by D.L. Chase, D.L. Sgontz, E.R. Blosser and W.M. Henry of Battelle Columbus Laboratories (94). It was reported that total mercury in ambient air can be collected in iodine monochloride, but the subsequent analysis is relatively complex and tedious and contamination from reagents and containers is a problem.

A silver wool collector, preceded by a catalytic pyrolysis furnace, gives good recovery of mercury and simplifies the analytical step. An instrumental method based on particle counting proved unreliable, but another instrument using the 253.7 nm. Hg optical absorption line proved to be quite accurate for the determination of elemental mercury in air.

Measurement in Water

Atomic Absorption Methods (AA) – Mercury concentrations in water are often too low for direct analysis using flame AA. To compensate, a large volume of sample may be taken or mercury may be concentrated by solvent extraction, amalgamation, absorption or ion-exchange.

In the widely used Hatch and Ott method, potassium permanganate is added to a water sample contained in a bottle to convert mercury to the inorganic mercuric form. Excess permanganate is removed by addition of hydroxylamine hydrochloride and mercury reduced to the elemental form by addition of stannous chloride. An air stream is directed through the sample solution to serve as a carrier gas for the mercury which is swept out as a vapor. The vapor is then passed into an absorption cell where the 253.7 nm. Hg line emitted by a ultraviolet lamp is absorbed by the vapor. The mercury in the sample is totally converted to the atomic form; the sensitivity is increased substantially over that of the conventional flame technique where only a small fraction of the mercury is converted to the atomic form.

The Hatch and Ott method was modified in the EPA method to include oxidation with potassium persulfate to ensure complete oxidation of some organomercury compounds (95). Other modifications utilize addition of sulfuric and nitric acids to destroy residual organics (96), and use of helium as the carrier gas (97).

In the method by Moffitt and Kupel, biological and aquatic samples are decomposed with nitric acid and mercury ions reduced to mercury metal using stannous chloride (58). The mercury is swept out of solution by an air stream and absorbed into charcoal. The charcoal is then introduced into a small boat-shaped tantalum vessel which is inserted directly into an oxidizing air-acetylene flame. The minimum detectable quantity of mercury by this method is 20 nanograms.

Chau and Saitoh used a combination of solvent extractions and flameless atomic absorption (97a). Mercury in water samples was extracted with dithizone dissolved in chloroform. Dithizone has been found to react with some organomercury compounds as well as inorganic forms. The dithizone extraction also serves to concentrate the mercury in the smaller chloroform volume.

Following dithizone extraction/concentration, the mercury was extracted back into an aqueous solution; stannous chloride was added to reduce mercuric mercury to the metal, and measurement made using flameless AA. A sensitivity of 0.008 ppb was achieved using this technique. Other methods for analyzing mercury in water utilize amalgamation of mercury onto copper or silver collectors. In one fell swoop, this method can be used to simultaneously separate, concentrate and convert various mercury forms from water samples (50)(98)(99)(100)(101). In these methods, the copper or silver wire is electrically heated to drive off collected mercury as a vapor. The mercury vapor is then passed into the optical cell where the change in intensity of the 253.7 nm. Hg line is measured. Collection of mercury on the wires is done either as a heterogeneous chemical reaction or as electrodeposition under an applied potential of 3 volts. Less time may be needed for electrochemical collection; 20 minutes versus 2 hours in one case (98). However, if other metals are present in the water, they may also plate out and interfere with the mercury analysis. The method can be applied to measuring concentrations at the nanogram level in laboratory prepared samples.

Other flameless atomic absorption methods of collecting and simultaneously concentrating mercury from aqueous samples onto solids include absorption of mercury vapor into charcoal (58), and evaporation on a tantalum boat (102). The latter procedure involves extraction of mercury from urine samples into an organic layer, followed by evaporation of the organic layer (methyl isobutyl ketone) onto a tantalum boat. The boat is heated in the flame and mercury measured as the vapor.

Emission Spectroscopy – Conventional emission spectroscopy is not sufficiently sensitive for direct measurement of mercury in water samples (103). Methods involving plasma excitation sources do have the required sensitivity, and have been applied (103). Another method utilizes a DC discharge on samples of mercury vapor which were generated from water samples by reduction with sodium borohydride (97). In this last method, volatile mercury compounds can be separated and analyzed using a gas stream and rubber diaphragm. Only volatilized mercury diffuses through the diaphragm, so that both free and total mercury can be analyzed in the same sample. The above methods can have sensitivities in the parts per trillion (10^{-3} μg./l.) range.

In the April and Hume procedure (103), 10 ml. of sample solution is constantly aerated with helium. After the base line is stable, 0.5 ml. of 2% titanous chloride in 1 M H_2SO_4 is added to reduce mercury to elemental form. The purging of elemental mercury by the helium stream into the helium plasma causes an emission maximum in about 20 seconds. The emission, monitored by a conventional monochromator-photomultiplier tube, is linearly proportional to the mercury concentration from 0.001 to 1 μg./ml. The detection limit might be as low as 0.0002 μg./ml. in the 10 ml. sample, lower than the 0.001 μg./ml. limit in the Hatch and Ott flameless atomic absorption procedure (104).

Ag, Pd and Au interfere if present in concentrations 100 times greater than mercury. Pb, Cd, Zn, Cu do not interfere in concentrations 1,000 times greater than mercury. Concentrations of Cl^- over 0.1 M cause a slight depression. Phenyl and methylmercury can be analyzed by the same procedure. Elemental and dimethyl mercury are immediately aerated by the helium stream before addition of reductant unless oxidized to less volatile forms. The plasma torch unit is not commercially available but could be constructed for under $1,000.

SpectraMetrics, Inc. has introduced argon plasma emission spectrometers which should be applicable to the April and Hume procedure for mercury. Detection limits have not been indicated but might be comparable to helium plasma instruments.

In a method by Braman for measuring mercury in solution (97), the mercury is reduced to elemental form with $NaBH_4$ solution and allowed to diffuse through a latex or Saran Wrap membrane into a helium carrier stream; the mercury is excited as it is carried through a 900 v., 31.5 watts/inch DC discharge. A monochromator and photomultiplier tube measure the 253.7 nm. emission intensity.

With 200 ml. of sample solution in an open beaker and a latex membrane area of 12 cm.^{2}, the limit of detection is 4×10^{-6} μg./ml. Hg. Analyses of natural water samples gave measurements of 1 to 16×10^{-5} μg./ml. Dimethyl mercury could be determined without reduction, with the same sensitivity as elemental Hg. Oxidation during pretreatment has been successfully done with $KMnO_4$-H_2SO_4. The method can be used to determine ng. quantities of As and Sb (105).

Emission spectroscopy is less susceptible than atomic absorption to interference from aromatics, since aromatics do not usually emit in the 253.7 nm. region. But as in atomic absorption, cobalt is a potential interference, as it has an emission line at 253.649 nm., close to the mercury line at 253.652 nm. (106). The cobalt problem is avoided if the mercury is selectively removed such as in the Braman or the cold vapor method.

Neutron Activation Analysis – This technique was used to measure the total mercury content in fish (107)(108), some common foods (109) and samples of fresh water (110). Data for the food analysis show that except for certain fish the major foods are essentially free of mercury.

In each of these analyses, mercury was separated using anion exchange or precipitation (108). For example, the mercury in fresh water (110) was converted to $HgCl_2$ by adding sufficient HCl to a four-liter sample contained in a polyethylene bottle to make the final solution 0.1 N in HCl. The acidified water sample is passed through anion-exchange resin loaded filter paper to concentrate the Hg. The filter paper is allowed to air dry, sealed in Mylar bags and irradiated. Irradiation time was 2 hours at a thermal neutron flux of 1.3×10^{13} n./cm.^{2}-sec. The samples were cooled three to four days, transferred to a polyethylene bag and counted. Mercury in samples of fresh waters was found to range between 0.03 to 6.6 ppb. The detection limit of the instrument is calculated to be 0.005 μg. Hg.

A sample of methylmercuriguanidine was analyzed by first treating for 30 minutes with chlorine gas to degrade the compound and convert it to an analyzable form (110). Anion exchange followed by irradition showed that recovery was about 100% for the methylmercuriguanidine sample. Only 0.36% was recovered for samples which were not chlorine treated. Other organomercury compounds may possibly be analyzable using this procedure (110).

The good sensitivity, precision and accuracy obtained by neutron activation analysis is shown in Table 34 (109). The potential of neutron activation/anion exchange technique to distinguish between inorganically bound and organically bound mercury is indicated by the low recovery of the latter when chlorine pretreatment is not used (110).

TABLE 34: ANALYSES OF MERCURY STANDARDS AND OF SAMPLES BY NEUTRON ACTIVATION

Material	Mercury (parts per million)	
	This work	Other work
NBS orchard leaves*	0.148 ± 0.010	0.155 ± 0.006 0.162 ± 0.010
International Atomic Energy Agency standard flour*	4.6 ± 0.5	4.87 ± 0.06 4.9 ± 0.3
Food and Drug Directorate flour No. 32573	0.011 ± 0.003	0.011 ± 0.004 0.007 ± 0.02
Swedish fish No. 410-30	1.29 ± 0.13	1.14 1.17
Swedish fish No. 410-28	2.16 ± 0.22	2.24 2.20
Bowen kale*	0.25 ± 0.03	0.23

*Standard material.

Source: J.T. Tanner, M.H. Friedman, D.N. Lincoln, L.A. Ford and M. Jaffee; *Science,* 177, 1102 (1972)

The precipitation method was used to measure the mercury content in bottom sediments as well as fish (108). Sediment levels from Lake Erie at the mouth of the Buffalo River ranged from 1.95 to 6.79 ppm, on a dry weight basis. Mercury waste had been discharged into the Buffalo River prior to April 1970.

Automated Laboratory Analyzers – Laboratory automation of atomic absorption spectrophotometers was achieved by using a Technicon Auto-Analyzer (111)(112)(113). One such system automates the Hatch and Ott method for mercury. The rate of flow of the air stream through the sample and to the absorption cell is critical; the sensitivity increased with increasing flow rate.

It is possible to analyze 22 to 50 samples per hour (111)(113). Comparative throughput for nonautomated operation is not given, but automation should free the laboratory operator to do other work. Analytical results obtained for mercury in several types of sample matrices are shown in Table 35 (111).

TABLE 35: RESULTS OF DUPLICATE ANALYSES BY AUTOMATED PROCEDURE AND COMPARISONS WITH MANUAL PROCEDURE

	Hg concentration, ppm		
Sample	I[a]	II[b]	III[c]
Brine 1	0.0077	0.0071	
2	0.0078	0.0076	
3	0.0041	0.0044	
Fish 1	0.07		0.06
2	0.4		0.4
3	0.8		0.6
4	1.2		1.0
Blood[d]	0.055		
	0.062		
Coal[d]	0.48		
	0.45		
	0.50		
	0.52		

[a] I. Automated procedure.
[b] II. Beckman Mercury Vapor Monitor (Industrial Lab).
[c] III. Beckman Mercury Vapor Monitor (Government Lab).
[d] Replicate analyses.

Source: B.W. Bailey and F.C. Lo, *Anal. Chem.,* 43, (1971)

Continuous or Periodic Analyzers – Continuous analyzers can be used in the field to sample and analyze mercury in water unattendedly. The Olin Mercury Monitor for Liquids and the Geomet Model 103 are commercially available for analysis of mercury in water. The Olin Mercury Monitor for Liquids measures total mercury. A programmer sets the following sequence of instrumental functions (see Figure 38).

A predetermined volume of 10% hydrazine solution is added to the scrubber via the reagent loop. A fixed volume of catalyst solution (0.5% cupric chloride dihydrate, 3% potassium persulfate and 2% sulfuric acid) is then added to the filtered sample in the sample loop. Mercury-free air is bubbled through the reagent mixture until the reading of the UV

FIGURE 38: SCHEMATIC DIAGRAM OF MERCURY IN LIQUID MONITOR (OLIN)

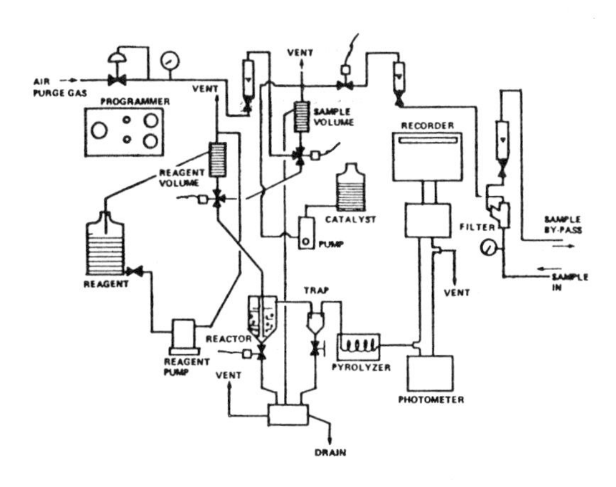

Source: I.A. Capuano, "Automatic Environmental Mercury Analyzers," 17th National ISA Analysis Instrumentation Symposium, Houston, Texas (April 1971)

detector has stabilized. An automatic servomechanism subsequently zeroes the recorder. The fixed volume of sample is then discharged into the reagent medium. The resulting elemental mercury is purged from the liquid by the air stream through the pyrolyzer and into the UV photometer tube for detection and measurement. The pyrolyzer is set at 900°C. and serves to destroy possible organic impurities that may absorb UV light. It also favors formation of halogen acids rather than nonabsorbing mercury halides. The detection limit of the Olin monitor for liquids is given as 0.3 ppb. The mean deviation was ± 0.178 parts per billion for sample concentrations at the 7.5 ppb level. The analysis cycle time is 6 or more minutes.

The Olin monitor has been described by I.A. Capuano in U.S. Patent 3,704,097; Nov. 28, 1972; and U.S. Patent 3,713,776; January 30, 1973; both assigned to Olin Corporation.

The Geomet Model 103 Mercury Air Monitor utilizes a modular attachment for applications to water-monitoring. The Geomet Module 201 provides for extraction of Hg from all forms of effluent waters, which can then be sampled by the Model 103 air monitor (see Figure 39). The sample from the water module is passed over two consecutive precious metal wire grids which collect the Hg.

FIGURE 39: AIRFLOW SCHEMATIC DIAGRAM (GEOMET 103)

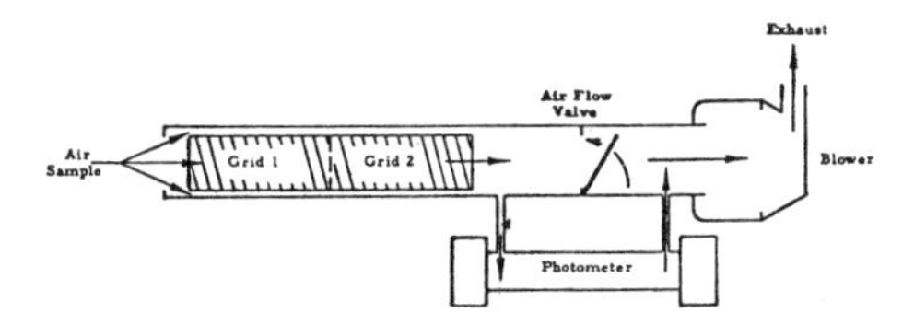

Source: Geomet, Inc., Rockville, Md.

At the completion of the sampling premeasurement concentration period, the grids are serially heated to vaporize the collected Hg into the UV detector cell.

Other – Nanogram quantities of mercury have been measured by flameless atomic fluorescence in a variety of samples including water, rock, wheat, flour and natural sediments (101). The advantages of atomic fluorescence over atomic absorption are said to include better sensitivity and lesser requirements for instrumentation. Sample premeasurement handling utilized mercury oxidation/reduction techniques, followed by collection of the vapor onto silver. The resulting silver amalgam was then heated to release the mercury vapor into the fluorescence cell for subsequent measurement.

Picogram (10^{-12} gram) quantities of mercury were measured in 1 to 10 mg. samples using an x-ray fluorescence procedure (116). In this method, liquid samples containing mercury are deposited on a tantalum strip that is then heated. The resulting mercury vapor is collected on a gold foil. A second heating releases the mercury to an inert gas mixture where it is measured by x-ray fluorescence.

Methylmercury present in fish tissue was analyzed by electron capture gas chromatography (52). The methylmercury in the fish tissues were isolated as methylmercury chloride and purified. The purified material was determined by the gas chromatographic technique, using electron capture monitoring.

A process developed by P.E. Coffey (117) permits accurate determination of the concentration, calculated as elemental mercury, of a sample gaseous, liquid or solid medium containing mercury, either in elemental or reducible form by (1) the reduction of mercury-containing compounds where necessary to elemental mercury in vapor form; (2) conversion of the mercury content to condensation nuclei; and (3) the measurement of the resultant condensation nuclei to yield an accurate indication of the concentration, calculated as elemental mercury, of mercury in the sample medium.

References

(1) Jones, H.R., *Mercury Pollution Control,* Park Ridge, N.J., Noyes Data Corp. (1971).
(2) Wallace, R.A., Fulkerson, W., Shults, W.D. and Lyon, W.S., "Mercury in the Environment - The Human Element," *Oak Ridge National Laboratory Report ORNL-NSF-EP-1,* National Technical Information Service, Springfield, Virginia (March 1971).
(3) Kothny, E.L., "The Three-Phase Equilibrium of Mercury in Nature," presented at the 162nd National American Chemical Society Meeting, Washington, D.C. E.L. Kothny, Air and Industrial Hygiene Laboratory, State of California Department of Public Health, 2151 Berkeley Way, Berkeley, California 94704.
(4) Friberg, L. and Vostal, J., *Mercury in the Environment - A Toxicological and Epidemiological Appraisal,* Chemical Rubber Company Press, Cleveland, Ohio 44128 (1972).
(5) Goldwater, L.J., "Mercury in the Environment," *Sci. Am.* 224, 5, 15 (1971).
(6) "Mercury in the Environment," *Geological Survey Professional Paper 713,* U.S. Government Printing Office, Washington D.C. 20402 (1970); compilation of papers on the abundance, distribution and testing of mercury in rocks, soils, waters, plants and the atmosphere.
(7) Hammond, A.L. "Mercury in the Environment: Natural and Human Factors," *Science* 71, 788 (1971).
(8) Wood, J.M., "A Progress Report on Mercury," *Environment* 14(1), 33 (1972).
(9) "Mercury in the Environment," *Env. Sci. Tech.* 4(1), 890 (1970).
(10) Aronson, T., "Mercury in the Environment," *Environment* 13(4), 16 (1971).
(11) "Mercury: Anatomy of a Pollution Problem," *Chem. Eng. News* 49(27), 22 (1971).
(12) Grant, N. "Mercury in Man," *Environment* 13(4), 2 (1971).
(13) Interagency Committee on Environmental Mercury, "Mercury in the California Environment," California State Department of Public Health, Environmental and Consumer Protection Agency, 2151 Berkeley Way, Berkeley, California 94704 (July 1971).
(14) Jonasson, I.R., "Mercury in the Natural Environment: A Review of Recent Work," *Geological Survey of Canada Paper 70-57,* Geological Survey of Canada, 601 Booth Street, Ottawa, Canada (1972).
(15) D'Itri, F.M., *The Environmental Mercury Problem,* Chemical Rubber Company Press, Cleveland, Ohio 44128; excellent review, contains 554 references. (1972).

(16) Jensen, S. and Jernelov, A., "Biological Methylation of Mercury in Aquatic Organisms," *Nature* 223, 753 to 754 (1969)
(17) Caton, G.M., Oliveira, D.P., Oen, C.J. and Ulrikson, G.U., "Mercury in the Environment - An Annotated Bibliography," *Oak Ridge National Laboratory Report ORNL-EIS-71-8*; Oak Ridge National Laboratory, Oak Ridge, Tennessee 37830 (May 1972).
(18) Stahl, Q.R., "Preliminary Air Pollution Survey of Mercury and Its Compounds," *EPA Publication APTD 69-40;* APTIC, EPA, Research Triangle Park, North Carolina 27711 (October 1969).
(19) "Background Information - Proposed National Emission Standards for Hazardous Air Pollutants: Asbestos, Beryllium, Mercury," *EPA Publication APTD-0753;* APTIC, EPA, Research Triangle Park, North Carolina 277-1 (December 1971).
(20) "Mercury in the Air," *Environment* 13(4), 28 (1971).
(21) Lawrence Berkeley Laboratory, "Instrumentation for Environmental Monitoring - Air," *Publ. LBL-1,* Volume 1, Berkeley, University of California (February 1, 1973).
(22) "Threshold Limit Values of Airborne Contaminants and Physical Agents with Intended Changes Adopted by ACGIH for 1971," American Conference of Government Industrial Hygienists, P.O. Box 1937, Cincinnati, Ohio 45201 (1971).
(23) Williston, S.H., "Mercury in the Atmosphere," *J. Geophy. Res.* 73(22), 7051 (1968).
(24) Mayz, E., Corn, M. and Barry, G., "Determinations of Mercury in Air at University Facilities," *Am. Ind. Hyg. Assoc. J.* 32(6), 373 (1971).
(25) Friberg, L. and Vostal, J., "Mercury in the Environment - A Toxicological and Epidemiological Appraisal," report by the Karolinska Institute, Stockholm prepared for the EPA, Office of Air Programs, Research Triangle Park, North Carolina 27711 under contract CPA 70-30 (Nov. 1971); differs from Reference (4).
(26) Dams, R., Robbins, J.A., Rahn, K.A. and Winchester, J.W., "Nondestructive Neutron Activation Analysis of Air Pollution Particulates," *Anal. Chem.* 42(8), 861 (1970).
(27) Harrison, P.R., Rahn, K.A., Dams, R., Robbins, J.A., Winchester, J.W., Brar, S.S. and Nelson, D.M., "Areawide Trace Metal Concentrations Measured by Multi-Element Neutron Activation Analysis," *JAPCA* 21(9), 563 (1971).
(28) Gray, D., McKown, D.M., Kay, M., Eichor, M. and Vogt, J.R., "Determination of the Trace Element Levels in Atmospheric Pollutants by Instrumental Neutron Activation Analysis," *IEEE Trans. Nucl. Sci.* NS-19(1), 194 (1972).
(29) Brar, S.S., Nelson, D.M., Kline, J.R. and Gustafson, P.F., "Instrumental Analysis for Trace Elements Present in Chicago Area Surface Air," *J. Geophy. Res.* 75(15), 2939 (1970).
(30) Blassen, E.R. and Thompson, R.J., "Elemental Analysis of Air Particulates Using Spark-Source Mass Spectrography," Paper No. 245, Pittsburgh Conference on Analytical Chemistry and Applied Spectroscopy, Cleveland, Ohio; E.R. Blassen, Battelle Memorial Institute, Columbus, Ohio 43201 (February 28 to March 5, 1971).
(31) Rhodes, J.R., Pradzynski, A.H. and Sieberg, R.D., "Energy Dispersive X-Ray Emission Spectrometry for Multi-Element Analysis of Air Particulates," presented at the 18th Analysis Instrumentation Symposium, San Francisco (May 3 to 5, 1972); published in the proceeding entitled *Analysis Instrumentation*, Volume 10, Instrument Society of America, Pittsburgh, Pennsylvania 15222.
(32) Lee, J. and Jervis, R.E., "Detection of Pollutants in Airborne Particulates by Activation Analysis," *Am. Nucl. Soc. Trans.* 11(1), 50 (1968).
(33) Foote, R.S., "Mercury Vapor Concentrations Inside Buildings," *Science* 177, 513 (1972).
(34) Long, L.H. and Cattanach, J., "Antoine Vapor Pressure Equations and Heats of Vaporization for the Dimethyls of Zinc, Cadmium and Mercury," *J. Inorg. Nucl. Chem.* 20, 340 (1961).
(35) Weiss, H.V., Koide, M. and Goldberg, E.D., "Mercury in a Greenland Ice Sheet: Evidence of Recent Input by Man," *Science* 174, 692 (1971).
(36) Joensuu, O.I., "Fossil Fuels As A Source of Mercury Pollution," *Science* 172, 1027 (1971).
(37) Ehrlich, P.R. and Holdren, J.P., "One-Dimensional Ecology," *Bul. At. Sci.* 28(5), 16 (1972).
(38) "Skids Under Mercury?," *Chem. Week* 110(10), 16; short column (1972).
(39) Hattman, E.A. and Schultz, H., "The Fate of Trace Mercury from the Combustion of Coal," paper No. 67 presented at the 14th National Meeting of the American Chemical Society.
(40) Patrick, D.R., Standards Development and Implementation Division, EPA, Research Triangle Park, North Carolina 27711.
(41) Billings, C.E. and Matson, W.R., "Mercury Emissions from Coal Combustion," *Science* 176, 1232 (1972).
(42) Bidstrup, P.L., *Toxicity of Mercury and Its Compounds,* Elsevier, Amsterdam (1964).
(43) "Cell Systems Keep Mercury from Atmosphere," *Chem. Eng. News* 50(7), 14; two-page review of mercury scavenging from vent gases from mercury chlor-alkali cells (1972).
(44) "Business Briefs," *J. Air Pollut. Contr. Ass.* 22(2), 134; product announcement (1972).
(45) Monsanto Enviro-Chem Systems Inc., 800 North Lindbergh Boulevard, St. Louis, Missouri 63166 (July 1972).
(46) High, K., Nonmetal Mine Inspector and Industrial Hygienist, U.S. Bureau of Mines, Alameda, Calif. 94501.

(47) National Industrial Pollution Control Council, *Mercury,* U.S. Government Printing Office, Washington, D.C. 20402 (October 1970).
(48) Beauchamp, I.L. and Tebbens, B.D., "Mercury Vapor Hazards in the University Laboratories," *Ind. Hyg. Quart.* 12(4), 171 to 174 (1951).
(49) "Part 1910 - Occupational Safety and Health Standards, National Consensus Standards and Established Federal Standards," *Federal Register* 36(105), 10466 (May 29, 1971); notice from the Occupational Safety and Health Administration, Department of Labor.
(50) Kalb, G.W. "The Determination of Hg in H_2O and Sediment Samples by Flameless Absorption," *At. Absorption Newsletter,* 9 (4), 84 (1970).
(51) Putnam, J.J., "Quicksilver and Slow Death," *National Geographic,* Volume 124, No. 4 (Oct. 1972).
(52) Sumner, A.K. and Saha, J.G., "Methylmercury Concentrations in Muscle Tissues of Fish from the Saskatchewan River," *Environ. Lettrs.,* 2 (3), 167 (1971).
(53) Gilmour, J.T., "Inorganic Complexes of Divalent Mercury in Natural Water Systems," *Environ. Lettrs.,* 2 (3), 143 (1971).
(54) Moore, F.L., "Solvent Extraction of Mercury from Brine Solutions with High-Molecular-Weight Amines," *Env. Sci. Tech.,* 6 (6), 525 (1972).
(55) Maass, W.B., "Water Pollution and Industrial Organic Finishing," *Metal Finishing,* page 61, (Jan. 1972).
(56) *The Merck Index,* 8th Edition, Merck & Co., Inc., Rahway, New Jersey (1968).
(57) McKee, J.E. and Wolf, H.W., *Water Quality Criteria,* 2nd Edition, California State Water Works Resources Control Board, Sacramento, California (1963).
(58) Moffitt, Jr., A.E. and Kupel, R.E., "A Rapid Method Employing Impregnated Charcoal and Atomic Absorption Spectrophotometry from the Determination of Hg in Atmospheric, Biological and Aquatic Samples, " *At. Abs. Newsletter,* 9 (6), 113 (1970).
(59) Berman, E., "Determination of Cd, Tl, and Hg in Biological Materials by Atomic Absorption," *At. Abs. Newsletter,* 6 (3), 57 (1967).
(60) Suchtelen, H., Warmoltz, N. and Wiggerinck, G.L., "A Method for Determining the Mercury Content of Air," *Philips Tech. Rev.* 11:91 (1949); *A.M.A. Arch. Ind. Hyg. Occupational Med.* 3:432 (1951).
(61) Woodson, T.T., "A New Mercury Vapor Detector," *Rev. Sci. Instr.* 10:303 (1939).
(62) McMurry, C.S. and Redmond, J.W., "Portable Mercury Vapor Detector," *U.S. Atomic Energy Commission Report Y-1188* (1958).
(63) Jacobs, M.B. and Jacobs, R., "Photometric Determination of Mercury Vapor in Air of Mines and Plants," *Am. Ind. Hyg. Assoc. J.* 26(3):261 (1965).
(64) Nelson, G.O., Van Sandt, W. and Barry, P.E., "A Dynamic Method for Mercury Vapor Detector Calibration," *Am. Ind. Hyg. Assoc. J.* 26:388 (1965).
(65) Hawkes, H.E. and Williston, S.H., "Mercury Vapor as a Guide to Lead-Zinc Deposits," *Min. Congr. J.* 48:30 (1962).
(66) James, C.H. and Webb, J.S., "Sensitive Mercury Vapor Meter for Use in Geochemical Prospecting," *Trans. Inst. Met.* 73:633 (1964).
(67) Barringer, A.R., "Interference-free Spectrometer for High Sensitivity Mercury Analyses of Soils, Rocks and Air," *Trans. Inst. Mining Met.* 75:B120 (1967)
(68) Vaughn, W.W. and McCarthy, Jr., J.H., "An Instrumental Technique for the Determination of Submicrogram Concentrations of Mercury in Soils, Rocks and Gas," *U.S. Geol. Surv. Profess. Papers 501-D, D123-7* (1964).
(69) Hemeon, W.C.L. and Haines, Jr., G.F., "Automatic Sampling and Determination of Microquantities of Mercury Vapor," *Am. Hyg. Assoc. J.* 22:75 (1961).
(70) Demidov, A.V. and Mokhov, L.A., "Rapid Methods for the Determination of Harmful Gases and Vapors in the Air," translated by Levine, B.S., *U.S.S.R. Literature on Air Pollution and Related Occupational Diseases* 10:114 (1962).
(71) Beckmann, A.O., McCullough, J.D. and Crane, R.A., *Anal. Chem.* 20 (1927).
(72) Nordlander, B.W., "Selenium Sulfide - A New Detector for Mercury Vapor," *Ind. Eng. Chem.* 19 (1927).
(73) Stitt, F. and Tomimatsu, Y., "Sensitized Paper for Estimation of Mercury Vapor," *Anal. Chem.* 23(8):1098 (1951).
(74) Grosskopf, K., *J. Ind. Hyg. Toxicol.* 20:21A (1938).
(75) Kusnetz, H.L., Saltzman, B.E. and Lanier, M.E., "Calibration and Evaluation of Gas Detecting Tubes," *Am. Ind. Hyg. Assoc. J.* 21(5):361 (1960).
(76) Magos, L., "Radiochemical Determination of Metallic Mercury Vapor in Air," *Brit. J. Ind. Med.* (London) 23:230 (1966).
(77) Ulfvarson, U., "Determination of Mercury in Small Quantities in Biologic Material by a Modified Photometric-Mercury Vapor Procedure," *Acta Chem. Scand.* 21(3):641 (1967).
(78) Ohta, Y., "Activation Analysis Applied to Toxicology; Measurement of the Concentration of Mercury in Hair by Neutron Activation Analysis," *Japan J. Ind. Health* (Tokyo) 8(5):12 (1966)

(79) Cortivo, D., et al, "Mercury Levels in Normal Human Tissue: I. Spectrographic Determination of Mercury in Tissue," *J. Forensic Sci.* 9:501 (1964).

(80) Cholak, J. and Hubbard, D.N., "Microdetermination of Mercury in Biological Materials," *Ind. Eng. Chem.* (Anal. Ed.) 18 (1946).

(81) Burke, W.J., Moskowitz, S. and Dolin, B.H., "Estimation of Mercury in Air," *Ind. Air Analysis,* 3M(8H2-85):22 (1948).

(82) *Manual of Analytical Methods - Recommended for Sampling and Analysis of Atmospheric Contaminants,* American Conference of Governmental Industrial Hygienists (1957).

(83) Kudsk, F.N., "Chemical Determination of Mercury in Air (An Improved Dithizone Method for Determination of Mercury and Mercury Compounds)," *Scand. J. Clin. Lab. Invest.* 16:1 (1964).

(84) Lugg, G.A. and Wright, A.S., "The Determination of Toxic Gases and Vapors in Air," *Commonwealth of Australia, Circular 14,* 2nd Edition (1955).

(85) Massmann, W. and Sprecher, D., "Die Toxikologische Analyse des Quecksilvers," *Arch. Toxikol.* 16 (1957).

(86) Razumov, V.A. and Aidarov, T.K., "Indirect Spectrophotometric Determination of Mercury Vapor Concentrations in the Air of Work Premises," *Hyg. Sanitation* 30(7):81 (1967).

(87) Aruin, A.S., "Determination of Mercury in Atmospheric Air," *U.S.S.R. Literature on Air Pollution and Related Diseases* 3:18 (1960).

(88) Sergeant, G.A., Dixon, B.E. and Lidzey, R.G., "The Determination of Mercury in Air," *Analyst* 82 (1957).

(89) Linch, A.L., Stalzer, R.F. and Lefferts, D.T., "Methyl and Ethyl Mercury Compounds - Recovery from Air and Analysis," *Am. Ind. Hyg. Assoc. J.* 29(1):79 (1968).

(90) Quino, E.A., "Determination of Dibutyl Mercury Vapors in Air," *Am. Ind. Hyg. Assoc. J.* 23(3):23 (1962).

(91) Hamilton, G.A. and Ruthven, A.D., "An Apparatus for the Detection and Estimation of Organo-Mercury Dusts and Vapors in the Atmosphere," *Lab. Pract.* (London) 15(9):995 (1966).

(92) Lawrence Berkeley Laboratory, "Instrumentation for Environmental Monitoring - Air," *Publ. LBL-1,* Volume 1, Berkeley, University of California (February 1, 1973).

(93) Moyer, R.H., Sibbett, D.J. and Sutherland, D.C., Anderson, H.H.; U.S. Patent 3,640,624; Feb. 8, 1972; assigned to Geomet Incorporated.

(94) Chase, D.L., Sgontz, D.L, Blosser, E.R. and Henry, W.M., *Report PB 210,822,* Springfield, Virginia, Nat. Tech. Info. Service (June 1972).

(95) *Methods for Chemical Analysis of Water and Wastes,* Environmental Protection Agency, Cincinnati, Ohio 45268 (1971).

(96) *Apparatus for Flameless Mercury Determinations,* Brochure from Perkin-Elmer Corp., Norwalk, Connecticut 06852 (1971).

(97) Braman, R.S., "Membrane Probe-Spectral Emission Type Detection System for Mercury in Water," *Anal. Chem.,* 43 (11), 1462 (1971).

(97a) Chau, Y.-K. and Saitoh, H., "Determination of Submicrogram Quantities of Hg in Lake Waters," *Env. Sci. Tech.,* 4 (10), 839 (1970).

(98) Brandenberger, H. and Bader, H., "The Determination of Mercury by Flameless Atomic Absorption," *At. Abs. Newsletter,* 7 (3), 53 (1968).

(99) Fishman, M.J., "Determination of Mercury in Water," *Anal. Chem.,* 42 (12), 1462 (1970).

(100) Doherty, P.E. and Dorsett, R.S., "Determination of Trace Concentrations of Mercury in Environmental Water Samples," *Anal. Chem.,* 43 (13), 1887 (1971).

(101) Muscat, V.I., Vichers, T.J. and Andren, A., "Sample and Versatile Atomic Fluorescence System for Determination of Nanogram Quantities of Mercury," *Anal. Chem.,* 44 (2), 218 (1972).

(102) Mesman, B.B. and Smith, B.S., "Determination of Hg in Urine by Atomic Absorption Utilizing the APDC/MIBK Extraction System and Boat Technique," *At. Abs. Newsletter,* 9 (4), 81 (1970).

(103) April, R.W. and Hume, D.N., "Environmental Mercury: Rapid Determination in Water at Nanogram Levels," *Science,* 170, 849 (1970).

(104) Hatch, W.R. and Ott, W.L., "Determination of Sub-Microgram Quantities of Mercury by Atomic Absorption Spectrophotometry," *Anal. Chem.,* 40 (14), 2085 (1968).

(105) Braman, R.S., Justen, L.L. and Foreback, C.C., "Direct Volatilization - Spectral Emission Type Detection System for Nanogram Amounts of Arsenic and Antimony," *Anal. Chem.,* 44 (13), 2195 (1972).

(106) Manning, D.C. and Fernandez, F., "Cobalt Spectral Interference in the Determination of Mercury," *At. Absorption Newsletter,* 7 (1), 24 (1968).

(107) Rottschafer, J.M., Jones, J.D. and Mark, Jr., H.B. "A Simple Rapid Method for Determining Trace Mercury in Fish via Neutron Activation Analysis," *Env. Sci. Tech.,* 5 (4), 336 (1971).

(108) Sivasankara Pillay, K.K., Thomas, Jr., C.C., Sondel, J.A. and Hyche, C.M., "Determination of Mercury in Biological and Environmental Samples by Neutron Activation Analysis," *Anal. Chem.,* 43 (11), 1419 (1971).

(109) Tanner, J.T., Friedman, M.H., Lincoln, D.N., Ford, L.A. and Jaffee, M., "Mercury Content of Common Foods Determined by Neutron Activation Analysis," *Science,* 177, 1102 (1972).
(110) Becknell, D.E., Marsh, R.H. and Allie, Jr., W., "Use of Anion Exchange Resin-Loaded Paper in the Determination of Trace Mercury in Water by Neutron Activation Analysis," *Anal. Chem.,* 42 (10), 1230 (1971).
(111) Bailey, B.W. and Lo, F.C., "Automated Method for Determination of Mercury," *Anal. Chem.,* 43 (11), 1525 (1971).
(112) Fishman, M.J. and Erdmann, D.E., "Automation of Atomic Absorption Analysis," *At. Absorption Newsletter,* 9 (4), 88 (1970).
(113) Stack, Jr., V.T., "Water Quality Surveillance," *Anal. Chem.,* 44 (8), 1972.
(114) Capuano, I.A., "Automatic Environmental Mercury Analyzers," 17th National ISA Analysis Instrumentation Symposium, Houston, Texas (April 1971).
(115) Manufacturer's Bulletin, Geomet Model 103, Geomet, Inc., 50 Monroe St., Rockville, Maryland 20850.
(116) *Chemical and Engineering News,* 50 (42), 11 (1972).
(117) Coffey, P.E.; U.S. Patent 3,711,248; January 16, 1973; assigned to Environment/One Corporation.

MOLYBDENUM

A limited amount of data suggests that molybdenum has a relatively low order of toxicity. Concentrations of molybdenum in ambient air may range from below detectable amounts to maximum values of the order of 0.34 μg./cu. m. of air. The tolerance for molybdenum in irrigation waters, according to Table 11, is only 0.005 mg./l. for continuous use, however. The tolerance on a short term use basis in fine textured soil is 0.05 mg./l.

Determination in Air

A tentative method for the analysis for molybdenum content of atmospheric particulates has been published (1). The method depends on the formation of a stable molybdenum thiocyanate complex. The complex is extracted with MEK and its absorbance is measured. A concentration of 0.055 μg./cu. m. of molybdenum can be measured by the method if a 20 cubic meter sample can be provided.

Determination in Water

The determination of trace amounts of molybdenum in water has been reported by R. Armstrong and C.R. Goldman of the Institute of Ecology of the University of California at Davis (2). They reported that molybdenum may be determined directly in solutions containing above 100 ppb. A spectrophotometric method using the extractable complex with diacetyldithiol was found to be the most convenient and the least subject to interference of a number of methods. Preconcentration of dissolved molybdenum by coprecipitation with hydrated manganese dioxide followed by spectrophotometry of the dithiol complex allows the determination of 1 μg. of molybdenum in a 5 liter sample (0.2 ppb) with a relative error of 3%.

Particulate matter may be removed from water samples by HA millipore filtration or by continuous flow (150 ml./min.) centrifugation at 28,000 *g.* The latter method is more convenient for processing the large (10 to 50 liter) water samples needed for estimating the trace metal content of suspended particles in unpolluted waters. Molybdenum is most conveniently released from organic materials by wet oxidation with perchloric acid, the residue being taken up in 3 N hydrochloric acid. Hydrofluoric acid digests of silicate samples were found to give higher recoveries than perchloric acid digests and were found to be more reproducible than fusion methods.

As pointed out earlier in Table 11, molybdenum may be determined by atomic absorption at a detection limit of 14 ppb.

References

(1) American Public Health Association, "Methods of Air Sampling and Analysis," Washington, D.C. (1972).
(2) Armstrong, R. and Goldman, C.R., Paper presented before ASTM 71st Annual Meeting, San Francisco, California (June 1968).

NICKEL

Nickel and its compounds are of concern as air pollutants because harmful effects of exposure to these materials have been observed among industrial workers. Exposure to airborne nickel dust and vapors may have produced cancers of the lung and sinus, other disorders of the respiratory system, and dermatitis. There is a substantially higher mortality rate among nickel workers due to sinus cancer, up to 200 times the expected number of deaths.

However, since other metal dusts have also been present in industrial exposures to nickel, it has not been possible to determine whether nickel is the carcinogen. Yet experiments have shown that nickel carbonyl and nickel dusts can induce cancer in animals. Nickel contact dermatitis was found in 77% of the females and 10% of the males suspected of having allergenic reactions to metals. No information on the effects of nickel air pollution on commercial and domestic animals, plants, or materials was found in the literature, according to Sullivan (1).

The most likely sources of nickel in the air appear to be emissions from metallurgical plants using nickel, engines burning fuels containing nickel additives and plating plants, as well as from the burning of coal and oil, and the incineration of nickel products. For example, in 1964, urban air concentrations of nickel averaging 0.032 $\mu g./m.^3$ and ranging up to a maximum of 0.690 $\mu g./m.^3$ were found in East Chicago, Ind.

Emission of nickel in particulate form can be controlled using normal control devices, such as precipitators, baghouses and scrubbers. Nickel carbonyl, which is gaseous, must first be decomposed by heat before it is removed as a particulate. No information has been found on the economic costs of nickel air pollution or on the costs of its abatement.

Wastewaters containing nickel originate primarily from metal industries; particularly plating operations (2). High levels of nickel have also been reported in wastes from silver refineries (3). In addition, basic steel works and foundaries, motor vehicle and aircraft industries, and printing have been mentioned as potential sources of nickel (4). Nickel concentrations typically found in plating and metal processing wastes are summarized in Table 36.

Golomb (5) states that generally, rinse waters may be considered to contain 1% of the plating bath concentration. Lowe (6) has reported that acid and sulfamate nickel plating baths generally contain nickel at 40,000 to 90,000 mg./kg. of solution, while electrodeless process baths are of weaker concentration, containing 7,000 to 7,500 mg./kg. of solution. Industries concerned primarily with copper or brass plating or processing may also contain nickel in plant wastewaters, but generally at low levels, 1 mg./l. or less, and are therefore of little concern. In addition, Hill (7) has reported nickel concentrations ranging from 0.46 to 3.4 mg./l. in acid mine drainage.

Nickel exists in waste streams as the soluble ion. In the presence of complexing agents such as cyanide, nickel may exist in a soluble complexed form. The presence of nickel cyanide complexes interferes with both cyanide and nickel treatment, and the presence of this species may be responsible for increased levels of both cyanide and nickel in treated wastewater effluents. The tolerance for nickel in irrigation waters is 0.05 mg./l. on a continuous use basis or 2.0 mg./l. on a short-term use basis in fine textured soil as shown

TABLE 36: SUMMARY OF NICKEL CONCENTRATIONS IN METAL PROCESSING AND PLATING WASTEWATERS

Industry	Nickel Concentration, mg/l		Reference
	Range	Average	
Tableware Plating			
Silver bearing waste	0-30	5	(8)
Acid Waste	10-130	33	(8)
Alkaline waste	0.4-3.2	1.9	(8)
Metal Finishing			
Mixed wastes	17-51	-	(9)
Acid wastes	12-48	-	(9)
Alkaline wastes	2-21	-	(9)
Small parts fabrication	179-184	181	(10)
Combined degreasing, pickling and Ni dipping of sheet steel	3-5	-	(6)
Business Machine Manufacture			
Plating wastes	5-35	11	(8)
Pickling wastes	6-32	17	(8)
Plating Plants			
4 different plants	2-205	-	(11)
Rinse waters	2-900	-	(12)
Large plants	up to 200	25	(13)
5 different plants	5-58	24	(14)
Large plating plant	88 (single waste stream)	-	(15)
	46 (combined flow)	-	
Automatic plating of Zinc base castings	45-55	-	(6)
Automatic plating of ABS type plastics	30-40	-	(6)
Manual barrel and rack	15-25	-	(6)

Source: PB 216,162

earlier in Table 5. The maximum allowable concentration of nickel in an effluent being discharged to a storm sewer or stream is 2.0 ppm. Nickel is one of the relatively nontoxic metals found in the tissues of man, ranking in this respect with the essential elements, iron, cobalt, copper and zinc, according to Cothern (16). Its physiological role, if any, has not been established. The toxicity for mammals is low, ranking with the essential metals and with chromium, barium and silver (17).

Nickel carbonyl vapors have been shown to cause pulmonary cancer (18), acute pneumonitis (19), alterations in ribonucleic acid (20) and appears to be related to acute mycardial infarction (21). Nickel dust has shown a correlation with lung cancer (22) and inhalation of powdered metallic nickel (15 mg./m.^{3} in air) has been shown to cause pulmonary lesions in guinea pigs and rats (23). Nickel carbonyl is highly toxic by inhalation. The LD_{50} values (the amount that kills 50% of the animals) for 30 minute exposure for mice, rats and cats are 10, 35 and 270 ppm by volume, respectively. It is suspected of causing a high incidence of carcinoma of the respiratory passages after long exposure to low concentrations (24). Little seems to be published concerning the effects of nickel in water supplies, although 6 ppm appears to be toxic for some aquatic insects and the lethal dose for fish is in the range of 0.08 to 1.0 ppm (25). Nickel is of low toxicity to fish. The 12 day LD_{50} is 0.5 ppm (26).

Measurement in Air

Sampling Methods: Dusts and fumes of nickel compounds may be collected by any method suitable for collection of other dusts and fumes; the impinger, electrostatic precipitator and filters are commonly used. The National Air Sampling Network uses a high volume filtration sampler (27). The sampling of $Ni(CO)_4$ requires a special sampling solution, and various sampling techniques are discussed with the quantitative methods.

Quantitative Methods: Emission spectroscopy has been used by the National Air Pollution Control Administration for nickel analysis of samples from the National Air Sampling Network (27(28). The samples are ashed and extracted to eliminate interfering elements. The minimum detectable nickel concentration by emission spectroscopy is 0.0064 $\mu g./m.^3$ for urban samples and 0.0016 $\mu g./m.^3$ for nonurban samples. The different sensitivities result from different extraction procedures required for urban samples (27).

Thompson et al (28) have reported that the National Air Pollution Control Administration uses atomic absorption to supplement analyses obtained by emission spectroscopy. The method has a minimum detectable limit of 0.004 $\mu g./m.^3$, based on a 2,000 $m.^3$ air sample.

The ring-oven technique has been adapted to the measurement of nickel particulates. Dimethylglyoxime is used as the coloring agent. The lower limit of identification is 0.08 $\mu g.$ and the normal range of measurement is 0.10 to 1.0 $\mu g.$ (29). In addition to the above test, a number of colorimetric procedures, all using dimethylglyoxime, have been described for the detection of nickel [(30) through (35)].

An atomic absorption method for determining nickel in biological materials has been reported by Sunderman (36). The nickel is extracted with dimethylglyoxime. This method permits measurement of the nickel levels in normal tissue or urine.

Bergmann et al (37) reported an x-ray fluorescence method for the determination of nickel in petroleum feed stocks. They claim a limit of detection of 0.05 $\mu g.$ of nickel per 10 grams of feed stock.

Measurement in Water

Nickel may be monitored in water by the following instrumental techniques (38): Polarographic, Spark Source Mass Spectrometry, and X-Ray Fluorescence. Nickel may be measured in water down to a detection limit of 6 ppb by atomic absorption as noted earlier in Table 11.

References

(1) Sullivan, R.J., "Air Pollution Aspects of Nickel and Its Compounds," *Report PB 188,070*, Springfield, Virginia, Nat. Tech. Info. Service (September 1969).

(2) Patterson, J.W. and Minear, R.A., "Wastewater Treatment Technology," *Report PB 216,162*, Springfield, Virginia, Nat. Tech. Info. Service (February 1973).

(3) Banerjee, N.G. and Banerjee, T., "Recovery of Nickel and Zinc from Refinery Waste Liquors: Part I - Recovery of Nickel by Electrodeposition," *Jour. Sci. Industr. Res.*, 11B:77-78 (1952).

(4) Dean, J.G., Bosqui, F.L. and Lanouette, K.H., "Removing Heavy Metals from Waste Water," *Env. Sci. Tech.*, 6:518-522 (1972).

(5) Golomb, A., "Application of Reverse Osmosis to Electroplating Waste Water Treatment. Part II. The Potential of Reverse Osmosis in the Treatment of Some Plating Wastes," *Plating,* 59:316-319 (1972).

(6) Lowe, W., "The Origin and Characteristics of Toxic Wastes with Particular Reference to the Metals Industries," *Wat. Poll. Control* (London), 1970:270-280.

(7) Hill, R.D., "Control and Prevention of Mine Drainage," presented at the Environmental Resources Conference on Cycling and Control of Metals; Battelle Columbus Laboratories, Columbus, Ohio (October 31 to November 2, 1972).

(8) Nemerow, N.L., *Theories and Practices of Industrial Waste Treatment,* Addison Wesley, Reading, Massachusetts (1963).

(9) Anderson, J.S. and Iobst, Jr., E.H., "Case History of Wastewater Treatment in a General Electric Appliance Plant," *Journal Water Pollution Control Fed.,* 40:1786-1795, (1968).

(10) McElhaney, H.W., "Metal-Finishing Wastes Treatment at the Meadville, Pa., Plant of Talon, Inc.," *Sew. Ind. Wastes,* 25:475-482 (1953).

(11) Wise, W.S., "The Industrial Waste Problem: IV. Brass and Copper Electroplating and Textile Wastes," *Sew. Industr. Wastes,* 20:96-102 (1948).

(12) Anonymous, *State of the Art: Review on Product Recovery,* Water Pollution Control Research Services 17-7 ODJW 11/69, U.S. Department Interior, Washington, D.C. (November 1969).

(13) Pinkerton, H.L., "Waste Disposal," in *Electroplating Engineering Handbook,* 2nd Edition, A.K. Graham, editor-in-chief, Reinhold Publishing Co., New York (1962).

(14) Battelle Columbus Laboratories, "An Investigation of Techniques for Removal of Cyanide from Electroplating Wastes," *Water Pollution Control Research Series 12010 EIE 11/71,* Environmental Protection Agency, Washington, D.C. (November 1971).

(15) Chalmers, R.K., "Pretreatment of Toxic Wastes," *Wat. Poll. Control* London, 1970:281-291.

(16) Cothern, C.R., "Determination of Trace Metal Pollutants in Water Resources and Stream Sediments," *Report PB 213,369;* Springfield, Virginia, Nat. Tech. Info. Service (1972).

(17) Schroeder, H.A., Balassa, J.J. and Tipton, I.H., *J. Chron. Dis.* 15, 51 (1961).

(18) Sunderman, F.W. and Sunderman, Jr., F.W., *Am. J. Clin. Pathol.* 35, 203 (1961).

(19) Sunderman, F.W., Ronge, C.S., Sunderman, Jr., F.W., Donnelly, A.S. and Lucszyn, G.W., *Am. J. Clin. Pathol.* 36, 477 (1961).

(20) Sunderman, F.W., *Amer. J. Clin. Pathol.* 39, 549 (1963).

(21) Sunderman, F.W., Nomoto, S. and Nechat, M., *Proc. Fourth Ann. Conference on Trace Subst. in Envir. Health,* p. 352, Columbia, Missouri (1970).

(22) Morgan, J.G., Brit. S. *Industr. Med.* 15, 224 (1958).

(23) Hygiene Guide Series, *Am. Ind. Hyg. Assoc. J.,* pages 6 to 10 (November - December 1964).

(24) Kincaid, S., Beckworth, C.H. and Sunderman, F.W., *Am. J. Clin. Pathol.* 26, 107 (1956).

(25) Warnick, S.L. and Bell, H.L., *J. Water Pollution Control Fed.* 41, 280 (1969).

(26) Her Majesty's Stationery Office, "Water Pollution Research" (1968).

(27) "Air Pollution Measurements of the National Air Sampling Network - Analyses of Suspended Particulates, 1957-1961, *U.S. Public Health Serv. Publ.* 978 (1962).

(28) Thompson, R.J., Morgan, G.B. and Purdue, L.J., "Analyses of Selected Elements in Atmospheric Particulate Matter by Atomic Absorption," Preprint. Presented at the Instrument Society of America Symposium, New Orleans, Louisiana (May 5 to 7, 1969).

(29) West, P.W., et al, "Transfer, Concentration and Analysis of Collected Airborne Particulates Based on Ring Oven Techniques," *Anal. Chem.* 32(8):943-6 (1960).

(30) Feigl, F. and Stern, R. "The Application of Spot Reactions in Qualitative Analysis," *Z. Anal. Chem.* 60:1 (1921).

(31) Kielczewski, W. and Supinski, J., "Determination of Microgram Quantities of Nickel by the Impregnated-Paper Method," *Chem. Anal.* (Warsaw), 10(4):677 (1965).

(32) Kobayashi, Y., "A Rapid Determination of a Small Amount of Nickel by Means of a Detector Tube," *J. Chem. Soc. Japan, Ind. Chem. Sect.* 58(10):728 (1955).

(33) Kutzelnigg, A., "New Reagent Papers for Porosity Testing and for the Determination of Metals," *Metalloberflaeche* 5:B 113 (1951).

(34) Marion, S.P. and Zlochower, I., "A Spot Test Analysis of the Group III Cations," *J. Chem. Educ.* 36:378 (1959).

(35) Tananaev, N.A., "Qualitative Analysis of the Elements of the First to Third Groups When Present Together, with Special Regard to Spot Tests," *Z. Anorg. Allgem. Chem.* 140:320 (1924).

(36) Sunderman, Jr., F.W., "Measurements of Nickel in Biological Materials by Atomic Absorption Spectroscopy, *Am. J. Clin. Pathol.* 44(2):182 (1965).

(37) Bergman, J.G., et al, "Determination of Sub-Ppm Nickel and Vanadium in Petroleum by Ion Exchange Concentration and X-Ray Fluorescence," *Anal. Chem.* 39:1259 (1967).

(38) Lawrence Berkeley Laboratory, "Instrumentation for Environmental Monitoring - Water," *Publ. LBL-1,* Volume 2, Berkeley, University of California (February 1, 1973).

NICKEL CARBONYL

Nickel carbonyl is a flammable liquid which boils at 43°C. Because it is volatile and easily decomposable into metallic nickel and carbon monoxide, it is very toxic. It is considered to be five times as toxic as carbon monoxide and can also bring a large concentration of active nickel into the body and deposit it in the lungs.

Measurement in Air

Brief et al (1) mention five methods which have been used for the determination of nickel carbonyl. (1) An air sample can be bubbled through a saturated solution of sulfur in trifluoroethylene. The sulfur reacts with nickel to form a precipitate. Spectrographic examination is sensitive to 0.0003 ppm nickel carbonyl.

(2) An air sample may be drawn through a tube containing red mercuric oxide at 200°C. and the liberated mercury may be determined spectrographically. A parallel stream of air is drawn through an oxidizing reagent to convert the CO to CO_2, and the stream is passed over mercuric oxide; the liberated mercury is again determined spectrographically. The difference in the amounts of mercury vapor measured corresponds to the nickel carbonyl content in the air. A sensitivity of 0.0014 ppm is reported.

(3) The nickel carbonyl may be absorbed in chloramine-B. The nickel determination is accomplished colorimetrically using dimethylglyoxime. For a 30 minute sample, at the suggested sampling rate of 0.5 liter per minute, a sensitivity of 0.01 ppm is obtained.

(4) Another colorimetric method uses iodine in carbon tetrachloride as the collection medium. The nickel is colored with dimethylglyoxime. A sensitivity of 0.1 ppm to nickel carbonyl is claimed.

(5) Nickel carbonyl may be collected in dilute sulfuric acid followed by spectrophotometry using sodium diethyldithiocarbamate as the coloring agent. Brief et al (1) collected a sample in dilute hydrochloric acid in a midget impinger (0.1 cfm for 30 minutes). The nickel was complexed with alpha-furildioxime and extracted with chloroform. The nickel content is determined spectrophotometrically, and the method is sensitive to 0.0008 ppm.

The American Industrial Hygiene Association indicates (2) that a direct field instrument is commercially available. This instrument is useful for continuous monitoring of operations involving nickel carbonyl. The instrument can detect concentrations as low as 10 and as high as 1,500 ppb.

Another field method has been described by Kobayashi (3). Air is drawn through a tube filled with silica gel impregnated with 0.5% gold chloride. In the presence of nickel carbonyl, the silica gel changes from a light yellow to bluish violet. The concentration of nickel carbonyl is a function of the length of the colored layer. The useful range of a 100 cubic centimeters sample is 200 to 600 ppm. By measuring the minimum volume of test gas needed to color the silica gel at a constant sampling rate, the concentration of nickel carbonyl, down to 3 ppm, can be determined.

In another method (4), the test air is drawn at 0.5 l./min. through an absorption tube containing one or two 15 mm. diameter filter papers and then through two absorption vessels with porous plates. Each plate contains 3 ml. of a 1.5% solution of chloramine B in alcohol. The chloramine B solution retains the nickel carbonyl vapor. The colored vapor is compared with standards. The sensitivity of the method is 1 μg. of nickel carbonyl; the error does not exceed 10%.

Sunderman et al (5) have developed a gas chromatographic method for the detection of nickel carbonyl in the blood and breath. A Carbowax 20 M column and electron capture detector were used. McCarley et al (6) describe a continuous monitor of nickel carbonyl. Polarized light is used to detect nickel carbonyl deposited on borosilicate glass.

Concentrations in the range of 0.05 to 4 parts per million can be measured.

References

(1) Brief, R.S., Venable, F.S. and Ajemian, R.S., "Nickel Carbonyl: Its Detection and Potential for Formation," *Am. Ind. Hyg. Assoc. J.* 26:72 (1965).
(2) "Nickel Carbonyl," Hygienic Guide Series, *Am. Ind. Hyg. Assoc. J.* 29:304 (1968).
(3) Kobayashi, Y., "Rapid Method for the Determination of Low Concentrations of Nickel Carbonyl Vapor," *Yuki Gosei Kagaku Kyokai Shi* 15:466 (1957).
(4) Belyakov, A.A., "The Determination of Microgram Quantities of Nickel, Nickel Tetracarbonyl, and its Solid Decomposition Products in Air," *Zavodsk. Lab.* 26:158 (1960).
(5) Sunderman, Jr., F.W., Roszel, N.O. and Clark, R.J., "Gas Chromatography of Nickel Carbonyl in Blood and Breath," *Arch. Environ. Health* 16:836 (1968).
(6) McCarley, J.E., Saltzman, R.S. and Osborn, R.H., "Recording Nickel Carbonyl Detector," *Anal. Chem.* 28:880 (1956).

NITRATES

Samples of outside atmospheres may contain 0.1 to 10.0 μg./m.^{3} of nitrates. Nitrates plus nitrites expressed as mg./l. of nitrogen should be less than 10 ppm in domestic water supplies and it is desirable that they be virtually absent as noted earlier in Table 3. The maximum allowable concentration of nitrate ion in effluent being discharged to a storm sewer or stream is 45.0 ppm.

Measurement in Air

A tentative method for the analysis of nitrates in atmospheric particulate matter has been published (1). It involves particulate collection on a glass fiber filter followed by extraction with water. The acidified extract is used to nitrate 2,4-xylenol. The nitrate is extracted with xylene, treated with caustic and determined colorimetrically. Usually 24 hour samples are employed. The method has a working range of 10 to 100 μg. of nitrate ion and a sensitivity of 5 μg.

Measurement in Water

Nitrates can be monitored in water by the following instrumental techniques (2): Activation Analysis; Automated Colorimetric; Colorimetric; and UV Absorption.

References

(1) American Public Health Association, "Methods of Air Sampling and Analysis," Washington, D.C. (1972).
(2) Lawrence Berkeley Laboratory, "Instrumentation for Environmental Monitoring - Water," *Publ. LBL-1,* Volume 2, Berkeley, University of California (February 1, 1973).

NITRIC ACID

Nitric acid is a direct emission product from nitric acid manufacturing, but a much more important source may be photochemical reactions. Very little of the nitric oxide emitted is ever accounted for in the form of particulate nitrate. Nitric acid has been shown to be an important product of photooxidation of nitric oxide in the presence of hydrocarbons in laboratory experiments (1). The kinetic mechanisms usually postulated for conversion of nitric oxide and nitrogen dioxide to products also favor nitric acid formation. Soviet air quality standards for nitric acid establish a single exposure (20 minute averaging time) limit of 0.15 ppm. The Western German engineering society, V.D.I. has established a long-term exposure value of 0.5 ppm and a shorter term value of 1.0 ppm (not to be exceeded more than once in any 2 hours.

Measurement in Air

Because of a lack of an acceptable technique for atmospheric analysis, one cannot assess the importance of nitric acid as an atmospheric pollutant. Optical techniques, including open path instrumentation, should offer possibilities for measurement of nitric acid, however.

References

(1) Gay, Jr., B.W. and Bufalini, J.J., *Envir. Sci. & Technol.* 5, 422 (1971).

NITRITES

Measurement in Water

A report by D. Sam (1) describes a convenient method of determining submicrogram concentrations of inorganic nitrites in natural stream water. This technique uses ion exchange concentration followed by spectrophotometric measurement. It is adapted from the procedure reported by Wada and Hattori for the analysis of nitrites in seawater.

Nitrites can be monitored in water by the following instrumental techniques (2): activation analysis; automated colorimetric; colorimetric; and ion selective electrode. According to R.C. Crawford et al (3), nitrite ion may be determined photometrically using Saltzman's reagent comprising N-(1-naphthyl)-ethylenediamine and sulfanilic acid which forms a dye specifically with NO_2^-. The determination of nitrites by automated chemical analyses has been reviewed by K.H. Mancy (4).

References

(1) Sam, D., "Analytical Method for Determination of Inorganic Nitrites in Natural Stream Water," *Report XC-755-697,* Springfield, Virginia, Nat. Tech. Info. Service (1972).

(2) Lawrence Berkeley Laboratory, "Instrumentation for Environmental Monitoring - Water," *Publ. LBL-1,* Volume 2, Berkeley, University of California (February 1, 1973).

(3) Crawford, R.W., Rigdon, L.P. and Thompson, R.J.; U.S. Patent 3,776,697; December 4, 1973; assigned to U.S. Atomic Energy Commission.

(4) Mancy, K.H., *Instrumental Analyses for Water Pollution Control,* Ann Arbor, Michigan, Ann Arbor Science Publishers (1971).

NITROGEN COMPOUNDS

Air quality standards have been established for aniline as an example of one organic nitrogen compound. Soviet standards call for a maximum of 0.5 ppm on either a single exposure (20 minute averaging time) or 24 hour averaging time basis. West German standards call for 0.6 ppm on a basis not to be exceeded more than once in any 4 hours and 0.2 ppm on a long-term exposure basis. The contrast is interesting because, as discussed under "Arsenic" in this volume, Soviet standards are usually much more rigorous than those of other countries.

Benzidine is an organic nitrogen compound which has been singled out as one of 9 materials cited in "Proposed Toxic Water Pollutant Effluent Standards" by the EPA.

According to this proposed standard, as published in the *Federal Register* for Dec. 27, 1973, no benzidine may be discharged into streams with a flow of less than 10 cu. ft./sec. or lakes with an area less than 500 acres. Faster flowing streams or larger lakes are permitted to receive 1.8 μg./l. of benzidine. However when receiving stream flow is less than ten times waste stream flow, the allowable limit is cut by a factor of ten.

Measurement in Air

A process has been developed by C.B. Murphy (1) which permits the detection and measurement of trace amounts of complex nitrogenous organic compounds in an oxygen-containing gas such as air. The gas stream is passed through a pyrolyzer where the nitrogeneous organic compound is oxidized to produce nitrogen dioxide, water vapor and carbon dioxide and the nitrogen dioxide hydrolyzed to produce nitric acid which is then reacted with ammonia to produce ammonium nitrate particles, or nuclei, which are then measured in a condensation nuclei detector. Filters are provided to remove background particles which may be present in the original gas sample or unoxidized carbon particles resulting from the pyrolysis. Amounts of organic compounds as low as 0.000001 g./l. of air have been readily detected and measured by this method.

A process developed by J.W. Belisle (2) permits detecting toluene diisocyanate or other aromatic isocyanates, or aromatic amines, in ambient air. The process consists of contacting a sample of the air with an acid solution of glutaconic aldehyde containing in admixture therewith an amount of cationic ion exchange resin.

The isocyanate is converted to the corresponding amines; amines react with the reagent to produce a yellow color which is concentrated on the surface of the resin. The method is rapid and of high sensitivity. A composition useful for preparation of the test solution is prepared by mixing 1-(4-pyridyl)pyridinium chloride hydrochloride with a cationic ion exchange resin in particulate form.

A process developed by H.P. Silverman and G.A. Giarrusso (3) provides a rapid means for detection of toxic reducing vapors found in rocket propellants, such as hydrazine and unsymmetrical dimethyl hydrazine (UDMH). By this method, measurement of reducing vapors can be made within a few minutes with a relatively simple, inexpensive, lightweight, highly sensitive and portable sensing instrument. The instrument may be in the form of a thin film sensitized badge that can be worn by personnel and can be carried in a pocket or worn on a belt to effect a portable warning device for personnel working in areas of possible contamination by toxic reducing vapors.

Such complete sensor assembly for detecting reducing vapors may be comprised of a vacuum deposited thin film of one of the noble metals to form a conductive electric pathway on an inert nonconductive substrate. The pathway may be constructed so as to form two legs of a Wheatstone bridge when connected to a suitable instrument. One leg of the film is coated with an inert and impervious coating or tape, such as silicone, rubber, vinyl tape or acrylic plastic.

The leg to be exposed is sensitized by coating it with an appropriate reducible metallic salt. The legs are connected to the measuring instrument, which is basically a Wheatstone bridge with an amplifier and readout circuits. When a reducing vapor contacts the senitized leg, the metallic sensitizing salt reacts to form metal which is deposited upon the noble metal film. The deposited metal decreases the resistance of that side of the bridge so as to unbalance the bridge circuit. This unbalance is detected by the instrument and is shown on an ammeter or galvanometer. The amount of unbalance shown is a measure of the amount of the reducing vapor present in the atmosphere.

A process developed by E.V. Crabtree, E.J. Poziomek and D.J. Hoy (4) provides a colorimetric method of detecting the presence of microgram amounts of isocyanide groups in mixtures. The method comprises the steps of placing the chemical compound or mixture upon an inert support medium and then contacting with the detector a benzidine reagent or tetra base reagent producing a blue color indicating the presence of the isocyanide group.

In a process developed by S.W. Benson, G.R. Haugen and R.S. Jackson (5) the presence of nitrogen-containing compounds is detected by sampling vapor in the vicinity of the suspected compounds, reacting the vapor under conditions to convert the compound to nitric oxide. The nitric oxide is reacted with atomic oxygen with the chemiluminescent emission of light. This light is detected to determine the presence of the suspected compound.

Measurement in Water

Organic nitrogen or total nitrogen can be monitored in water by the following instrumental techniques (6): Activation Analysis; Automated Colorimetric; and Colorimetric. A nitrogen compound detector developed by F.L. Boys (7) is a high sensitivity electronically controlled analytical system capable of measuring nitrogen compound concentrations of less than 1,000 parts per million in sample test fluids.

Prior analytical methods of measuring nitrogen compound concentrations less than 1,000 parts per million generally have been based upon either the Kjeldahl or Dumas methods, both of which are time consuming and relatively expensive in operation averaging approximately several hours per test. The Kjeldahl method generally is predicated upon the principle of digesting the nitrogen compound in sulfuric acid, distilling the neutralized acid and subsequently titrating the liberated nitrogen compounds while the Dumas method frees the nitrogen by combustion of the nitrogen organic compounds and subsequently volumetrically determines the analytic ratio of the freed components.

Because such methods of determining nitrogen content are deficient economically, new experimental methods for nitrogen analysis have been under investigation by scientists in attempts to provide a high sensitivity, economically feasible nitrogen analyzer. The method of nitrogen analysis presented in this process includes the vaporizing of the sample test fluid and the subsequent reduction of the nitrogen compounds therein to ammonia by passing a stream of hydrogen gas into contact with the vaporized sample. The gaseous product of the hydrogenation is subsequently passed through a material, e.g., Ascarite, to remove the hydrogen sulfide present in the gaseous product before the formed ammonia is absorbed by a stream of deionized water. A measurement of the conductivity of the ammonia solution by a highly sensitive electronic detection circuit provides an indication of the nitrogen concentration in the sample test fluid.

A process developed by J.A. McNulty and R.T. Moore (8) permits the determination of nitrogen in water and in oxygenated hydrocarbon matrices. The sample is volatilized and thereafter pyrolyzed by the use of a catalyst. After pyrolysis, acidic type gases including CO_2 are removed and thereafter an electrochemical determination is made to determine whether a nitrogen compound is present in the residual gas. In the apparatus, a specific type of scrubber is utilized for removing the acidic type gases including CO_2 from the products of pyrolysis.

References

(1) Murphy, C.B.; U.S. Patent 3,410,662; November 12, 1968; assigned to General Electric Company.
(2) Belisle, J.W.; U.S. Patent 3,533,750; October 13, 1970; assigned to Minnesota Mining and Manufacturing Company.
(3) Silverman, H.P. and Giarrusso, G.A.; U.S. Patent 3,549,329; December 22, 1970; assigned to TRW Inc.
(4) Crabtree, E.V., Poziomek, E.J. and Hoy, D.J.; U.S. Patent 3,567,382; March 2, 1971; assigned to U.S. Secretary of the Army.
(5) Benson, S.W., Haugen, G.R. and Jackson, R.S.; U.S. Patent 3,647,387; March 7, 1972; assigned to Stanford Research Institute.
(6) Lawrence Berkeley Laboratory, "Instrumentation for Environmental Monitoring - Water," *Publ. LBL-1,* Volume 2, Berkeley, University of California (February 1, 1973).
(7) Boys, F.L.; U.S. Patent 3,497,322; February 24, 1970; assigned to Sinclair Research, Inc.

(8) McNulty, J.A. and Moore, R.T.; U.S. Patent 3,565,583; February 23, 1971; assigned to Dohrmann Instruments Company.

NITROGEN OXIDES

Of the various oxides of nitrogen, the most important as air pollutants are nitric oxide (NO) and nitrogen dioxide (NO_2), nitrogen sesquioxide (N_2O_3), nitrogen tetroxide (N_2O_4), and nitrogen pentoxide (N_2O_5), but only N_2O is present in the atmosphere in appreciable concentrations. The term NO_x is used to represent the composite atmospheric concentration of nitric oxide and nitrogen dioxide.

Nitric oxide (NO) and a comparatively small amount of NO_2 are formed under high temperature conditions such as those that accompany the burning of fossil fuels. They are emitted to the atmosphere from automobile exhausts, furnace stacks, incinerators and vents from certain chemical processes. Although vast quantities are produced by natural biological reactions, the resultant concentrations of atmospheric NO_x are low because these reactions take place over wide areas.

Most of the NO_x produced technologically is in the form of NO, which is subsequently oxidized in the atmosphere to the more toxic and irritant NO_2. Normally, at low-NO concentrations of 1.2 mg./m.^{3} (1 ppm) or less, the direct reaction with oxygen of the air proceeds slowly. The oxidation of NO to NO_2 is speeded up enormously, however, by photochemical processes involving reactive hydrocarbons.

Nitrous oxide (N_2O) is not considered an air contaminant. There is no evidence to suggest N_2O is involved in photochemical reactions in the lower atmosphere. Formation of N_2O is largely due to the decomposition of nitrogen compounds by soil bacteria and to reactions between nitrogen and atomic oxygen or ozone in the upper atmosphere. The mean atmospheric concentration of N_2O is about 449 μg./m.^{3} (0.25 ppm). The oxides, N_2O_4, N_2O_3, and N_2O_5, are not found in appreciable concentrations under urban atmospheric conditions. Nevertheless, it is feasible that even in small concentrations, N_2O_3 and N_2O_5 could be involved in the reactions leading to photochemical air pollution as noted by the EPA (1).

A variety of air quality standards have been promulgated for nitrogen oxides. California has set a limit of 0.1 ppm based on a 1 hour averaging time for nitrogen dioxide at an adverse level; the limit is 3.0 ppm for NO_2 at a serious level. Combined nitrogen oxides are restricted to 0.1 ppm for consideration at a serious level by the State of Colorado.

Soviet air quality standards call for a maximum of 0.045 ppm of NO_2 on either a single exposure (20 minutes) or 24 hour basis. East German standards specify 1.0 ppm of nitrogen oxides on a basis not to be exceeded more than once in 8 hours and 0.5 ppm on a long-term basis. Canadian (Ontario) air quality standards specify 0.1 ppm on a 24 hour averaging basis. Emission standards for nitrogen oxides from nitric acid plants have been set at 2,560 ppm in the U.K. and in Australia, according to a review by A.C. Stern (1a).

Measurement in Air

The most accurate, reliable and useful methods for measurement of the atmospheric nitric oxide (NO), nitrogen dioxide (NO_2), and nitrogen oxides (NO_x) employ standard techniques to collect the sample in an absorbing solution. The analytical determination is made colorimetrically or spectrophotometrically. Analytical methods based on physical properties such as absorption or emission of radiation, oxidation or reduction and methods based on gas-chromatographic separation and detection are not yet in common use but will probably receive more attention in the future. The colorimetric Griess-Saltzman method is deemed the most suitable manual method generally applicable to the measurement of NO_2 in the atmosphere (1).

This method is based on the reaction of NO_2 with sulfanilic acid to form a diazonium salt, which couples with N-(1-naphthyl)-ethylenediamine dihydrochloride to form a deeply colored azo dye. Air is sampled through a fritted bubbler into a solution of the Griess-Saltzman reagent for periods of 30 minutes or less, and the color is allowed to develop for an additional 15 minutes after completion of sampling. The color produced, which is proportional to the amount of NO_2 sampled, is measured at 550 nanometers (nm.). The method can be used to determine concentrations of NO_2 in the air from 40 to 1,500 micrograms per cubic meter (0.02 to 0.75 ppm).

Ordinarily, interferences are not a problem, although high ratios of sulfur dioxide to nitrogen dioxide (about 30:1) can cause bleaching and give misleadingly low values. Interference from other oxides of nitrogen or ozone are negligible at concentrations found in polluted air. Peroxyacetylnitrate (PAN) can give a response of up to 35% of an equivalent molar concentration of NO_2, but in ordinary ambient air PAN concentrations are too low to cause significant error. The color formed is reasonably stable, but samples should be read within 1 hour of completion of sampling.

In the original method Saltzman used solutions of sodium nitrite as calibration standards. He reported that, under laboratory conditions, 0.72 mol of nitrite produces the same color as 1 mol of NO_2 gas and incorporated this factor into his calculations. In recent years this stoichiometric factor has been the source of considerable controversy. Values ranging from 0.5 to 1.00 have been reported, and one recent work reconfirmed Saltzman's original findings. The method can be standardized by using an accurately known concentration of NO_2 gas and thus eliminate the stoichiometric factor from the calculations.

The Griess-Saltzman method for NO_2 measurement cannot be used successfully when the delay between sample collection and color measurement is more than 4 to 6 hours or when sampling periods of longer than 1 hour are required. In such situations the Jacobs-Hochheiser method is preferred. With this method, NO_2 can be measured in a network where a central laboratory is used for analyzing field samples. With a later modification of the procedure, NO_2 sampling periods can be as long as 24 hours, and the delay of analysis can be at least 2 weeks, if necessary.

In the Jacobs-Hochheiser procedure, polluted air is passed through aqueous sodium hydroxide, in which nitrite ion is formed from the NO_2. The resulting solution, which contains sodium nitrate in addition to sodium nitrite, is treated with hydrogen peroxide to remove possible sulfur dioxide interference, and acidified. The nitrous acid produced is measured by the Griess-Saltzman diazotizing-coupling procedure, except that sulfanilamide is used in place of sulfanilic acid. In the development of the Jacobs-Hochheiser procedure, the ratio of nitrite to nitrate was determined and an average stoichiometric factor of 0.63 was selected for converting nitrite ion to NO_2 concentrations.

Because the sampler adopted by the National Air Surveillance Network (NASN) employs a 24 hour sampling period, it precluded the use of the Griess-Saltzman method and all data obtained by this network since its inception have been derived by the Jacobs-Hochheiser method. Because of discrepancies concerning the NO_2-nitrite stoichiometric factor, all NASN computations are now based on the assumption that one molecule of NO_2 produces one molecule of nitrite and no nitrate.

The absorption efficiency for NO_2 varies with the sampling system used. Detailed studies have shown that, under laboratory conditions, the NASN sampling system has an analytical efficiency of approximately 35%, a correction that is applied in NASN calculations of NO_2 concentrations. Other investigators have found different efficiencies. The EPA designated reference method for nitrogen dioxide in air was the Jacobs-Hochheiser method until recently.

Research in analytical chemistry led the Environmental Protection Agency in July 1972 to postpone for one year the regulation of nitrogen oxides (NO_x) emissions by stationary sources as noted by the American Chemical Society (2). EPA chemists had found flaws in the method for measuring nitrogen dioxide (NO_2) in the air. Use of the method, they concluded, may have exaggerated the NO_2 problem in some of the nation's 247 Air Quality Control Regions (AQCR's). The persistent need to define the quality of the environment precisely, routinely, and at reasonable cost bred steady activity elsewhere in analytical chemistry as well. Progress in 1972 included work on particulates and organic compounds. New data emerged also on global sources of carbon monoxide.

Late in 1971, EPA had classified 45 AQCR's in 29 states as Priority I for NO_2. The action was based in large part on data obtained by the method found later to be faulty. To meet air quality standards in Priority I regions, states may have to control NO_x emissions from stationary sources. The NO_x standard to protect human health is a maximum of 100 μg. per cubic meter of air (annual arithmetic mean). The stationary sources include nitric acid plants and certain types and sizes of steam generating plants. EPA at any rate had planned to implement the NO_x control procedures July 31, 1972.

Scientists at EPA's National Environmental Research Center, Research Triangle Park, North Carolina, meanwhile, had been studying the reliability of the analytical method for NO_2 (the reference method). They found that when the actual concentration of NO_2 in air is 30 to 60 $\mu g./m.^3$, the method shows it to be much higher. When actual concentration is above 130 $\mu g./m.^3$, the method shows it to be lower. The EPA chemists found also that nitric oxide in the air can cause the NO_2 method to read incorrectly high.

These findings suggested that some Priority I regions might not in fact have an NO_x problem. They cast doubt also on the validity of some years of data acquired by the reference NO_2 method. To avoid costly, perhaps needless controls on NO_x emissions, EPA postponed the implementation date to July 1, 1973. At the same time the agency launched an intensive study of several analytical methods for NO_2, including the current reference method. The methods are being tested and cross-checked at some 200 sites of the National Air Surveillance Network and the Continuous Air Monitoring Program. The work is being done by the Quality Assurance and Environmental Monitoring Laboratory at the National Environmental Research Center.

EPA will use the resulting data to decide, first, whether to reclassify any of the Priority I regions. The agency will use the data also to try to work out a sound correction factor for the current reference method for NO_2. Such a factor would allow continued use of the method (if necessary), which has certain practical advantages in air monitoring; and it would restore the value of the mass of NO_2 data from previous years. If the research does not produce a sound correction factor, EPA will have to develop and adopt a new reference method for NO_2 and may lose much of the value of existing data on NO_2 in the air. The agency expected to be able to resolve these questions early in 1973. (These analytical difficulties do not affect the air quality standard for NO_2. Nor do they affect NO_x emission regulations for moving sources, such as the automobile.)

In fact, as noted recently (2a), the Griess-Saltzman method has now been validated by the ASTM in contrast to the Jacobs-Hochheiser method which was originally specified in EPA regulations and later cast in doubt.

As noted by Altshuller (3), the measurement of nitric oxide in the ambient atmosphere has involved use of analyzers utilizing the colorimetric Griess-Ilosvay reaction for nitrogen dioxide (4), in which the nitric oxide is oxidized to nitrogen dioxide before analysis. A number of efforts have been made to overcome difficulties in developing an oxidizing substrate capable of providing high conversion efficiencies over a range of atmospheric conditions.

The stoichiometry of the colorimetric nitrogen dioxide reaction used has been in dispute (5). In addition, the nitric oxide and nitrogen dioxide colorimetric analyzers have been difficult to keep in proper operation. No electroanalytical or UV analyzer has yet been provided that is adequate for ambient air monitoring of nitrogen oxides. The need for improved instrumentation is particularly urgent for nitrogen dioxide.

The promising approach involves the use of chemiluminescent emissions from the electronic transition, NO to NO_2, produced in the reaction of nitric oxide with ozone (6). The emission spectrum extends from 0.6 to 3 microns, with maximum intensity near 1.2 microns. This method for nitric oxide by determination is linear from 0.004 up to 10,000 ppm. Photomultipliers responding from the outset of emission at 0.6 micron to their cutoffs at 0.8 to 0.9 micron have been evaluated.

To obtain a sensitivity down to 0.004 ppm, choice of photomultipliers must be optimized and the operating temperature of the photomultiplier reduced to -20°C. Since such sensitivity is not needed in measurement of source emissions, less sensitive photomultipliers operating at room temperature can be used. For a wide range of gases present in polluted air, quenching of the chemiluminescent reaction has not been observed at reduced pressures. No interfering species have been identified. This same reaction has been mentioned as one method for ozone measurement.

The reaction of nitric oxide with atomic oxygen produces excited states that emit weakly, with some line structure occurring between 0.35 and 0.45 micron, followed by a more intense unstructured band with a maximum near 0.65 micron (7). Since atomic oxygen reacts with nitrogen dioxide to form nitric oxide on a one-to-one basis, this reaction can be used to measure NO + NO_2. This reaction can be used as the basis of a rapid response instrument for measuring nitrogen oxides. Use of the emission in the 0.65 micron region only provides discrimination against the chemiluminescent emissions produced by the corresponding reactions of atomic oxygen with carbon monoxide and sulfur dioxide (7). Instruments have been fabricated for monitoring nitric oxide in ambient air by reaction with O_3, and NO_x by reaction with atomic oxygen (8)(9).

Nitrogen oxides may be monitored in air by the following instrumental techniques (10): colorimetric (reference method of NO_2); coulometric; chemiluminescence; electrochemical transducers; infrared spectroscopy; and lasers.

Nitric oxide may be measured after NO_2 removal from the sample by oxidation with a supported chromic oxide catalyst to NO_2 followed by determination by the Griess-Saltzman method. The conduct of the Griess-Saltzman reaction for NO_2 determination is then reviewed in detail (10a).

Combined nitrogen oxides including nitric and nitrous acid mists (but not nitrous oxide, N_2O) may be measured by a colorimetric technique using phenoldisulfonic acid (10a). Improved chemical methods for the sampling and analysis of nitrogen oxide pollutants from the combustion of fossil fuels have been reviewed by J.N. Driscoll and A.W. Berger of Walden Research Corporation of Cambridge, Massachusetts (11). A comprehensive review of the available types of instrumentation for the determination of nitrogen oxides in stationary source emissions has been made by L.P. Parts, P.L. Sherman and A.D. Snyder of Monsanto Research Corporation (12).

An indicator for nitrogen oxides has been developed by L.W. Brauer (13). It consists of an inert granular solid carrying diphenylbenzidine-decasulfonic acid or an alkali metal salt thereof and is responsive to oxidizing nitrogen oxides and acids by changing color from white to grey-blue.

A device developed by R.F. Rakowski (14) is a dosimeter device capable of giving an estimate of the concentration time product of an exposure to nitrogen dioxide in the range of toxicological interest without requiring an operation by the user other than visual inspection. The device will determine the degree of an accidental exposure to nitrogen dioxide so that appropriate medical action can be taken and will warn a user of the presence of hazardous levels of nitrogen dioxide gas.

The dosimeter strips of this device have as a base material a sheet comprising a uniform layer of silica gel bound to polyethylene terephthalate with a polyvinyl alcohol binder. The reactive reagents of this process are diphenylamine and oxalic acid.

Dosimeter strips prepared by dipping a base sheet in a methanol solution of the reactive reagents develop a forest green color when exposed to nitrogen dioxide. Exposure to other oxidizers produces a blue color.

A device developed by A.B. Smith (15) is a small pocket type of field use indicator of the colorimetric type for detecting and indicating nitrogen dioxide contamination in an atmosphere. In this device, a powder compact of diazotizing, coupling and stabilizing agents is mechanically supported in suitable manner to provide a self-contained, one-time indicator unit that is discarded after use.

A reagent for the determination of NO_2 in air has been developed by N.A. Lyshkow (16). It is a Griess-type colorimetric reagent for the analysis of nitrogen dioxide in air, the reagent containing 1 to 4 grams sulfanilamide or sulfanilic acid per liter of reagent, 0.025 to 0.75 gram N-(1-naphthyl)ethylenediamine dihydrochloride per liter of reagent, 0.025 to 0.075 gram 2-naphthol-3,6-disulfonic acid disodium salt per liter of reagent and the balance tartaric acid or acetic acid in an amount such that the reagent has a pH of less than four. The disodium salt is described as improving both the rate and intensity of color development, as well as the absorption efficiency of the reagent even at extremely low nitrogen dioxide levels.

A special colorimetric indicator solution for the detection of NO_2 has been developed by A.B. Smith, R.H. Hennig, A.J. Kurtz and G. Pantchenko (17). Soluble tablets suitable for providing such an indicator solution of the colorimetric type for quantitatively measuring the nitrogen dioxide contamination in an atmosphere are disclosed.

The indicator solution, which has a pH of not more than 4.0, is conveniently prepared by dissolving a tablet containing a diazotizing agent mechanically supported in a suitable manner and a second tablet containing a coupling agent also mechanically supported in a suitable manner in a predetermined amount of water. A measured solution is transferred to an absorption bubbler of the type customarily employed in colorimetric analysis, the reagent solution is exposed to metered quantities of air containing nitrogen dioxide as a contaminant thus resulting in the formation of red-violet diazo dye. On examination of the sample thus exposed in a colorimeter, the nitrogen dioxide content is readily measured.

A procedure for continuous NO_2 analysis has been developed by F. Schulze (18). It involves the use of 1-amino-2,5-benzenedisulfonic acid as the diazotizable compound together with N-(1-naphthyl)ethylenediamine dihydrochloride as the coupling compound and makes possible a highly improved system for continuously analyzing air for nitrogen dioxide which, in turn, makes possible the employment of more compact and more efficient conventional apparatus.

Because of its inherent sensitivity and specificity the Griess reaction, originally described in 1879 and, involving the formation of an azo dye, has become the accepted reaction for determining trace amounts of nitrite ion and hence nitrogen dioxide. The general reactions involve formation of nitrous acid by reaction of nitrogen dioxide with water, diazotization of an aromatic amine by nitrous acid, and coupling the diazo compound with a coupling reagent to form an azo dye.

Various aromatic amines have been examined for speed of reaction and sensitivity but in current practice the choice has narrowed down to either sulfanilic acid or sulfanilamide. 1-naphthylamine was early proposed as a coupling reagent but today one of its derivatives, N-(1-naphthyl)ethylenediamine dihydrochloride, is the preferred coupling reagent because of its more rapid coupling rate and deeper color of the resultant azo dye.

The most frequently used azo dye reagent in continuous analyzers is a formulation containing 5.0% acetic acid, 0.5% sulfanilic acid (diazotizing compound) and 0.005% N-(1-naphthyl)ethylenediamine dihydrochloride (hereafter referred to by the abbreviation NEDA) (the coupling compound) in water which is known as "modified Saltzman reagent." The procedure employed follows the tentative test method issued by the American Society for

Testing and Materials, ASTM D2012-63T, which may be summarized as follows. Nitrogen dioxide is absorbed by passing the air sample and modified Saltzman reagent, in volume ratio of from 100:1 to 1,000:1, concurrently through a special absorber that provides adequate contact for quantitative absorption of this gas. Following diazotization and coupling, the pink dye is measured continuously in a recording colorimeter by comparison with unreacted reagent in a similar reference cell.

The ASTM procedure lacks nothing in sensitivity or specificity for trace amounts of nitrogen dioxide but otherwise, however, the procedure is deficient in a number of respects as has been pointed out in the technical literature. The shortcomings are to a considerable extent due to defects in the chemistry of the analytical system, however, some of these chemical deficiencies require mechanical concessions in the apparatus which further contribute to the lack of smooth and efficient operability. The various deficiencies and proposals that have been tried for avoiding them include the following.

It has been pointed out in the technical literature that the use of acetic acid in the modified Saltzman reagent is a corrosion and hygienic nuisance. The acetic acid corrodes metal parts, it softens some plastic parts and its odor is objectionable. To avoid the acetic nuisance, it has been proposed to substitute other organic acids such as citric or tartaric acid for adjusting pH of the reagent. While this eliminates the odor and corrosion problem it introduces a new and just as bothersome problem. Citric and tartaric acid solutions, in concentrations suitable for this use, are nutrient media for slimes and molds. Mold spores floating in the air are drawn into the apparatus with the sample air and the resultant growths clog the absorber and contaminate the reagent.

Instrumental response to rapid fluctuations in nitrogen dioxide concentration is limited by the speed of the chemical reaction used in measurement. This deficiency is objectionable in air monitoring and it is a very serious defect in analyzer-controllers which must have rapid response to maintain uniform concentrations of nitrogen dioxide in environmental chambers. It has been determined and reported in the literature that presently used reagents require at least 6 minutes for development of full color response. This necessitates use of oversize gas absorbers and long lag times in the apparatus between the gas absorber and the optical measuring cell which results in slow instrumental response of 10 to 15 minutes to changes in concentration.

In addition to causing slow instrumental response, this time lag also causes peaks and dips in concentration to be averaged out due to reciprocal blending of successive increments of reagent. Various reagent formulations, some based on the use of sulfanilamide in place of sulfanilic acid, plus activators for accelerating the rate of reaction have been proposed but experience with these formulations demonstrates little if any practical increase in reaction rate.

In any type of gas liquid absorber, there is an initial interface between gas and reagent. Because of capillary effects, the reagent creeps back along the walls of the absorption apparatus into the gas zone. If the gas is not saturated with moisture, the reagent in the capillary film will evaporate and deposit any solids it may carry in solution. The solid film thus deposited then acts as a wick, carrying more solution into the evaporation zone with the result that a crust of solid will form which in time may clog the apparatus. Reaction between nitrogen dioxide and reagent occurs in this crust and the resultant dye cannot reach the photometric system and low results are obtained.

At other times, the colored crust may flake off and discolor the reagent thereby yielding false high results. It has been proposed to presaturate the air by bubbling through water but this expedient is impractical because nitrogen dioxide is quite soluble in water, thus serious errors in analysis occur. In conventional practice, large volumes of reagent are required for continuous operation. Normally between 1 and 5 ml./min. of reagent are supplied to the gas absorption system. There is thus required from 10 to 50 liters of reagent per week. Preparation and storage of these quantities is inconvenient especially when outlying field monitoring stations must be serviced on a weekly basis. Reagent storage in these orders of magnitude also militate against easy portability of the analyzer. To minimize the reagent storage and handling problem, the expedient of using low reagent flows has been practiced. Unfortunately, this expedient introduces mechanical problems such as inefficient gas absorption, difficulty in maintaining constant reagent metering rate, as well as even slower instrumental response.

To avoid the need for frequent replenishment with large volumes of fresh reagent, it is a practice to recover spent colored reagent by filtering through activated carbon. The clarified reagent is then rejuvenated by addition of NEDA which the activated carbon removes along with color. While this eliminates the need for periodic preparation of large volumes of reagent, it does not eliminate the storage of large batches of recovered as well as spent reagent.

The shelf life of conventional reagents is poor due to darkening caused by decomposition of NEDA. It is known that acid solutions of NEDA decompose in a few days. The diazotizable reagent is acidic. Thus, when optimum amount of NEDA is added to fresh or recovered reagent, there is no assurance that the optimum amount will still be there after several days or a week's operation. Analytical errors may thus be introduced due to darkening of reagent and depletion of NEDA content.

Sulfur dioxide, a frequent copollutant in urban atmospheres, produces negative interference in conventional methods of nitrogen dioxide analysis. This interference is due to two effects. Sulfur dioxide is absorbed by reagent and reacts with nitrogen dioxide; it thus competes with reagent for the available nitrogen dioxide. Because the reaction between reagent and nitrogen dioxide is relatively slow, there is ample opportunity for the sulfur dioxide to remove some of the nitrogen dioxide.

The second source of interference is the bleaching effect of sulfur dioxide on the azo dye. These forms of interference have been recognized but have been accepted as more or less unavoidable in normal air monitoring practice. While it is only moderate with low concentrations of sulfur dioxide, it can cause appreciable errors in highly polluted atmospheres. In environmental chambers, where exceptionally high concentrations of sulfur dioxide may be employed, this form of interference can cause serious negative errors.

Conventional apparatus for nitrogen dioxide analysis was designed to accommodate the slow reaction time of reagents. There would be little advantage in improving the chemistry of the procedure if improved apparatus were not available. There is, therefore, need for improved apparatus for capitalizing on any improvements effected in the chemistry of the process.

Another problem with conventional apparatus is calibration of the solution metering pump, necessary to insure correct air to reagent ratio. Still another problem is loss of water from reagent due to evaporation into the air sample stream. When reagent is recovered and recycled continuously, some provision must be made for replacing the water lost by evaporation otherwise the system will ultimately run dry.

Figure 40 shows a suitable form of apparatus for the conduct of this process. Blocks **G**, **K** and **I** represent reagent vessels of about 1 liter capacity. Blocks **J** and **L** represent metering pumps. Vessel **G** is normally kept about half filled with spent reagent from the analyzer. Spent reagent flows from the bottom of **G** into the bottom of **I** which is a one liter column about three-fourths filled with 30 mesh activated carbon.

A plug of nylon fiber in the top of this column acts as a filter for the carbon. Metering pump **J** pumps clear recovered reagent from the top of **I** to the analyzer at a constant rate of about 1 cc/min. At the same time, metering pump **L** pumps a solution of NEDA from storage vessel **K** into the suction side of pump **J** at a constant rate of 0.025 cc/min., or 250 cc/week. Since the optimum concentration of NEDA in the complete reagent is 0.05 gram per liter, the stock solution in **K** should contain 0.5 gram NEDA per 250 cc. Operation of the analyzer is started by filling reagent reservoir **G** with about 1 liter of the new reagent having the following composition: 15 grams ABDSA (70% technical grade); 100 to

FIGURE 40: IMPROVED COLORIMETRIC ANALYZER FOR NO_x MEASUREMENT

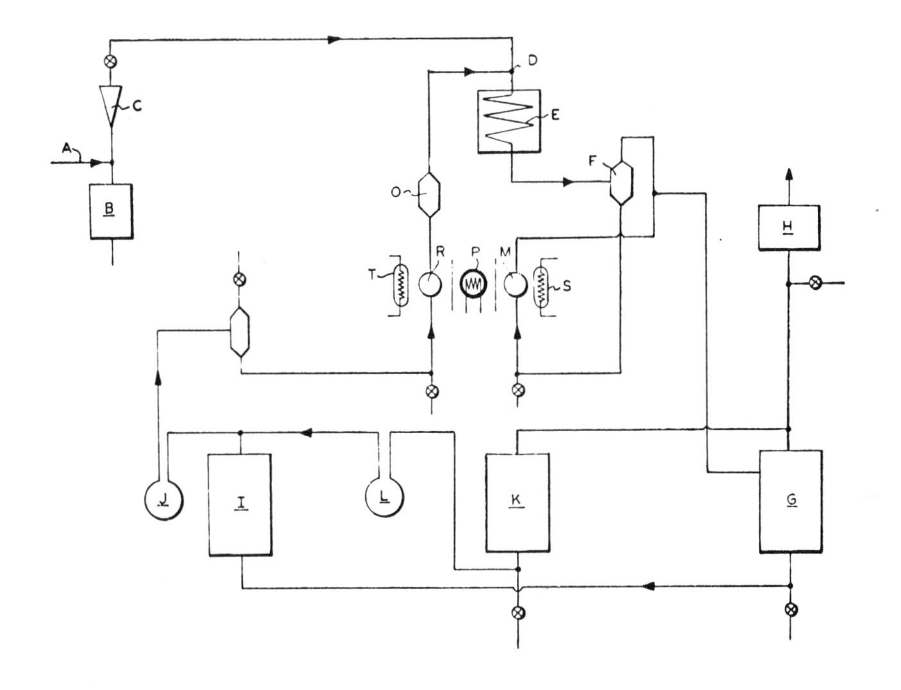

Source: F. Schulze; U.S. Patent 3,512,937; May 19, 1970

to 200 cc ethylene glycol, glycerol or diethylene glycol; and water to make 1 liter. Normally, the reagent is first decolorized by treating with 5 grams of powdered activated carbon for several hours at room temperature and then filtering. This, however, is not essential because the reagent will be decolorized in passing through activated carbon filter **I**. Reagent is allowed to flow from **G** into **I** until all air has been displaced from the activated carbon filter. To reservoir **K** is added 250 cc of coupling reagent having the following composition: 0.5 gram NEDA; 0.33 gram sodium bicarbonate and 250 cc water.

Both metering pumps **J** and **L** are then started in order to pump complete reagent into the analyzer. An air trap **N** is located in the reagent line to entrap any air that might have been included in the reagent stream due to incomplete deaeration of the activated carbon filter. Clear reagent then flows through reference flow cell **R**.

A 5 cc pipette **O** located in the reagent line beyond the flow cell is for the purpose of calibrating the metering pumps. To calibrate the pumps, the pipette is emptied, the pumps are started and filling time is measured by means of a stop watch and pumping rate in cubic centimeters per minute is calculated.

Normally when analyzing for nitrogen dioxide in the 0 to 20 pphm range a 1,000 to 1 air to liquid volume ratio is used. For reagent pumping rate of about 1 cc/min. the air sample rate is adjusted, by means of rotameter **C**, to 1,000 times the reagent flow. In this manner it is unnecessary to adjust the pump to an exact rate of delivery, which is a tedious operation. Knowing the reagent delivery rate, it is a simple matter to adjust airflow rate by means of the rotameter.

Sample air from valve **A** (which is a 3 way valve for switching from purifying column **B** to air to be analyzed) meets reagent **D** and both flow concurrently down wetted wall absorber **E.** Column **B** containing activated carbon and soda lime is employed for purifying air for the purpose of zeroing the analyzer. **E** is a 10 turn, 3 inch diameter helix constructed of 4 to 5 mm. i.d. glass tubing. Reaction between nitrogen dioxide and reagent takes place in the helix; colored reagent is separated from air in separator **F.** Reagent flows through measuring cell **M** from whence it flows to spent reagent reservoir **G.** Motive power for the air stream is furnished by vacuum pump **H.**

The color measuring system consists of reference and measuring flow cells, **R** and **M,** photocells **S** and **T,** and a light source, **P.** Recorder and other accessories of the photometer are not shown because the type of equipment used for that purpose is unrelated to the chemistry or apparatus of the analyzer proper. The analyzer as described is quite flexible with respect to concentration of nitrogen dioxide being analyzed. The above outlined specification describes the procedure for analyzing gases containing between 0 to 20 pphm nitrogen dioxide. Range of the analyzer can be extended by reducing the air to reagent ratio, which can be done in two ways.

Airflow can be reduced from about 1,000 to 100 cc/min. to obtain a tenfold increase in range. On the other hand, reagent flow can be increased from about 1 to 5 cc/min. When this is done, it is necessary to raise the NEDA content of the coupling reagent to 2.5 grams per 250 cc to maintain the optimum concentration of 0.05 gram of NEDA per liter in the complete reagent. The analyzer may be calibrated by placing a standard color solution in the measuring flow cell in which the color has been generated by adding a known quantity of sodium nitrite to the complete reagent. Alternately, the instrument can be dynamically calibrated by the method described in ASTM method D2012-63T.

A device for nitric oxide analysis has been developed by J.N. Harman, III and R.M. Neti (19). The nitric oxide in the gas stream is reacted with an excess amount of ozone so as to completely convert the nitric oxide into nitrogen dioxide. The resulting gas containing nitrogen dioxide and residual ozone is passed through a scrubber which removes the ozone and does not affect the nitrogen dioxide content of the gas. The resulting gas is then analyzed for its NO_2 content, which is a function of the nitric oxide content of the sample gas. The method and apparatus may also be utilized for determining the total amount of nitrogen dioxide and nitric oxide in a sample gas. The device is particularly applicable to the monitoring of air for air pollution control.

A scrubber which is capable of removing interferents from an NO_2-containing stream to be analyzed without affecting the NO_2 content of the stream has been developed by J.N. Harman, III and R.M. Neti (20). The scrubber contains argentic oxide (AgO) which effectively removes both positive and negative interferents from the sample gas stream, but does not affect the NO_2 content of the stream. The scrubber is particularly applicable to the monitoring of the NO_2 content of polluted air.

A process developed by N.A. Lyshkow (21) for determining the concentration of NO and NO_2 in gas sample is one in which a gas sample containing NO and NO_2 is first passed through a scrubber in which the sample is contacted with a reagent solution capable of development of color in response to contact with NO_2. The intensity of the color thus developed is a measure of the NO_2 content of the gas sample. The unabsorbed gases are treated to remove residual amounts of NO_2 and the NO in the sample is oxidized to NO_2 which is absorbed in a second scrubber with a reagent solution capable of development of color in response to contact with NO_2, with the intensity of color developed in the second scrubber being a measure of the NO content of the gas sample.

A device developed by N.A. Lyshkow (22) is a colorimeter for determining the concentration of gaseous pollutants such as NO_x in air. The device incorporates improved scrubber and photocell assemblies. The scrubber includes a helical coil into which an air sample is drawn and admixed with a liquid absorbent whereby the liquid flows gravitationally through the coil in the form of wave fronts to provide complete contact between the air sample and the liquid absorbent with minimum agitation. The photocell assembly includes a pair of spaced photoresistors and a light source spaced therebetween, with the light source having a lamp housing enclosing the light source and defining a pair of spaced openings whereby the light source projects optical spots to the photoresistors to illuminate the photoresistors with light, the relative intensity of which is dependent upon the color developed in solutions passed in front of the photoresistors as an indication of the pollutant gas concentration in the air sample.

A device developed by R.M. Neti and C.C. Bing (23) permits the measurement of the nitric oxide concentration in a gas stream such as polluted air or automobile exhausts. The nitric oxide measured may be the pollutant itself or if nitrogen dioxide is the pollutant, it may be converted to nitric oxide for measurement.

Nitric oxide absorbs light either in the far ultraviolet or in the infrared regions of the electromagnetic spectrum. Spectrophotometers have therefore been built to determine the nitric oxide levels. However, the usefulness of these devices decreases if the detection limits are a few parts per million, as they frequently are in polluted air. A second approach has been to oxidize the usually nonreactive nitric oxide to the reactive nitrogen dioxide by either gas phase oxidation or liquid phase oxidation and measure the nitrogen dioxide either coulometrically or colorimetrically. These methods also lack wide dynamic range and are restricted in utility to concentrations of a few parts per million.

Another approach has been to study some of the gas phase reactions of nitric oxide. One of these involves the reaction of nitric oxide with ozone upon mixing nitric oxide and ozone. Light is emitted as a by-product of this chemical reaction. The amount of light so emitted can be used as a measure of the nitric oxide. According to the prior art, such devices can be used if this reaction is observed under vacuum. Thus the working limit range of devices of this type is usually about 10^{-3} to 2 mm. mercury pressure as the lower and upper limits. The reaction chambers must either be of stainless steel or heavy glass flasks.

These are usually coated with white barium sulfate and/or magnesium oxide to improve the light collection efficiency. The emitted light is detected by a suitable photomultiplier tube together with a current measuring device. A typical device of this type is described in U.S. Patent 3,528,779. The chemiluminescent technique possesses the advantage of having a wide dynamic range and is useful for monitoring from the ambient levels, i.e., parts per billion, to the source levels, i.e., parts per million. The disadvantages of the known chemiluminescent techniques are that they require a very high vacuum, careful and precise control of the working pressure and efficient light collection and collimating systems.

It is the object of this development to provide an apparatus for the measurement of the nitric oxide concentration in a gas stream which, although it uses the chemiluminescent technique, has none of the disadvantages of the prior art systems. A very high vacuum is not necessary in the device. Further, careful and precise control of the working pressure is not necessary. Yet another advantage is that the highly efficient light collection and collimating systems of the known devices are not necessary. Much simpler light collection and collimating systems can be utilized.

This method consists of measuring the nitric oxide concentration in a gas stream with ozone at the entrance to a reaction chamber so that the mixture enters the reaction chamber as a point source. The intensity of the light emitted as a result of the chemiluminescent reaction between the ozone and the nitric oxide at the point source is measured. This light intensity is a function of the nitric oxide concentration in the gas stream, whereby the nitric oxide concentration is measured.

A device useful for determining trace amounts of NO has been developed by A. Fontijn and P.H. Vree (24). The device is actually suitable for detecting trace amounts of volatile substances such as carbon oxides, nitrogen oxides, sulfur oxides and oxygen. The method involves taking a measured sample of the gaseous mixture containing the gas to be detected, separating the gas being detected from the mixture where necessary, forming a first component which includes the gas being detected and, preferably, a carrier gas, providing a second component which contains a substance to be reacted with the first component to produce chemiions, activating at least one of the substances to produce the species necessary to form chemiions, mixing the components to form the chemiions, passing the mixture between electrodes and measuring an electrical current produced by the chemiions formed.

Figure 41 shows the form of apparatus which may be displayed in the conduct of this process. Referring to the figure, the apparatus comprises a chromatographic gas separation column **2** of conventional type, having an inlet tube **4** connected to one end and an outlet tube **6** connected to the opposite end. A sample-injecting inlet port **8** is connected to the inlet tube **4**. The outlet tube **6** leads into the measuring apparatus which will now be described.

The measuring apparatus includes a discharge gas inlet tube **10** passing through a microwave cavity **12**, of conventional design, to a Y-shaped gas divider unit **14**. One leg of the Y comprises a sample arm tube **16** containing a pair of electrodes **18** and **20**. The other leg of the Y comprises a reference arm tube **22**, also containing a pair of electrodes **24** and **26**. Both the sample arm tube **16** and the reference arm tube **22** are joined at their outlet ends to a line **28** leading to a vacuum pump (not shown).

Connected into the sample arm tube **16** is a continuation of the outlet tube **6** from the gas chromatograph column **2**. Connected into the reference arm tube **22** is a tube **30** from a source of ballast gas (not shown). This last-mentioned tube **30** first passes through the microwave cavity **12** as does the tube **6**. Also connected to the gas discharge inlet tube **10** may be another tube **32** for introducing a gaseous catalyst in a modification of this technique.

FIGURE 41: APPARATUS FOR DETERMINATION OF TRACE AMOUNTS OF NO

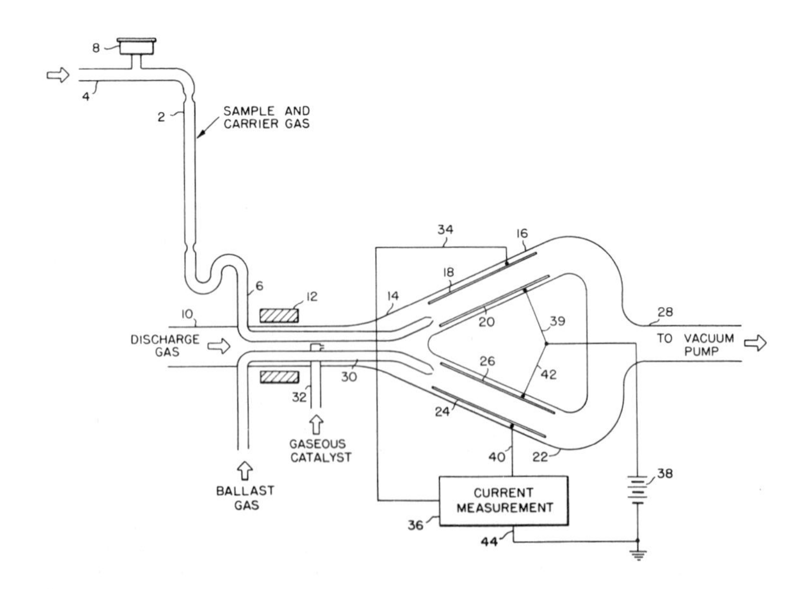

Source: A. Fontijn and P.H. Vree; U.S. Patent 3,713,773; January 30, 1973

One of the electrodes **18** in the sample arm **16** is connected by a lead **34** to a current measuring device **36** such as a vibrating reed electrometer. The other electrode **20** in the sample arm **16** is connected to one side, e.g., the positive side, of a battery **38** by a lead **39.** One of the electrodes **24**, in the reference arm **22**, is also connected to the current measuring device **36** by a lead **40**. The other electrode **26**, in the reference arm **22**, is connected to the same side of the battery **38** as is electrode **20** by a lead **42**. The other (negative) side of the battery **38** is connected to ground. The current measuring device **36** also has a terminal **44** connected to ground.

The following is an example of the use of the apparatus in carrying out the method of detecting nitric oxide (NO). First, a flow of carrier gas, e.g., helium, is started through the inlet tube **4** and through the chromatograph column **2** to establish equilibrium conditions. At about the same time, a flow of ballast gas, also helium, is started through the ballast gas inlet tube **30** and into the reference arm **22** of the measuring apparatus, and a mixture of nitrogen and helium containing about one volume percent nitrogen is passed into the discharge gas tube **10.**

Meanwhile, the vacuum pump is also started to establish a vacuum of about 1 mm. of Hg in the sample arm **16** and the reference arm **22**. Then the microwave cavity **12** is energized. This causes the gaseous molecules to be broken down into atoms: $N_2 \rightarrow N + N$. The flow rate of the nitrogen-helium mixture is about 1 cc/sec.

A 1 cc sample of the gas mixture which contains the NO to be measured is then injected into the sample injection port **8** and permitted to flow through the gas chromatograph column **2.** It is assumed that previous test runs have been made to determine the exact time interval it will take for the NO to appear at the exit end of the column. In the apparatus shown, the sample is passed through the microwave cavity to produce the oxygen atoms needed for the production of ions. However NO does not need to be activated in order to obtain a measurement of its concentration.

The current measurement apparatus **36** includes a conventional electrometer connected to a recorder. When the atoms from the NO of the injected sample pass through the sample arm **16**, a peak will be observed in the recorded reading of the electrometer. The reason for the appearance of the peak is that in mixtures of nitrogen atoms and oxygen atoms, ions are produced by reactions involving excited N_2 and/or excited NO molecules in reactions such as:

$$N_2^* + NO^* \rightarrow N_2 + NO^+ + e^-$$

$$N + N + NO^* \rightarrow N_2 + NO^+ + e^-$$

in which N_2^* and NO^* are excited molecules formed by N-N and N-O atom association reactions. The area under the peak of the recorded curve is measured to give a relative measurement of the concentration of NO in the sample. To get an actual reading of concentration, this value must be compared with the area under a peak recorded with a known sample of NO. The relative values are different for each apparatus setup.

Another technique for NO-NO_2 analysis has been developed by J.A. Williamson, Jr. (25). The method involves introducing the sample gas into a leak-tight pressure sample cell; passing visible radiation through to detect for NO_2; introducing an oxygen-containing gas under pressure; detecting again with visible radiation for the combined amounts of NO and NO_2; and controlling the cell temperature.

A process developed by A. Warnick and A.D. Colvin (26) is one whereby concentrations of nitric oxide in gaseous mixtures are determined by measuring the chemiluminescence of the reaction between the nitric oxide and ozone. The gaseous mixture is introduced with virtually laminar flow into a reaction chamber at a location proximate to the inner surface of a light transmitting element. Reaction chamber pressure preferably is maintained above about 5 torr.

A technique developed by G.W. Wooten, R. Iltis and V.L. Johnson (27) is a chemiluminescent method for detecting NO_x (NO and NO_2), SO_2, CO and other gases which react with atomic oxygen to produce chemiluminescence. Atomic oxygen generated by pulse electrical discharge is added at a controlled and known flow rate to a reaction chamber to which is also added at a controlled and known flow rate a gas mixture to be analyzed. The reaction chamber is maintained at a sufficient pressure to support chemiluminescence, normally above about 0.1 torr and in the range of about 0.5 to 5.0 torr.

Photoelectric means is used to measure the chemiluminescence, and the photoelectric means is synchronized with the pulse discharge so that photoelectric means is turned off during the time of a pulse and for sufficient time thereafter to allow the light generated by the pulse discharge to disappear, whereby the only light the photoelectric means sees is chemiluminescence. The reaction zone and the oxygen generating means are compact and are located adjacent one another to allow miniaturization of the equipment.

A technique developed by R.T. Menzies (28) is based on the discovery that a laser can provide emissions which are absorbed by pollutants, such as NO, NO_2, N_2O_4, SO_2 and CO. The laser emission is used to resonantly excite the pollutant molecules and then to act as a local oscillator for a superheterodyne radiometer to detect the fluorescence occurring in response to the stimulation.

A process developed by R. Chand (29) is one whereby nitrogen oxide and sulfur dioxide concentrations present in a gaseous mixture are rapidly and continually monitored by measuring the current passing between an inert metallic sensing electrode and a counter electrode which electrodes are in contact with an aqueous electrolyte solution and at which sensing electrode the oxides are electrooxidized.

The sensing electrode is composed of an inert metal, whereas the counter electrode is composed of an electroactive lead sulfate which is electrochemically reduced when electrically interconnected with the sensing electrode in the presence of the aqueous electrolyte solution. A variable voltage source maintains the potential at the sensing electrode at the desired level.

References

(1) Environmental Protection Agency, "Air Quality Criteria for Nitrogen Oxides," *Publication No. AP-84,* Washington, D.C., U.S. Government Printing Office (January 1971).

(1a) Stern, A.C. in *Industrial Pollution Control Handbook,* H.F. Lund, Editor, McGraw-Hill Book Co. (1971).

(2) American Chemical Society, *Chemistry in 1972,* Washington, D.C., A.C.S. (1973).

(2a) Anonymous, *Chem. & Eng. News,* pages 13 to 16 (March 5, 1973).

(3) Altshuller, A.P., "Analytical Problems in Air Pollution Control," in NBS Special Publication 351, *Analytical Chemistry: Key to Progress on National Problems,* Washington, D.C., U.S. Government Printing Office (1972).

(4) Chand, R. and Marcote, R.V., *Development of Portable Electrochemical Transducers for the Detection of SO_2 & NO_x,* Contract CPA 22-69-118, U.S. Department of Health, Education and Welfare, Public Health Service, National Air Pollution Control Administration, Durham, North Carolina (September 1970).

(5) Mueller, P.K., Kothny, E.L., Pierce, L.B., Belsky, T., Imada, M. and Moore, H., *Anal. Chem.* 43, 1R (1971).

(6) Fontijn, A., Sabadell, A.J. and Ronco, R.J., *Anal. Chem.* 42, 575 (1970).

(7) Snyder, A.D. and Wooten, G.W., *Feasibility Study for the Development of a Multifunctional Emission Detector for NO, CO and SO_2,* Contract CPA 22-69-8, U.S. Department of Health, Education and Welfare, Public Health Service, National Air Pollution Control Administration, Durham, North Carolina (August 1969).

(8) Ronco, R.J. and Fontijn, A., *Prototype Continuous Chemiluminescent NO Monitor,* Contract CPA-70-79, Environmental Protection Agency, Air Pollution Control Office, Durham, North Carolina (May 1971).

(9) Iltis, R. and Mueller, J.C., *Oxides of Nitrogen Analyzer,* Contract CPA 70-70, Environmental Protection Agency, Air Pollution Control Office, Durham, North Carolina (June 1971).

(10) Lawrence Berkeley Laboratory, "Instrumentation for Environmental Monitoring - Air," *Publ. LBL-1,* Volume 1, Berkeley, University of California (February 1, 1973).

(10a) American Public Health Association, *Methods of Air Sampling & Analysis,* Washington, D.C. (1972).
(11) Driscoll, J.N. and Berger, A.W., *Report PB 209,268,* Springfield, Virginia, Nat. Tech. Info. Service (July 1971).
(12) Parts, L.P., Sherman, P.L. and Snyder, A.D., *Report PB 204,877,* Springfield, Virginia, Nat. Tech. Info. Service (October 1971).
(13) Brauer, L.W.; U.S. Patent 3,388,075; June 11, 1968.
(14) Rakowski, R.F.; U.S. Patent 3,574,552; April 13, 1971; assigned to the U.S. Secretary of the Air Force.
(15) Smith, A.B.; U.S. Patent 3,681,027; August 1, 1972; assigned to Resource Control, Inc.
(16) Lyshkow, N.A.; U.S. Patent 3,375,079; March 26, 1968; assigned to Precision Scientific Company.
(17) Smith, A.B., Hennig, R.R., Kurtz, A.J. and Pantchenko, G.; U.S. Patent 3,704,098; November 28, 1972; assigned to Resource Control, Inc.
(18) Schulze, F.; U.S. Patent 3,512,937; May 19, 1970.
(19) Harman, III, J.N. and Neti, R.M.; U.S. Patent 3,652,227; March 28, 1972; assigned to Beckman Instruments, Inc.
(20) Harman, III, J.N. and Neti, R.M.; U.S. Patent 3,677,708; July 18, 1972; assigned to Beckman Instruments, Inc.
(21) Lyshkow, N.A.; U.S. Patent 3,667,918; June 6, 1972; assigned to Pollution Monitors, Inc.
(22) Lyshkow, N.A.; U.S. Patent 3,712,792; January 23, 1973; assigned to Pollution Monitors, Inc.
(23) Neti, R.M. and Bing, C.C.; U.S. Patent 3,692,485; September 19, 1972; assigned to Beckman Instruments, Inc.
(24) Fontijn, A. and Vree, P.H.; U.S. Patent 3,713,773; January 30, 1973; assigned to Aerochem Research Laboratories, Inc.
(25) Williamson, Jr., J.A.; U.S. Patent 3,718,429; February 27, 1973; assigned to E.I. du Pont de Nemours and Company.
(26) Warnick, A. and Colvin, A.D.; U.S. Patent 3,746,513; July 17, 1973; assigned to Ford Motor Co.
(27) Wooten, G.W., Iltis, R. and Johnson, V.L.; U.S. Patent 3,749,929; July 31, 1973; assigned to Monsanto Research Corporation.
(28) Menzies, R.T.; U.S. Patent 3,761,715; September 25, 1973; assigned to California Institute of Technology.
(29) Chand, R.; U.S. Patent 3,763,025; October 2, 1973; assigned to Dynasciences Corporation.

ODOROUS COMPOUNDS

Offensive odors in the air are a major air pollution problem in some areas. These malodors cause many complaints, provoking emotional disturbances, mental depression and irritability. In some instances health effects such as nausea, vomiting, headache, loss of sleep, loss of appetite, and impaired breathing are induced. Contact with odorants may cause varying degrees of reactions in allergic individuals, particularly children, according to Sullivan (1).

Sociologically, odor pollution can interfere with human relations in many ways. It can damage personal and community pride, discourage capital investment and lower the socio-economic status of both the individual and the community. Some state, county and city governments have enacted laws that prohibit the emission of air pollutants which unreasonably interfere with the enjoyment of life and property. However, no odor standard has been established.

No information has been found on the effects of odor air pollution on animals. Odors per se have no effect on plants or materials. However, some odorants such as hydrogen sulfide and sulfur dioxide may affect animals, plants and materials.

The sources of odors are numerous and include pulp and paper mills, animal rendering plants, sewers and sewage treatment plants, garbage dumps and incinerators, chemical plants, petroleum refineries, metallurgical plants and internal combustion engines, particularly diesel and aircraft engines. The most offensive odors come from plants or processes which produce low molecular weight sulfur and nitrogen compounds, such as ethyl and methyl mercaptans, hydrogen sulfide, ammonia and dimethylamine. Environmental air concentrations of obnoxious odorants frequently exceed the odor threshold concentration in some local areas, and the odor has on occasions been recognized 20 miles from the source. The most generally accepted method of abatement of odors is incineration at the source. However, improper incineration may in itself be a source. Other abatement methods include adsorption, absorption, particulate removal, source elimination, process changes, chemical control, containment, odor masking, odor counteraction, biological control and dilution.

Economically, noxious odors may stifle the development and growth of a community. Both people and industry desire to locate in a place where it is pleasant to work, live and play. Tourists shun polluted areas, and rental and real estate property values may decrease. The control of odor pollution is often very costly to an industry, depending on the odor problem and the type of industry. This cost may be reduced by economic benefits derived from recovered heat or waste products. About $75 million have been spent for air pollution control in the kraft paper industry alone. The general topic of *Odors and Air Pollution* has been the subject of a comprehensive bibliography with abstracts prepared by the Environmental Protection Agency (2).

Measurement in Air

The overall topic of measurement of odors in air has been discussed by Cooper (3). Methods of odor analysis may be divided into two groups: organoleptic and chemical or instrumental as noted by Sullivan (1). The organoleptic methods, which rely on detection with the human nose, are completely subjective, but other methods are available to convert the subjective measurements into some meaningful objective results. Chemical or instrumental methods for analyzing odorants, which are numerous, usually suffer from lack of sensitivity. Sensitive noses can detect odors in quantities impossible to identify and monitor with commercially available instrumentation or chemical methods.

Sampling Methods: Samples may be collected in 250 ml. Pyrex gas collecting tubes. The air sample is aspirated with a rubber squeeze bulb into the tube and isolated with stopcocks at both ends of the tube (4)(5).

Qualitative Methods: Only the nose can measure odor quality, and even then, results are strictly qualitative. The odor surveys that have been conducted are examples of qualitative odor analyses. In these surveys, panelists have been asked to sniff ambient air or air samples and describe the odor quality, strength and acceptability (4)(5).

Quantitative Methods (Organoleptic): The most common method used is the vapor dilution technique. With this method, a sample is usually taken at the sampling station (in ambient air, a plant wastegas stream, or any other desired sampling point) with a gas sampling tube. The sample is then returned to the laboratory, where it is diluted, usually by means of a syringe, and presented to a panel of observers for evaluation of the odor threshold dilution (4)(6)(7)(8).

A modification of this method is the syringe dilution technique. The sample is collected in a syringe and removed in part to another syringe for dilution to produce a test dilution for human appraisal. Sensitivity limits this method to use with nonambient odors. However, it has the advantage of being simple and easily portable. Benforado et al (4) consider removal of the samples to the laboratory for analysis an advantage, but Gruber et al (5) believe this to be a disadvantage.

The vapor dilution method may be a static method, a continuous method, or a volatilization technique. Some instruments that have been based on the vapor dilution method, using the human nose as the detector, are listed as follows: (1) static method - Checkovich-Turner Osmometer (7); Barail Osmometer (9); Elsberg-Levy Olfactometer (10)(11); and Fair-Wells Osmoscope (12); (2) continuous method - Allison-Katz Odorimeter (13); Zwaardemaker Olfactometer (14); Procter and Gamble Osmo (15); Scentometer (5); and Nader Odor Evaluation Apparatus (16); (3) volatilization technique - Flask Dilution Method (17); and Enclosed Sniff-Blotter Technique (7).

Of these instruments, the scentometer requires special mention because it is portable, inexpensive and requires only one man for its operation. However, this last advantage may become a disadvantage when it is desirable to have the opinion of an odor panel rather than a single person. The instrument has several ports which allow air to pass through activated charcoal to provide clean air for dilution with the odorous air sample. By opening and closing the ports, the operator can adjust the dilution threshold concentration. Moreover, he can breathe clean air to allow his nose to recover from olfactory fatigue, the main problem associated with sniffing.

Other methods, based on such properties as vapor adsorption, liquid dilution and diffusion, are the following. The vapor adsorption and breakthrough method is based upon the time required for odor to break through an adsorber column of known volume. The Moncrieff Adsorption Unit is based on this technique (18). The liquid dilution method uses an odorless solvent to dilute the odorous material and the human appraisal is made on either the flask of diluent or on fractions of the diluent. The Elsberg-Levy Olfactometer (10)(11) and Foster-Smith-Scofield Stimulator (13) use this technique.

The rate of diffusion method requires the odorant to be placed on an adsorptive surface at the end of a diffuser column which encloses odorless, static air. Rates of diffusion may be measured by determining the time required by the odorant to diffuse through the full length of the tube as in the Ramsey Unit (19) or the diffusion time may be detected as the odorant passes sniff ports along the length of the tube as in the Snell Laboratory Air Force Unit (20).

Turk (21) has described a method for determining the intensity and character of diesel exhaust odors. In this method an odor panel is screened by giving each person a triangle test and intensity test. The triangle test consists of allowing each person to sniff five sets of three samples. Two of three samples are identical, while the third is different. He must detect which sample is different. The intensity test requires the person to rank in intensity a solution of odorant with a series of dilutions of the same odorant.

Panelists selected are then asked to compare diesel exhaust gases with standards. In Turk's method, the standards were 32 liquids contained in polyethylene bottles. The head gas expelled by squeezing the bottles served as the reference odors. Overall exhaust odor intensity was rated on a 1 to 12 scale, and the qualities burnt, oily, pungent and aldehyde-aromatic were each rated on a 0 to 4 scale representing the following intensities: none, slight, moderate, strong and extreme.

Duffee (22) reports that Battelle Institute has developed a sniff kit for rendering plant odors. Methyl disulfide is presented to a human odor panel at five concentrations, ranging from 0.001 to 10%, for comparison with the rendering odors. He claims that the kit may be used by a single untrained observer to determine the effectiveness of odor control systems for rendering odorants or to compare odorant sources within or between plants.

Quantitative Methods (Instrumental): Gas chromatography has been exploited by several investigators (23)(24)(25)(26) as a means of measuring odorants in the range of the odor threshold concentration of the mercaptans. Applebury and Schaer (23) have reported successful results. They used a 40 ml. sample and a Porapak Q column (¼" x 6') at 90°C. The detector was a coulometric cell with platinum electrodes similar to a design recommended by Adams et al (27). The reported minimum detectable concentrations were the following: 0.1 ppm or 150 μg./m.^{3} for hydrogen sulfide; 0.5 ppm or 1,000 μg./m.^{3} for methyl mercaptan; and 0.5 ppm or 650 μg./m.^{3} for sulfur dioxide.

Stevens et al (28) have developed a gas chromatography method which they claim can be used to determine the concentration of sulfur dioxide and other odorous gases produced in kraft paper mills. Polyphenyl ether was coated (4%) on 30 to 40 mesh Teflon powder containers and packed in 24 feet of Teflon tubing. A small amount (0.05%) of phosphoric acid was also added. It was found that sulfur dioxide, hydrogen sulfide, methyl mercaptan, and carbon disulfide could be separated by this column with very little loss. A flame photometric detector was used to measure concentrations down to 0.01 ppm. An evaluation of measurement methods and instrumentation for odorous compounds from stationary sources has been made by H.J. Hale of Esso Research and Engineering Company (29). As pointed out by Hall and Salvesen (29), the instrumental measurement of odorants rather than odors is a valid approach only when there is prior agreement as to what odorants are primarily responsible for odor. No reliable correlations exist in the general case between odor and the physical or chemical properties of odorants, and the extent to which satisfactory correlations have been made to date varies greatly from one industry to another. This was examined for three specific industries which have been the subject of major odor complaints: kraft pulp mills, petroleum/petrochemical refineries and animal rendering plants.

The degree to which instrumentation is available for field use in emission measurements can be related directly to the complexity of the odorant mixture to be measured. The nature of the odorant mixture in kraft mill emissions is well defined, and instrumentation to measure it is commercially available. The situation in petroleum refineries is more complex: a part of the odor problem is in stack emissions which can be instrumented directly, but another part is due to small amounts of material from diffuse sources which can only be measured at present by ambient methods. The odor problem in animal rendering plants is even less well defined: there is no agreement yet as to the identity of the specific odorants most responsible for the odors observed, and instruments for their detection and measurement are just now being developed.

The odorants emitted in kraft pulping consist primarily of four reduced sulfides: H_2S, methyl mercaptan, dimethyl sulfide and dimethyl disulfide. The H_2S and methyl derivatives are considered equally objectionable in odor complaints and they are frequently measured together as total reduced sulfide (TRS). They are present in varying proportions in different streams, together with moderate amounts of SO_2 which is not malodorous.

The literature provides ample evidence that these four are the key components both for odor emissions and for process changes to control them. The much smaller amounts of higher sulfides which are present have a secondary effect and they respond similarly to the methyl homologs in odor controls.

Petroleum/petrochemical refinery emissions also contain H_2S and mercaptans, as key components. The problem of odorant measurements is much different from kraft mills, however, since a primary object of petroleum refining is the removal and disposal of large amounts of sulfur present in the original feed stock. The refining process handles many streams rich in H_2S and other organic sulfides. These are converted to nonodorous products, chiefly elemental sulfur and SO_2.

The principle emission point for residual sulfides in the refinery is the burner stack of the sulfur plant. This normally consists chiefly of SO_2 and other combustion gases. It may contain H_2S or other sulfides during periods of upset conditions. Small amounts of carbonyl sulfide are also present, and they normally exceed the amount of H_2S. The sulfur plant in a modern refinery processes all collectible gas streams to remove residual H_2S and sulfides, before the gas is vented to the atmosphere.

The odorants emitted from small diffuse sources throughout the plant are more varied in composition. Key components include H_2S, alkyl and aromatic mercaptans, all of which will be included in an ambient measurement of total reduced sulfur. The massive amounts of SO_2 compared to reduced sulfur in refinery emissions are a complicating factor, and the contribution of COS requires further study.

The odorant picture for animal rendering plants is much less clear. The composition of the odorants released is directly related to the nature and quality of the material being processed. Odorants such as trimethylamine which are highly characteristic of the emissions from a plant processing feathers, hair or fish meal are minor components or missing in plants processing animal flesh. Other objectionable amines such as skatole, putrescine and cadaverine are present only if the feed to the plant is not properly washed, or if it contains badly decayed flesh, which can be avoided. The odorants recognized as generally characteristic of animal waste rendering include aliphatic aldehydes, free fatty acids, ammonia and variable amounts of amines and H_2S depending on the specific feed stock. The laboratory instruments available for the analysis of these recognized odorants are reviewed herein, to indicate preferred approaches for a prototype for field use.

The continuous automatic instrumentation available for the direct measurement of total reduced sulfide depends upon two types of sensors: bromine coulometric titration, and the flame photometric detector. Coulometric titration can be carried out continuously in a flowing stream, or in a microcoulometer which accepts very small discrete samples such as the fractions eluted from a gas chromatographic column.

Both approaches are very useful for differential measurements, but they are subject to difficulties with zero stability which require frequent calibration if absolute values are desired. Coulometric cells can also be based on titration with iodine or with silver ions, which respond to different portions of the total sulfur present.

Flame photometry responds equally to all sulfur atoms. It can be used as a total measurement, or more desirably in a GC/FPD combination. This permits a distinction between S types such as H_2S, SO_2, total S, or some other selected component such as methyl mercaptan. Automatic programmed equipment for such measurements is available. Continuous measurements can also be made using lead acetate tape samplers or similar automatic instruments which are sensitive only to H_2S, or to H_2S and mercaptans as key components.

An alternate approach is to convert all sulfur compounds present to SO_2, with or without a preliminary separation of the SO_2 initially present, and use this as an analysis of sulfides in the sample. This combination has been applied particularly with a conductivity cell or a microcoulometer for the detection of SO_2. It could also be used with any of the very large number of SO_2 detection systems available. The analogous approach of catalytically or thermally converting all sulfur compounds to H_2S has also been used, particularly with the silver cell microcoulometer or lead acetate tape as the sensor for H_2S.

Wet chemical methods which consume reagents are considered intrinsically less desirable for field measurements than the use of electrical or optical sensors, but they are much more flexible when analyses of varying composition are required. Most of the standard methods for product quality measurements in petroleum refining are based on wet chemistry.

Many of these measurements can be automated, if enough samples are to be measured in a given time. Nevertheless, they have been excluded as far as kraft mills and refineries are concerned in this review, which is aimed primarily at the evaluation of physical methods of measurement. Automated wet chemistry may well be a preferred approach to the analysis for odorants in animal rendering plants, where both qualitative and quantitative base data are still required.

Measurement in Water

The problem of odor and odor detection with particular reference to wastewaters has been discussed by K.H. Mancy (30). He pointed out that odor is a human physiological response to odorant volatile matter that stimulates smell in man. Determination of odor is based solely on the olfactory senses of the analyst, or on those of a group of individuals, and on the ability of the analyst (or group) to distinguish between different levels and kinds of odors. The testing is based entirely on arbitrary comparison since no absolute units or base for odor exist (31)(32).

Several authors have attempted to characterize and classify the origin of odor in natural and wastewaters [(32) through (37)]. Most of these studies treat taste and odor as closely connected human responses. Taste determinations using human subjects are generally not recommended especially in cases of wastewater or untreated industrial effluents, and thus are excluded from this discussion.

Odor can always be related to the presence of volatile organic and/or inorganic species present in water. Odor intensity is a function of the volatility and the concentration of the odor-causing species, as well as of certain environmental factors such as temperature, ionic strength and pressure. It has been claimed that there are only four basic types of odor: sweet; sour; burnt; and goaty, realizing that the many odors are in fact combinations of two or more of these groups.

Odors often can be related to the presence of certain biological forms in the wastewater, such as algae and actinomycetes. Such odor-causing organisms are believed to secrete characteristic volatile oils during growth, and upon decomposition and decay. Such poetic terms as musty, earthy, woody, moldy, swampy, grassy, fishy and wet leaves have been used to describe odors (31)(38).

Recent studies of odor characteristics and human response have led to a proposal of a steriochemical theory of odor (33)(38). This theory relates the response to odor to the geometry of molecules. It has been postulated that the olfactory system is composed of receptor cells of certain different types, each representing a distinct primary odor, and that odorous molecules produce their effects by fitting closely into receptor sites on these cells. This concept is similar to the lock and key theory used to explain certain biochemical reactions; e.g., enzyme with substrate, antibody with antigen, and desoxyribonucleic acid with ribonucleic acid in protein synthesis.

Seven primary odors are distinguished (39), each of them by an appropriately shaped receptor at the olfactory nerve endings. The primary odors, together with reasonably familiar examples are (a) camphoraceous, e.g., camphor or moth repellent; (b) musky, e.g., pentadecalactone as in angelica root oil; (c) floral, e.g., phenylethyl methyl ethyl carbinol as in roses; (d) pepperminty, e.g., methone as in mint candy; (e) pungent, e.g., acetic acid as in vinegar; and (g) putrid, e.g., butyl mercaptan as in rotten eggs. A classification of odor by chemical type is shown in Table 37 (34). It has been claimed that every known odor can be made by mixing the seven primary odors in certain combinations and proportions (34).

Odors resulting from mixtures of two or more odoriferous substances are extremely complex. The mixture may produce an odor of greater or lesser intensity than might be expected from summing the individual odors, or a completely different kind of odor may be produced (33)(34)(35). Accordingly, it is frequently necessary to characterize the odor of the wastewater and that of the receiving stream both separately and in combination if the actual relationship and effect are to be determined.

Odor intensity is expressed in terms of the threshold odor number (32)(33). By definition, the threshold odor number is the greatest dilution of the sample that still leaves a preceptible residual odor. The test procedure is based on successive dilution of a sample with odor-free water, disregarding any suspended matter or immiscible substances, until a dilution is obtained which has a barely perceptible odor. It has been recommended that odor tests be run at 25° and 60°C. or 40° and 60°C. (32).

In all cases the sampling and test temperature should be reported, since the threshold odor is a function of temperature. A given sample, under fixed conditions, will emit a characteristic odor stimulus, but the response to this stimulus and the judgement based upon this response are purely subjective matters, and their interpretation may vary considerably from individual to individual (33)(36)(37). Consequently, it is desirable to use a panel or group of judges, rather than a single analyst for both qualitative and semiquantitative evaluation of odors in water or wastewater samples (31).

Quantitative measurement of odor-causing matter seems to be the only exact approach for odor measurement. Gas chromatographic procedures have been used to identify and measure the quantities of odor-causing substances in water samples (38). A positive identification is accomplished once a correlation is established between the isolated substance and its human odor sensation. Additional research work is needed to reduce these techniques to a practical level for day to day operations and automatic measurement procedure applicable for monitoring purposes.

TABLE 37: ODORS CLASSIFIED BY CHEMICAL TYPES

Odor class	Chemical types included	Odor characteristics				
		Fragrance	Acidity	Burntness	Caprylicness	Algae and fungi
Estery	Ethers Esters Lower ketones	high	medium	low to medium	medium	----
Alcoholic	Phenols and cresols Alcohols Hydrocarbons	high	medium to high	low to high	medium	Asterionella Coelosphaerium
Carbonyl	Aldehydes Higher ketones	medium	medium	low to medium	medium	Mallemonas
Acidic	Acid anhydrides Organic acids Sulfur dioxide	medium	very high	low to medium	medium	Anabaena
Halide	Quinones Oxides (including ozone) Halides Nitrogen compounds	High	medium to high	medium to high	low to high	Dinobryon Actinomycetes
Sulfury	Selenium compounds Arsenicals Mercaptans Sulfides and hydrogen sulfide	medium	medium	very high	very high	Aphanizomenon
Unsaturated	Acetylene derivatives Butadiene Isoprene Vinyl monomers	high	medium	medium	high	Synura
Basic	Higher amines Alkaloids Ammonia and lower amines	high	medium	low to medium	high	Uroglenopsis Dinobryon

Source: G.M. Fair; *J. New England Water Works Association;* 1933.

References

(1) Sullivan, R.J., "Air Pollution Aspects of Odorous Compounds," *Report PB 188,089,* Springfield, Virginia, Nat. Tech. Info. Service (September 1969).

(2) Environmental Protection Agency, "Odors and Air Pollution," *Publ. No. AP-113,* Research Triangle Park, North Carolina, Office of Air Programs (October 1972).

(3) Cooper, Jr., H.B.H., *Hydrocarbon Processing,* pages 97 to 101 (October 1973).

(4) Benforado, D.M., Rotella, W.J. and Horton, D.L., "Development of an Odor Panel for Evaluation of Odor Control Equipment," *J. Air Pollution Control Assoc.* 19(2): 101 (1969).

(5) Gruber, C.W., Jutze, G.A. and Huey, N.A., "Odor Determination Techniques for Air Pollution Control," *J. Air Pollution Control Assoc.* 10(4): 327 (1960).

(6) *Manual on Sensory Testing Methods,* American Society for Testing Materials Special Tech. Publ. No. 434 (May 1968).

(7) Matheson, J.F., "Olfactometry: Its Techniques and Apparatus," *J. Air Pollution Control Assoc.* 5(3): 167 (1955).

(8) "Standard Method for Measurement of Odor in Atmosphere (Dilution Method)," American Society for Testing Materials Standard Method D 1391-57.

(9) Barail, L.C., "Odor Measurement," *Soap and Sanitary Chem.* 25(6):133 (1949).

(10) Elsberg, C.A. and Levy, I., "The Sense of Smell, I. A New and Simple Method of Quantitative Olfactometry," *Bull. Neurol. Inst.* 4:5 (1935).

(11) Elsberg, C.A., Levy, I. and Brewer, E.D., "A New Method for Testing the Sense of Smell," *Science* 83:211 (1936).

(12) "Odor Determinations on a Numerical Basis ... The Fair-Wells Osmoscopes," *Bull. No. 524,* Eimer & Amend, New York, N.Y.

(13) Foster, D., Smith, L.A. and Scofield, E.H., "A New Dirhinic Olfactory Stimulator," *Am. J. Psychol.* 60:272 (1947).

(14) Zwaardemaker, H., *Fortschritte der Medicin* 19:721 (1889).

(15) Heller, H., "A Critical Discussion of Teudt's Theory," *Am. Perfumer* 14:365 (1920).

(16) Nader, J.S., "An Odor Evaluation Apparatus for Field and Laboratory Use." Presented at the 1957 Annual Meeting of the American Industrial Hygiene Association, St. Louis, Missouri.
(17) Jerome, E.A., "Olfactory Thresholds Measured in Terms of Stimulus Pressure and Volume," *Arch. Psychol.* 39(274):5 (1942).
(18) Moncrieff, R.W., "The Characterization of Odors," *J. Physiol.* 125:453 (1954).
(19) Ramsey, W., *Nature* 26:187 (1882).
(20) Gee, A.H., *Organoleptic Appraisal of Three Component Mixtures,* American Society for Testing Materials Reprint 105b (1954).
(21) Turk, A., "Selection and Training of Judges for Sensory Evaluation of the Intensity and Character of Diesel Exhaust Odors," City College of the City of New York. *U.S. Public Health Service Publ. 999-AP-32* (1967).
(22) Duffee, R.A., "Appraisal of Odor-Measurement Techniques," *J. Air Pollution Control Assoc.* 18(7): 472 (1968).
(23) Applebury, T.E. and Schaer, M.J., "Analysis of Kraft Pulp Mill Gases by Process Gas Chromatography," Department of Chemical Engineering, Montana State University, Bozeman, Montana (1968).
(24) Brooman, D.L. and Edgerley, Jr., E., "Concentration and Recovery of Atmospheric Odor Pollutants Using Activated Carbon," *J. Air Pollution Control Assoc.* 16:25 (1966).
(25) Feldstein, M., Balestrieri, S. and Levaggi, D.A., "Studies on the Gas Chromatographic Determination of Mercaptans," *J. Air Pollution Control Assoc.* 15(5):215 (1965).
(26) Sableski, J.J., "Odor Control in Kraft Mills, A Summary of the State of the Art," U.S. Public Health Service, National Center of Air Pollution Control, Cincinnati, Ohio (May 10, 1967).
(27) Adams, D.F., et al, "Improved Sulfur Reacting Microcoulometric Cell for Gas Chromatography," *Anal. Chem.* 38(8):1094 (1966).
(28) Stevens, R.K., O'Keeffe, A.E., Mulik, J.D. and Krost, K.J., "Gas Chromatography of Reactive Sulfur Gases in Air at the Parts-Per-Billion Level. 1. Direct Chromatographic Analysis," National Air Pollution Control Administration, Cincinnati, Ohio (1969).
(29) Hall, H.J. and Salvesen, R.H., "Evaluation of Measurement Methods and Instrumentation for Odorous Compounds in Stationary Sources, I. State of the Art," *Report PB 212,812,* Springfield, Virginia, Nat. Tech. Info. Services (July 1972).
(30) Mancy, K.H. in *NBS Spec. Publ. 351,* "Analytical Chemistry - Key to Progress on National Problems," Washington, D.C., U.S. Government Printing Office (1972).
(31) Amer. Publ. Health Association, Amer. Water Works Association and Water Pollution Control Federation, *Standard Methods for the Examination of Water and Wastewater,* 12th Edition (1965).
(32) American Society for Testing Materials, *Manual on Industrial Water and Industrial Wastewater,* 2nd Edition, Philadelphia, Pennsylvania (1964).
(33) Baker, R.A., *J. Water Pollution Control Fed.* 34, 582 (1962).
(34) Fair, G.M., *J. New England Water Works Association* 47, 248 (1933).
(35) Rosen, A.A., Skeel, R.T. and Ettinger, M.B., *J. Water Pollution Control Fed.* 35, 777 (1963).
(36) Rosen, A.A., Peter, J.B. and Middleton, F.M., *J. Water Pollution Control Fed.* 34, 7 (1962).
(37) Spaulding, C.H., *J. American Water Works Association* 34, 877 (1942).
(38) Gerald, F.A., *The Human Senses,* New York, Wiley (1953).
(39) Cleary, E.J., *J. American Water Works Association* 54, 1347 (1962).

OIL AND GREASE

Oily waste materials in wastewaters are measured in terms of their hexane solubility for purposes of pollution evaluation, according to Patterson and Minear (1). Hexane is an organic solvent employed to separate oily organic compounds from wastewaters. Oily wastes include greases, as well as many types of oils. Grease is not a specific chemical compound, but a rather general group of semiliquid materials which may include fatty acids, soaps, fats, waxes and other similar materials extractable into hexane.

Unlike some industrial oils which represent precise chemical composition, greases are, in effect, defined by the analytical method employed to separate them from the water phase of the waste (2). A waste treatment manual recently published by the American Petroleum Institute (3) suggests the following classification for types of oily wastes.

(1) Light hydrocarbons - including light fuels such as gasoline, kerosine and jet fuel, and miscellaneous solvents used for industrial processing, degreasing

or cleaning purposes. The presence of waste light hydrocarbons may make removal of other, heavier oily wastes more difficult.

(2) Heavy hydrocarbons, fuels and tars - includes the crude oils, diesel oils, No. 6 fuel oil, residual oils, slop oil, asphalt and road tar.

(3) Lubricants and cutting fluids - oil lubricants generally fall into two classes; nonemulsifiable oils such as lubricating oils and greases, and emulsifiable oils such as water-soluble oils, rolling oils, cutting oils and drawing compounds. Emulsifiable oils may contain fat, soap, or various other additives.

(4) Fats and fatty oils - these materials originate primarily from processing of foods and natural products. Fats result from processing of animal flesh. Fatty oils for the most part originate from the plant kingdom. Quantities of these oils result from processing soybeans, cottonseed, linseed and corn.

There are many industrial sources of oily wastes. Table 38 presents the major categories of industry producing oil and grease laden waste streams, and lists characteristic types and sources of oily wastes associated with each category. Table 39 summarizes reported oily waste concentrations for many of these industries. By far the three major industrial sources of oily waste are petroleum refineries, metals manufacture and machining, and food processors. Petroleum refineries produce large quantities of oil and oily emulsion wastes.

TABLE 38: INDUSTRIAL SOURCES OF OILY WASTES

Industry	Waste Character
Petroleum	Light and heavy oils resulting from producing, refining, storage, transporting and retailing of petroleum and petroleum products.
Metals	Grinding, lubricating and cutting oils employed in metal-working operations, and rinsed from metal parts in clean-up processes.
Food Processing	Natural fats and oils resulting from animal and plant processing, including slaughtering, cleaning and by-product processing.
Textiles	Oils and grease resulting from scouring of natural fibers (e.g., wool, cotton).
Cooling and Heating	Dilute oil-containing cooling water, oil having leaked from pumps, condensers, heat exchangers, etc.

Source: American Petroleum Institute; *Industrial Oily Waste Control*

TABLE 39: CONCENTRATIONS OF OILY MATERIAL IN INDUSTRIAL WASTEWATERS

Industrial Source	Oily Waste Conc., mg/l	Reference
Petroleum Refinery*	40-154	6
Petroleum Refinery*	35-178	7
Petroleum Refinery*	20	8
Steel Mill		
Hot Rolling	20	3
Cold Rolling	700 (500 free oil)	9
Cold Rolling	60-500	10
Cold Rolling Coolant	2,088-48,742 (2,035-36,664 free oil)	11
Cold Rolling Rinse	113-3,034 (83-2,284 free oil)	11
Food Processing	3830	12
Food Processing (Fish)	520-13,700	13
Metal Finishing	100-5,000	14
Metal Finishing	665	15
Oil Field Brine	25-50	16
Paint Manufacture	1900	12
Aircraft Maintenance	500-1200 (250-500 free oil)	12

* API oil separator effluent.

Source: J.W. Patterson and R.A. Minear; *Report PB 216,162;* February 1973

Because of the long history of pollution problems associated with petroleum refining, the American Petroleum Institute (API) has exerted a good deal of effort in developing and publishing methods of oily waste control (3)(4). Nevertheless, comprehensive federal industrial waste profile on the petroleum refining industry reported no values for oil content of refinery wastewater, and stated that "... data concerning the amounts of oil ... (in wastewater) ... are not complete enough to justify inclusion ..." in that report (5). Indicative of the oil content of refinery waste, however, is a reported value of 154 mg./l. of oil, after preliminary skimming to partially remove floating oils (6).

In the metals industry the two major sources of oily wastes are steel manufacture and metal working. Oily wastes include both emulsified and nonemulsified or floating oils. In steel manufacture, steel ingots are rolled into desired shapes in either hot or cold rolling mills. Oily wastes from hot rolling mills contain primarily lubricating and hydraulic pressure fluids. In cold strip rolling, however, the steel ingot is usually oiled prior to rolling, to lubricate and to reduce rusting. Additional oil-water emulsions are sprayed during rolling to act as coolants.

After shaping, the steel is rinsed to remove the adhering oil. Rinse and coolant waters from the cold rolling mills may contain several thousand mg./l. of oil, of which 25% or more may be emulsified and thus difficult to separate from the wastewater. Emulsified oil in hot rolling mill effluents rarely exceeds 20 mg./l. (3). More concentrated oily wastes, such as from batch dumps of spent coolant or lubricating fluid must also be treated. Metal working produces shaped metal pieces such as pistons and other machine parts. Oily wastewaters from metal working processes contain grinding oils, cutting oils, and lubrication fluids. Coolant oil-water emulsions are also employed in many metal working processes. Soluble and emulsified oil content of wastewaters may vary from 100 to 5,000 mg./l. (14).

The third major source of oily wastes and particularly greases, is the food processing industry. In the processing of meat, fish and poultry, oily and fatty materials are produced primarily during slaughtering, cleaning and by-product processing. The major grease sources are the rendering areas, in particular from the wet (or steam) rendering process which gives the highest levels, pound per pound of scrap processed, of hexane extractables in food processing waste streams (3). Grease content in meat packinghouse waste streams may run several thousand mg./l. (17), and it has been reported that waste from a fish processing plant contained 520 to 13,700 mg./l. of fish oil (13). The maximum allowable concentration of oil in an effluent to be discharged to a storm sewer or stream should be 10 ppm.

Reverting now from oily and greasy wastewaters from industrial processing to a consideration of oil spills, one finds that the grounding of the supertanker Torrey Canyon (18)(19) off the coast of England with the release of 30 million gallons of crude oil on March 18, 1967, set in motion a chain of events which constitutes a watershed in man's awareness of the hazards of aquatic oil pollution and his concern for its prevention and control (20).

On May 28, 1967, President Johnson directed the Departments of Interior and Transportation to examine how the resources of the nation could best be mobilized against oil spill pollution. As a result, *Oil Pollution - A Report to the President* was issued in February 1968 (21). Oil spilled on water was characterized as "... one of the most devastating substances in the environment..." Of even greater concern, the country was found "... not fully prepared to deal effectively with spills of oil or other hazardous materials, large or small, and much less with a Torrey Canyon type disaster..."

Since issuance of the above report, some progress has been made in the detection of oil spills (22)(23)(24), the identification of the source of the spill [(25) to (30)], and the cleanup of oil spilled [(31) to (43)]. More progress will be required, particularly in prevention and cleanup, before our aquatic environment can be assured of minimal adverse effects from oil pollution [(19)(44) to (48)].

The problems created by the Torrey Canyon and other large disasters are fairly obvious and more readily studied than the immense number of daily small discharges of oil to the aquatic environment. So long as the developed nations of the world, particularly the U.S., depend upon petroleum to fuel and lubricate their technologically oriented societies, the problem of controlling aquatic oil pollution will exist.

A survey of water quality instrumentation related to oil pollution has been published by the Lawrence Berkeley Laboratory of the University of California (49). The objective of this survey is to describe analytical instrumental procedures suitable for detecting and characterizing aquatic oil pollution sufficiently well to trace the pollution to its source for remedial action and impact study.

Most of the attention to the problems of oil in water since Torrey Canyon, concerns oil pollution on the open seas and coastal areas (owing to heavy maritime petroleum traffic and increased offshore exploration and production). The scope of this survey of water quality instrumentation is not directly involved with marine applications; however, since much of the technical discussion in the technical and mass media on marine oil pollution also applies to oil pollution of inland waterways, e.g., estuaries, lakes, canals, rivers, etc., material concerning marine oil pollution is also included.

A number of commendable references are available. An excellent, general discourse on oil pollution appeared in the second edition of *Water Quality Criteria* (50). More recently a number of conferences on oil pollution have been held in this country. The proceedings have been published (31)(32)(33)(51). Other informative general references include publications of the Federal government (21)(46)(52)(53) and bibliographies [(54) to (59)].

The effects of aquatic oil pollution on man himself are quite minimal for the simple reason that he finds the prospect of oily water so aesthetically displeasing that he avoids personal contact with it. For further information see the National Technical Advisory Committee on Water Quality Criteria Report of 1968, under the Subcommittee for Recreation and Aesthetics (53).

Contamination of domestic water supplies in particular must be avoided. The Subcommittee for Public Water Supplies (53) made the following statement: "... It is very important that water for public water supply be free of oil and grease. The difficulty of obtaining representative samples of these materials from water makes it virtually impossible to express criteria in numerical units. Since even very small quantities of oil and grease may cause troublesome taste and odor problems, the Subcommittee desires that none of this material be present in public water supplies ..."

Aesthetic considerations predominate over toxicological considerations, as the following excerpt (references omitted) from McKee and Wolf (50) serves to illustrate: "... An exhaustive search of literature to determine the toxic components of refinery wastewaters toward humans revealed a paucity of toxicological data on the ingestion of such substances by humans or by test animals. From the approximate order of magnitude based on data from the fields of occupational health and industrial hygiene, it was concluded that any tolerable health concentrations for oily substances far exceed the limits of taste and odor. It appears, therefore, that hazards to human health will not arise from drinking oil polluted waters, for they will become esthetically objectionable at concentrations far below the chronic toxicity level..."

Removing oil contaminants from domestic water supplies may be costly (50): "... Floating or emulsified oil in raw water supplies, will complicate the coagulation, flocculation, and sedimentation processes at a treatment plant. Oil coated floc may not settle properly. If free or emulsified oil reaches sand filters or ion exchange beds it will coat the grains, decrease the effectiveness of filtration and interfere with backwashing. Taste- and odor-producing compounds will require the greater use of activated carbon or heavy chlorination. Again, however, the taste and odor factor will control the threshold or limiting concentration of oily material acceptable in a domestic water supply..."

The Subcommittee on Water Quality for Agricultural Uses (53) made no specific recommendation in regard to contamination by oily substances. Some data are available on agricultural effects due to oil. Limited application of crude petroleum has been reported to improve rather than interfere with crops (50). Farm animals have been reported to develop a fondness for crude oil and suffer adverse effects, including death, as a result (60). An effluent standard of 30 ppm of oil emulsified in water was set by the Ohio Department of Health for a creek, due to its use by grazing cattle (61).

The Subcommittee for Water Quality Requirements for Industrial Water Supplies noted (53): "... Steam generation and cooling are unique water uses in that they are required in almost every industry..." Both functions are adversely affected by contamination with oil (50): "... In steam production, the presence of oil in boiler feed water may cause foaming, priming, overheating of tubes resulting in blistering or failure, and poor transmission of heat from the metal to the water ..."

As a result, the American Boiler Manufacturers Association in its standard guarantee on steam purity specifies that "... the total quantity of oil or grease, or substance which is extractable either by sulfuric ether or by chloroform, shall not exceed 7 ppm in the boiler water when the sample being tested is acidified to 1% hydrochloric acid, or 7 ppm in the feed water when the sample being tested is first concentrated at low temperature and pressure to the same ppm total solids as the boiler water..." Additional examples of adverse industrial effects of oily water are reported (50).

Examples of wildlife flora and fauna which have benefited from the release of oily substances into their environment are few indeed. They appear to be limited to bacteria and fungi which feed on components of the oily matter, or to kelp which flourished apparently as a result of oil implicated extermination of natural predators (62). Otherwise, evidence is abundant that contamination of the aquatic environment with oily material is an adverse event, particularly for fauna. Numerous graphic examples are described in lay terms by Marx in his chapter "Oil vs. Marine Life" (45). A more technical review is presented by Revelle et al (47).

McKee and Wolf (50) summarize: "... Oil films may interfere with gas exchange, coat bodies of birds and fish, impart a taste to fish flesh, exert a direct toxic action on some organisms as a result of water-soluble components, and interfere with fish food organisms and the natural food cycle. Oil from surface films becomes adsorbed on clay particles, settles to the bottom, and there remains a source of pollution, for it may be stirred up to float again or may leach toxic principles..."

Most of the available toxicity data (50)(53) are reported as the median tolerance limit (TL_m), the concentration that kills 50% of the test organisms within a specified time span, usually in 96 hours or less. In many cases, the differences are great between TL_m concentrations and concentrations that are low enough to permit reproduction and growth. Aromatic hydrocarbons are generally implicated as the toxic components present in petroleum and its products.

Pickering and Henderson (61) studied the toxicity of several important petrochemicals to fathead minnows, bluegills, goldfish and guppies in both soft and hard water. Standard bioassay methods were used. The petrochemicals tested were benzene, chlorobenzene, o-chlorophenol, 3-chloropropene, o-cresol, cyclohexane, ethyl benzene, isoprene, methyl methacrylate, phenol, o-phthalic anhydride, styrene, toluene, vinyl acetate and xylene. These were all similar in their toxicities to fish, with 96 hour TL_m values ranging from 12 to 368 mg./l.

Except for isoprene and methyl methacrylate, which were less toxic, values for all four species of fish for the other petrochemicals ranged from 12 to 97 mg./l., a relatively small variation. In general, o-chlorophenol and o-cresol were the most toxic. Cairns (63) reports the following 96 hour TL_m values of naphthenic acid for bluegill sunfish, 5.6 mg./l.; pulmonate snail, 6.1 to 7.5 mg./l. (in soft water); and diatom, 41.8 to 43.4 mg./l. in soft water and 28.2 to 79.8 mg./l. in hard water. The term "naphthenic acid" refers to a family of cyclohexane carboxylic acids present in petroleum.

Surface oil films interfere with the respiration of aquatic insects (a fact used for control of the mosquito). Oil is also a well-known hazard to waterfowl (21) by destroying the buoyancy and insulation of their feathers. The mortality rate of waterfowl which encounter oil slicks approaches 100%. The effects of marine oil contamination on flora and fauna have been even more striking. Several oil spills have been studied in extensive detail (18)(33)(48)(62).

Measurement in Water

In the previous section it has been shown that it is of utmost importance to detect, identify, and measure discharges of oil and grease into the aquatic environment to determine the nature of the offending material so that its source can be located and timely corrective measures instituted. This includes not only earliest possible termination of oil discharge, but also a thorough determination of the extent and amount of aquatic contamination to surface water, ground water, sediment, aquifer, fauna and flora to assess the environmental damage and the effectiveness of the cleanup.

Oil processing and handling facilities would also benefit from the development of oil detection and identification techniques to pinpoint sources of spillage or leakage to surface or ground water or to determine the efficiency of oil recovery from processing water prior to release to receiving waters. With the assistance of modern analytical chemical technology, considerable progress toward this goal has been made.

Revised standard manuals on the analysis of water and wastewater (64)(65)(66) describe elementary methods for sampling, extracting (if necessary) and weighing oily or greasy (insoluble organic) material which may be present in water or wastewater. Quantitation of such material is important in the treatment of sewage, because oily matter can interfere with bacterial degradation and plug carbon filters or render trickling ponds or sludge digesters ineffective. Quantitation is also essential to determine the levels of oil contamination dispersed throughout the aquatic environment.

Unfortunately, these elementary laboratory methods for measuring the bulk of insoluble organic matter in water and wastewater offer little insight into the identity of the source of an unobserved oily discharge or its probable environmental impact. To determine these aspects, it is necessary to develop information on the composition of the oil, i.e., its constituent organic compounds.

Because petroleum in particular is such a gross mixture of organic compounds, extensive detailed analysis of its composition has posed formidable difficulties in the past. In recent years a number of qualitative tests and analytical techniques have appeared in petroleum research and chemistry publications (67)(68)(69) which apply either directly or with modification to oil spill identification.

Because the various techniques employed in oil spill analysis are scattered throughout the technical literature, a standard manual setting forth analytical procedures in a generally accepted, systematized manner is a highly desired goal. As an initial response to this situation, a booklet entitled *Laboratory Guide for the Identification of Petroleum Products* (70) was issued in 1969 by the Federal Water Pollution Control Administration (now the Water Quality Office, EPA). Besides describing additional sampling techniques, it provides water pollution analysts with a series of specific analytical procedures. The purpose of these procedures is to characterize an unobserved oil spill sufficiently well to identify its probable source when compared with suspect samples.

Offering suggested procedures on a trial basis exposes these procedures to the broad, general usage necessary for their acceptance as standardized methods. Although the booklet is incomplete by current standards and regional EPA laboratories already employ alternate approaches, it constitutes a useful initial step toward devising a systematic analysis of oily material in the aquatic environment. Parallel efforts by the American Society for Testing and Materials (ASTM), Subcommittee D-19.10 are underway (71).

The choice of specific techniques from among the multiplicity of analytical procedures applicable to petroleum and its products depends on factors such as size, sampling methods, personal analytical preferences, and, perhaps most important, the availability of instrumentation. The more elementary bulk properties of an oil spill sample, such as density, viscosity, solubility in selected solvents, ash content, sulfur content, refractive index, distillation range, etc. can be used to characterize an unobserved oil spill for general classification purposes (70).

Gas chromatography (GC), atomic absorption spectrophotometry (AA), thin layer chromatography (TLC), and infrared spectroscopy (IR) are examples of moderately sophisticated laboratory instruments which have been applied successfully to oil spill source identification by EPA technical support laboratories, among others. These latter instruments play an even more important role when assessment of the environmental impact of oil spills is the goal, since the questions of "what sort of" and "how much" noxious or toxic organic compounds are present require investigation into the molecular composition of the oily mass.

Because crude petroleum and many of its refined products represent an extremely complex and variable mixture of chemical compounds, mostly organic, no single analytical technique presently available is capable of analyzing this material in toto. However, the still developing field of GC, in conjunction with ancillary techniques, most closely approaches this ideal; therefore, GC is the most important single technique. The sample requirements for a complex mixture like crude petroleum seldom exceed 10 mg. for a dual detector GC run. Generally GC analysis is complemented with additional analytical data in order to establish conclusively the identity of an oil spill sample from among several similar suspected samples. Atomic absorption spectrophotometry (AA), thin layer chromatography (TLC) and infrared spectroscopy (IR) are currently the most utilized of the sensitive complementary techniques. All four techniques are limited in field applications to laboratory batch-type analysis as opposed to continuous, stream monitoring.

Although perhaps desirable, the outlook for continuous, remote sensing, oil spill monitors out in the environment is not promising. This is due to a number of factors such as the tendency of oil to float on the surface where it is subject to wind drift, wave action, fluctuation with the water level, and adhesion to solid surfaces contacting it. Devices already suitable for refinery process monitoring (72)(73)(74) become prohibitively expensive when adaptations for general remote environmental monitoring are considered.

Airborne oil spill detection devices have been studied (75)(76)(77). It must be kept in mind, however, that unless the oil slick detected is obviously linked, either through direct, unambiguous contact or strongly suspected recent contact, to the source of the discharge, detailed laboratory comparative analysis between the spill samples and suspect samples will be necessary to identify the source.

Oil and grease may be monitored in water by the following instrumental techniques (49): atomic absorption spectrophotometry; gas chromatography; IR spectrophotometry reflectance sensors; thin layer chromatography; and UV spectrophotometry. Other schemes for oil/water pollution monitoring have been proposed. For example, ultrasonic wave attenuation in water as a function of degree of fuel oil concentration was studied by F.K. McGrath (78). The development of an airborne remote laser fluorosensor for use in oil pollution detection has been reported by R.M. Measures and M. Bristow of the University of Toronto (79).

A device suitable for the simple determination of the quantity of oil in a wastewater has been developed by C.J. Overbeck and J.C. Means (80). With increasing involvement in pollution control and wastewater analysis, there is an express need for a procedure for determining the amount of oil-in-wastewater, suitable for field use. For this to be convenient, the equipment must be sturdy and the procedure must be relatively quick. A detection limit of 10 ppm is desired. Several approaches have been tried for the analysis of oil-in-wastewater. Included are spectrophotometric methods, optical diffusion methods, gravimetric methods, volumetric methods, colorimetric methods and chemical methods. Most procedures involve solvent extraction.

Spectrophotometric methods depend on infrared or ultraviolet absorption. These approaches naturally require instrumentation and this method is incapable of detecting lower homologs of the aromatic series. Ultraviolet absorption has the advantage of being able to be done directly on the water sample. In addition, rugged, UV monitoring instruments are available for this purpose.

Dependence upon transmissions, however, subjects the method to a few interferences, primarily turbidity. Variations in the type of hydrocarbon present will also affect the results. For a fairly consistent type of water, the instrument has useful application, but for a wide variety of wastewater analysis, the accuracy of the results would be doubtful.

Other photometric methods depend on fluorescence. Measurement of fluorescence can be done with a fluorometer or by visual comparison with a standard. Oils, however, have a wide range of fluoroscence characteristics ranging in color and intensity. Again, a procedure based on this property can be devised for a specific water, but serious errors arise when attempting to do it to a wide variety of samples. Previous work done in this field indicated no reliable correlation between fluorescence and the amount of oil present in the water. Several optical methods for measuring oil in water use three kinds of instruments: turbidimeter, nephelometer, and the tyndalmeter, which depend on apparent absorbence, reflectance, and the scattering of light respectively. Each of these is subject to interference with other suspended material, and there is the problem of free oil which floats or adheres to the sides of the sample holder. Classical approaches for oil-in-water analysis are refinements of the basic principle of extracting the oil from the water sample with a suitable solvent, evaporating the solvent to dryness, then weighing the residue. Solvents commonly used are hexane, ether, petroleum ether, chloroform, and carbon tetrachloride. For field use this approach is unsatisfactory because a sensitive analytical balance is required. In addition prolonged drying time is involved.

A remaining possibility is volumetric measurement of the oil after extraction and evaporation. An approach used in the past for a volumetric measurement of residual oil after extraction and evaporation of solvent is given in API methods 732-53, *The Determination of Non-Volatile Oily Matter.* The problem with this method is that a centrifuge was necessary. Since centrifuging is not practical in the field, other approaches had to be tried. Furthermore, occasionally the evaporation was too rapid, and the entire sample would blow out the tube.

The device of this process comprises a volumetric measurement of oil-in-wastewater. The apparatus consists of a bulb and a measuring device in an inverted position. A syringe is the measuring device. After extraction of the sample, the solvent is evaporated to approximately 10 ml. It is then transferred to the apparatus having a special bulb or funnel on top and an inverted syringe fitted to the bulb or funnel.

The evaporation is completed in the funnel using a hot water bath and the residual oil drawn into the syringe and measured. It is better to draw the oil into the syringe while a little solvent remains so that sufficient fluidity is retained. Evaporation is completed in the syringe with the plunger fully extended. The oil is collected by pushing in the plunger until the oil is near the top of the measuring area. The reading is then made.

A technique developed by D.M. Zall (81) permits detecting a hydrocarbon contaminant in an aqueous solution by applying chromic-sulfuric acid to filter paper wetted by the aqueous solution. A positive test is indicated by the development of a bluish-green color.

References

(1) Patterson, J.W. and Minear, R.A., "Wastewater Treatment Technology," 2nd Edition, *Report PB 216,162,* Springfield, Virginia, Nat. Tech. Info. Service (February 1973).
(2) American Public Health Association, *Standard Methods for the Examination of Water and Wastewater,* 13th Edition, APHA, New York (1971).
(3) American Petroleum Institute, *Industrial Oily Waste Control,* API, New York.
(4) American Petroleum Institute, *Manual on Disposal of Refinery Wastes,* 7th Edition, API, New York (1963).
(5) Anonymous, *The Cost of Clean Water: Vol. III, Industrial Waste Profiles, No. 5 - Petroleum Refining,* U.S. Department of the Interior, Washington, D.C. (1967).
(6) Wigren, A.A. and Burton, F.L., "Refinery Wastewater Control," *Jour. Water Pollution Control Federation,* 44:117-128 (1972).
(7) Anonymous, "Humble Oil Treats Wastes at Baytown," *Water Sew. Wks.,* 118:1W4-5 (1971).
(8) Rose, W.L. and Gorringe, G.E., "Activated Sludge Plant Handles Loading Variations," *Oil Gas Journal,* 70:40:62-65 (1972).
(9) Symons, C.R., "Treatment of Cold Mill Wastewaters by Ultrahigh-Rate Filtration," *Journal Water Pollution Control Federation,* 43:2280-2286 (1971).
(10) Donovan, Jr., E.J., "Treatment of Wastewater for Steel Cold Finishing Mills," *Water Waste Engng.,* F22-F25 (November 1970).
(11) Anonymous, "Treatment of Wastewater - Waste Oil Mixtures," U.S. EPA Report 12010 EZV 02/70 (1970).
(12) Boyd, J.L. and Shell, G.L., "Dissolved Air Flotation Application to Industrial Wastewater Treatment," presented at 45th Annual Conference Water Pollution Control Federation (1972).
(13) Chun, M.J., Young, R.H.F. and Burbank, Jr., N.C., "A Characterization of Tuna Packing Waste," *Proc. 23rd Purdue Industrial Waste Conference,* 33:786-805 (1968).
(14) Brink, R.J., "Operating Costs of Waste Treatment in General Motors," *Proc. 19th Purdue Indust. Waste Conference,* 19:12-16, (1964).

(15) Germain, J.E., Vath, C.A. and Griffin, C.F., "Solving Complex Waste Disposal Problems in the Metal Finishing Industry," presented at Georgia Water Pollution Control Association Conference (September 1968).
(16) Wallace, Jr., J.T. and Brown, J.S., "Deep Bed Filtration of Oilfield Produced Water," presented at 27th Indust. Waste Conference, Purdue University (1972).
(17) Garrison, V.M. and Geppert, R.J., "Packinghouse Waste Processing Applied Improvement of Conventional Methods," *Proc. 15th Purdue Indust. Waste Conference,* 15:207-217 (1960).
(18) Smith, J.E., *Torrey Canyon Pollution and Marine Life,* Cambridge University Press, Cambridge, England (1968).
(19) Ludwigson, J.O., "Oil Pollution at Sea" (in *Oil Pollution: Problems and Policies,* S.E. Degler, Ed.) Bureau of National Affairs, Washington, D.C. (1969).
(20) Price, R.I., "International Activity regarding Shipboard Oil Pollution Control," in *Proceedings of Joint Conference on Prevention and Control of Oil Spills,* American Petroleum Institute, New York (December 15 to 17, 1969).
(21) *Oil Pollution - A Report to the President* (a report on pollution of the nation's waters by oil and other hazardous substances by the Secretary of the Interior and the Secretary of Transportation), U.S. Government Printing Office, Washington, D.C. (February 1968).
(22) Allen, A.A. and Estes, J.E., "Detection and Measurement of Oil Films," in *Santa Barbara Oil Symposium,* University of California, Santa Barbara, California (1970).
(23) Swaby, L.G. and Forziati, A.F., "Remote Sensing of Oil Slicks;" in *Proceedings Joint Conference on Prevention and Control of Oil Spills,* American Petroleum Institute, New York, N.Y. (Dec. 15-17, 1969).
(24) Catoe, C.E. "Remote Sensing of Controlled Oil Spills;" in *Proceedings of Joint Conference on Prevention and Control of Oil Spills,* American Petroleum Institute, Washington, D.C. (June 15-17, 1971).
(25) Kreider, R.E., "Identification of Oil Leaks and Spills;" in *Proceedings of Joint Conference on Prevention and Control of Oil Spills,* American Petroleum Institute, Washington, D.C. (June 15-17, 1971).
(26) Kawahara, F.K., *Laboratory Guide for the Identification of Petroleum Products,* Analytical Quality Control Laboratory, EPA, Cincinnati, Ohio (1969).
(27) Brunnock, J.V., Duckworth, D.F. and Stephens, G.G., "Analysis of Beach Pollutants;" in *Scientific Aspects of Pollution of the Sea by Oil,* P. Hepple, Editor; Elsevier Publishing Company Ltd., New York, N.Y. (1968).
(28) Ramsdale, S.J. and Wilkinson, R.E., "Identification of Petroleum Sources of Beach Pollution by Gas-Liquid Chromatography;" in *Scientific Aspects of Pollution of the Sea by Oil,* P. Hepple, Editor; Elsevier Publishing Company Ltd., New York, N.Y. (1968).
(29) Adlard, E.R., Creaser, L.F. and Matthews, P.H.D., "Identification of Hydrocarbon Pollutants on Seas and Beaches by Gas Chromatography," *Anal. Chem.* 44, 64 (1972).
(30) Kawahara, F.K., "Gas Chromatographic Analysis of Mercaptans, Phenols, and Organic Acids in Surface Waters with Use of Pentafluorobenzyl Derivatives," *Environ. Sci. and Technol.* 5, 235 (1971).
(31) *Proceedings of Joint Conference on Prevention and Control of Oil Spills,* American Petroleum Institute, Washington, D.C. (June 15-17, 1971).
(32) *Proceedings Joint Conference on Prevention and Control of Oil Spills,* American Petroleum Institute, New York, N.Y. (December 15-17, 1969).
(33) *Santa Barbara Oil Symposium,* University of California, Santa Barbara, California (1970).
(34) *Chemical Treatment of Oil Slicks,* Federal Water Pollution Control Administration, Washington, D.C. distributed by Water Quality Office, EPA, Washington, D.C. (1969).
(35) *Oil Dispersing Chemicals,* Federal Water Pollution Control Administration, Washington, D.C., distributed by Water Quality Office, EPA, Washington, D.C. (1969).
(36) *Cleaning Oil Contaminated Beaches,* Federal Water Pollution Control Administration, Washington, D.C., distributed by Water Quality Office, EPA, Washington, D.C. (1969).
(37) *Recovery of Oil Spills Using Vortex Assisted Airlift System,* Water Quality Office, EPA, Washington, D.C. (1970).
(38) *Evaluation of Selected Earthmoving Equipment for the Restoration of Oil Contaminated Beaches,* Water Quality Office, EPA, Washington, D.C. (1970).
(39) *Oil/Water Separation System with Sea Skimmer,* Water Quality Office, EPA, Washington, D.C. (1970).
(40) *Feasibility Analysis of Incinerator Systems for Restoration of Oil Contaminated Beaches,* Water Quality Office, EPA, Washington, D.C. (1970).
(41) *Floating Oil Recovery Device,* Water Quality Office, EPA, Washington, D.C. (1971).
(42) *Concept Development of a Hydraulic Skimmer System for Recovery of Floating Oil,* Water Quality Office, EPA, Washington, D.C. (1971).
(43) *Recovery of Floating Oil Rotating Disk Type Skimmer,* Water Quality Office, EPA, Washington, D.C. (1971).

(44) Biglane, K.E., "A History of Major Oil Spill Incidents," in *Proceedings Joint Conference on Prevention and Control of Oil Spills,* American Petroleum Institute, New York, N.Y. (December 15-17, 1969)
(45) Marx, W., *Oilspill,* Sierra Club, San Francisco, California (1971).
(46) *Clean Water for the 1970's - A Status Report,* Federal Water Quality Administration, U.S. Government Printing Office, Washington, D.C. (1970).
(47) Revelle, R., Wenk, E., Ketchum, B.H. and Corino, E.R., "Ocean Pollution by Petroleum Hydrocarbons", in *Man's Impact on Terrestrial and Oceanic Ecosystems,* W.H. Matthews, F.W. Smith and E.D. Goldberg, Editors; M.I.T. Press, Cambridge, Massachusetts (1971).
(48) Blumer, M., Sanders, H.L., Grassle, J.F. and Hampson, G.R., "A Small Oil Spill," *Environment* 13, 2 (1971).
(49) Lawrence Berkeley Laboratory, "Instrumentation for Environmental Monitoring - Water," *Publ. LBL-1,* Volume 2, Berkeley, University of California (February 1, 1973).
(50) McKee, J.E. and Wolf, H.W., *Water Quality Criteria,* State Water Resources Control Board, Sacramento, California (1963).
(51) Hoult, D.P., *Oil on the Sea,* Plenum Press, New York, N.Y. (1969).
(52) *National Oil and Hazardous Substances Pollution Contingency Plan,* Council on Environmental Quality, Executive Office of the President, Washington, D.C. (August 1971).
(53) *Water Quality Criteria,* Report of the National Technical Advisory Committee to the Secretary of the Interior, U.S. Government Printing Office, Washington, D.C. (1968).
(54) Yee, J.E., *Oil Pollution of Marine Waters, Bibliography No. 5,* Department of the Interior, Washington, D.C. (1967), distributed by National Technical Information Service, Springfield, Virginia.
(55) *Oil Spillage Study: Literature Search and Critical Evaluation for Selection of Promising Techniques to Control and Prevent Damage,* Battelle Memorial Institute, Richland, Washington; distributed by National Technical Information Service, Springfield, Virginia (1968).
(56) *Water Pollution by Oil Spillage,* Industrial Information Services Search No. 460, Science Library, Southern Methodist University, Dallas, Texas (1969).
(57) Radcliffe, D.R. and Murphy, T.A., *Biological Effects of Oil Pollution - Bibliography,* Federal Water Pollution Control Administration, Washington, D.C.; distributed by Water Quality Office, EPA, Washington, D.C. (1969).
(58) Holmes, R.W., Paterman, H.M. and Walstead, K., *Oil Pollution Index-Catalog,* Volumes 1 to 4, University of California, Santa Barbara, California (1972).
(59) Walstead, K. and Hearth, F.E., *Oil Pollution in the Santa Barbara Channel: A Comprehensive Bibliography,* University of California, Santa Barbara, California (1972).
(60) Ferguson, H.F. (Chairman), "Oil Pollution and Refinery Wastes," Committee on Oil Pollution, Conference of State Sanitary Engineers, *Sewage Works, Jour.* 7, 104 (1935).
(61) Pickering, Q.H. and Henderson, C., "Acute Toxicity of Some Important Petrochemicals to Fish," *J. Water Poll. Control Federation,* 38 (9), 1419 (1966).
(62) North, J.W., *Successive Biological Changes Observed in a Marine Cove Exposed to Large Oil Spillage,* Ref. 61-6, Institute of Marine Resource, University of California, San Diego, California (1961).
(63) Cairns, Jr., J., "Environment and Time in Fish Toxicity," *Industrial Wastes,* 2 (1), 177 (1957).
(64) *Standards Methods for the Examination of Water and Wastewater,* 13th Edition, American Public Health Association, Washington, D.C. 20036 (1971).
(65) *Methods for Chemical Analysis of Water and Wastes,* Analytical Quality Control Laboratory, EPA, Cincinnati, Ohio 45268 (1971).
(66) *Annual Book of ASTM Standards,* Part 23, American Society for Testing and Materials, Philadelphia, Pennsylvania 19103 (1972).
(67) "Petroleum," *Anal. Chem.,* 39, 157R (1967).
(68) "Petroleum," *Anal. Chem.,* 41, 152R (1969).
(69) "Petroleum," *Anal. Chem.,* 43, 162R (1971).
(70) Kawahara, F.K., *Laboratory Guide for the Identification of Petroleum Products,* Analytical Quality Control Laboratory, EPA, Cincinnati, Ohio 45268 (1969).
(71) Dewling, R.T., Director, Surveillance and Analysis, EPA, Edison, New Jersey 08817 (Nov. 1972).
(72) Klemas, V., "Detecting and Measuring Oil on Water," *Instrumentation Technol.,* 19 (9), 54 (1972).
(73) Goolsby, A.D., "Continuous Monitoring of Water Surfaces for Oil Films by Reflectance Measurements," *Environ. Sci. Technol.,* 5 (4), 357 (1971).
(74) Pust, H.W., Kreider, R.E. and Gardiner, K.W., "An Automated Instrumental Approach to the Analysis of Oil in Water," in *Analysis Instrumentation, Vol. 9: Proceedings of the 17th Annual ISA Analysis Instrumentation Symposium,* A.M. Bartz, R.L. Chapman and L. Fowler, Editors, Instrument Society of America, Pittsburgh, Pennsylvania 15222 (1971).
(75) Swaby, L.G. and Forziati, A.F., "Remote Sensing of Oil Slicks," in *Proceedings Joint Conference on Prevention and Control of Oil Spills* (December 15-17, 1969, New York), American Petroleum Institute, Washington, D.C. 20006; distributed by National Technical Information Service, Springfield, Virginia 22151.

(76) Guinard, N.W. and Purves, C.G., "The Remote Sensing of Oil Slicks by Radar," *Naval Research Laboratory Report No. R 07-02* (April 28, 1970) to the Commandant, U.S. Coast Guard, distributed by National Technical Information Service, Springfield, Virginia 22151.
(77) Allen, A.A. and Estes, J.E., "Detection and Measurement of Oil Films," in *Santa Barbara Oil Symposium,* University of California, Santa Barbara, California 93106 (1970).
(78) McGrath, F.K., *Report AD-747,084,* Springfield, Virginia, National Technical Information Service (June 1972).
(79) Measures, R.M. and Bristow, M., *Report N72-20479,* Springfield, Virginia, National Technical Information Service (December 1971).
(80) Overbeck, C.J. and Means, J.C.; U.S. Patent 3,626,751; December 14, 1971; assigned to Nalco Chemical Company.
(81) Zall, D.M.; U.S. Patent 3,700,409; October 24, 1972; assigned to the U.S. Secretary of the Navy.

ORGANICS

Measurement in Water

The term "organics" covers a multitude of products but "petrochemicals" can be monitored in water by the following instrumental techniques (1): gas chromatography; gas chromatography/mass spectrometry/computer; and thin layer chromatography.

As pointed out by the American Chemical Society (2), the EPA in 1972 expanded sharply its ability to identify specific organic chemicals in the environment. The primary tool is the tandem gas chromatograph-mass spectrometer (GC/MS). Twenty-three of the agency's laboratories were thus equipped, up from only six in mid 1971. The number should rise to about 50 by 1975. Six of the installations are tied into a computerized GC/MS data bank at Battelle Memorial Institute, Columbus, Ohio, and others will join them. Overall the system will be the nation's first integrated sampling and analysis network for organic pollutants.

The gas chromatograph in GC/MS equipment separates the unknown organic compounds in the sample; the mass spectrometer produces their spectra. The compounds are identified by comparing their spectra with those of known compounds, a process speeded greatly at Battelle by a computer matching system. The data bank there contains more than 11,000 spectra for organic compounds, and the number is growing steadily.

Organic pollutants, particularly in water, have been analyzed as a rule by collective methods such as biochemical oxygen demand. These methods do not pick out individual compounds. Some such compounds, however, may exert marked effects in addition to depleting the oxygen in a stream as they degrade biologically. Thus GC/MS, by detecting specific compounds, should do much to enhance scientists' ability to assess the environmental impacts of organic pollutants.

GC/MS also is turning up unexpected pollutants. In December 1972, experts at EPA's Southeastern Environmental Research Laboratory (SERL), Athens, Georgia, drew up a list of about 250 organic pollutants that actually have been detected in water. John McGuire, L.H. Keith, and their colleagues at SERL believe that the number of potential ones lies somewhere between 10,000 and 100,000. Keith reported recently the detection of 37 different organic compounds in one sample of water downstream from a petrochemical plant.

An apparatus developed by L.G. von Lossberg (3), permits measuring the organic contaminants in a flowing stream of water. The device is especially adapted for use in the measurements of organic contaminants in substantially mineral-free water such as condensate or demineralized water, though by slight modification it is adaptable also for use with raw or partially treated water. It has been found that organic contamination of condensate and the demineralized water may result and frequently does result from various causes such as process leaks, improper cross-connections, vapor contamination, heat exchanger leaks, etc. Also, for example, demineralized water produced from surface sources generally contains organic traces due to the fact that most pretreatment is not capable of completely removing organic constituents at all times. Also, for instance, organic contamination can be carried by steam generated by boilers having organic constituents entering the steam-water cycle.

Organic matter present in otherwise highly pure water used in manufacturing processes is frequently highly undesirable as it may result in damage for instance to textiles, chemicals, pharmaceuticals and certain foods undergoing processing. It has been customary to determine the existence of organic matter in the water by taking spot samples and subjecting these to chemical tests during which the organic constituents are oxidized by various chemical reagents. However, such testing techniques are time consuming and further their utility is extremely limited due to the difficulty of correlating the contamination and the counteracting or quality control measures.

With these considerations in mind it is a primary object of this technique to provide a continuous method for measuring the organic contaminants in a flowing stream of water whereby it makes it possible to correlate the quality control measures with the degree of contamination present at a given time within the stream.

In accordance with this process a comparatively small proportion of the flow of the main stream or flowing supply is diverted into a separate flow path to provide a flowing test sample. This sample is continuously vaporized within an enclosed portion of its flow path and pure gaseous oxygen is continuously delivered into it and combined with it, following which at a later stage in its flow path the sample stream is condensed, the resulting condensate containing dissolved carbon dioxide in proportion to the amount of organic contaminants plus such inorganic salts as may have existed in the main stream.

Where the main stream is of a mineral-free nature the resulting amounts of carbon dioxide in the condensate will constitute a measure of organic contamination. Similarly, if the flowing test sample is made mineral-free as by being passed through an ion exchange bed to remove the interfering inorganic salts, the same result will follow.

In the event the flowing sample itself consists of raw or only partially treated water, the process still may be carried out to secure an accurate indication of organic impurities by taking parallel samples of the main stream, one of which consists of the test sample earlier mentioned, while the other consists of a monitoring sample. In the monitoring sample, the amount of carbon dioxide present is constantly measured, without first subjecting the sample to the vaporizing and oxidizing steps. It will be apparent that the difference in the amount of carbon dioxide in the two samples at any given time will indicate the amount of organic matter present.

A technique developed by P. Moyat (4) permits the determination of organics in water. The apparatus used comprises a container for the polluted water under test and provided with electrodes. The water is treated with an acid electrolyte and current passed between the electrodes. Elemental or nascent oxygen at the anode combines with carbon of the organic substances and is evolved as carbon dioxide, the latter being measured as the indication of pollution with respect to the amount or flow of the water.

A device developed by W. Juda and M.S. Frant (5) provides for the detection of intermediate and high molecular weight organic compounds in aqueous electrolytic solutions, as well as other traces of chemical agents, through measurement of the rate of variation in overvoltage at one electrode in an electrolyzing circuit.

A device developed by M.M. Zuckerman and A.H. Molof (6) involves measuring the soluble organic content of a water sample by employing a refractometer which compares the refractive index of the water sample to the refractive index of a known or related water sample to obtain a measure of the total soluble content. The technique of this process then corrects this value for the content of the soluble inorganic material by use of an apparatus which compares the electrical conductivity of the water sample to the electrical conductivity of a known or related water sample. The system has the distinct advantage over existing methods of requiring minimal supervision, and short delay time before obtaining analytical results. Thus the system is particularly adapted for performance monitoring of a water treatment facility or for qualitatively monitoring the soluble organic content of a water-stream.

A process developed by G. Claus (7) is one where the amount of dissolved organic material in a solvent medium is measured by passing light radiation of 280 millimicron wavelength through the medium. The amino acid Tyrosine within the medium absorbs radiation of this wavelength. By measuring the absorbence of the radiation, the concentration of Tyrosine is determined and the amount of organic material can be calculated by the known proportional relationship of Tyrosine to total organic matter.

A process developed by F.T. Eggertsen (8) involves estimating the amount and volatility of organic materials in contaminated water samples wherein a sample of the contaminated water is moved into a heated furnace and the vapors swept directly into a flame-ionization detector.

References

(1) Lawrence Berkeley Laboratory, "Instrumentation for Environmental Monitoring - Water," *Publ. LBL-1,* Volume 2, Berkeley, University of California (February 1, 1973).
(2) American Chemical Society, *Chemistry in 1972,* Washington, D.C., Department of Public and Member Relations (1973).
(3) von Lossberg, L.G.; U.S. Patent 3,205,045; September 7, 1965.
(4) Moyat, P.; U.S. Patent 3,224,837; December 21, 1965; assigned to Hartmann & Braun AG, Germany.
(5) Juda, W. and Frant, M.S.; U.S. Patent 3,586,608; June 22, 1971; assigned to Prototech Incorporated.
(6) Zuckerman, M.M. and Molof, A.H.; U.S. Patent 3,635,564; January 18, 1972; assigned to Envirotech Corporation.
(7) Claus, G.; U.S. Patent 3,751,167; August 7, 1973.
(8) Eggertsen, F.T.; U.S. Patent 3,753,654; August 21, 1973; assigned to Shell Oil Company.

OXYGEN DEMAND

Oxygen demand is closely related to the question of organic waste content, whether the wastes be fecal wastes or petrochemical wastes. As pointed out by C.F. Gurnham (1), such wastes consume the naturally dissolved oxygen (DO) in streams. Total depletion of DO produces bad odors, offensive appearance and even partial depletion may cause fish kills. It is very difficult to measure each individual pollutant and hence general organic pollution is usually measured by determining oxidizable carbon or measuring oxygen consuming power.

The dissolved oxygen content of a stream may approach saturation value under favorable, nonpolluted conditions. At best, however, this concentration is low; it is only 14 mg./l. at the freezing point and decreased with increasing temperatures. Photosynthesis helps to maintain its level. If about 2 mg./l. of DO can be maintained at all parts of a stream, nuisance conditions can be avoided, according to Gurnham. A higher concentration, perhaps 4 to 6 mg./l. is necessary to insure a healthy population of game fish. Dissolved oxygen (DO) can be monitored in water by using a membrane probe (2).

As regards oxygen demand the original method was measurement of biological oxygen demand (BOD) which relates to the ability of streams to purify themselves and to the fact that municipal sewage was often treated by biological processes. Several years ago it was generally believed that if only biodegradable organics were put into a body of water, that bacteria would consume the organics and thereby remove them as contaminants.

Subsequent studies have shown that if the concentration of the biodegradable organics is high enough, the bacteria will deplete the oxygen in the water during their consumption of the organics. The oxygen-depleted water will fail to sustain animal life and the water becomes even more seriously contaminated.

Various methods were devised for measuring the biochemical oxygen demand (BOD) of water contaminated with biodegradable organics. One such method requires taking a sample of the water and incubating it (in a closed container) until the bacteria have consumed all the organics they can, then measuring the amount of oxygen used up by the bacteria. It is generally believed that in about 20 days, the bacteria will have consumed all the organics which require oxygen during consumption, assuming there was enough oxygen in the water to allow them to consume all the biodegradable organics.

In an article titled "Graphical Determination of BOD Curve Constants" by H.A. Thomas (3), a method was set forth which allows one to compute the ultimate BOD from BOD tests of only 5 days. While the BOD measurements are widely employed, they require tests of several days duration. Some attempts have been made to run accelerated BOD determinations in a few hours, but environmental conditions limit the accuracy of this method. The 5 day BOD is useful in describing the past demands of the stream but useless for pollution control because of the 5 day delay.

Hence, various rapid oxidation and combustion techniques have been developed to measure chemical oxygen demand (COD). The results generally have little relationship to biological oxygen demand which is a relatively slow oxidation process and takes place under much different environmental conditions.

Measurement in Water

Oxygen demand can be monitored in water by the following instrumental techniques (2): biochemical oxygen demand (BOD) - DO probe and respirometry/manometric; and chemical oxygen demand (COD) - colorimetric.

A device developed by F. Poepel and H. Steinecke (4) permits measuring oxygen absorption more especially measuring biochemical oxygen absorption, for example, when ascertaining the BOR (biological oxygen requirement).

In conventional methods used hitherto for determining the biochemical oxygen consumption, the oxygen is produced constantly by electrolytic methods commensurate with the instantaneous consumption and the current intensity recorded in conjunction with the time taken. The biochemical oxygen consumption may be ascertained graphically by means of integrating the area between the recorded curve and abcissa. The expenditure in electrical circuit members is considerable and the evaluation for ascertaining intermediate values particularly cumbersome. The apparatus can therefore only be operated by trained personnel.

In contrast, the method of this process simply involves determining the drop in pressure created by the oxygen absorption in a measuring system closed to the outer atmosphere and maintained at a constant temperature, in which method a test vessel is associated with an oxygen storage container and a pressure gauge, and in which the absorbed oxygen quantities are supplied intermittently from the storage container to the test vessel, and are directly indicated on an instrument in known units, for example, milligrams per liter.

A process developed by I. Lysyj and K.H. Nelson (5) involves the analysis of wastewater for biochemical oxygen demand using a pyrolysis chamber functionally connected to a hydrogen flame ionization detector so that first a value indicative of total organic content of a sample is determined and then that value is converted by suitable factors to a value indicative of biochemical oxygen demand.

A method developed by R.D. Mikesell (6) utilizes a continuous respirometer apparatus for measuring the rate at which oxygen in a fluid, such as sewage, is consumed to thereby obtain an indication of the respiration rate of the bacterial population included in such sewage. The apparatus includes a closed airtight container having an inlet and an outlet. Inlet oxygen sensing means is provided for sensing the amount of dissolved oxygen in the fluid entering the container and outlet oxygen sensing means is provided for sensing the amount of oxygen in the sewage exhausted out the outlet. Pump means is provided for circulating the sewage through the closed container at a constant flow rate whereby the amount of dissolved oxygen in the sewage may be determined at both the inlet and outlet to thereby enable the operator to determine the rate at which such oxygen is consumed or respired by the bacteria included in the sewage.

A process developed by R.M. Arthur (7) involves an automatic recording apparatus and method for measuring the accumulated amount of oxygen absorbed by a liquid culture, e.g., of a bacteria microorganism suspension. The apparatus can be used for batch or continuous flow methods. In biological applications, however, the batch method can be advantageously employed to provide a continuous graphical record of oxygen utilization by a culture.

The apparatus is sensitive, simple to use, inexpensive to construct and highly accurate in measuring the gas absorption and expiration characteristics of a substance and in this connection has been found particularly useful in measuring oxygen utilization by bacteria cultures. Although the measurement is automatic, this process affords a number of advantages other than its automatic recording features, for instance, the apparatus can be provided with an aeration chamber large enough to contain a large amount of culture. This is often necessary so that a truly representative sample of large volumes of waste is tested. The large sample volume also provides for better accuracy in determination of a low oxygen demand.

A method developed by V.A. Stenger and C.E. Van Hall (8) is a method for determining the oxygen demand of a material containing oxidizable components, which method involves the combustion of a small sample of the material to be analyzed in a heated, continuous stream of carbon dioxide. The carbon monoxide produced as the result of combustion with carbon dioxide relates directly to the total oxygen demand (TOD) of the sample. Hence, quantitative analysis of the combustion gases for carbon monoxide yields a measure of the oxygen demand of the sample.

As used herein, "combustion" refers to the reaction or equilibration of carbon dioxide with an oxidizable material in the sense that the oxidant (carbon dioxide) is reduced and the oxidizable material is oxidized. By TOD is meant the net oxygen demand of the sample. Thus oxygen dissolved in the sample and other oxidant source materials contained in the sample lower the oxygen demand of the sample as it is measured.

A refinement of the above process is described by C.E. Van Hall (9). In this modification, the carbon dioxide feed gas plus the sample are introduced into a combustion tube, moisture is then removed from the effluent, and the moisture reduced effluent then is passed through a suitable carbon monoxide detector. The effluent then is processed to convert carbon monoxide therein to carbon dioxide and the thus processed effluent is then pumped at a controlled flow rate and is reintroduced to the combustion chamber.

A method developed by C.J. Overbeck and J.J. Hickey (10) permits the determination of COD in water by a procedure which reduces time and effort normally required for such an analysis by standard methods. The improvement comprises the use of an oxidizing composition. The oxidizing composition consists essentially of an equimolar solution of sulfuric and phosphoric acid, also containing therein a water-soluble hexavalent chromium salt.

A method developed by J.D. Laman and G.R. Wessels (11) involves continuously monitoring the amount of biodegradable organics in aqueous material. The method comprises adding caustic to precipitate any magnesium, calcium, copper or iron that may be present, filtering or otherwise separating out the precipitate, adding a permanganate solution (usually potassium permanganate) to the filtrate (precipitate-free material), passing the filtrate through the sample cell of a colorimeter, and measuring the color change (in millivolts) on a continuous recorder as a function of the amount of biodegradable material in the aqueous sample being tested. This method is especially useful in determining the chemical oxygen demand in wastewater streams, and performs such measurement in less than one hour as compared to a matter of 6 to 8 hours or more for other known methods.

A process developed by J.L. Teal, C.E. Hamilton and D.A. Clifford (12) permits determining the oxygen demand of combustible materials in aqueous dispersions. A feed gas stream composed of an inert gas containing oxygen is passed at a constant rate through a combustion supporting, porous catalyst bed in the combustion zone which is heated to a combustion supporting temperature and then is fed into a detector for small amounts of free oxygen.

A small amount of a dilute aqueous dispersion of a combustible material is injected into the feed gas stream within the combustion zone upstream from the catalyst bed. The resulting gaseous product is then swept from the combustion zone into the oxygen detector by the continuing pressure of the feed gas stream. Since the feed gas stream contains only a small amount of oxygen, very small amounts of combustible materials in the aqueous dispersion produce a large percentage deviation in the oxygen content of the effluent gas from the combustion zone.

A variation of the above described process which permits measuring oxygen demands of aqueous dispersions in the parts per billion range has been developed by J.L. Teal, C.E. Hamilton and D.A. Clifford (13). To achieve accurate measurements at such low levels of oxygen demand, the method is operated with the feed gas stream containing oxygen within the range from about 0.5 up to about 10 ppm by volume. A catalyst bed at least 8 cm. long assures complete and reproducible combustion. It is also essential that the aqueous sample to be analyzed be substantially free of dissolved oxygen upon injection into the combustion zone.

References

(1) Gurnham, C.F., *Industrial Wastewater Control,* New York, Academic Press (1965).
(2) Lawrence Berkeley Laboratory, "Instrumentation for Environmental Monitoring - Water," *Publ. LBL-1,* Volume 2, Berkeley, University of California (February 1, 1973).
(3) Thomas, H.A., *Water & Sewage Works,* pages 123 and 124 (March 1950).
(4) Poepel, F. and Steinecke, H.; U.S. Patent 3,282,803; November 1, 1966; assigned to J.M. Voith GmbH, Germany.
(5) Lysyj, I. and Nelson, K.H.; U.S. Patent 3,567,391; March 2, 1971; assigned to North American Rockwell Corporation.
(6) Mikesell, R.D.; U.S. Patent 3,731,522; May 8, 1973; assigned to Robertshaw Controls Company.
(7) Arthur, R.M.; U.S. Patent 3,740,320; June 19, 1973.
(8) Stenger, V.A. and Van Hall, C.E.; U.S. Patent 3,421,856; January 14, 1969; assigned to The Dow Chemical Company.
(9) Van Hall, C.E.; U.S. Patent 3,567,385; March 2, 1971; assigned to The Dow Chemical Company.
(10) Overbeck, C.J. and Hickey, J.J.; U.S. Patent 3,540,845; November 17, 1970; assigned to Nalco Chemical Company.
(11) Laman, J.D. and Wessels, G.R.; U.S. Patent 3,558,277; January 26, 1971; assigned to The Dow Chemical Company.
(12) Teal, J.L., Hamilton, C.E. and Clifford, D.A.; U.S. Patent 3,560,156; February 2, 1971; assigned to The Dow Chemical Company.
(13) Teal, J.L., Hamilton, C.E. and Clifford, D.A.; U.S. Patent 3,679,364; July 25, 1972; assigned to The Dow Chemical Company.

OZONE (OXIDANTS)

Oxidants are defined as those substances, with the exception of NO_2, which liberate iodine from a buffered KI solution (1). In the atmosphere, the largest component of the oxidant category is ozone. Smaller components include peroxyacetylnitrate (PAN) which is discussed separately in this volume. The contributions of NO_2 will have to be subtracted in an analytical determination of oxidants.

The concentration of oxidants in air as measured by the KI method (see text to follow) has been set on an air quality basis by several states because of the contribution of such oxidants to smog formation. On a 1 hour averaging time basis, the California state limit is 0.15 ppm, the Colorado state limit is 0.1 ppm and the limit for the St. Louis metropolitan area is 0.15 ppm. The New York state overall limit is 0.15 ppm. On a 24 hour averaging time basis, the Canadian (Ontario) limit is 0.1 ppm.

Measurement in Air

The KI evolution technique can be applied in various ways. Both manual and continuous colorimetric methods have been described in detail (1). The reaction may also be utilized in an amperometric method for continuous oxidant monitoring (1). Photochemical oxidants such as ozone, may be monitored in air by the following instrumental techniques (1a): chemiluminescence (reference method); coulometric; colorimetric and ultraviolet absorption. The EPA designated reference method for photochemical oxidants such as ozone in air is the gas phase chemiluminescent method.

However, according to Altshuller (2) the present oxidant analyzers are unsatisfactory monitoring tools. The oxidant has no exact meaning since the response obtained depends on the presence of various interfering substances. The substrate used to eliminate sulfur dioxide interference oxidizes nitric oxide to nitrogen dioxide and thus increases the signal caused by nitrogen dioxide (3). Recently, it has been shown that hydrogen peroxide is an atmospheric oxidant, but the response of commercial analyzers to this oxidant is poor (4).

It should be evident that each of the major oxidants should be measured separately. Several ozone analyzers have been developed and evaluated under field conditions. All of these instruments are based on utilization of chemiluminescent reactions (5). The three types of reactions involved are as follows: (1) reaction of ozone with Rhodamine B absorbed on silica-gel disc (Regener Method), with emission at 0.59 micron; (2) mixing of ozone with excess nitric oxide, resulting in chemiluminescence from excited nitrogen dioxide that extends from 0.6 to 3.0 microns; and (3) mixing of ozone with excess ethylene, resulting in a chemiluminescence peak near 0.43 micron (Nederbragt's method).

All of these systems provide a specific and sensitive means of analyzing ozone rapidly over an adequate linear range. However, both the Regener (6)(7)(8) and the ozone-nitric oxide techniques (5)(9) have several disadvantages, compared to the Nederbragt technique, for use in a routine air monitoring instrument (5)(10)(11).

The Regener approach requires preparation and calibration of the disc. Because of the activation and decay characteristics of the chemiluminescent surface of the disc, a 4 min. mode of operation is preferable; the mode involves a calibration cycle, purge cycle, sample cycle and purge cycle (7)(8). This type of operation also complicates methods for obtaining signal read-out. The ozone-nitric oxide reaction requires use of reduced pressure to avoid quenching the chemiluminescent reaction (5)(9). Detection of emissions in the spectral region involved dictates the use of photomultipliers near the edge of the chemiluminescent response above 0.6 microns; thus cooling to -20°C. is required to obtain a sensitivity down to 0.01 ppm.

The Nederbragt type detector can operate at atmospheric pressure without quenching and with a sensitivity down to 0.003 ppm at 5°C. with a less expensive photomultiplier (5)(10)(11). The ozone-ethylene reaction can be used with a detector of less complicated design, excellent operation and the good cost and size characteristics.

A device developed by H.H. Anderson, R.H. Moyer, D.J. Sibbett and D.C. Sutherland (12) permits detecting and analyzing pollutant gases in the atmosphere, particularly sulfur dioxide, ozone, nitrogen dioxide, and nitric oxide utilizing the catalyzed chemiluminescence reaction of luminol (5-amino-2,3-dihydro-1,4-phthalazinedione) with hydrogen peroxide. Sampled air streams, after appropriate treatment by adsorption column, are reacted with surface films of luminol-hydrogen peroxide solutions to give continuous, real time analysis of pollutant gases.

The chemiluminescence method of monitoring air pollutants utilizes five or six microreactors (channels) simultaneously which are monitored sequentially by a single photomultiplier. Channel monitoring is controlled by a rotary shutter which moves discretely from channel (microreactor) to channel. Quantitative analysis of the gaseous components of the atmosphere is obtained by comparison of the signals obtained from the separate channels with calibrated standards for each channel. Signal processing may utilize simple computer circuitry.

A device developed by N.A. Lyshkow (13) permits analysis of a gas stream of air and ozone in which the gaseous reactant is passed as a thin layer through a tortuous path to generate a highly turbulent flow for contact with a coreactant in the form of a solid, liquid or gas in a surface reaction to produce flashes of radiation by chemiluminescence. The flashes of radiation can then be measured as relating to the amount of ozone contained in the gas stream.

References

(1) American Public Health Association, *Methods of Air Sampling and Analysis,* Washington, D.C. (1972).

(1a) Lawrence Berkeley Laboratory, "Instrumentation for Environmental Monitoring - Air," *Publ. LBL-1,* Volume 1, Berkeley, University of California (February 1, 1973).

(2) Altshuller, A.P., in NBS Special Publication 351, *Analytical Chemistry: Key to Program on National Problems,* Washington, D.C., U.S. Government Printing Office (August 1972).

(3) Saltzman, B.E. and Wartburg, Jr., A.F., *Anal. Chem.* 37, 779 (1965).

(4) Gay, Jr., B.W., paper presented before 161st National A.C.S. meeting, Los Angeles, California (Mar. 28 to April 2, 1971).

(5) Hodgeson, J.A., Martin, B.E. and Baumgardner, R.E., "Comparison of Chemiluminescent Methods for Measurement of Atmospheric Ozone," in *Progress in Analytical Chemistry,* Volume 5, New York, Plenum Press (1971).

(6) Regener, V.H., *Geophys. Research* 69, 3795 (1964).

(7) Tommerdahl, J.B., Final Report on Contract CPA-22-69-7, Durham, North Carolina, National Air Pollution Control Administration (December 1969).

(8) Hodgeson, J.A., Krost, K. J., O'Keeffe, A.E. and Stevens, R.K., *Anal. Chem.* 42, 1995 (1970).

(9) Fontijn, A., Sabadell, A.J. and Ronco, R.J., *Anal Chem.* 42, 575 (1970).

(10) Nederbragt, G.W., Vander Horst, A. and Van Duijan, J., *Nature* 206, 87 (1965).

(11) Warren, G.J. and Babcock, G., *Rev. Sci. Instr.* 41, 280 (1970).

(12) Anderson, H.H., Moyer, R.H., Sibbett, D.J. and Sutherland, D.C.; U.S. Patent 3,659,100; April 25, 1972; assigned to Geomet, Incorporated.

(13) Lyshkow, N.A.; U.S. Patent 3,712,793; January 23, 1973.

PARTICULATES

As pointed out by A.C. Stern (1) a number of individual states in the U.S. have air quality standards for suspended particulate matter. These typically vary from 0.1 to 0.2 mg./m.^{3}. There are also state air quality standards for deposited particulate matter expressed in tons per square mile per month with figures ranging from 10 to 15 for residential areas to 30 for industrial areas.

Canadian (Ontario) air quality standards call for a maximum of 0.06 mg./m.^{3} in residential-rural areas over a 1 year averaging time and 0.11 mg./m.^{3} in industrial-commercial areas on the same time basis. Emission standards for particulate matter have been set in the U.K. (1) as 460 mg./m.^{3} for particles greater than 10 microns and 115 mg./m.^{3} for particles less than 10 microns. As pointed out by S.K. Friedlander (1a), the national ambient air quality standard for primary particulate matter is 75 $\mu g./m.^3$, annual geometric mean. By primary matter is meant particles introduced into the atmosphere in particulate form. Secondary particulates are those formed in the atmosphere from certain gases by chemical and physical processes; the secondary standard is 60 $\mu g./m.^3$. These standards are now exceeded in almost all major cities as pointed out by Friedlander (1a).

Measurement in Air

Particulates may be monitored in air by the following instrumental techniques (1)(1b): particulates (mass loading) - gravimetric (high volume sampling) (reference method), beta-radiation absorption, piezoelectric microbalance, resonant frequency and capacitance; particulates (opacity) - opacity light meter; particulates (size distribution) - dry impingement, sedimentation, electrostatic precipitation, microscopy (optical and electron), light scattering, holography, condensation nuclei counters, and light transmission; particulates (velocity) - Pitot tubes, and Doppler-shift technique.

The EPA designated reference method for determination of the mass loading of particulates in air is the high volume sampling method. Thus, a number of sophisticated techniques have replaced such early day techniques as the Rengelmann test which simply measured the density of black smoke by visual comparison with a hand-held card having areas of varying degrees of blackness.

Particulate analysis is more complex and in some ways less advanced than analysis of gaseous air pollutants, as noted by Altshuller (2). Certainly, there is a clear lack of instruments for analyzing particulates by mass, size and chemical composition. The great bulk of available results have been obtained by collecting samples on filters in the field and then weighing and analyzing the collected sample in the laboratory.

Particles larger than 10 microns settle readily and they are usually associated with settled dust and dirt. Particles in this size range are deposited in the nasopharyngeal region of the respiratory tract and do not tend to penetrate effectively into the pulmonary and tracheobronchial regions (3). Consequently, these large particles have not been associated with toxicological action. Particles above 10 microns also do not have significant effects on light scattering and on visibility.

Sampling for these large particles has been accomplished with dustfall jars, adhesive coatings, cyclonic collectors (for higher volume samples), long horizontal tunnels (as fractional elutriators), and various impactors (3). Dustfall jars appear as satisfactory as more complex and expensive techniques are, although the dustfall jar has poor time resolution.

The justification for the use of dustfall jars has decreased, however, since they have become increasingly poor measures of particulate pollution as the level of atmospheric particles smaller than 10 microns has increased. In view of these circumstances, little justification exists for directing effort toward development of instruments that can measure particles larger than 10 microns.

Particles between 0.1 and 10 microns probably are those of the greatest concern in air pollution. The particles in this size range contribute most of the particle mass. These particles penetrate into the pulmonary and tracheobronchial regions. The proportion of the particles that are between 0.1 and 1 micron is particularly important, as noted in *Fine Dust and Particulates Removal* by H.R. Jones (4).

Particles in this size range are largely responsible for reduction in visibility and for haze and turbidity. Studies of physiological response to particulate matter indicate that particles smaller than 1 micron can have greater irritant potency than larger particles (3). Measuring techniques for particles in the 0.1 to 10 micron range have included the Volz sun photometer, the integrating nephelometer, various other forward-scattering and right angle-scattering instruments, cascade impactors, electron microscopes and tape samplers (3).

None of these techniques is adequate for providing a quantitative measurement of mass concentration of 0.1 to 10 micron particulates. Several of these techniques have been research tools not primarily intended to give an overall measure of mass concentration; nonetheless, they have given valuable specialized measurements on particular characteristics of particulates in this size range.

The tape sampler has received considerable use in the measurement of suspended particulates. Visual color of the spots on the tape has been compared with a standard gray scale. More frequently, reflectance or transmittance measurements have been used. Transmittance measurements, however, have been shown to relate poorly to the mass of particulates. Reflectance also is not ordinarily related to total suspended particulate but to dark suspended matter, inasmuch as reflectance is also a function of absorbance, and absorbance is a function of color.

Both techniques suffer from a number of complications related to variability in the characteristics of the deposits. Neither technique can be considered satisfactory for measurement of absolute concentrations of particulates, but tape samplers can be used to obtain relative values. Careful standardization is critical. Changes in the characteristics of the particulates over periods of years at a site because of fuel changes or control efforts may limit the usefulness of these measurements even for obtaining relative values.

The integrating nephelometer has received considerable evaluation in recent years in an attempt to relate its response to mass concentration of suspended particulates [(5) through (8)]. This instrument was designed to measure meteorological range, which in turn correlates well with visual range. However, a close correspondence between nephelometer response and mass concentration is obtained only if the particle size distribution and other particle characteristics remain constant. Unfortunately, the experimental measurements themselves show this not to be the case (3).

A visual range of 7.5 miles can be associated with mass concentrations ranging all the way from 50 to 200 μg./m.^{3}. Such a range of mass concentrations represents the entire range of annual geometric mean concentrations of suspended matter in urban areas. Similarly, a visual range of 5 miles can be associated with suspended particulate loadings ranging from 75 to 300 μg./m.^{3}. Such results clearly indicate that nephelometer measurements are completely inadequate as a means of determining whether particulate emissions comply or tend to comply with air quality standards for particulates.

This instrument certainly is useful, however, in its original purpose, which is the measurement of meteorological or visual range. Since light scattering is associated primarily with particles in the 0.1 to 1 micron range, it would seem more useful to attempt to relate nephelometer measurements to mass concentration in the 0.1 to 1 micron range than to total suspended particulates.

The equipment that has received the greatest use in particulate measurement has been the high volume sampler (3). This device necessitates the replacement of filters that must be transported to the laboratory for weighing and chemical analysis. Ordinarily, a 24 hour average sample is obtained. When such integrated total weights of suspended particulate are required, the high volume sampler is an adequate device. However, this sampler can collect particles well above 10 microns and may not be too efficient in collecting particles approaching 0.1 micron. In addition, the sampling rate is sensitive to the mass of material collected.

It is in the utilization of high volume samples for subsequent chemical analysis that a multitude of problems occur. The glass fiber filter medium as ordinarily used gives high blank readings for a considerable number of cations and anions. Furthermore, reactions can occur on the filter medium between collected materials. Oxidation or volatilization of some collected substances can and does occur. Other filter materials can be used, particularly for special analytical applications, but such filter materials usually have limitations associated with their use as well.

It should be apparent that these sampling problems cause greater obstacles to overall analytical techniques themselves. Conversely, it seems futile to expend much effort on improving analytical techniques if the greater source of uncertainty results from the sampling technique itself.

The Anderson sampler has received considerable use as a particle-sizing device. Although materials collected on the various stages must be returned to the laboratory for weighing or analysis, this sampler can serve as a useful interim approach for both mass and composition measurements.

A number of types of instruments ought to be developed for the measurement of particulates. A continuous or periodic monitoring instrument capable of measuring the mass concentration of aerosols below 10 microns is needed. An instrument has been developed that uses an electrostatic precipitator to deposit aerosol particles directly onto a piezoelectric quartz crystal microbalance (9). The balance is sensitive enough to permit measurements of incremental mass for time periods of less than 1 minute.

Collection can be significantly influenced by sorption or desorption of water vapor, but this effect can be reduced by use of a dual crystal detector. Adhesion of larger particles depends on the composition of the particulate and the relative humidity. Addition of a surface active agent to the flow samples can help minimize this difficulty. Because of these possible limitations, such equipment requires careful field evaluation and comparison with normal particle-sizing collection techniques (10)(11)(12).

Equipment capable of periodic monitoring of particle size distribution by mass concentration also is required. An investigation is underway of a new cascade impactor concept for particle size fractionation, in which by beta-ray attenuation is used to measure material collected on a filter tape.

In view of the difficulties in filter collection and chemical analysis, development of on-site chemical analyzers for various chemical species in particulate matter may be justified. Aside from monitoring needs, improvement of our knowledge of gas aerosol transformation in the ambient atmosphere requires instruments capable of the diurnal monitoring of such species as sulfuric acid, total particulate sulfate, particulate nitrate, and organic aerosol. All of these substances contribute to the loading of particulates in the 0.1 to 1 microns range. Research studies using particle-sizing samplers also have identified lead, sulfate, nitrate, carbon-containing species and ammonium and chloride ions as constituents in the 0.1 to 1 micron range [(10) through (14)].

To properly associate visibility with air pollution control effectiveness, both the total mass concentration within the 0.1 to 1 micron range and the contribution of these chemical species to the mass concentration must be known. Of course, the direct association with visibility also involves computing the on-site contributions of individual species based not only on mass and size distribution but also on optical characteristics of the particles.

Despite emphasis on automated field methods, laboratory procedures will continue to be useful in particle size and morphological analysis of airborne particles (15)(16). For example, optical and electron microscopy can be used for particle sizing. These techniques, along with x-ray diffraction and other measurements of physical properties, have been used to characterize a wide variety of types of particles. An atlas of photomicrographs of various airborne particles is available to aid in microscopic identification (17).

In addition to determination of the physical concentration and type of particulate matter, there is the question of the chemical composition of the individual particles. A variety of methods have been published as the result of work by an Intersociety Committee (17a), covering the analysis of the following materials in atmospheric particulates: benzo(2)pyrene; polynuclear aromatics; arsenic; beryllium; iron; manganese; molybdenium; selenium; nitrates; plutonium; and strontium. The electron microprobe has been used for identification and for estimation of relative concentration of a variety of elements including Al, Si, P, S,

Cl, K, Ca, V, Fe, Ni, Cu, Zn, Br, La, Ce, and Pb in airborne particles (18)(19). In addition, Milliport filters and beryllium disks have been scanned by area for individual elements. Lead has been found associated with bromide and chloride, but samples also have been scanned in which the lead is present with sulfate. Particles can be located in which no elements can be identified in the x-ray distribution patterns, probably because they contain carbon and other lower atomic weight elements.

The monitoring of a variety of hazardous or potentially hazardous elements, including Be, Hg, As, Cd, V, and Mn may require the development on on-site analyzers. For this purpose x-ray fluorescence techniques should have potential. Work is underway to estimate better the potential of x-ray fluorescence for analysis of airborne particulate fractions. Atomic absorption already has been shown to have potential for continuous analysis of certain elements in air, such as lead (20).

Such equipment would allow sampling each 24 hours or oftener, 7 days a week, instead of the more frequent analysis ordinarily associated with filter collections and laboratory analysis. A monitoring technique for various elements also may be desired for analysis of some types of stationary emissions.

Elemental Analysis – Particulate samples are usually analyzed for the elements after collection on a filter. Contamination of the filter media with trace quantities of the elements is a common limitation on analysis. The glass fiber filter most commonly used for collection is not satisfactory for trace analysis for a number of metals because of such contamination. Other substrates, such as polystyrene membrane of Millipore filters, have been used but still are far from completely adequate. For a few elements such as mercury, other collection techniques are used, such as amalgamation or use of gas bubblers.

Methods of analysis utilized included colorimetry for a few elements. Usually physical techniques including optical emission spectroscopy, atomic absorption spectrophotometry, neutron activation analysis, x-ray fluorometry, spark-source mass spectroscopy, and stripping voltammetry have been used for elemental analysis. The status of techniques available through 1966 has been reviewed with emphasis on wet chemical techniques, ring-oven methods and atomic absorption spectrophotometry (21).

Optical emission spectroscopy has been used for elemental analysis over the past 15 years in Federal monitoring activities. This spectrographic technique has been applied to analysis of 16 elements: As, Be, Bi, Cd, Co, Cr, Cu, Fe, Mn, Mo, Ni, Pb, Sb, Sn, Ti, V and Zn (22). This method involves considerable analytical skill but it is still only semiquantitative, lacks sensitivity for a number of elements, and is limited by blanks in the filter substrate.

Atomic absorption is a more quantitative and specific technique and is very sensitive for some elements. This technique can be used for As, Be, Na, K, Ca, Mg, Ba, Cd, V, Co, Cu, Zn, Ag, Ni and Pb (11)(23)(24)(25). The method is limited by the availability of hollow-cathode lamps. A number of these elements have been measured after particle-sizing of airborne particulates (11). Atmospheric precipitation samples have been analyzed for Pb, Zn, Cu, Fe, Mn and Ni (26).

The thermal neutron-activation analysis of particulate matter in air has received increased use in recent years (27)(28). Scintillation counting using a thallium-doped NaI crystal on samples on cellulose fiber was utilized to determine Al, V, Mn, Na, Cl, and Br. In much more ambitious use of neutron activation analysis, airborne particulate samples were collected on polystyrene filters and counted by means of a lithium-doped De detector.

The procedure permits determination of up to 33 elements provided that 5 minute irradiations are utilized with 3 and 15 minute cooling periods and 2 to 5 hour irradiation with a 20 to 30 hour radiation [20 to 30 day cooling period (29)]. Even with this scheme, most of the low molecular weight elements cannot be analyzed. The sensitivities for a number of the elements are marginal or inadequate because of various limitations, including interference of other substances in the sample or a high blank value. Included in the group of elements analyzed by this method are Mg, S, Cl, Ti, Ni, and Ag. Determination of Li, Be, B, C, N, O, F, Si and P could not be made. Therefore, only a few elements lighter than K could be analyzed. In addition Pb and Cd, biologically important elements that occur in airborne particulate matter, are not included in the scheme. A number of the elements that can be measured are not presently of concern biologically nor do they contribute significantly to the mass of particulates. These elements include Ga, La, Sm, Eu, Au, Sc, Ce, Co and Th. Elements such as Cr, F, Ni, Zn, Se, Sb and Hg are determinable only after a 20 to 30 day decay period. Therefore, although this neutron-activation analysis scheme in principle appears very attractive, in practice it can only be applied with one or more other analytical techniques if determination of a range of biologically important elements or a mass balance is desired.

A particularly interesting study of elemental analysis of particulates included use of optical emission spectrography, parts-source mass spectrography, x-ray fluorescence, atomic absorption, CHN analysis and chlorine analysis by a colorimetric technique (19). By this means, 77 elements could be determined or estimated in particulates collected in Cincinnati, Denver, St. Louis, Washington, Chicago and Philadelphia. Bound oxygen could not be included in this analytical scheme. This study is one of the few that permits any evaluation of analytical results from two or more methods on the same samples.

The agreement was good between the analyses by atomic absorption and x-ray fluorescence for Ca, Fe, Cu and Zn, often being within 20% or less of each other. The agreement was not as close for K and Pb in some samples. When the analyses for Cu and Zn by optical emission spectrographic technique were compared with those by atomic absorption of x-ray fluorescence, the values obtained by emission spectrography averaged a factor of two to three lower. The Fe analyses by the optical emission spectrographic technique also averaged appreciably lower than those by atomic absorption or x-ray fluorescence.

It is likely that even with additional improvements in these techniques, at least two if not three different analytical techniques will be required to cover the range of elements of interest with adequate sensitivity. It also would seem of considerable importance to conduct concurrently some additional comparisons of each of the several techniques for elemental analyses of the same sets of airborne particulate samples. If confidence can be placed in these analyses being correct, on an absolute basis, within 20 to 30%, the reliability of each of the methods applicable to each element of interest must be ascertained.

Anion Analysis – Analyses for sulfate and nitrate are routinely made on bulk particulate samples collected on the high volume sampler at urban and nonurban sites (22)(30)(31). Sulfate is determined by the methylthymol blue method by means of an autoanalyzer (31). Nitrate is assayed colorimetrically following reduction to nitrite by alkaline hydrazine (31).

In investigations on the particle size distribution of sulfate, several analytical procedures have been used. In one of these investigations, high temperature reduction of sulfate to hydrogen sulfide was utilized, followed by an iodometric, microcoulometric titration for the hydrogen sulfide (14). In other work (10)(32), a modified turbidimetric procedure was used to analyze the size-fractionated material from an Anderson sampler (33). Nitrate was also determined colorimetrically after reduction following particle sizing (12). Phosphate and chloride also have been analyzed colorimetrically after particle sizing (12).

Organic Particulates – A considerable amount of analytical effort has gone into the measurement of various types of organic substances in particulate matter. Polynuclear aromatic hydrocarbons, particularly benzo(a)pyrene, have received the most attention. Certain of the procedures involving column chromatography followed by spectrophotometric analysis have interference problems associated with them.

Paper chromatography and thin layer chromatography have received attention as separation techniques. Spectropolynuclear aromatic hydrocarbons, azaheterocyclics and polynuclear ring carbonyls are determined after use of column and thin layer chromatography. Gas chromatography has been applied to n-alkanes in airborne particulates as well as to polynuclear aromatic hydrocarbons. However, adequate separation of benzo(a)pyrene from benzo(e)pyrene has been a problem in applying gas chromatographic procedures to the analysis of airborne particulates. The details of these developments have been reviewed in several publications (34)(35). The status of work on pesticides, nonvolatile fatty acids and phenols is also considered in one of these reviews (35).

More recently, a simple and rapid procedure has been developed for the determination of benzo(a)pyrene, benz(c)acridines and 7H-benz(de)anthracen-7-one (36). Benzene extracts were separated by one-dimensional thin layer chromatography followed by analysis of benzo(a)pyrene by spectrophotometry or spectrophotofluorometry and the benz(c)acridines and 7H-benz(d)anthracen-7-one in trifluoroacetic acid, by spectrofluorometry.

The procedure was applied to 6 month composites from 52 cities. Most analyses for polycyclic organic materials are applied to 24 hour integrated samples or composites representing even longer integration times. A sensitive procedure, requiring less than 20 minutes, that involves thin layer chromatography and direct fluorometric measurement was applied to 3 hour sequential air samples for analysis of 7H-benz(de)anthracen-7-one and phenalen-1-one (37).

As pointed out by the American Chemical Society (38), the Environmental Protection Agency should soon be using x-ray fluorescence for large-scale routine analysis of particulates in the environment. L.S. Birks and J. Gilfrich at the Naval Research Laboratory, Washington, D.C., helped work out the specifications for a suitable instrument in collaboration with J. Wagman of EPA. EPA planned to install the device in 1973 at its laboratory at Research Triangle Park, North Carolina. There it will analyze contaminants from EPA sampling stations around the nation.

X-ray fluorescence is used mainly today in production control in steel, cement and other industries. For pollutant analysis, particulates are filtered from air or water, and the filter is inserted directly into the x-ray machine. It then reads out, simultaneously, the amounts of as many chemical elements in the sample as it is designed to accommodate. EPA's new instrument will analyze for 15 to 20 elements at a time.

Among the problems tackled by Birks and his colleagues was the sensitivity of the instrument. At first it could detect only about 1 microgram of an element per square centimeter of filter material. The machine that evolved, however, can detect as little as one hundredth of that amount, or about 10 nanograms per square centimeter.

Future research on x-ray fluorescence likely will be aimed at assessing the chemical combinations of elements in the sample, as oxides, organic compounds, and the like. Such information not only will tell more about the environment than does analysis only for individual elements, it will be needed eventually in biomedical evaluation of the effects of pollutants.

The state of the art in instrumentation for the measurement of particulate emissions from combustion sources has been reviewed by G.J. Sem et al of Thermo-Systems Inc. of St. Paul, Minnesota (39).

A sampling device for the analysis of furnace gases from glass or other furnaces has been developed by S. Krinov (40). The device consists of a portable sampling train that can be moved to test samples from various parts of a furnace for analysis of furnace gas. Further, it has excellent resistance to abrasion, heat, chemicals and thermal shock, thereby permitting operation at the same temperature as that of the furnace gas, thereby avoiding premature cooling and collection of condensate.

A device developed by C.B. Moore and B. Vonnegut (41) consists of a monitor and spectrometer device for indicating the total concentration in the atmosphere of particles below any desired maximum mobility level. This monitor is, for example, particularly useful in the determination of concentrations and the distribution of sizes of suspended particles in the atmosphere. U.S. Patent 2,986,923 by B. Vonnegut describes an apparatus for detecting and measuring aerosols. In that apparatus, no special provision is made for the removal of the fast charging ions. In the apparatus of this process, however, an electric field is imposed in an ion filter that causes the charging ions to move in a controlled fashion. Movement of the fast ions is thus controlled independently. This apparatus therefore makes it possible to obtain the approximate spectra of particle size, in the sample of gaseous fluid being examined, from the mobility of the charged particles.

The device of this process operates by electrically charging incoming airborne particulate matter to the maximum that can be accepted (under the set conditions), removing all excess charging ions and then measuring the resulting charge concentrations. The apparatus consists of (a) a suction fan or other suitable means for introducing air into the intake; (b) a charging section in which large quantities of fast, gaseous ions are produced in a potential gradient so that passing particles acquire a net charge of one polarity; (c) an ion filter section where all fast ions and charged particles with mobilities greater than some preset value are removed by electrostatic precipitation and (d) a Faraday cage in which the potential produced by the net mean charge concentration on the remaining particles is measured continuously with an electrometer.

A device developed by E. Schreiber and O. Skatsche (42) is a device for determining the soot content of flue gases by the degree of blackening of a filter traversed by a predetermined amount of flue gases taken from the flue gas current via a sampling pipe. In a conventional device of this kind the soot content is determined by means of a number of appropriately marked small filter paper plates. To perform the measurement, the operator has to move close to the source to be tested for the purpose of taking the sample and then to effect the zero-white-level adjustment of the visual reading instrument by hand for the subsequent evaluation procedure. This method is, however, comparatively complicated and time-consuming, apart from the risk of confusing the blackened filter paper plates. Additional individual sources of error are inherent in the zero-white-level adjustment of the visual reading instrument by hand.

It is the object of this device to eliminate individual influences in connection with the determination of the soot content of flue gases to a considerable extent in order to be able to successfully exploit the valuable clues on the combustion pattern of the tested source provided by this method. The device comprises a filter tape capable of performing a preferably rhythmic motion in transverse relation to the sampling pipe, the blackening of the filter being measurable by means of an adjacent photoelectric measuring instrument.

Such a device permits the continued surveillance of the soot content of the flue gases without interrupting the operation of the source to be tested and the avoidance of delays due to the time required for the analysis of the test filters. It is both possible with this method to take samples at short intervals during operation, evaluating the blackened filter tape in another place or at a later date, and to analyze every single sample by the immediate measurement of the relative light transmittance of the filter tape on the test bench.

A device developed by A. Goetz (43) permits the collection and analysis of the particulate content of defined volumes of gaseous samples. The device is particularly adapted for use with particulates of submicron diameter and is particularly adapted for use with samples having a particle concentration less than that which can be measured by scattered light photometry of airborne particulates. Specifically, the device may be used with samples where the particulate mass concentration is in the order of less than a 10^{-7} fraction of the mass of the supporting gas phase, i.e., concentrations so low that chemical tracing methods become ineffective.

The particulates from the sample are collected on a polished opaque surface which is then illuminated at a low angle of incidence to provide only scattered light at a high angle of incidence with the deposited particulates producing the light scattering such that the scattered light intensity is a measure of the particulate concentration.

A device developed by A.R. McFarland and C.M. Peterson (44) is an aerosol sampler which combines the inertial impact separation principle with a liquid collection substrate.

In the past, aerosol samplers have utilized the inertia impact principle to separate aerosol from the gas to be sampled. However, in such samplers, a solid or jell substrate was generally used to collect the separated aerosol particles. One of the disadvantages of the samplers utilizing a solid or jell collection substrate was that the operation of the sampler had to be periodically discontinued so that the collection substrate and separated aerosol could be removed from the sampler.

Other previous aerosol samplers have used a liquid substrate to collect the separated aerosol particles. These samplers, however, have not utilized the inertia impact principle to separate the aerosol from the gas, but have used other means, such as electrostatic deposition, to separate the aerosol. However, such samplers were generally more expensive to manufacture and maintain than the inertial impactor samplers.

The reason for not heretofore using a thin liquid film as a collection substrate with an inertial impactor sampler was that it was thought that the relatively high velocity gas stream, such as is required for inertial separation of the liquid film and would cause aerosolization of the liquid, thereby producing inefficient and inaccurate results.

The device of this process effects a combination of the inertial impactor principle with liquid collection means by utilizing a nozzle assembly plate in which the nozzles are arranged and disposed with respect to the radii of the rotating collection disc so that the high velocity gas streams directed at the disc by the nozzles do not significantly interrupt the integrity of the liquid substrate film and do not cause substantial aerosolization of the liquid. Also, an aspirator is used to remove the liquid and hydrosol from the disc as it reaches the periphery of the disc, whereby high collection efficiencies can be achieved.

A device for the determination of the concentration of airborne dust particles in an atmosphere such as that of a clean room has been developed by B.O. Lindahl and V. Lamme (45). A slide having a dark, accurately flat top surface is exposed to air to be sampled. A beam of light is directed across the surface, parallel thereto. Only accumulated dust particles on the surface reflect light in directions nonparallel thereto and the light they thus reflect is concentrated on a photoresponsive element. Its response affords an indication of quantity of particles on the slide.

A dust monitoring method developed by S. Badzioch and P.G.W. Hawksley (46) involves passing a stream of gas containing dust into a centrifugal flow separating device, maintaining constant one characteristic of gas flow through the device, allowing the dust concentration in the device to increase, and monitoring another characteristic of gas flow through the device.

For example, the characteristic monitored may be the difference between the pressure of the gas at the inlet to the device and the pressure of the gas at the outlet, the characteristic kept constant being the rate of flow through the device. Alternatively, the characteristic monitored may be the rate of flow of gas through the device, the difference in pressure between the inlet and outlet of the device being then kept constant.

The method can readily be carried out repeatedly at regular intervals and in each instance monitoring performed after the same period of time has elapsed to allow dust to collect in the device, which must have means for enabling the dust collected to be removed at regular intervals, i.e., periodically before the next period of dust collection begins.

If a record were kept of the monitored characteristic with such repetition of the method, the record would be of saw-tooth pattern, the envelope of which would represent the variation of dust concentration with time. The method is therefore applicable to the control of manufacture in a plant so as to control, in turn, the dust concentration in a gas stream leaving the plant. If absolute determination of dust concentration in terms of weight per unit volume of gas is required then calibration of the monitoring equipment, using samples of dust which is to be monitored is necessary.

A device developed by K.E. Noll (47) involves a rotary means for evaluating the quantity and size distribution of particulate matter in the atmosphere. Instruments employing the process of inertial impaction for the collection of particles have been used for many years. Most of these instruments were designed for specific sampling conditions and are not ideally suited for collecting the complete size, density and concentration range of particulate matter present in the atmosphere.

Sampling techniques and information on particulate matter in the atmosphere have been provided by meteorological research concerned with condensation nuclei and raindrop size distributions, and biological research concerned with spore formation and transport. However, these technologies have been interested in restricted types and size ranges of particulate matter and no attempt has been made to collect a representative size distribution of all particulate matter present. Air pollution measurements of atmospheric particulate matter have generally provided only the total weight of material present.

Little is known about the variation of these particles under different conditions and because of their considerable polydispersity, a mean size or total weight is not adequate to describe them. Thus, the determination of both the size distribution of particulate matter and the quantity present in urban and rural environments would provide much valuable information which could be used to better understand the role of these particles in the atmosphere. Furthermore, a sampling technique for determining the quantity and size distribution of particles present in the atmosphere is necessary for the monitoring and control of air pollution in urban and rural areas.

An instrument which would provide this information would be of particular use in monitoring industrial pollution problems. Such a device must meet particular requirements for sampling of particles in the atmosphere due to the small numerical concentration of these particles and their peculiar inertia properties. The device must sample at a high volumetric flow rate and not be affected by variations in windspeed and direction during the sampling period. Versatility of the instrument is necessary due to the wide range of particle sizes over which a high sampling efficiency must be maintained and a frequent large concentration change over the range of particle sizes. Furthermore, the instrument must be sufficiently portable to permit this use at a number of sampling sites.

It is therefore an object of this development to provide a portable multistage rotary inertial impaction sampler for the collection of particulate matter in the atmosphere by impaction, which device has a great degree of versatility for collecting suspended particles of all sizes.

Figure 42 shows the configuration of such a rotary inertial impactor device. As shown there, a motor **9** or other suitable rotary drive means transmits rotative motion to shaft **15** through a V belt **12** and a pair of pulleys **11** and **13**. The shaft rotates in an upper bearing **14** and a lower bearing **26** which are attached to the supporting frame **23**. A rotor arm **17** is attached firmly to the shaft for rotative motion.

In the form shown, four impactor stages **24** are attached to the rotor arm. Each impactor stage **24** is made up of four collector elements **19, 20, 21** and **22** of different widths. Suitable sizes for these sample elements have been found to be 1/32 of an inch for element **22**; 1/8 for element **21**; ½ inch for element **20** and 2 inches for element **19.** The widths can vary in any desired ratio. The rotor arm was designed to operate at approximately 450 rpm, but may be caused to rotate at any desired rate of speed. The four collector stations are located on the rotor arm at various distances from the shaft providing a velocity variation and thus a variety of impingement conditions for collecting atmospheric particulate matter.

Each impactor stage **24** is removably attached to the rotating beam **17** by use of machine screws or other suitable fastener means passed through holes provided in section **25** of element **19.** The individual elements of the impactor stages are fastened together as shown in the detail at the base of the figure, by use of setscrews **31** or other suitable fastener means. The two larger stages **19** and **20** are constructed of flat stock material and have rounded edges **30** with a radius of approximately ¼ inch to prevent turbulent breakaway

FIGURE 42: PERSPECTIVE ELEVATION OF ROTARY INERTIAL IMPACTOR FOR SAMPLING

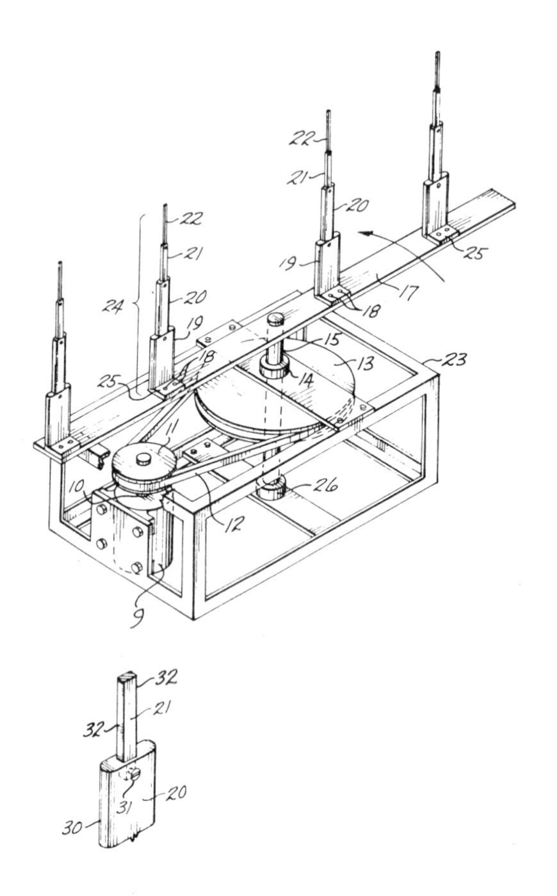

Source: K.E. Noll; U.S. Patent 3,633,405; January 11, 1972

at these points. The two upper stages **21** and **22**, which are a part of each impactor stage **24**, are similarly made of a flat stock material but have a sharp filed edge **32**. The elements of the individual impactor stages are made from aluminum stock with highly polished surfaces which have been black anodized to prevent backscatter of light when viewed in a microscope. Provision may be made at each end of rotor arm **17** for the addition of weights to balance the apparatus under various conditions.

For example, if only one impactor stage is to be used it would be necessary to provide a counterbalancing weight on the opposite side of the rotor arm to prevent rotational imbalance. Similarly when all four sample plates are in place and located off center as shown, an imbalance will be present unless a counterbalancing weight is provided. In operation, the leading face of each sample element is coated with a tacky or adhesive material. White petroleum jelly, for example, may be advantageously used due to its stability, ready availability and simplicity of application to the leading face of the sample elements. Samples may be collected by bolting sets of assembled collectors to the rotor blade **17** and rotating the complete assembly at any desired rate of rotation to provide the desired impaction rate. Particles are impacted on each collector face as the collector moves through the air due to inertial forces of the particles. The particles which are captured by the impactor stages may be sized and counted by visual methods well-known in the art. Surface illumination of the samples being examined may be provided by overhead lighting thus allowing the sampler face to be viewed directly without any cover glass or other coating needed between the particles and the microscope objectives. Particle size may be determined by comparison to a known standard by techniques well-known in the art.

A device developed by P. Coubon (48) is one employing a thin filter element mounted integrally with a centrifugal fan-like rotor so that when the rotor is rotated air will be drawn through the filter. This assembly is mounted within a fairly spaceous enclosure with limited ingress and egress orifices so that air will tend to be recirculated within the enclosure. Solid particles in the air entering the enclosure will tend to be collected in a spot at the center of the filter element, the size of the spot being a measure of the amount of dust in the air.

An apparatus developed by E.B. Childs (49) permits determination of the quantity of soot and other solid, nonwhite contaminants present in the gases being discharged from a combustion chamber. There are several testing devices in use in industry today from which some indication of the degree of contamination of waste gases can be ascertained. These devices generally fall into two distinct types which are briefly discussed hereinafter.

A first type can be referred to as a continuous smoke tester which operates on the principle of determining the degree of transparency of the gas by the use of photoelectric devices. A recording device is used in conjunction with photoelectric equipment so that a permanently visible record of the study is provided.

The main disadvantage inherent in this type of equipment is the poor sensitivity to smoke below the visibility level. This disadvantage prevents the use of the device except in the study of gases containing an established minimum of contamination. Other factors limit their utility such as unsuitability for being transported for various applications and exacting installation requirements which render its adaptation to specific situations expensive and time consuming. Furthermore, devices of this character are delicate and require considerable attention if they are to be maintained in proper working order.

Another type of testing device is commonly known as the spot tester. Spot testers provide a paper record from the smoke itself by causing smoke to deposit a portion of its contaminant upon a suitable paper. The means employed for doing this are of varying types. Some devices utilize an intermittently operated filter tape, the operation being either manual or mechanical. For various reasons, equipment of the spot testing type is unsatisfactory.

In the first place, no understanding or indication can be derived from a single spot test as to whether or not the smoke condition that is made apparent by the spot has been continuous and constant during the taking of the record or has continuously varied from one degree of density to another. Therefore, in order to determine what conditions are taking place in a combustion chamber over a given interval of time, it is necessary to make several spot tests to indicate the changing conditions. This, of course, requires a considerable amount of time. Furthermore, it is not possible to make immediate adjustments to an engine and to observe their effects during the taking of the record.

The record must first be taken and studied before any adjustments can be made, and after the adjustments are made, additional spot tests must be run in order to determine the effect of the adjustment to the engine. Oftentimes several series of tests and adjustments are necessary in order to effect the necessary corrections to the engine condition. The adjustment is difficult to make becuase the engine conditions are continuously changing, especially if the engine has just been started in operation because at such time the combustion chamber is cold, as is the incoming fuel, and thus combustion efficiency is lower than when the engine and fuel have become heated. Therefore, in order to obtain reliable data and to make accurate adjustments, it is necessary to allow the engine to heat up for many minutes before spot tests are completed. Additionally, in using the spot test, if the smoke that is being examined is heavily laden with contaminants, the initial spot that is deposited is likely to be so dense as to be meaningless except to indicate that the gas is quite heavily laden with contaminants. Oftentimes, a spot test of half the time of a previously made one under substantially the same conditions will produce a smoke spot of substantially the same color and density. Therefore, it is necessary to run several spot tests of progressively shorter duration before a spot can be produced which will have a definite meaning when compared with its standard.

The main disadvantage of the spot type test is that it does not afford a record corresponding to the changes that take place in a combustion chamber of an engine either as the changes are effected by reasons of the physical conditions of the engine or as effected by adjustments made thereto by the individual who is examining it, as a result of which disadvantages, accurate records and adjustments are extremely difficult to obtain, if not in many cases, impossible.

There is also a good deal of condensation trouble. If spot testers are not warmed up (on a furnace, or in another makeshift manner) frequently first spots are wet and soggy and must be repeated. Spot testers have been brought to a fair degree of perfection within the limits of their applicability and for those purposes usually produce satisfactory results. However, there is considerable need for an instrument of a more sensitive nature so as to permit a higher degree of perfection in the building and operation of combustion chambers and engines.

U.S. Patent 2,667,779 discloses a device for drawing off and filtering a contaminated gas through a moving filter tape so as to permit an observation of the contaminant content, and the changes in the degree of contamination. However, the device of U.S. Patent 2,667,779 requires that the filter tape record be compared with prestandardized test tapes to determine the quantity of contaminants in the gas under test.

In the device of this process, means are provided for indicating the amount of solid, nonwhite contaminant suspended in a gas and for moving a filter tape through the stream. The apparatus also includes means for generating a signal representative of the amount of contaminants filtered from the stream by the tape, and means for indicating the number of times the contaminants signal exceeds a predetermined value for a predetermined period of time.

Thus, the apparatus provides for fast interpretation of results of the analysis by indicating a numerical value representative of a degree of amount of contaminants in the gas. By providing for fast interpretation, the device is particularly useful in field investigations where compliance with smoke regulations is involved.

Figure 43 is a schematic diagram showing such an apparatus. Filter paper **13** from a paper tape feed **10** is pulled through a tape head **11, 12** by a tape pull motor **14.** The tape **13** passes from the motor to a tape rewind spool **15.** The tape head comprises an inflow plate **11** and an outflow plate **12.** A conduit **21** interconnects an exhaust probe (not shown) in an exhaust system such as that used with a diesel engine and a bore or conduit **11'** formed in the inflow plate. A bore or conduit **12'** formed in the outflow plate is connected to a conduit **20** having a filter **17** therein. The conduit is connected at its downstream end to a vacuum pump **19.** A vacuum gauge **16** and a vacuum regulator **18** are also connected to the conduit to provide a constant vacuum in the conduit.

The vacuum pump acts to draw exhaust from the exhaust probe (not shown) through the conduit **21,** the inflow conduit or bore, the filter tape and the outflow conduit or bore. The amount of light from a lamp **23** which is reflected from the solid, nonwhite contaminent smoke trace on the filter tape is sensed by a photocell **22.** The photocell provides an output representative of the darkness of the smoke trace to a recorder **24.** The recorder, the photocell, the lamp and the tape pull motor are connected to a regulated 12 v. DC power supply **25.**

FIGURE 43: MOVING TAPE FILTER DEVICE FOR ANALYZING SOLIDS IN SMOKES

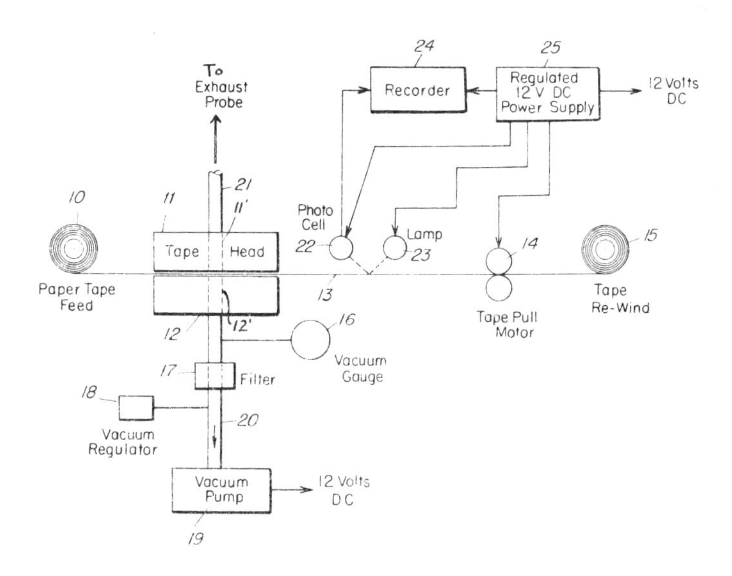

Source: E.B. Childs; U.S. Patent 3,653,773; April 4, 1972

The regulated power supply and the vacuum pump are connected to a 12 volt DC source, such as a vehicle battery. To monitor a diesel engine exhaust, the exhaust probe (not shown) is suitably aligned along the axis of the exhaust stack or tail pipe at the outlet end of the system. The probe may be constructed of stainless steel and include a heat radiator to dissipate heat. The conduit **21** may be constructed from small tubing, such as $^3/_{16}$ inch inside diameter to reduce sample flow time between the probe and the tape head **11, 12.** The vacuum pump is provided to further reduce the time lag due to the transfer of the sample from the probe to the tape head **11, 12.** To stabilize the flow of exhaust through the system, the vacuum regulator **18** may be suitably set to provide a 4 inch Hg vacuum indication on the gauge **16.**

The tape **13** is considerably wider than the mouth of the inflow and outflow conduits **11'**, **12'** so that the tape acts as a gasket and seals the mouth of the outflow conduit **12'**. The inflow plate **11** is fitted sufficiently tightly on the outflow plate **12** to substantially eliminate leakage into the outflow conduit **12'** from the atmosphere. A suitable diameter for the inflow and outflow conduits **11'**, **12'** is $^3/_8$ inch.

The tape pull motor **14** is powered by the regulated power supply **25** to provide a constant tape speed. A suitable tape speed is 10 inches per minute for diesel engine exhaust monitoring. The lamp **23** and the photocell **22** are enclosed in a bored head such that the photocell senses only the light reflected from the tape. The head may be made from black Bakelite board to provide a light path, as indicated by the dash line from the lamp to the tape, and reflected to the photocell at approximately a 60° angle.

A device developed by R.W. McIlvaine (50) provides a method of determining the difficulty of removing pollutants from gas by wet scrubbing. It includes the steps of passing a sample of the gas through a portable test wet scrubber stage for removing some of the pollutants therefrom, measuring the amount of pollutants remaining in the gas sample after passage through the wet scrubber stage and adjusting the pressure drop across the scrubber stage to obtain the desired level of pollutants remaining after passage therethrough. The results obtained from such a testing technique would be a direct indication of the difficulty of removing the particular contaminants from the tested gas and would give vendors, purchasers, and engineers a reliable index as to the size of equipment needed and the operational characteristics and power required to produce results in compliance with the applicable codes and ordinances.

A device developed by N.S. Smith, Jr. and G.E. Fasching (51) is one whereby size and flow rate of dust particles entrained in a flowing gas stream are measured by an electrogasdynamic dust monitor. A multiple segment collector section generates currents with interdependent values related to dust properties.

An apparatus developed by M.J. Pilat (52) is a cascade impactor for measuring the quantity and size distribution of suspended particles in stacks, ducts and other pollution sources. The impactor which is adapted to be placed within the duct carrying the particle-laden fluids comprises a tubular body containing a plurality of serially spaced apart impactor plates interspaced between serially spaced apart jet stages, each succeeding jet stage having a smaller gas flow cross-section than the preceding jet stage.

A sized portion of the particulate matter suspended in gases flowing through the cascade impactor is captured by collection surfaces on the impactor plate placed below the specific jet stage which imparts a sufficient amount of inertia to the particle for it to impinge upon the collection surface. Dry collection surfaces are used for liquid aerosols whereas collection surfaces covered with a sticky substance such as grease are utilized for dry particulate matter. The size distribution of the particulate matter is reflected by the amounts of particulate matter adhering to the various impactor plates.

A device developed by J.F. Vaneldik, R.I. Wilson, G.R. Kampjes and W.J. Lavender (53) embodies an electrostatic precipitator especially suited for use in a gas sampling system for removing solid particulate material from a sample gas stream withdrawn from an ore reduction roaster for passage to a gas chromatography unit. The precipitator is designed to remove residual and very small particulate material from such a gas stream after such a stream has first passed through a relatively coarse screen filter.

The electrostatic precipitator includes a housing divided by a downwardly extending partition into an ionization chamber containing an ionizing wire electrode and a collector chamber containing a plate-type collector electrode. The partition terminates upwardly of the floor of the housing to provide a dust collection chamber through which the gas stream flows on passing from the ionization chamber to the collector chamber.

In the overall gas sampling system, a blowback gas system is provided for dislodging solid material from the screen filter and from the upstream portion of the sampling conduit and for returning dislodged material to the roaster to prevent cumulative plugging of the system. Operation of such a blowback system is usefully controlled by a sampling programmer of the gas chromatography unit to which the system is coupled so that such blowback system operates automatically between successive analyses.

An apparatus developed by R.L. Chuan (54) is an apparatus for sensing airborne particulate matter which combines the impaction action of a decelerated small air jet with a quartz crystal microbalance to detect, by mass sensing, the presence of micron size airborne particulate matter.

A quartz crystal, coated with a tacky material and oscillating in a shear mode as part of a resonant circuit, responds to very small changes of mass on its surface by changes in the resonant frequency. This frequency is mixed with that of a reference oscillator and the differential frequency, which is proportional to collected mass, may be counted or converted to an analog signal suitable for direct recording or display, or for the performing of control functions.

A device developed by M.C. Gourdine and S.E. Law (55) is an electrostatic device for measuring the mass of particulate matter entrained in a gaseous flow per unit volume of such flow. The gaseous flow is passed down a bounded flow path and the entrained particles are exposed to and charged by a corona discharge. The charged entrained particles are carried downstream to a collection section where the particles are repelled by a charged deposition electrode and driven toward a grounded electrode. A portion of the charged particles are collected on a dielectric surface positioned in front of the grounded electrode. The accumulated charge on the dielectric surface is then measured by an induction electrode and an electrometer to show the particle mass per unit volume of the flow.

According to a process developed by N.R. Whetten (56) the size distribution of particles including particulate matter in a gas sample is determined by generating a three-dimensional alternating current electric field in an expansion chamber for mass selective sorting of the particles whereby condensible vapor saturated charged particles in the gas sample which are in a particular range of charge-to-mass ratio are trapped within the electric field and the particles outside the range are swept away.

An expansion of the gas in the expansion chamber causes condensation of droplets on the trapped particles and a condensation nuclei counter may be used for detecting the number of droplets which corresponds to a like number of the trapped particles in the particular range of charge-to-mass ratio occurring in the gas sample.

A device developed by W.A. Riggs (57) is an opacity monitor for measuring the light transmissibility of gases in a furnace stack. A collimated light source is directed through the stack, a pair of photocells are disposed to measure respectively the intensity of the source and the intensity of the light after it passes through the stack, the two measurements being balanced for normal conditions and the ratio of the two measurements being used as an indication of the opacity of the stack gas.

A device developed by J.O. Hanson and R.M. Ross (58) is an apparatus for monitoring suspended aerosols and solid particulates in a gas. The apparatus consists basically of a filter in a spring-mounted holder, means such as a pump for moving gas through the filter, means for causing the filter assembly to oscillate substantially in a horizontal plane, means for measuring the period of oscillation of the filter assembly and for comparing the period measured to the tare period of oscillation, and means for converting the period difference into an indication of the mass added to the filter during the sampling period. In its preferred form the apparatus includes means for assuring that the filter and matter collected thereon have sufficiently low moisture content during the weighing operation.

References

(1) Stern, A.C. in *Industrial Pollution Control Handbook,* H.F. Lund, Editor, New York, McGraw-Hill Book Company (1971).

(1a) Friedlander, S.K., *Envir. Sci. & Tech.* 7, (13), 1115 to 1118 (December 1973).

(1b) Lawrence Berkeley Laboratory, "Instrumentation for Environmental Monitoring - Air," *Publ. LBL-1,* Volume 1, Berkeley, University of California (February 1, 1973).

(2) Altshuller, A.P., in NBS Special Publication 351, *Analytical Chemistry - Key to Progress on National Problems,* Washington, D.C., U.S. Government Printing Office (August 1972).

(3) U.S. Public Health Service, "Air Quality Criteria for Particulate Matter," *Publ. No. AP-49,* Washington, D.C., National Air Pollution Control Administration (January 1969).

(4) Jones, H.R., *Fine Dust and Particulates Removal,* Park Ridge, New Jersey, Noyes Data Corporation (1972).

(5) Charlson, R.J., Horvath, H. and Pueschel, R.F., *Atm. Environ.* 1, 469 (1967).

(6) Charlson, R.J., Ahlquist, N.C. and Horvath, H., *Atm. Environ.* 2, 455 (1968).

(7) Noll, K.E., Mueller, P.K. and Imada, M., *Atm. Environ.* 2, 465 (1968).

(8) Charlson, R.J., Ahlquist, N.C., Selvidge, H. and MacCready, P.B., *J. Air Pollution Control Assoc.* 19, 937 (1969).

(9) Olin, J.G., Sem. G.J. and Christenson, D.L., Report on Contract CPA-22-69-83, Durham, North Carolina, National Air Pollution Control Administration (May 1970).

(10) Wagman, J., Lee, R.E. and Axt, C.J., *Atm. Environ.* 1, 479 (1967).

(11) Lee, R.E., Patterson, R.K. and Wagman, J., *Envir. Sci. Technol.* 2, 288 (1968).

(12) Lee, R.E. and Patterson, R.K., *Atm. Environ.* 3, 249 (1969).
(13) Robinson, E. and Ludwig, F.L., *J. Air Pollution Control Association* 17, 664 (1967).
(14) Ludwig, F.L. and Robinson, E., *Atm. Environ.* 2, 13 (1968).
(15) Giever, P.M., in *Air Pollution,* Volume 2, A.C. Stern, Editor, New York, Academic Press (1968).
(16) McCrone, W.C., in *Air Pollution,* Volume 2, A.C. Stern, Editor, New York, Academic Press (1968).
(17) McCrone, W.C., Draftz, R.G. and Delly, J.G., *The Particle Atlas,* Ann Arbor, Michigan, Ann Arbor Science Publishers (1971).
(17a) American Public Health Association, *Methods of Air Sampling and Analysis,* Washington, D.C. (1972).
(18) Landstorm, D.K. and Kohler, D., Final Report on Contract CPA-22-69-33, Durham, North Carolina, National Air Pollution Control Administration (December 1969).
(19) Henry, W.M. and Blosser, E.R., Final Report on Contract CPA-22-69-153, Durham, North Carolina, National Air Pollution Control Administration (August 1970).
(20) Loftin, H.P., Christian, C.M. and Robinson, J.W., *Spec. Lett.* 3, 161 (1970).
(21) West, P.W., "Chemical Analysis of Inorganic Pollutants," *Air Pollution,* Stern, A.C., Editor, Vol. 2, Academic Press, New York (1968).
(22) Morgan, G.B., Ozolins, G. and Tabor, E.C., "Air Pollution Surveillance Systems," *Science* 170, 289 (1970).
(23) Burnham, C.D., "Determination of Lead in Airborn Particulates in Chicago and Cook County, Illinois, by Atomic Absorption Spectroscopy," *Environ. Sci. Technol.* 3, 472 (1969).
(24) Thompson, R.J., Morgan, G.B., Purdue, L.J., "Analysis of Selected Elements in Atmospheric Particulate Matter by Atomic Absorption," *At. Absorption Newsletter* 9, 53 (1970).
(25) Sachder, S.L. and West, P.W., "Concentration of Trace Metals by Solvent Extraction and their Determination by Atomic Absorption Spectrophotometry," *Environ. Sci. Technol.* 4, 749 (1970).
(26) Lazrus, A.L., Lovange, E. and Lodge, J.P., "Lead and Other Metal Ions in United States Precipitation," *Environ. Sci. Technol.* 4, 55 (1970).
(27) Brar, S.S., Nelson, D.M., Kanabrocki, E.L., Moore, C.E., Burnham, C.D. and Hattori, D.M., "Thermal Neutron Activation Analysis of Particulate Matter in Surface Air of the Chicago Metropolitan Area," *Environ. Sci. Technol.* 4, 50 (1970).
(28) Hoffman, G.L., Duce, R.A. and Zoller, W.H., "Vanadium, Copper and Aluminum in the Lower Atmosphere Between California and Hawaii," *Environ. Sci. Technol.* 3, 1207 (1969).
(29) Dams, R., Robbins, J.A., Rahn, K.A. and Winchester, J.W., "Nondestructive Neutron Activation Analysis of Air Pollution Particulates," *Anal. Chem.* 42, 861 (1970).
(30) "Air Quality Data from the National Air Sampling Networks and Contributing State and Local Networks," *Publication No. APTD 68-9,* 1966 Edition, U.S. Department of Health, Education and Welfare, Public Health Service, National Air Pollution Control Administration, Durham, North Carolina (1968).
(31) McMullen, T.B., Faoro, R.B. and Morgan, G.B., "Profile of Pollutant Fractions in Non-Carbon Suspended Particulate Matter," *J. Air Pollution Control Assoc.* 20, 369 (1970).
(32) Lee, Jr., R.E. and Wagman, J., "A Sampling Anomaly in the Determination of Atmospheric Sulfate Concentration," *J. Amer. Ind. Hyg. Assoc.* 27, 266 (1966).
(33) Jacobs, M.B., *The Chemical Analysis of Air Pollutants,* Interscience, New York (1960).
(34) Sawicki, E., "Airborne Carcinogens and Allied Compounds," *Proc. Arch. Environ. Health* 14, 46 (1967).
(35) Hoffman, D. and Wynder, E.L., "Organic Particulate Pollutants," *Air Pollution,* Stern, A.C., Editor, Volume 2, Academic Press, New York (1968).
(36) Stanley, T.W., Morgan, M.J. and Grisby, E.M., "Application of a Rapid Thin-Layer Chromatographic Procedure to the Determination of Benzo(a)pyrene, Benz(c)acridines and 7H-Benz(de)anthracene-7-one in Airborne Particulates from Many American Cities," *Environ. Sci. Technol.* 2, 699 (1968).
(37) Stanley, T.W., Morgan, M.J. and Meeker, J.E., "Rapid Estimation of 7H-Benz(de)anthracene-7-one and Phenalen-1-one in Organic Extracts of Airborne Particulates from 3-Hour Sequential Air Samples," *Environ. Sci. Technol.* 3, 1198 (1969).
(38) American Chemical Society, *Chemistry in 1972;* Washington, D.C., Department of Member and Public Relations (1973).
(39) Sem, G.J., *Reports PB 202,665 and 202,666,* Springfield, Virginia, National Technical Information Service (1971).
(40) Krinov, S.; U.S. Patent 3,070,990; January 1, 1963; assigned to Pittsburgh Plate Glass Company.
(41) Moore, C.B. and Vonnegut, B.; U.S. Patent 3,178,930; April 20, 1965; assigned to Arthur D. Little, Inc.
(42) Schreiber, E. and Skatsche, O.; U.S. Patent 3,464,257; September 2, 1969.
(43) Goetz, A.; U.S. Patent 3,475,951; November 4, 1969; assigned to California Institute Research Foundation.
(44) McFarland, A.R. and Peterson, C.M.; U.S. Patent 3,518,815; July 7, 1970; assigned to Environmental Research Corporation.
(45) Lindahl, B.O. and Lamme, V.; U.S. Patent 3,526,461; September 1, 1970; assigned to SAAB Aktiebolag, Sweden.

(46) Badzioch, S. and Hawksley, P.G.W.; U.S. Patent 3,605,485; September 20, 1971; assigned to Coal Industry (Patents) Limited, England.
(47) Noll, K.E.; U.S. Patent 3,633,405; January 11, 1972; assigned to The Battelle Development Corp.
(48) Courbon, P.; U.S. Patent 3,590,629; July 6, 1971; assigned to Charbonnages De France, France.
(49) Childs, E.B.; U.S. Patent 3,653,773; April 4, 1972; assigned to Mobil Oil Corporation.
(50) McIlvaine, R.W.; U.S. Patent 3,668,825; June 13, 1972; assigned to National Dust Collector Corp.
(51) Smith, Jr., N.S. and Fasching, G.E.; U.S. Patent 3,679,973; July 25, 1972; assigned to the U.S. Secretary of the Interior.
(52) Pilat, M.J.; U.S. Patent 3,693,457; September 26, 1972; assigned to The Battelle Development Corp.
(53) Vaneldik, J.F., Wilson, R.I., Kampjes, G.R. and Lavender, W.J.; U.S. Patent 3,705,478; December 12, 1972; assigned to Sherritt Gordon Mines Limited.
(54) Chuan, R.L.; U.S. Patent 3,715,911; February 13, 1973; assigned to The Susquehanna Corporation.
(55) Gourdine, M.C. and Law, S.E.; U.S. Patent 3,718,029; February 27, 1973; assigned to Gourdine Systems Incorporated.
(56) Whetten, N.R.; U.S. Patent 3,740,149; June 19, 1973; assigned to General Electric Company.
(57) Riggs, W.A.; U.S. Patent 3,743,430; July 3, 1973; assigned to Shell Oil Company.
(58) Hanson, J.O., and Ross, R.M.; U.S. Patent 3,744,297; July 10, 1973; assigned to General Electric Company.

PEROXYACETYLNITRATE (PAN)

When reactive organic substances and nitrogen oxides accumulate in the atmosphere and are exposed to the ultraviolet component of sunlight, the formation of such compounds as ozone and peroxyacetylnitrate occurs. These photochemical oxidants, as they are known, are believed to have adverse effects on animals, humans and vegetation.

Measurement in Air

Peroxyacetylnitrate has been analyzed by gas chromatography with electron-capture detectors (1). This method has also been described in detail as a result of a study by an Intersociety Committee (2). Because nitrate is unstable, grab sampling is not satisfactory; however, monitoring on a given site can be conducted. Calibration problems and the lower stability of the electron-capture detector considerably limit the utilization of this technique by monitoring networks, as noted by Altshuller (3).

References

(1) Davley, E.F., Kettner, K.A. and Stephens, E.R., *Anal. Chem.* 35, 589 (1963).
(2) American Public Health Association, *Methods of Air Sampling and Analysis,* Washington, D.C. (1972).
(3) Altshuller, A.P. in NBS Spec. Publication 351, *Analytical Chemistry: Key to Progress on National Problems,* Washington, D.C., U.S. Government Printing Office (August 1972).

PESTICIDES

Pesticides include a spectrum of chemicals used to control or destroy pests which may cause economic damages or present health hazards. These are employed in agriculture, forestry, food storage, urban sanitation and home use. Since hundreds of such chemicals are presently available, this discussion is limited to those synthetic organic pesticides which currently are used in the greatest volume and are potential health hazards to humans, domestic and commercial animals and fish and wildlife because of the pesticides' inherent toxicity or persistence.

Pesticides can cause poisoning by ingestion, absorption through the intact skin, or inhalation. In cases of accidental occupational poisonings, it has usually been impossible to determine if the exposure was predominantly respiratory or dermal as noted by Finkelstein (1).

Of all the pesticides, the chlorinated hydrocarbon and organophosphorus insecticides are of major concern because of their health hazard. The acute toxicity of the organophosphates, on the average, is somewhat greater than that of the chlorinated hydrocarbons. However, the latter group is considerably more persistent because of their greater stability. Some members of the chlorinated hydrocarbon group, especially DDT, dieldrin, and BHC, have been found as residues in human fat tissue in all parts of the world.

Acute poisonings of commercial and domestic animals have usually been accidental and involved the more toxic organophosphorus insecticides. Animals also store the chlorinated hydrocarbon residues in fat tissue, and as with humans, the significance of this storage is not completely known. When ingested, as little as 7 to 8 ppm of DDT residue on hay will result in 3 ppm being excreted in cow's milk, and butter made from such milk will contain 65 ppm.

Fowl, fish, and many forms of wildlife have been adversely affected by pesticides, especially the chlorinated hydrocarbons. Birds are affected by DDT resulting in thin-shelled eggs and a decrease in hatchability. Wildlife in general have been affected in various parts of the country.

Herbicides may cause damage to other than the target plants if the dosage is too great. Some insecticides have produced undesirable flavors in plants used as food. Translocation of DDT and other insecticides into crops from the soil has been observed, but apparently this does not result in a high residue level. There have been no reports of damage to inanimate materials from the pesticides as such, but some of the solvents used in spraying applications could have a damaging effect on paint and other surfaces.

The primary source of pesticides in the air is the process of application. Even under the most ideal conditions, some amount will remain in the air following the application. However, under certain meteorological conditions, the pesticide spray or dust does not settle and can drift some distance from the area of application. Many episodes have occurred in which these drifting pesticide clouds have caused inhalation poisonings as well as toxic residues on croplands.

It is known that pesticides will volatilize into the air from soil, water, and treated surfaces. It has also been observed that rain and snow can wash pesticides from the air back to soil and water surfaces. There is good evidence that pesticide-containing dust originating from soil can enter the ambient air and be transported for considerable distances before falling back to the earth. The full significance of this with respect to effects on the total environment is still not known.

The economic and social benefits gained by the use of pesticides have been great. Pesticides have contributed to the eradication or reduction of a number of human diseases both in the United States and in other parts of the world. It has been estimated that nationally about five dollars are saved for every dollar invested in chemical pesticide usage. Although episodes of damage caused by pesticide drift resulting from agricultural treatment have been reported, no tabulated data on costs of damage due to air pollution from pesticides have been found. The costs of pesticide air pollution damage to humans, wildlife, and other animals cannot be estimated. The general topic of data management for monitoring pesticides and related compounds has been reviewed by G.B. Wiersma and H. Tai of the Office of Pesticide Programs of EPA (2).

In a successful and well managed ambient monitoring system the first phase is the planning and the collection of raw data from the field. Certain basic principles are involved in data collection. First, the objectives must be defined clearly. Precisely what is one monitoring for? For example, in pesticides, one defines what pesticides, chemical contaminates and heavy metal residues are of interest. Another fundamental objective is what one is trying to do with the data. The National Pesticide Monitoring Panel wrestled with this subject for a good deal of time. They came to the conclusion that, in pesticide monitoring, the fundamental objectives should be to determine the presence, levels and changes in time of pesticide residues in a media. Other monitoring systems will have different objectives, but these must be defined prior to implementation in the field (3). Second, one must define the sampling population. One cannot send a field team out with just the objective of simply collecting environmental information or samples. They have to be told precisely what kind of samples to collect, where to collect them and how to collect them. In addition, so that the data will make some kind of sense, a certain amount of homogeneity has to be established about the sampling population. For example, to facilitate data organization and analysis in monitoring of soils for pesticides, the United States was divided into two general categories, cropland and noncropland (4).

The third principle of data collection is to employ a sound, statistically based sampling design. This is of critical importance, particularly if some estimate is required on the reliability of data about the occurance, the amount and the change through time of the pollutants detected.

The fourth principle is to maintain consistency in collection of samples. They should be collected using the same techniques and procedures at each site. This requires that the people handling data collection be properly trained in the correct procedures and maintain sufficient caution to eliminate the possibility of cross-contamination of the samples. Once the sample has been collected in the field, it should be sent, as rapidly as possible, to the laboratory.

Care should be taken to insure that the samples arrive in a condition closely approximating their natural state. Once a sampling team is in the field, they should obtain as much information as possible about the site while there. In a monitoring program, in addition to soil and crop samples, the inspectors should collect the following information on:

(a) The crops that are grown on the sampling area;

(b) How much irrigation was used and how many inches of water were involved; and

(c) The pesticide or fertilizer used in the year of sampling. This includes the amount applied, the crop it was applied to, what it was used for, its formulation and method of application.

This information has been invaluable in the evaluation and interpretation of information generated by monitoring studies. Once the samples have been properly collected and sent to the laboratory, the second phase of data management begins. This is the extraction and analysis of the raw environmental samples to determine contaminant levels. In a typical case, these include pesticides, heavy metals, PCB's and other chemical contaminants.

When one looks at the general principles of the analytical technology of the pesticide residues, the important ones are the sensitivity and the selectivity of the method. The general trend has been to develop analytical methods specific for a particular pesticide or a particular group of pesticides, and sensitive to a level of 0.1 ppm, or even down to 0.1 ppb. The basic operations of the analytical procedures are:

(a) Subsampling in the analytical laboratory;

(b) Separation of the residues from the sample matrix, or extraction;

(c) Removal of interfering substances or cleanup; and

(d) Identification and determination of the residue.

In case the identity of the pesticide to be analyzed is not known in advance, the confirmation procedure may become a distinct operation all by itself. This operation can be elaborate and costly, in terms of both the manpower and the analytical equipment required for the identification of a component at a concentration of 1.0 ppm.

Specific chemical reaction and spectrophotometry were, in the early days, the principal analytical methods of identifying pesticides. As both the type and the use of pesticides have grown more and more complicated, the development of analytical methodology has inevitably evolved around a technique known as chromatography, a separation technique by which a multicomponent mixture, passing through a separating medium, can be divided into individual components according to, generally speaking, their molecular weight, size or structure. Column chromatography which generally refers to separation of liquid or dissolved substances has been particularly useful in the cleanup steps.

Gas chromatography, which deals with the separation of volatile compounds, has been developed as the mainstay of the pesticide analytical procedures. Due to the availability of a specific and sensitive detecting system, most, if not all, the recent multidetection or multiresidue analyses have utilized gas chromatography. The essential analytical procedures described are the following.

Extraction – The first step in preparing a sample is to extract the pesticide residue from the material. Since no single solvent can dissolve or extract all known pesticides, the selection of solvent will naturally depend upon the type of pesticides and the matrix material. A great variety of solvents or mixed solvents have been used. The most common ones are pure normal hexane, or normal hexane mixed with isopropanol, acetonitrile, acetone, ether or methylene chloride. The methods of extraction include shaking, rotating, or the use of an extractor, such as Soxhlet extractor.

Cleanup – This is a step of separating the pesticide residues from the bulk of coextracted materials. Two basic techniques are commonly applied:

(a) Partition: the extract is further extracted with an immiscible solvent. Ideal separation would be that the pesticides go into one layer while all other materials remain in another layer. The knowledge of the solubility of pesticides in various solvents, and its partition coefficient, or p-value, between solvents is of prime importance for a successful and quantitative partition.

(b) Column chromatography: some materials, especially lipids, peptides, fats and oils, have similar solubility as pesticides, and are difficult to be separated by partition only. The so-called column cleanup has become the essential step in handling biological samples.

The extract, or the partitioned extract, is introduced onto a column containing a separating medium, then eluted with a solvent, pure or mixed. The most commonly used column material is made up of mixture of magnesium and silicon oxides. The eluate is then concentrated to a definite volume, and ready for the final step; i.e., qualitative or quantitative analysis.

Separation and Determination of Pesticides – More often than not, the extract contains a group of pesticides and their decomposition products. All forms of chromatography, including paper, thin-layer and gas, have been used for this final stage of analysis. Among these chromatographic techniques, gas-liquid partition chromatography (GLC), or gas chromatography (GC), has been the most widely applied. Extensive research has been carried out on its application to quantitative analysis, especially in the field on the specific detectors. A detailed discussion on gas chromatography is beyond the scope of this presentation. However, two fundamental operating parameters are briefly described as follows:

(a) Column materials usually have a thermally stable compound, the stationary phase, coated on an inert material, the support. The most commonly used stationary phases are polysiloxanes such as DC-200, SE-52, OV-17 or its fluorinated derivatives such as QF-1; polyethylene glycol polymer such as Carbowax 20M; or hydrocarbons, such as Apiezon. The support materials are usually diatomaceous earth, fire brick powder, Teflon or glass beads.

(b) Detectors most frequently used include: microcoulometric detector, for halogen-containing or nitrogen-containing compounds; electron capture detector, for compounds containing electronegative atoms, especially chlorine; alkali-metal modified flame ionization detector, for selective detection such as phosphorus; and flame photometric detector, for selective detection such as phosphorus or sulfur.

The complexity of pesticides, i.e., the type, the use, the matrix and the reaction and decomposition, has been a challenge to analytical technology in both qualitative and quantitative analysis. While the exact application of general analytical principle varies with specific circumstance, the accuracy, or the authenticity of analytical results of a certain procedure has been of great concern to the users.

Collaborative study to establish the validity has been the practice. The general trend in the pesticide analysis is the development of well-defined analytical manuals as an authoritative guidance. Agencies such as FDA (5), Water Quality Office (6), Perrine Laboratory of EPA (7) and Food and Drug Directorate of Canada (8) have published comprehensive texts on pesticides analysis.

In the laboratory it is necessary to maximize information obtained from each sample. One tries to analyze each sample for as many pesticide classes, chemical compounds and heavy metals as practical. This requires that one schedule chemical analysis and the flow of samples through the laboratory to take advantage of the different degradation rates among pesticides and chemical compounds. Those most likely to degrade rapidly in storage will be analyzed first.

One should maintain a sample library. Whether this would be practical for all media and monitoring programs is debatable. In a laboratory which collects primarily soil and crop samples, it is a highly valuable and important part of the operation.

After the data has been collected and chemically analyzed, the raw data is transferred to a central staff. Their responsibility is to analyze and interpret the data so that it is useful to decision makers and others. This is the third phase of data management. The first principle governing data analysis is the function and structure of the central staff.

It is important that this staff originally plans and manages the flow of information from the field to the laboratory. Naturally an effective system requires extremely close coordination and liaison between the laboratory staff, the field personnel and the planning and analysis staff. This staff should consist of people who have expertise in the areas of primary concern, as well as sound ecological training.

A second principle of data analysis is that one should not attempt to extract more from the data than was set forth in the original objectives. Many monitoring programs fall short at this point. It is critical that one maximize information from a sample collected in the field, but, in attempting to do this, one should never attempt to draw conclusions and inferences from data that are not justified by the original sampling objectives and designs.

A third principle of data analysis is the proper use of computer support. The critical first step is that there is a careful edit of raw data sheets by not only a technician but, where appropriate, by professionals. It has been found that several man days of professional level review have paid off in a reduction of errors and an increase in the confidence placed on the information. When establishing a computer support program, it is necessary to make certain that the program, as established, is flexible and responsive to changes that are bound to occur in future handling of the data.

The programmers should be allowed to develop their techniques with a minimum of interference. To aid in the development of a viable program, the analysis of data should be kept as simple as possible and still be consistent with stated objectives. The addition of new ideas and the institution of too many changes should be minimized. The only changes initiated during the developmental stage should be those that could not be instituted after completion of the program. When a program is completed, one tends to initiate changes and additions cautiously, finding that many times ideas that seemed pertinent at the time were not quite as valuable several weeks later.

The value of the programmable calculator should be emphasized. These machines are, in practically every sense a desk top computer unit. Their use for mathematical and statistical analyses has saved a great deal of time in analyzing data. As an example, a regression analysis at the computer center would take two weeks including data preparation, key punching and analysis. This same amount of data was analyzed using a programmable calculator in less than an hours' time. The presence of such a strong analytical tool, readily available, allows one to conduct data analyses in great depth because it frees one from the time consuming and monotonous calculations required on machines of lesser capacity.

A fourth principle of data analysis is the proper application of statistical analysis to the data. Statistical analysis actually begins when planning field studies. That is the implementation of a statistically sound sampling plan, and it is the foundation of all future data analyses.

The next step in statistical analysis is to define the distribution form of the data. In pesticide residue data, the normal distribution is not appropriate because the data has been found to be severely skewed to the right. This can be somewhat alleviated by transforming the pesticide residues to logarithms. Tests have shown that while this transformation does not truly meet all the requirements of the normal distribution, it is close enough, and it has been used in lieu of better transformation.

A second problem in pesticide analysis is what to do with zeros. Somewhere between 25 and 75% of the data collected in the field either has pesticide or contaminant residues below the detectable limits of chemical assays or else residues are truly absent from the sample. In environmental monitoring work a miss is as important as a hit. How are these zeros to be considered in the analysis of data? Unfortunately, the same success has not been encountered in handling the zero problem as was the case with transforming the distributions.

One technique used is to determine the frequency distribution for a particular land area, either a biome, state or group of states, and calculate this frequency distribution either using probit analysis or a transformation involving standard deviates. This type of analysis has provided a good picture of the distribution of a pesticide residue through a medium, and it has allowed one to place confidence intervals about the levels detected.

Despite the obvious use of statistics in data handling, sometimes the best method is simply to forego any statistical analysis. In the fifth principle of data analysis it is necessary to attempt, wherever possible, to correlate the monitoring data collected in one system with that data collected in other systems. This correlation will begin with the central operational staff mentioned earlier, but the more sophisticated correlation of this data should probably take place at a higher echelon. In EPA, for all monitoring data, this would most logically be conducted by the Office of Monitoring.

All this effort is worthless if the information is not put into a usable form and presented both to the administrators and decision makers within the organization. In addition, it should be published in open literature so that other segments of the public interested in the results and needing the information can have it in a reasonable length of time.

Up to this point the discussion has centered on sensor data management for systems which do not utilize remote sensing techniques. One might well now consider the question: What are the potentials for remote sensing of pesticides? How will they influence existing monitoring networks? Remote sensing for pesticides, the techniques and possibilities, will be considered through the following information.

A review of the literature (2) reveals that remote sensing techniques have been applied to detecting a wide variety of environmental parameters. Multispectral photography taken from satellites has been used to identify soil, salt flats and water surfaces. Microwave radiometry has been used to detect soil moisture levels in the upper 2 to 3 meters of the soil. A variety of techniques are being tested for the remote sensing of various pollutants in the air.

Some include using shifts in sound frequencies (Doppler effects) to measure wind profiles in the boundary layer. Another technique is to use optical correlation methods to identify certain gaseous pollutants and the use of Raman spectroscopy to identify NO and SO_2 (but with limited success) has also been reported.

Aerial photographs have been used to trace pulp mill effluents in marine waters and success is claimed. Mixed results have been reported when trying to identify pollution zones in estuaries using infrared Ektachrome color film with a dark red filter and Kodachrome X with a UV filter. Microwave radiometry has been tested for possible use in identifying oil slicks in bad weather or at night. The cited examples are hardly exhaustive but are given as examples of current attempts to utilize remote sensing techniques to monitor some environmental parameters.

None of the above studies were conducted to determine the feasibility of identifying pesticide residues using remote sensing. To do so using present techniques would certainly be stretching the state of the art. However, one is encouraged by the possibilities inherent in present remote sensing techniques that could make remote sensing of pesticide residues a reality within the relatively near future.

One area of particular promise is the use of induced luminescence using UV and visible lasers. The use of laser beams to identify a wide variety of surface materials has been reported. The concept basically is that the laser beams induce momentary luminescence in a material and that different materials have different signatures which are detectable at very low levels.

Using this method they were able to obtain characteristic signatures for a variety of surfaces including Teflon, epoxy, light and dark phenolic surfaces, a variety of different kinds of leaves and certain minerals and oils. It is very speculative at this time to predict the usefulness of a method like this for identifying pesticide compounds. Certain basic research would have to be undertaken to test feasibility of this approach. They are as follows:

(a) The relative merits of passive versus active sensing will have to be investigated. If active methods are used it will be necessary to determine what frequencies are most efficient for detecting pesticide residues.

(b) Both the infrared and Raman spectra for various pesticide compounds will have to be identified and selections made of the characteristic identifying bands.

(c) The background spectrum data in natural environments will have to be studied and variations noted due to media, season, time of day, overhead cover, etc.

(d) The type of sensors best suited to detecting pesticide residues will have to be identified and developed.

(e) Actual field tests will have to be made to test the procedures.

(f) In the beginning, remote sensing will have to be complimented by extensive ground-truth studies.

The use of pesticides has become a routine practice in modern agriculture. While these compounds have great advantages in the control of predatory insects, they represent a possible danger to the aquatic environment when present in even trace concentrations.

The National Technical Advisory Committee on Water Quality Criteria has recommended "... that environmental levels ... not be permitted to rise above 50 nanograms/liter..." Many of the states have incorporated pesticide criteria in their water quality standards. Therefore, the monitoring of surface waters for pesticides is an essential part of the measurement of water quality.

As regards pesticides in wastewaters, R.M. Santaniello of Gulf Degremont, Inc. (9) has presented data on permissible and desirable limits for pesticides in public water supplies.

Compound	Permissible (mg./l.)	Desirable
Aldrin	0.017	None
Chlordane	0.003	None
DDT	0.042	None
Dieldrin	0.017	None
Endrin	0.001	None
Heptachlor	0.018	None
Lindane	0.056	None
Methoxychlor	0.035	None
Phosphates plus carbamates	0.1	None
Herbicide mixture of 2,4-D; 2,4,5-T; and 2,4,5-TP	0.1	None

Several pesticides (four in all) were named recently by EPA in the proposed toxic pollutant effluent standards as published in the *Federal Register* for December 27, 1973. For discharge into streams having a flow rate of less than 10 cu. ft./sec. or into lakes having an area of less than 500 acres, it is proposed that there be no discharge of aldrin-dieldrin, DDT (or DDD or TDE), endrin or toxaphene. The limitations on discharges to faster flowing streams, larger lakes or salt water are tabulated below.

	Fresh Water	Salt Water
Aldrin-Dieldrin	0.5 μg./l.	5.5 μg./l.
DDT (and DDD and TDE)	0.2 μg./l.	0.6 μg./l.
Endrin	0.2 μg./l.	0.6 μg./l.
Toxaphene	1.0 μg./l.	1.0 μg./l.

With the above, there is stipulated a condition that when the receiving stream flow is less than 10 times the waste stream flow, these concentrations are reduced by a factor of 10.

Measurement in Air

The analytical procedures for the determination of pesticides in the environment generally involve four steps:

(a) A sampling method that collects a sufficient quantity of the material to permit analysis;

(b) An extraction procedure to remove the specific pesticide(s) from the the bulk of nonpesticide environmental material;

(c) Separation or cleanup to remove nonpesticidal interfering materials carried along during the extraction; and

(d) Detection and identification.

The analysis of pesticides has been handicapped by the low concentrations present in the ambient air, which in the past have made the completion of the above four steps difficult. Only in recent years has instrumentation become sufficiently sophisticated, especially for detection and identification, so that valid information may be obtained. Table 40 presents the analytical sensitivity which has been acquired over the years as understanding of the pesticide residue subject has increased.

TABLE 40: INCREASING ANALYTICAL SENSITIVITY (MINIMUM DETECTABILITY) FOR PESTICIDES

Year	Sensitivity
1930	10 ppm
1935	10 ppm
1940	10 ppm
1945	1 ppm

(continued)

TABLE 40: (continued)

Year	Sensitivity
1950	0.1 ppm
1955	0.02 ppm
1960	1 ppb
1965	0.1 ppb

Source: H. Finkelstein; *Report PB 188,091;* September 1969

Sampling Methods: In addition to the low concentration of airborne pesticides, sampling is also complicated by the coexistence of nonpesticidal materials in both aerosol and vapor phases. A sampling method has been developed by the Midwest Research Institute (10) using a three-section sequential collection train consisting of a glass cloth filter; an impinger containing 2-methyl-2,4-pentanediol; and an adsorption tube containing alumina. The maximum air sampling rate is approximately 29 l./min., and about 24 hours are generally required to collect enough sample for analysis.

Tabor (11) sampled only particulate DDT by the collection of particulate samples on glass-fiber filters. In another study, air samples were collected at rooftop level in Pittsburgh with a two-stage sampling system to separate airborne dust into two fractions (12). The large particles were collected by sedimentation on 71 horizontal trays, and particles which penetrated through this section of the sampler were collected on an MSA 1106B glass-fiber filter. Each sample of particulates was obtained by continuous sampling at an average flow rate of 1.22 $m.^3$/min. (12).

An air sampling system for the differential collection of aerosol and gaseous fractions of airborne herbicides has been reported (13). It consists of a rotating disk impactor for collecting aerosol droplets down to approximately 3μ in diameter, followed by a midget impinger to collect the gaseous fraction. The impactor was specially designed and constructed of glass, Teflon, and stainless steel to prevent contamination of the collection fluid with substances that interfere with electron capture gas chromatography.

Incoming air impinges on the impaction disk that rotates slowly through a fluid well containing n-decane. The impacted droplets wash off into the collection fluid. The disk then passes through a Teflon squeegee to remove the adhering droplets, thus presenting a smooth surface containing a fluid film upon which the air stream impinges.

There are other sampling techniques used to measure operators' hazards in the field. The level of exposure may be determined by the amount of toxicant trapped on the filter of a respirator worn by the operator. Samples are also collected in a special respirator which is modified to simulate nasal breathing characteristics.

Contact samples are collected on pads attached at suitable points on the operators' clothing. Additional samples may be taken by means of suction-operated equipment placed in the breathing zones. The mass to size ratios of airborne particles are evaluated by sampling the air in the breathing zones through cascade impactors and also by collecting the fallout on slides set at different heights in the working area (14).

Air samples were collected at tractor operators' breathing zones using all-glass fritted absorbers and electric or hand-operated suction pumps. Exposures were also determined by attaching filter pads to double-unit respirators (15).

Quantitative Methods: Although methods have been developed to determine the concentration of some pesticides or a component of the pesticide, e.g., phosphorus, the methods are very tedious and time consuming. More research is needed to reduce the methods to procedures that can be used economically for routine analyses in the National Sampling Network.

The collected pesticide must first be extracted from the particulate matter and generally needs a cleanup treatment to remove other interfering substances before the final analysis can be performed.

The extraction and cleanup procedure for the Midwest Research Institute sampling train (10) is as follows. The filter cloth is washed with methanol. The alcoholic mixture is poured over an alumina adsorbent which has previously been transferred to a chromatography tube. The treated alumina column is extracted with hexane. The impinger solution (2-methyl-2,4-pentanediol) is diluted with water and this solution is extracted with hexane. The combined hexane extracts are concentrated by evaporation.

The treated hexane solution is passed through a Florisil column and the pesticides eluted from the column by first adding 0.5% dioxane in hexane to remove the chlorinated hydrocarbon pesticides followed by 5.0% dioxane in hexane to remove the organophosphate pesticides. The two solutions are concentrated before final analysis by gas chromatography.

Tabor (11) used pentane followed by benzene to extract chlorinated and thiophosphate pesticides from particulate matter. The residues of extracts were analyzed by gas chromatography without further treatment.

Pesticide analysis is generally accomplished by some form of chromatography. Three types of chromatography can be used for the quantitative determination of pesticide residues: gas chromatography; thin-layer chromatography; and paper chromatography. The gas chromatographic technique has been used to separate complex pesticide mixtures in a single operation.

A highly sensitive detector for chlorinated pesticides is the electron-capture detector. It is capable of measuring some chlorinated pesticides in concentrations as low as the picogram range (10^{-12} grams). Another detector widely used with gas chromatography of chlorinated hydrocarbons is the microcoulometric detector. This detector operates on the following principles:

(a) As each chlorinated pesticide emerges from the chromatographic column, it passes through a combustion tube where the pesticide is burned with oxygen to yield hydrogen chloride, water and carbon dioxide.

(b) The gas stream then flows through a titration cell containing silver ion, which is maintained electrochemically at a constant concentration.

(c) Hydrogen chloride precipitates the silver ion stoichiometrically and the current required to regenerate it from a silver electrode is recorded as a chromatographic peak.

The detector can measure as little as 10^{-8} grams of chlorine or sulfur. Other detectors that have been used include hydrogen flame detectors (sensitive to carbon-containing compounds), sodium thermionic detectors (sensitive to thiophosphates), and flame-photometric detectors (sensitive to phosphorus).

The Midwest Research Institute (10) method of analysis uses gas chromatography. The chlorinated pesticides were determined by using two different columns with an electron-capture detector. The organophosphate pesticides were determined by using two different columns with a flame photometric detector. Tabor (11) used an electron-capture detector and a sodium thermionic detector for determining chlorinated and thiophosphate pesticides.

Westlake and Gunther (16) have reviewed gas chromatographic detection systems used in pesticide residue evaluations. "Detectability" as used by them, refers to Sutherland's definition (17); the detectable level is the concentration of pesticide above which a given sample of material can be said, with a high degree of assurance, to contain the chemical analyzed. Westlake and Gunther (16) in their review have discussed each of the available methods, including illustrations of the devices and literature references to their use. Also included are a table listing more than 100 pesticides for which infrared spectra have been published, and a table listing more than 50 pesticides for which mass spectal data have been published. Thomson and Abbott (18) have also described two methods termed Chemical Group Analysis and Biological Test Methods.

Chemical Group Analysis – By comparison with a certain class of substances, the pesticide can be identified and determined. This method does not give precise identification and cannot be applied without purification of the extracted material. In applying the Chemical Group Analysis technique to organophosphorus pesticide residues, phosphorus is usually determined quantitatively, not the pesticide compound. The essence of the method is the extraction and cleanup of the residue to insure the absence of natural phosphorus compounds.

The phosphorus in the subdivided extract is eventually converted to phosphoric acid by wet oxidation. The subsequent addition of ammonium molybdate and reduction with stannous chloride produces a heteropoly blue color which is compared against standards produced from solutions of known phosphorus content. This method has only limited selectivity and is sensitive down to 5 μg. of any one pesticide, i.e., 0.1 ppm in a 50 gram sample.

A simple screening method for the rapid estimation of organophosphorus pesticides using the above chemical-end-method of analysis has recently been introduced. In this method the compounds are extracted from the sample, cleaned up on a silica gel chromatoplate, and oxidized with ammonium persulfate or a nitric-perchloric acid mixture for phosphorus determination. Colorimetric and esterase inhibition methods of estimating total phosphorus have been used to develop a method of automatic wet chemical analysis.

In the analysis of chlorinated pesticide residues, the pesticides are extracted and after a limited cleanup process, they are spotted onto a filter paper flag, which is burned in a flask of oxygen. The chloride formed is absorbed in dilute sulfuric acid and can be estimated colorimetrically by the addition of ferric ammonium sulfate and mercuric thiocyanate; the sensitivity of this method of estimation is approximately 5 μg. of organochlorine pesticide.

A more sensitive method is to measure the quantity of chloride produced potentiometrically. This makes the method sensitive down to 0.5 μg. of pesticide. Recently a new continuous chloride ion system has been developed for use with a completely automated combustion apparatus to determine organochlorine pesticides and their residues.

Biological Test Methods – These methods show the presence or absence of toxicologically significant residues. They are basically useful as sorting methods or for confirming the presence of pesticide residues. The basis of the bioassay methods is the comparison of the response of selected insects to pretreated or unknown samples with the insects response to a series of standard pesticides under the same test conditions.

These methods are very sensitive and can easily detect pesticide levels of 0.1 ppm, but the methods do not distinguish between pesticides of similar toxicity. Some of the common insects and other organisms that are used in such bioassays are vinegar fly, housefly, mosquito larvae, mites, brine shrimp, daphnia, guppies and goldfish (19).

A process for detecting phosphorus insecticide vapors in the air by a condensation nuclei technique has been described by S.J. Fusco (20). The condensation nuclei measuring technique detects extremely minute quantities of a gas or vapor from airborne particles of the material being detected on which a fluid such as water, for example, will condense to form droplets. A commercially available condensation nuclei counter measures particles as small as 0.001 micron in radius, and detects mass concentrations of one part in 10^{14}.

The invisible nuclei are grown into microscopic particles which scatter light in the dark field optical system of a cloud chamber in the instrument. The number of microscopic fluid droplets formed are directly related to the original concentration of toxic gas present in the sample atmosphere and the amount of light scattered by the droplet as measured by electrooptical means in the detection circuit employed.

Some gases have been converted to submicroscopic particles for detection by a condensation nuclei technique. One conversion method is described in U.S. Patent 3,198,721. The gas to be detected is reacted with a second gas in the sample atmosphere by using a converter having a hot platinum wire, a source of ultraviolet light, a spark cap or some other suitable element. A different conversion is carrier out in U.S. Patent 3,117,841.

The conversion of a particular gas or vapor to be detected is carried out by reacting the material with an aqueous medium selected from the group consisting of water and aqueous solutions of a volatile reactant to form the nuclei. An acid-base chemical reaction or hydrolysis takes place wherein the reaction product is a salt or hydrolyzate, respectively, having the proper physical form.

It can be seen from the above conversion processes that formation of a product is needed to serve as a nucleating medium for water or some other fluid. It is a principle object of this process, therefore, to provide a method and apparatus for detecting gases or volatile chemical compounds containing a phosphoryl or phosphate grouping or like chemical compounds which can be converted in the same manner.

A process developed by E.J. Poziomek and E.V. Crabtree (21) whereby electrophile compounds used as intermediates or pesticides can be detected by forming volatile derivatives of the compounds by treatment with N-alkylformamides producing isocyanide ions which contact an inert support medium, then treating the medium and its contents with a detector agent for the ions thus producing the color detector signal for the ions.

A technique developed by R.L. Fick and A.H. Markey (22) permits the determinations of a contaminant such as 2,4-D herbicide in a gaseous atmosphere. The device functions by passing a contaminated gas flow over a surface means and simultaneously flowing a liquid on the exterior of the surface means so that the gas contacts the liquid. The liquid entraps and, preferably, at least partially dissolves the airborne contaminant as the liquid flows across the surface means. The liquid receiver collects the contaminated liquid, the liquid passing therefrom to a measuring means where the presence of contaminant in the liquid can be determined.

Measurement in Water

Methods for Organic Pesticides in Water and Wastewater is the title of a publication by the Environmental Protection Agency (6). The Food and Drug Administration's *Pesticide Analytical Manual* (5), the Canadian Department of Agriculture's *Guide to the Chemicals Used in Crop Protection* (23), and *Official Methods of Analysis* of the Association of Official Analytical Chemists (24), should also be consulted by pesticide residue analysts. Other helpful references for general practice and analytical quality control are ASTM Part 30 *Tentative Recommended Practice for General Gas Chromatography Procedures* (25), and the Environmental Protection Agency manual *Control of Chemical Analyses in Water Pollution Laboratories* (26).

The techniques used in measuring pesticides in water closely follow those described above for measuring pesticides in air. Chromatography is followed by such detection systems as election capture detectors, microcoulometric titration detectors and flame photometric detectors.

The electron capture detector (EC) is extremely sensitive to electronegative functional groups, such as halides, conjugated carbonyls, nitriles, nitrates, and organometallics (27) (28). It is virtually insensitive to hydrocarbons, amines, alcohols and ketones. The selective sensitivity of halides makes this detector particularly valuable for the determination of organochlorine pesticides. It is capable of detecting picogram (10^{-12} gram) quantities of many organochlorine pesticides. Organophosphorus pesticides containing nitro groups are also detected, although with much less sensitivity. Electron capture detectors may be of parallel plate and concentric tube or concentral design and employ one of two ionization sources: tritium (H^3) or radioactive nickel (Ni^{63}).

The tritium detector has a temperature limit of 225°C. (it should not be operated above 210°C.) which makes it susceptible to a buildup of high boiling contaminants which reduce its sensitivity and require frequent cleanup. The nickel detector, on the other hand, can be operated or baked out up to 400°C. to reduce contamination and cleaning problems.

The microcoulometric detector (MC) (29) is selective for halogen containing compounds, except fluorides, when used with the halogen cell. Under optimum conditions, this detector is capable of detecting 5 to 20 ng. of organochlorine pesticides. Although the sensitivity of this detector is not as great as that of electron capture, the high degree of specificity makes it a very valuable instrument for qualitative identification as well as for minimizing sample cleanup. Under the proper oxidative-reductive conditions, the system can be made specific for sulfur, phosphorus and nitrogen compounds.

The electrolytic conductivity detector (ECD) has a sensitivity 2 to 3 times greater than the microcoulometric system. Although perhaps slightly less selective than the MC, it is, nonetheless, effective for qualitative identification, and cleanup appears to be less of a problem (30). Use of the electrolytic conductivity detector in the reductive mode with a platinum catalyst is recommended when determining halogen compounds. If the oxidative mode is used, a scrubber must be employed to remove SO_2, which also responds to the detector.

The flame photometric detector (FPD) is selective for sulfur and phosphorus (31). With the use of a dual head, the detector is capable of simultaneously measuring both sulfur and phosphorus as well as a normal flame ionization response. Using a single head, either sulfur or phosphorus and normal flame ionization response is measured. The characteristic optical emissions of sulfur and phosphorous are measured using filters with transmission at 394 mμ (sulfur) and 526 mμ (phosphorus). The FPD is capable of determining subnanogram quantities of both sulfur (4×10^{-11} grams) and phosphorous (10^{-11} grams).

References

(1) Finkelstein, H., "Air Pollution Aspects of Pesticides," *Report PB 188,091;* Springfield, Virginia, National Technical Information Service (September 1969).

(2) Wiersma, G.B. and Tai, H. in Proceedings of EPA Environmental Quality Sensor Workshop, Las Vegas, Nevada (November 30 to December 2, 1971).

(3) Council on Environmental Quality, Annual Report of Working Group on Pesticides, Washington, D.C. (1971).

(4) Wiersma, G.B., Sand, P.F. and Cox, E.L., *Pesticides Monitoring Journal* 5, (1), 63 to 66 (1971).

(5) Food and Drug Administration, *Pesticides Analytical Manual,* Washington, D.C., Department of Health, Education and Welfare (1968).

(6) Water Quality Office, *Methods for Organic Pesticides in Water and Wastewater,* Washington, D.C., Environmental Protection Agency (1971).

(7) Perrine Laboratory, *Analytical Manual,* Perrine, Florida, Environmental Protection Agency (1971).

(8) Food and Drug Directorate of Canada, *Analytical Manual,* Ottawa, Canada (1969).

(9) Santanello, R.M., in *Industrial Pollution Control Handbook,* H.F. Lund, Editor, New York, McGraw-Hill (1971).

(10) Midwest Research Institute, Report published for the Division of Pesticides, U.S. Food and Drug Administration.

(11) Tabor, E.C., "Pesticides in Urban Atmospheres," *J. Air Pollution Control Association* 13:415 (1965).

(12) Antommaria, P., Corn, M. and DeMaio, L., "Airborne Particulates in Pittsburgh: Association with p,p'-DDT," *Science* 150:1476 (1965).

(13) Bamesberger, W.L. and Adams, D.F., "An Atmospheric Survey for Aerosol and Gaseous 2,4-D Compounds" in *Organic Pesticides in the Environment, Advan. Chem. Ser.* 60 (1966).

(14) Lloyd, G.A. and Bell, G.J., "The Exposure of Agricultural Workers to Pesticides Used in Granular Form," *Ann. Occup. Hyg.* 10:97 (1967).

(15) Jegier, Z., "Hazards of Insecticide Applications in Quebec," *Can. Journal Public Health* 56:233 (1965).

(16) Westlake, W.E. and Gunther, F.A., "Advances in Gas Chromatographic Detectors Illustrated from Applications to Pesticide Residue Evaluations," *Residue Rev.* 18:175 (1967).

(17) Sutherland, G.L., "Residue Analytical Limit of Detectability," *Residue Rev.* 10:8 (1965).

(18) Thomson, J. and Abbott, D.C., *Pesticide Residue,* London, The Royal Institute of Chemistry (1967).

(19) Sun, Y.-P., "Bioassay--Insects," in *Analytical Methods for Pesticides, Growth Regulators and Food Additives,* Volume II, G. Zweig, Editor, New York, Academic Press (1963).
(20) Fusco, S.J.; U.S. Patent 3,607,085; September 21, 1971; assigned to General Electric Company.
(21) Poziomek, E.J. and Crabtree, E.V.; U.S. Patent 3,645,693; February 29, 1972; assigned to The U.S. Secretary of the Army.
(22) Fick, R.L. and Markey, A.H.; U.S. Patent 3,751,967; August 14, 1973; assigned to The Dow Chemical Company.
(23) Canada, Department of Agriculture, *Guide to the Chemicals Used in Crop Protection,* Ottawa, Canada, Queen's Printer (1968).
(24) Association of Official Agr. Chemists, *Official Methods of Analysis,* Washington, D.C. (1965).
(25) American Society for Testing Materials, *Tentative Recommended Practice for General Gas Chromatography Procedures,* Philadelphia, Pennsylvania (1968).
(26) Environmental Protection Agency, *Control of Chemical Analyses in Water Pollution Laboratories,* Cincinnati, Ohio (1972).
(27) Lovelock, J.E. and Lipsky, S.R. *J. Am. Chem. Soc.* 82, 431 (1960).
(28) Lovelock, J.E., *Anal. Chem.* 33, 162 (1961).
(29) Challacombe, J.A. and McNulty, J.A., *Residue Reviews* 5, 57 (1964).
(30) Coulson, D.M., *J. Gas Chromatography* 4, 285 (1966).
(31) Brody, S. and Chaney, J., *J. Gas Chromatography* 4, 42 (1966).

PHENOLS

Phenols may occur as air pollutants, as emissions from glass-fiber manufacture, for example. Soviet air quality standards for phenol are 0.0026 ppm both on a single-exposure and 24 hour averaging time basis. West German standards are 0.15 ppm on a basis not to be exceeded more than once in 4 hours and 0.05 ppm on a long-term exposure basis. The general problem of phenols in wastewaters has been reviewed by Patterson and Minear (1).

McKee and Wolf (2) have listed the following industries as characteristic sources of phenolic pollutants in wastewaters:

Gas works (production)
Wood distillation
Oil refineries
Sheep and cattle dip
Chemical plants
Photographic developers
Explosives
Coal tar distilling
Mine flotation wastes
Insecticides
Resin manufacture
Coke ovens

In addition, aircraft maintenance (3)(4), foundry operations (5), Orlon manufacture (6), caustic air scrubbers in paper processing plants (4), rubber reclamation plants (7)(8), nitrogen works (9), fiberboard factories, plastic factories, glass production (10), stocking factories (11) and fiber glass manufacturing (12)(13)(14) have been reported as contributing phenols to wastewaters. Table 41 summarizes the levels of phenol found in wastes of various industries.

Although described in the technical literature simply as phenols, this waste category may include a variety of similar chemical compounds among which are various phenols, chlorophenols and phenoxyacids. In terms of pollution control, reported concentrations of phenol are thus the result of a standard analytical methodology which measures a general group of similar compounds rather than being based upon specific identification of the single compound, phenol (hydroxybenzene).

TABLE 41: LEVELS OF PHENOL REPORTED IN INDUSTRIAL WASTEWATERS

Industrial Source	Phenol Concentration, mg/l	Reference
Coke Ovens		
Weak ammonia liquor, without dephenolization	3,350-3,900	15
	1,400-2,500	16
	2,500-3,600	17
	3,000-10,000	18
	580-2,100	19
	700-12,000	10
	600-800	20
Weak ammonia liquor, after dephenolization	28-332	21
	10	18
	10-30	16
	4.5-100	9
Wash oil still wastes	30-150	22
Oil Refineries		
Sour water	80-185 (140 ave.)	22
General waste stream	40-80	23
Post-stripping	80	24
General (catalytic cracker)	40-50	25
Mineral oil wastewater	100	9
API Separator effluent	0.35-6.8 (2.7 ave.)	26
General wastewater	30	27
General wastewater	10-70	28
General wastewater	10-100	29
Petrochemical		
General petrochemical	50-600	30
Benzene refineries	210	9
Nitrogen works	250	9
Tar distilling plants	300	9
Aircraft maintenance	200-400	3, 4
Herbicide manufacturing	210	31
(includes chloro derivaties and Phenoxy acids).	239-524	32
Other		
Rubber reclamation	3-10	7, 8
Orlon manufacturing	100-150	6
Plastics factory	600-2,000	10
Fiberboard factory	150	10
Wood carbonizing	500	10
Phenolic resin production	1,600	33
Stocking factory	6,000	11
Synthetic phenol, plastics, resins	12-18	34
	369 (after cooling water separation)	
Fiberglass manufacturing	40-400	13

Source: J.W. Patterson and R.A. Minear; Report PB 216,162; February 1973

Measurement in Air

A tentative method for the determination of phenolic compounds in the atmosphere has been published (34A) as the result of work by an Intersociety Committee. The air is scrubbed with an alkaline solution. Phenolic particulates are collected by passage of the air through a fiber glass filter. The phenolic materials are separated from other compounds by distillation from an acidified system. The phenols are then determined by coupling with 4-amino-antipyrine in an alkaline medium containing an oxidant. A cubic meter of air containing 1.3 ppb of phenol will give a measurable color.

Measurement in Water

Phenolics may be monitored in water by the following instrumental techniques (35): Gas Chromatography, Thin-Layer Chromatography (Fluorescence) and UV Spectrophotometry.

Phenols may be measured using the Technicon CSM-6 in the nominal range of 0-5 ppm as noted earlier in Table 12. The detection limit using that instrument is given as 0.05 ppm.

A phenol detection process has been described by R.J. Nadalin (36). It comprises the steps of (a) admixing a sample of the fluid suspected of containing a phenol with a small volume of an organic solvent, such as chloroform or carbon tetrachloride, that is, a good solvent for phenol, not significantly miscible with the fluid, and which itself does not react with the reagent in step (b), and (b) admixing the solution of organic solvent containing a relatively high concentration of the phenol with a reagent composed of a nonaqueous solution of sulfuric acid and an oxidizing agent, composed of at least one compound selected from a group consisting of a vanadate ion and a chromate ion, which reacts with a phenol to produce a colored reduction product, whereby to produce an intense coloration depending on the quantity of the phenol.

The determination of phenolic compounds in water by instrumental analysis has been discussed in some detail by Prof. R.F. Christman of the University of Washington (37).

References

(1) J.W. Patterson and R.A. Minear, "Wastewater Treatment Technology," 2nd Ed., *Report PB 216,162,* Springfield, Va., Nat. Tech. Info. Service (Feb. 1973).

(2) J.E. McKee and H.W. Wolf, *Water Quality Criteria,* 2nd Ed., California State Water Quality Control Board, Publication No. 3-A, 1963.

(3) G.W. Reid, R. Daigh and R.L. Wortman, "Phenolic Wastes from Aircraft Maintenance," *Jour. Wat. Poll. Control Fed.,* 32:353-391, 1960.

(4) G.W. Reid and R.W. Libby, "Phenolic Waste Treatment Studies," *Proc. 12th Purdue Industrial Waste Conf.,* 12:250-258, 1957.

(5) R.P. Barzler, D.J. Giffels and E. Willoughby, "Pollution Control in Foundry Operations," in *Industrial Pollution Control Handbook*, Herbert F. Lund, ed., McGraw-Hill Book Co., New York, 1971.

(6) H.A. Schesinger, E.F. Dul, and T.A. Fridy, Jr., "Pollution Control in Textile Mills," in *Industrial Pollution Control Handbook*, Herbert F. Lund, ed., McGraw-Hill Book Co., New York, 1971.

(7) W.A. Parsons, *Chemical Treatment of Sewage and Industrial Wastes*, National Lime Association, Washington, 1965.

(8) W.D. Sechrist and N.S. Chamberlin, "Chlorination of Phenol Bearing Rubber Wastes," *Proc. 6th Purdue Industrial Waste Conf.,* 6:396-412, 1951.

(9) H.J. Wurm, "The Treatment of Phenolic Wastes," *Proc. 23rd. Purdue Industrial Wast Conf.,* 23:1054-1073, 1968.

(10) W. Noack, Formal Discussion to Biszysko and Suschka, "Investigations on Phenolic Wastes Treatment in an Oxidation Ditch," in *Advances in Water Pollution Research, Munich Conference,* Vol. 2, 285-295, Pergamon Press, New York, 1967.

(11) T. Ide, Formal Discussion to Biszysko and Suschka, "Investigations on Phenolic Wastes Treatment in an Oxidation Ditch," in *Advances in Water Pollution Research, Munich Conference,* Vol. 2, 285-295, Pergamon Press, New York, 1967.

(12) J.M. Baloga, F.B. Hutto, Jr., and E.I. Merrill, "A Solution to the Phenolic Problem in Fiberglas Plants," *Wat. Sew. Works,* 118:7-13, 1971.

(13) G.W. Fletcher, S.H. Thomas, and D.E. Cross, "Development and Operation of a Closed Wastewater System for the Fiberglas Industry," presented at the 45th Annual Water Pollution Control Federation Meeting, 1972.

(14) Johns-Manville Products Corporation, "Phenolic Waste Reuse by Diatomite Filtration," *U.S. EPA Report 12080EZF* 09/70, 1970.

(15) P.D. Kostenbader and J.W. Flecksteiner, "Biological Ozidation of Coke Plant Weak Liquor," *Jour. Wat. Poll. Control Fed.,* 41:199-207, 1969.

(16) C.W. Fisher, "Coke and Gas," in *Chemical Technology Volume 2, Industrial Wastewater Control,* F. Fred Gurnham, ed., Academic Press, New York, 1965.

(17) W.E. Carbone, R.N. Hall, H.R. Kaiser, and C.G. Bazell, "Commercial Dephenolization of Ammoniacal Liquors with Centrifugal Extractors," *Proc. 5th Ontario Indust. Waste Conf.*, pp. 42-58, 1958.
(18) Resource Engineering Associates, *State of the Art Review on Product Recovery,* U.S. Dept. Interior, Washington, D.C., 1969.
(19) J. Biszysko and J. Suschka, "Investigations of Phenolic Wastes Treatment in an Oxidation Ditch," in *Advances in Water Pollution Research, Munich Conference,* Vol. 2, 285-294, Pergamon Press, New York, 1967.
(20) G.F.S. Clough, "Biological Oxidation of Phenolic Waste Liquor," *Chem. Proc. Eng.*, 42 (1):11-14, 1961.
(21) T.W. Lesperance, "Biological Treatment of Phenols," *Proc. 8th Ontario Industrial Waste Conf.*, pp. 59-66, 1961.
(22) B.S. Graves, "Biological Oxidation of Phenols in a Trickling Filter," *Proc. 14th Purdue Industrial Waste Conf.*, 14:1-6, 1959.
(23) W.T. McPhee and A.R. Smith, "From Refinery Waste to Pure Water," *Proc. 16th Purdue Industrial Waste Conf.*, 16:311-326, 1961.
(24) M. Benger, "The Disposal of Liquid and Solid Effluents from Oil Refineries," *Proc. 21st Purdue Industrial Waste Conf.*, 21:759-767, 1966.
(25) W. Steck, "The Treatment of Refinery Waste Water with Particular Consideration of Phenolic Streams," *Proc. 21st Purdue Industrial Waste Conf.*, 21:783-790, 1966.
(26) R.F. Peoples, P. Krishnan, and R.N. Simonsen, "Nonbiological Treatment of Refinery Wastewater," *Jour. Wat. Poll. Control Fed.*, 44:2120-2128, 1972.
(27) A.A. Wigren and F.L. Burton, "Refinery Wastewater Control," *Jour. Wat. Poll. Control Fed.* 44:117-128, 1971.
(28) E.F. Mohler, Jr., H.F. Elkin and L.R. Kumnick, "Experience with Reuse and Biooxidation of Refinery Wastewater in Cooling Tower Systems," *Jour. Wat. Poll. Control Fed.*, 36:1380-1392, 1964.
(29) B.A. Carnes, J.M. Eller, and J.C. Martin, "Reuse of Refinery and Petrochemical Wastewaters," *Ind. Wat. Eng.*, 9:25-29, 1972.
(30) B.W. Dickenson and W.T. Laffey, "Pilot Plant Studies of Phenol Waste from Petrochemical Operations," *Proc. 14th Purdue Industrial Waste Conf.*, 14:780-799, 1959.
(31) T.B. Henshaw, "Adsorption/Filtration Plant Cuts Phenols from Effluent," *Chem. Eng.*, 78:47-49, 1971.
(32) A.E. Sidwell, "Biological Treatment of Chlorophenolic Wastes," *U.S. EPA Report 12130EGK 06/71,* 1971.
(33) Anonymous, *The Cost of Clean Water, Vol. III, Industrial Waste Profile No. 10, Plastics Materials and Resins,* U.S. Dept. Interior, Washington, D.C., 1967.
(34) T.P. Schumaker and R.H. Zanitsch, "Physical/Chemical Treatment: A Solution to a Complex Waste Problem," presented at the 45th Annual Water Pollution Control Federation Meeting, 1972.
(34a) American Public Health Assoc., *Methods of Air Sampling and Analysis,* Wash., D.C. (1972).
(35) Lawrence Berkeley Laboratory, "Instrumentation for Environmental Monitoring-Water," *Publ. LBL-1,* Vol. 2, Berkeley, Univ. of Calif. (Feb. 1, 1973).
(36) Nadalin, R. J., U.S. Patent 3,544,271; December 1, 1970; assigned to Westinghouse Electric Corp.
(37) Christman, R.F., in *Instrumental Analysis for Water Pollution Control,* K.H. Mancy, Ed., Ann Arbor, Mich., Ann Arbor Science Publishers, Inc. (1971).

PHOSGENE

It has been determined that concentration levels of phosgene gas in air in excess of about 0.1 part per million is a potential health hazard and constitutes a serious problem, particularly in industrial plants where, because of the nature of the work and the chemical processes and reagents involved, relatively small but potentially dangerous amounts of phosgene gas may be inadvertently liberated in the atmosphere to which plant personnel are exposed. Additionally, even in cases where workers may not be exposed to such an atmosphere, phosgene gas is relatively highly reactive to some organic materials and may cause undesirable effects such as, for example, relatively rapid rates of deterioration of certain industrial plastics or elastomers.

Measurement in Air

A process developed by C.B. Murphy (1), permits the detection and analysis of low levels of concentration of phosgene gas in other nonreactive gases such as air. Concentrations of phosgene as low as 0.1 part per million may be detected by exposing the gas to a source of ammonia to produce small nucleogenic solid reaction particles which may be measured in a condensation nuclei detector.

Reference

(1) C.B. Murphy; U.S. Patent 3,416,896; December 17, 1968; assigned to General Electric Co.

PHOSPHATES

The permissible concentration of phosphorus (presumably in the form of phosphates) in domestic water supplies is given as 10-50 μg./l. in Table 3. The desirable concentration is 10 μg./l. according to that same source.

Measurement in Water

Phosphates can be monitored in water by continuous colorimetric techniques (1). Phosphates can be measured by the Technicon CSM-6 in the nominal range of 0-8 ppm as noted earlier in Table 12. The detection limit is 0.08 ppm using that instrument.

A process has been developed by V.A. Stenger and D.N. Armentrout (2) for directly determining inorganic phosphorus in water samples. It comprises adding a very small amount of an alkali metal salt to the water solution containing the phosphorus compound, atomizing the water solution pneumatically, injecting the atomized particles into a flame ionization detector where interaction between the alkali metal and the phosphorus produces in the flame an enhanced ionization that is proportional to the phosphorus concentration.

An automated water monitoring instrument for phosphorus content of wastewaters has been described by M.J. Prager (3). The analytical principle employed was flame emission photometry. Phosphorus compounds burned in a hydrogen flame emit at about 525 millimicrons.

Conditions were established for the sensitive measurement of phosphorus in water. Operating parameters investigated included fuel and air flow rates, burner configuration, operating temperature, method of sample aerosolization, etc.

Using an ultrasonic nebulizer to aerosolize samples of triethylphosphate in water, it was possible to detect phosphorus at a concentration of less than 2 parts per billion. A procedure was worked out for distinguishing between organic and inorganic phosphorus with ion exchange resins. In measurements designed to determine interference by sodium and calcium, it was observed that the method is about 1000 times more sensitive towards phosphorus than towards calcium. A prototype instrument was designed, fabricated, tested, by NUCOR Corp. of Denville, New Jersey and delivered to EPA, Southeast Environmental Research Laboratory in Athens, Georgia.

References

(1) Lawrence Berkeley Laboratory, "Instrumentation for Environmental Monitoring-Water," *Publ. LBL-1, Vol. 2,* Berkeley, Univ. of Calif. (Feb. 1, 1973).
(2) Stenger, V.A. and Armentrout, D.N., U.S. Patent 3,607,070; Sept. 21, 1971; assigned to Dow Chemical Co.
(3) Prager, M.J., "Automated Water Monitoring Instrument for Phosphorus Contents," *Report EPA-R4-73-026,* Wash. D.C., Enviromental Protection Agency (June 1973).

PHOSPHITES

Measurement in Water

A process has been developed by A.W. Grobin, Jr. (1) which permits determining the phosphite ion content of a sample in which hypophosphite ion is also present by adding to the to-be-tested sample an acid buffered metal anion solution capable of forming colored heteropoly acids with phosphorous-containing compounds and then adding a phosphite ion selective reducing agent. The intensity of the resulting colored heteropolyphosphite complex is then measured by colorimetry and the measured intensity compared with a standard to determine phosphite content of the sample.

References

(1) Grobin, A.W., Jr., U.S. Patent 3,425,805; February 4, 1969; assigned to International Business Machines Corp.

PHOSPHORIC ACID

Measurement in Air

The air pollution aspects of phosphoric acid have been reviewed by Athanassiadis (1).

Sampling: Sampling of phosphoric acid mist in gas effluents from stacks of phosphoric acid plants (thermal process) can be done by a method based on that used by the U.S. Public health service for particulate sampling. The sampling train consists of a probe, cyclone, filter, four impingers, pump, dry-gas meter, calibrated orifice, and manometer, in the order listed (2). The cyclone is designed so that particulate samples can be separated into two fractions, one having particle diameters less and the other having particules greater than 5μ. The first two impingers contain water, the third one is dry, and the fourth contains silica gel. When phosphoric acid mist is sampled, the fritted-glass and paper filters are removed, so that all particles passing through the cyclone are collected in the impingers.

Quantitative Methods: Determination of phosphorus in phosphoric acid emitted from stacks of phosphoric acid plants can be made with the ammonium phosphomolybdate colorimetric method, which is based on the spectrophotometric determination of the yellow ammonium phosphomolybdovanadate complex formed when orthophosphate reacts with the reagent in an acid medium. The method is applicable to materials in which phosphorus compounds can be quantitatively oxidized to the orthophosphate form (3)(4)(5).

Interference comes form (1) certain ions which reduce the color to molybdenum blue, (2) oxalates, tartrates, and citrates which tend to bleach the color, (3) high concentrations of iron and (4) of dichromate, resulting from the close resemblance of the color of the dichromate ion to the yellow complex ammonium phosphomolybdovanadate. This method is applicable to the determination of total phosphates in the concentration range of about 50 μg. to 2,000 μg., with a replication prescision of ±1.0% (6). Another method used for the determination of phosphoric acid in stack gas samples is acid-base titration.

References

(1) Athanassiadis, Y.C., "Air Pollution Aspects of Phosphorus and Its Compounds," *Report PB 188,073,* Springfield, Va., Nat. Tech. Information Service (Sept. 1969).
(2) *Atmospheric Emissions from Thermal-Process Phosphoric Acid Manufacture,* U.S. Dept. of Health, Education, and Welfare, Public Health Service, U.S. Govt. Printing Office, Washington, D.C. (Oct. 1968).
(3) Guinland, K.P., et al., "Spectrophotometric Determination of Phosphorus as Molybdovanadophosphoric Acid," *Anal. Chem.* 27:1626 (1955).

(4) Rogers, R.N. "Determination of Phosphate by Differential Spectrophotometry," *Anal. Chem.* 32:1050 (1960).
(5) Talvitic, N.A., et al., "Spectrophotometric Determination of Phosphorus as Molybdovanadophosphorus Acid, Application to Airborne Particulate Matter," *Anal. Chem.* 34:866 (1962).
(6) Striplin, M.M., "Development of Processes and Equipment for Production of Phosphoric Acid," *Chemical Engineering Report No. 2,* Tennessee Valley Authority (1948).

PHOSPHORUS AND PHOSPHORUS COMPOUNDS

Some forms of phosphorus are toxic to humans and animals. Yellow phosphorus is a protoplasmic poison, while red phosphorus is comparitively nontoxic. Some of the phosphorus inorganic compounds, such as phosphine gas, are highly toxic at concentrations of 400 to 600 ppm, while others, such as tricalcium phosphate, are only mildly so. Many of the organic compounds of phosphorus are extremely toxic. However, very few of the known phosphorus compounds have been investigated with respect to their toxic effects on humans, animals, and plants, as discussed by Y.C. Athanassiadis (1).

The effect of some phosphorus compounds is the inhibition of cholinesterase, a process which is slow and sometimes cumulative and irreversible. The nervous system is affected by this type of adverse action. Nevertheless, the potential hazard represented by the ambient air concentrations of phosphorus and its compounds is unknown.

The source of almost all phosphorus and phosphorus products is phosphate rock. The major uses of phosphorus compounds are the manufacturing of fertilizers, pesticides, and industrial chemicals. These processes are the major sources of phosphorus emissions in the ambient air. Boilers and furnaces burning some crude oils and coals which contain phosphorus represent secondary emission sources of phosphorus compounds. The production of phosphorus and its primary products has been increasing annually.

No national data exist on the ambient air concentrations of phosphorus. The few local data available from special studies indicate an average ambient air concentration, in 1954, of 1.43 μg./m.^{3}.

Abatement of phosphorus emissions includes the use of scrubbers, fiber mist eliminators, high-energy wire mesh contactors, and electrostatic precipitators.

No information has been found on economic costs of phosphorus air pollution. Investment in air pollution control equipment within the phosphate industry has been relatively high, especially in Florida, where the phosphate fertilizer industry is concentrated. Operating costs of control equipment is this section of the fertilizer industry have been estimated at six million dollars annually; this also includes the control of fluoride emissions.

Measurement in Air

Sampling Methods: One of the problems in the determination of free phosphorus in air is separating free phosphorus from the many phosphorus compounds found in the urban air. A recently developed sampling method draws air through a macro impinger containing 100 ml. of xylene at a rate of 1 cfm for 15 minutes. A filter paper attached to the exit catches any fumes which pass through the xylene (2).

In a spectrographic method developed recently for the rapid determination of phosphine, air samples are collected at the rate of 0.5 l./min. (for about 10 minutes) in a fritted-glass bubbler. The bubbler has a collection efficiency of 86.2% at 0.5 l./min. (75% at 1 l./min.) and contains silver diethyldithiocarbamate (3).

Quantitative Methods: Phosphorus is usually determined and estimated as phosphoric acid expressed as phosphorus pentoxide (P_2O_5). This determination may be done gravimetrically by forming a magnesium pyrophosphate ($Mg_2P_2O_7$) precipitate, titrimetrically, or colorimetrically. (4).

A number of colorimetric methods (5) are available, most of which are based on the development of a blue color when molybdate is treated with a reducing agent. The ring-oven colorimetric method (6) uses as a reagent orthodianisidine molybdate dissolved in glacial acetic acid which contains sodium molybdate. The limit of identification is 0.002 μg. and the range 0.05 to 2.0 μg.

The interference of phosphate in fluorimetric procedures became the basis of various studies aimed at developing a highly sensitive method of determining trace amounts of phosphate (5). One such method, based upon fluorescence quenching by phosphate of aluminum-morin chelate, permits the determination of 0.5 to 10 μg. of phosphate, or 0.5 to 0.1 μg. per ml. in solution. Ions of a number of elements, especially metals, were found to interfere; and for the method to be specific, phosphate is separated from such ions (5).

Recently, a method has been developed for the determination of free phosphorus and phosphorus vapor in air that depends on trapping the element and its compounds in xylene. Separation of phosphorus compounds from free phosphorus is done by dissolving the compounds in water and separating the phases, converting the free phosphorus to silver phosphide, oxidizing the phosphide to phosphate, and estimating the phosphate by the molybdenum blue method. Standards are prepared in the range of 0 to 35 μg. of phosphorus, and the colorimetric portion of the procedure is carried out with each standard (2).

Phosphine in air can be determined by colorimetric methods. Recently a spectrographic method (3) has been developed for the rapid determination of phosphine in the range of 0.1 to 1 ppm. The method uses silver diethyldithiocarbamate as reagent and depends on the formation of a complex of phosphine and the reagent with an absorption maximum at 465 mμ. It is necessary to calibrate the procedure by a second method analyzing phosphine as phosphate, since no primary phosphine standards are presently available.

For the detection of organophosphorus pesticides, gas chromatography has been used in conjunction with a flame-photometric (7), thermionic (8), or electron-capture detector. See preceding section on pesticides for more details on analytical methods of organophosphorus pesticides.

A process developed by A. Karmen (9) is one whereby the presence of phosphorus in a gaseous material is detected by monitoring the rate of vaporization of a metallic material, which rate increases when a probe containing such metallic material is heated in the presence of a sample containing phosphorus. The rate of vaporization of the metallic material is measured by a flame ionization detector or flame photometry device. The metal could be an alkali metal, barium, strontium, calcium and mixtures of the same, and could be in the form of a pure metallic material, a metallic amalgam, a metallic salt and mixtures of the same.

A process developed by C.E. van der Smissen (10) involves the detection of the presence of phosphorus and/or sulfur-containing compounds in air and other gases. In particular, this process involves the quantitative determination of phosphorus or sulfur-containing compounds by means of the coloration of a hydrogen flame.

The gas to be analyzed either contains oxygen or oxygen is added thereto, and then the gas is mixed with hydrogen, with the amount of hydrogen being in excess of that required for the complete reaction of the oxygen contained in the gas. Finally, the mixture of gas with oxygen is ignited to obtain a color produced by the resulting flame. The process has the advantage of giving a great sensitivity of detection.

If the gas to be analyzed does not contain oxygen, a separate oxygen stream is introduced into the hydrogen stream along with the gas to be analyzed. In this mixture, the quantity of hydrogen is again greater than that needed for the complete reaction of the oxygen. This modification is also of great sensitivity for detecting the presence of phosphorus or sulfur in the gas. Figure 44 shows a suitable form of apparatus for the conduct of the process.

FIGURE 44: APPARATUS FOR DETECTING PHOSPHORUS (OR SULFUR) COMPOUNDS IN A GAS BY FLAME COLORATION

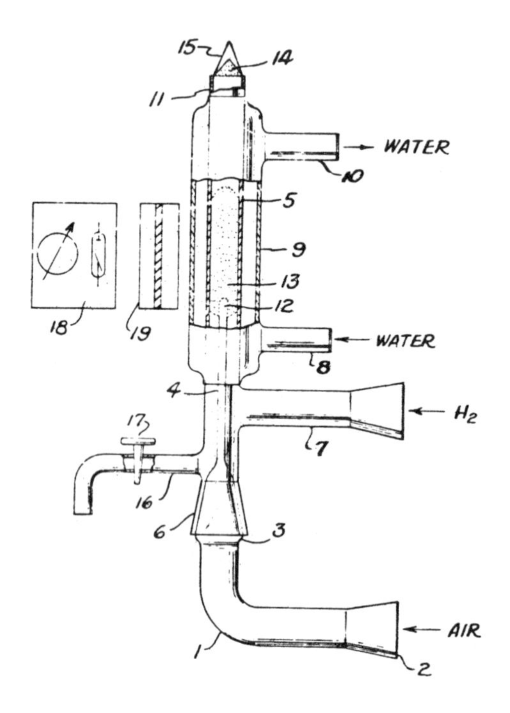

Source: C.E. van der Smissen; U.S. Patent 3,213,747; October 26, 1965

The burner is composed of two parts. The first part is a quartz tube **1** having inlet **2** for the gas to be tested and a conical portion **3** drawn out into a fine bore tube **4**, the upper end of which is open. Tube **4** is partially telescoped within a quartz, Pyrex or high temperature resistant glass tube **5**, the lower end **6** of which is seated on the conical portion **3**. Hydrogen is introduced adjacent the lower end of tube **5** through inlet pipe **7**. Tube **5** is cooled by a cooling agent such as water entering pipe **8** into water jacket **9** which surrounds tube **5**, the cooling agent being exhausted through pipe **10**. The upper end **11** of tube **5** is an external burner tube for the formation of an upper flame, while a lower first flame **12** is formed at the outlet end of tube **4** and this flame produces a glow zone **13**.

In operation, the gas such as air containing either phosphorus or sulfur compounds flows through into inlet **2** of pipe **1** and through tube **4**. At the same time hydrogen enters through pipe **7**. The velocity of the air in pipe **1** is about from 0.3 to 0.6 liters per minute and the velocity of the hydrogen is from about 0.1 to 0.4 liters per minute. The hydrogen is ignited at the upper end **11**, and the flame strikes back to ignite the first lower flame **12.** Tube **5** is cooled by the water jacket **9** and intense cooling causes the water formed by the flame **12** to collect on the inner wall of tube **5**, which water drops to the drain pipe **16** closed by the stopcock **17**. When the gas being tested is free from sulfur and phosphorus compounds, the flame **12** is bluish-white. This flame **12** has about the diameter of tube **4** and a length of from about 3 to 5 mm. It has a type of semicircular to elliptical shape. The upper flame has an inner coloration **14** and an outer conical shape **15**. In the absence of phosphorus or sulfur, it has only a slight bluish, reddish, or yellowish glow depending upon the color of the glass tube **5**.

When the gas being tested contains phosphorus compounds, the flame **12** has an intensified whitish coloration and the edge of flame **15** is colored somewhat whitish to yellowish. In the glowing zone **13** above flame **12**, a green color appears at a distance of from about one to two times the length of flame **12** and which has a decreased intensity in an upward direction. This green coloration appears a second time at the end **11** of tube **5** in the form of a very sharply defined cone **14** within the flame **15**.

When the gas being tested contains a sulfur compound, no change occurs in the coloration of flames **12** and **15**. However, a bluish coloration appears in the glow zone **13** instead of the greenish phosphorus color. This bluish color appears somewhat higher above the tip of flame **12** than the green phosphorus color. Again, cone **14** in flame **15** has a blue sulfur color instead of the green phosphorus color.

When both phosphorus and sulfur compounds are found in the gas being tested, a blue color appears above a green color in the glow zone **13**. The cone **14** is colored either blue or green according to the preponderance of either sulfur or phosphorus compounds in the gas.

The colors were determined in zone **13** with the aid of a spectrophotometer **18**. The green phosphorus color was measured on the 520 millimicron line, and the blue sulfur color was measured on the 380 millimicron line. The colors in cone **14** can be determined in a like manner.

According to the concentration of either phosphorus or sulfur in an amount of about 10^{-6} grams per liter of air, the characteristic glow of both flames **12** and **13** is substantially unaffected by slight changes in the intensity of flame 12. Therefore, both colored zones **13** and **14** are well suited for measuring purposes. When the concentration drops to about 10^{-8} grams per liter of air, the intensity of the color is determined by the first or lower flame **12**. This concentration is very sensitive to the fluctuations of the quantities of air and hydrogen so that it is not constant with a change in the intensity of flame **12** at the maximum sensitivity obtainable. Therefore, only the glowing zone **13** between flames **12** and **15** is usuable.

The optical axis of the spectrophotometer was set to transverse tube **5** about from five to fifteen times the length of flame **12** above flame **12** for measuring the blue sulfur coloration. When measuring the green phosphorus color, the axis was positioned from about one to two times the length of flame **12** above flame **12**.

The size of the burners can be selected to give the velocity of the gas and hydrogen flows as required. The water jacket **9** can be replaced by air cooling, and also the flame temperatures can be lowered by diluting the combustion gases with nitrogen or other inert gas. By setting the spectrophotometer upon the corresponding two wave lengths, namely 520 and 380 millimicrons, the combined presence of both phosphorus and sulfur can be determined.

Also, the determination can be carried out at other wave lengths since the spectral bands employed are wide and the given wave lengths of 520 and 380 millimicrons designate only the sharpest lines. The detection of phosphorus or sulfur in air is shown by the following example.

The air to be tested and suspected of containing a phosphorus compound, such as tricresyl phosphate, phosphorus oxychloride and sulfur compounds such as sulfur dioxide, dimethyl sulfoxide or thiophene, was introduced through tube **1** and the hydrogen was introduced through tube **7**. The gas mixture was ignited in the burner tip **11**. Thereupon, the flame struck back and ignited the lower or first flame **12** inside tube **5**. The water cooling was turned on and the spectrophotometer **18** set to take readings on the wave lengths 520 and 380 millimicrons. The optical axis of the photometer was set to be selectively centered at either from about one or two times or from five to fifteen times the length of flame **12** above flame **12**.

When only a few micrograms of phosphorus or sulfur-containing compounds were in a cubic meter of air, the photometer showed the presence of a distinct color in the glow zone **13**, this color being green for phosphorus and blue for sulfur. The observation of a specific color was facilitated by using a light filter **19** which filtered only pure green light for the phosphorus observation and a pure blue light for the sulfur observation.

A device developed by R.L. Wilburn (11) is a sampler for the continuous and near instantaneous testing of air for the presence of organophosphorus compounds and other oxidants. These compounds are often dispersed as aerosols for insect control and their presence in the atmosphere should be known in order that the necessary precautions may be taken.

The sampler utilizes the chemiluminescent property of luminol (3-amino-phthalhydrazide) to determine the presence of these oxidants in the air. The air sample is taken in continuously and mixed with the reagent solution. The solution then passes a photomultiplier after which it is exhausted from the sampler. The current from the photomultiplier is amplified and may be displayed on a chart recorder. This process permits of analysis of concentrations of organophosphorous compounds in concentrations as low as 3 gammas per liter of air.

A process developed by L.H. Goodson, W.B. Jacobs and A.W. Davis (12) permits continuous monitoring of airstreams for the presence of phosphorus compounds which are cholinesterase inhibitors.

The process is carried out by absorption of the pollutants in an absorbing solution by creating turbulent flow of the airstream and the absorbing solution prior to passing the absorbing solution containing the pollutants to an electrochemical cell. Any electrical change produced in the electrodes of the electrochemical cell denotes the presence of pollutants, the electrical change being proclaimed by conventional means.

A process developed by C.N. La Rosa, M.J. Prager and J.L. Kalinsky (13) involves the detection of toxic organophosphorus airborne substances by frustrated multiple internal reflection spectroscopy.

The apparatus used for the conduct of this process, employs a thin germanium crystal prism in the form of an isosceles trapezoid or parallelogram whose base surface and opposite parallel surface are coated with platinum about 70 millimicrons thick, the coating having been applied in a vacuum on the order of 10 millimicrons for effective adsorptiveness. The sampled atmosphere is flowed across the base and the opposed surface of the crystal while infrared of wavelength band 9 to 10 microns is directed through the crystal for multiple internal reflection and the amplitude of infrared that passes through the crystal is monitored.

References

(1) Athanassiadis, Y.C., "Air Pollution Aspects of Phosphorus and Its Compounds," *Report PB 188,073,* Springfield, Va., Nat. Tech. Information Service (Sept. 1969).

(2) Rushing, D.E., "A Tentative Method for the Determination of Elemental Phosphorus in Air," U.S. Dept. of Health, Education, and Welfare, Public Health Service, Salt Lake City, Utah (1968).

(3) Dechant, R., et al., "Determination of Phosphine in Air," *Am. Ind. Hyg. Assoc. J.* 27:75 (1966).

(4) Jacobs, M.B., *The Analytical Toxicology of Industrial Inorganic Poisons* (New York: Interscience, 1967).

(5) Land, D.B., et al., "A Fluoremetric Method for Determining Trace Quantities of Phosphate," *Mikrochim. Acta* (Vienna) 6:1013 (1966).

(6) Stern, A.C. (Ed.), *Air Pollution,* Vol. I (New York: Academic Press, 1968).

(7) Midwest Research Institute. Report for Division of Pesticides, U.S. Food and Drug Administration.

(8) Ives, N.F., et al., "Pesticide Residues, Investigation of Thermionic Detector Response for the Gas Chromatography of P, N, As, and Cl Organic Compounds," *J. Assoc. Off. Analyt. Chem.* 50(1):1 (1967).
(9) Karmen, A., U.S. Patent 3,425,806; Feb. 4, 1969; assigned to the U.S. Secretary of Health, Education and Welfare.
(10) van der Smissen, C.E., U.S. Patent 3,213,747; Oct. 26, 1965; assigned to O.H. Drager.
(11) Wilburn, R.L., U.S. Patent 3,287,089; Nov. 22, 1966; assigned to U.S. Secretary of the Army.
(12) Goodson, L.H., Jacobs, W.B. and Davis, A.W., U. S. Patent 3,715,298; Feb. 6, 1973; assigned to U.S. Secretary of the Army.
(13) La Rosa, C.N., Prager, M.J. and Kalinsky, J.L., U.S. Patent 3,582,209; June 1, 1971; assigned to U.S. Secretary of the Navy.

POLLENS

Aeroallergens are airborne materials which elicit a hypersensitivity or allergic response in susceptible individuals. The major effects of aeroallergens on human health are the production of allergic rhinitis and bronchial asthma. If the symptoms of allergic rhinitis occur during a particular season of the year, it is commonly called hay fever. It has been estimated that there are 10 to 15 million hay fever sufferers in the United States and that 5 to 10% of the untreated patients will develop bronchial asthma, as discussed by Finkelstein (1).

The common aeroallergens affecting human health are pollens of wind-pollinated plants, molds, house dust, and a miscellaneous group of vegetable fibers, cosmetics, paints, and others. The pollens are the most important of the entire list, and ragweed provides the most common of the pollens. More than 90% of the pollinosis occurring in this country is due to ragweed pollen.

Laboratory animals are used routinely in allergy studies, but exposure is usually by injection; most animals do not exhibit allergenic reaction to inhalation of aeroallergens. There is no evidence that aeroallergens have adverse effects on plants or materials.

Insufficient information exists to establish environmental air standards for the aeroallergens. Daily pollen counts are taken and pollen indexes derived in many local areas of the country by the use of a standardized procedure. However, because many variables are involved, these values are used more as guidlines than as standards. Generally, indexes of 5 to 15 are considered moderate, and acute hay fever symptoms last only a few days. Indexes above 15 are indicative of heavy pollen concentrations, and an index of 25 or more on any given day will usually cause severe symptoms of hay fever in most of the susceptible population. These values are relative, however, and may vary considerably between local areas.

Ragweed establishes itself readily in freshly turned soil, and therefore is found in abundance both in farmland and in urban areas in most parts of North America. Of the other aeroallergens, the molds are ubiquitous; their usual habitat is the soil and dust, and they become airborne through local air disturbances. House dust consists of small organic particulates. Because it is found in every indoor environment, house dust is probably the most common aeroallergen after pollens. Danders and other similar aeroallergens are found in the air close to their source, and their concentration in the air is therefore limited. They are allergenic to humans when the source is in close proximity to the susceptible individual.

The potential of other air pollutants to act synergistically with the natural allergens has become a new area of study in recent years. Several investigators have observed an increase in hospital admissions for bronchial asthma on days of high air pollution.

There are many materials which are aeroallergenic to sensitized individuals. However, some of the allergens are incorporated into products in such a way that their presence cannot always be recognized. Stuffing in pillows, mattresses, and toys may be of feathers, kapok, or other materials that can be highly allergenic.

The emission and dispersal of ragweed pollen have been studied in much detail. It has been found that pollen release occurs primarily in the early morning, and once the pollen is airborne, its dispersal is dependent upon horizontal and vertical air movements. If there is little air movement, dispersal of the pollen from a given source may be negligible. However, upward air flow can carry pollen up to high elevations, whereas horizontal air movements can carry the pollen great distances in all directions. During the ragweed season, daily pollen concentrations over much of the Eastern and Central United States commonly reach 350 to 1,000 grains per cubic meter of air.

The abatement and control of aeroallergens have been concentrated on ragweed. Considerable money and effort have been expended by local municipalities in attempting to reduce the pollen concentration in the air by reducing the ragweed plant density. Herbicides such as 2,4-D have been used extensively for this purpose. However, many of the eradication programs have had little success, primarily because windborne pollen from outside the control area usually has entered the city in sufficient quantities to cause pollinosis in the local susceptible population.

The economic costs incurred by the effects of and the control of aeroallergens cannot be adequately estimated. Insufficient data are available regarding the costs of allergic illnesses, and there are no estimates for the cost of abatement on the regional scale that would be required for adequate control.

Measurement in Air

The methods used for the analysis of aeroallergen pollution are based primarily on microscopic observations of collected samples from the air. Basically, these procedures are qualitative, but a relative degree of quantitation is introduced by standardization of the procedure. Some quantitative procedures are used that sample a given volume of air, and the results are expressed in terms of a count per unit volume of air. Most of the procedures have been concerned primarily with pollen and molds; little attempt has been made to sample for the other aeroallergens.

Qualitative Methods: The gravity slide method for pollen sampling, first used by Durham (2) in 1946, was accepted as the standard procedure by the Pollen Survey Committee of the American Academy of Allergy (3) in the same year. This standard air sampling device consists of two circular parallel planes of polished steel 9 inches in diameter and 3 inches apart, with a slide holder raised 1 inch above the lower plane. It is supported by a 30 inch metal rod on a tripod laboratory stand. A petrolatum coated slide is placed in the slide holder and exposed to the air on an unobstructed seven to eight story rooftop for 24 hours. The entire exposed area of the slide (4.84 $cm.^2$) is examined microscopically and the pollen counted.

The count is divided by 4.84 and expressed as a count per square centimeter, or simply as a number. The count can be converted into short ragweed pollen grains per cubic yard by multiplying by the factor of 3.6; to giant ragweed by 3.87; to timothy by 1.14; to corn by 0.17; and so forth (2).

The pollen count is determined (by many local authorities) by daily exposing a series of these Durham gravity slides at various sites in and about an area. Some slides may be exposed at ground level. The number of particles trapped on the slide is dependent upon wind conditions during the sampling period, and therefore, it is difficult to relate the counts to actual concentrations in the air. However, the pollen counts thus obtained, after several years, show a pattern of pollen concentration increase and decrease and correlate to some degree with the general incidence of hay fever in a given local area.

The gravity slide method for pollen determination has a number of limitations. In particular, the sampling is for a 24 hour period and does not give any indication of peak concentrations which might have existed at any time during the sampling period. Also, Ogden and Raynor (4) demonstrated that slides placed parallel to the airflow collect much more pollen than those placed at right angles, and this difference becomes greater as the higher wind speeds prevalent at greater heights. An increase of 3 to 4 miles per hour in wind speed may result in a 50% increase in the amount of pollen trapped (5). Therefore, the gravity slide method yields values which may not be entirely comparable to neighboring sampling sites because sampling heights and wind speed and direction cannot be standardized.

Quantitative Methods: Several volumetric devices are available for drawing a measured amount of air (using a vacuum pump) into a sampler. The intent here is to determine as accurately as possible the actual concentration present in the air at any given time. A photoelectric, continuous-recording particle sampler has been used by Smith and Rooks (6) for studying the diurnal fluctuations of airborne ragweed pollen. Raynor (7) made use of a membrane filter device with an attached timer and measured air intake to obtain a series of sequential pollen samples. The membrane filters were then viewed through a microscope and the counts determined.

The Hirst Spore Trap draws a measured amount of air through an orifice, and the pollen is impacted on a microscope slide moved past the orifice at a rate of 2 mm./hr. by a clock mechanism. A 24 hour sample thus can be obtained, but the deposition has been spaced in time along the slide (8).

Volumetric sampling devices have presented the problem of isokinetic sampling. That is, with volumetric samplers, the intake opening must be continuously oriented into the wind, and the airflow through the sampler must be equal at all times to the wind speed in the free air approaching the intake. If these conditions are not met, a true representative sample cannot be obtained for particles the size of pollens (9).

Because the use of volumetric samplers is too difficult in routine pollen sampling, the simple Durham gravity slide method has remained the standard technique in spite of its deficiencies. Although it is inaccurate for short-term (1 day or less) measurements, it has been satisfactory for determining seasonal patterns. However, a number of devices have been devised which attempt to retain simplicity but yet improve upon the gravity slide method.

The simplest sampler has been a vertically-oriented wire of about 1 mm. in diameter which is placed in the air stream containing pollen. The air can go around the wire but the pollen is impacted on the surface. The wire is then examined through a microscope and the pollen grains counted. An improvement upon this has been the flag sampler (10). It consists of an ordinary household pin set in a glass bearing in which it moves freely. It has a flag of transparent tape wound about it that works like a weather vane to keep the coated leading edge of the pin facing into the wind. Particles unable to follow the air stream around the curved surface of the pin impact upon it and are counted by means of a microscope. A similar device uses a larger wind vane to keep the edge of a microscope slide facing into the wind to act as the trapping surface.

Such samplers are inexpensive and are suitable for use when a large number of samples are to be taken. However, their disadvantages are that they are efficient only when there is some wind (at least 5 mph), and the impaction surfaces are quickly covered and require frequent changes. Also, the wind velocity and fluctuations need to be known, which requires the use of a separate recording anemometer.

Another approach to impaction sampling has been to mechanically move an adhesive-coated surface through the air to be sampled. The rotorod sampler (11) consists of two vertical rods (plastic or metal) rotated about a vertical axis approximately 2 inches away at a speed of about 2,000 rpm. The coated collecting surface of the rods moves at a tangential speed of approximately 25 mph, which is higher than most air velocities sampled and thereby has a relatively high collection efficiency independent of wind speed. The rods are examined through the microscope and pollen counts made. Modifications of this device have been the rotobar (10), with a bar-shaped surface used instead of a rod, and the rotoslide (4), which uses a microscope slide.

The main disadvantage of these samplers has been that a high concentration of pollen can build up on the impaction surface in a short time (an hour or less) and, therefore, frequent changing is required when continuous sampling is desired. To obviate this difficulty, the rotodisk sampler (12) has been devised, which substitutes disks for the rods. The vertical edge of the disk is covered except for a small slit, and a timing mechanism automatically shifts the slit to expose a fresh sampling surface.

Fungi have also been sampled by the methods given above. As the pollen are being counted, some investigators may also count mold spores. However, fungi lend themselves to other sampling procedures that utilize growth of the organisms as a means of measurement. The basic methods are:

(1) Sedimentation (13): In this simple method of sampling airborne organisms, the suspended particulates are allowed to settle or plain surfaces or on surfaces coated with a nutrient growth medium. This method yields information on the total number of viable particles that have settled out during the given sampling period of time.
(2) Impingement into liquids (14) through (17): Air is drawn through a small jet and is directed against a liquid surface, the suspended fungi being collected in the liquid. Because of the agitation of the particles in the collecting liquid, aggregates are likely to be broken up. Therefore, the counts obtained by this method tend to reflect the total number of individual cells in the air and are higher than the value obtained by other methods.
(3) Impaction onto solid surfaces (18)(19): Air is drawn through a small jet(s) and particles are deposited on dry or coated solid surfaces, or on an agar nutrient. This method has been used to determine total cellular numbers, size distribution, total viable numbers, and variation in concentration per unit of time during a long sampling period.
(4) Filtration (20) through (23): The particulates are collected by passage of the air through a filter which can be cellulose-asbestos paper, glass wool, cotton, alginate wool, gelatin foam, or membrane material. The particulates are washed from the filters and assayed by appropriate microbiological techniques. Since the viability of the organisms can be detrimentally affected by dehydration in the air stream, the results may be biased in this method.
(5) Centrifugation (24)(25): The particulates are propelled by centrifugal force onto the collecting surface, which can be glass or an agar nutrient. Particulate size and particulate concentration can be obtained by this method.
(6) Electrostatic precipitation (26): Particles are collected by drawing air at a measured rate over an electrically charged surface of glass, liquid, or agar. The number of particles or viable number is then determined.
(7) Thermal precipitation (27): The organisms are collected on surfaces by means of thermal gradients. The design is based on the principle that airborne particles are repelled by hot surfaces and are deposited on colder surfaces by forces proportional to the temperature gradient. The particle size distribution can then be determined.

References

(1) Finkelstein, H., "Air Pollution Aspects of Aeroallergens (Pollens)," *Report PB 188,076;* Springfield, Va., Nat. Tech. Information Service (Sept. 1969).
(2) Durham, O.C., "The Volumetric Incidence of Atmospheric Allergens. IV. A Proposed Standard Method of Gravity Sampling, Counting, and Volumetric Interpretation of Results," *J. Allergy* 17:79 (1946).
(3) "Preliminary Report of the National Pollen Survey Committee of the American Academy of Allergy on Proposed Standardization of Pollen Counting Techniques," *J. Allergy* 17:179 (1946).
(4) Ogden, E.C., and G.S. Raynor, "A New Sampler for Airborne Pollen: The Rotoslide ," *J. Allergy* 40:1 (1967).
(5) Albert, M.M., "Significance of Pollen Counts. Interpretations and Misinterpretations," *N.Y. State J. Med.* 66:2409 (1966).
(6) Smith, R.D., and R. Rooks, "The Diurnal Variation of Airborne Ragweed Pollen as Determined by

a Continuous Recording Particle Sampler and Implications of the Study," *J. Allergy* 25:37 (1954).
(7) Raynor, G.S., "An Automatic Programming Filter Sampler," *J. Air Pollution Control Assoc.* 7:122 (1957).
(8) Hirst, J.M., "An Automatic Volumetric Spore Trap," *Ann. Appl. Biol.* 39:257 (1952).
(9) Watson, H.H., "Errors due to Anisokinetic Sampling of Aerosols," *Am. Ind. Hyg. Assoc. Quart.* 15:21 (1954).
(10) Harrington, J.B., Jr., G.C.Gill, and B.R. Warr, "High-Efficiency Pollen Samplers for Use in Clinical Allergy," *J. Allergy* 30:257 (1959).
(11) Cole, A.L., and J.B. Harrington, Jr., "Air Pollution by Ragweed Pollen. III. Atmospheric Dispersion of Ragweed Pollen," *J. Air Pollution Control Assoc.* 17:654 (1967).
(12) Cole, A.L., and A.W. Stohrer, "Sequential Roto-Disk Pollen Sampler," *Public Health Repts. (U.S.)* 81:577 (1966).
(13) Richardson, J.F., and E.R. Wooding, "The Use of Sedimentation Cell in the Sampling of Aerosols," *Chem. Eng. Sci.* 4:26 (1955).
(14) Cown, W.B., T.W. Kethley, and E.L. Fincher, "The Critical-Orifice Liquid Impinger as a Sampler for Bacterial Aerosols," *Appl. Microbiol.* 5:119 (1957).
(15) Ferry, R.M., L.E. Farr, and M.G. Hartman, "The Preparation and Measurement of the Concentration of Dilute Bacterial Aerosols," *Chem. Rev.* 44:289 (1940).
(16) Greenburg, L., and J.J. Bloomfield, "The Impinger Dust Sampling Apparatus as Used by the United States Public Health Service," *Public Health Repts. (U.S.)* 47:654 (1932).
(17) Moulton, S., T.T. Puck, and H.M. Lemon, "An Apparatus for Determination of the Bacterial Content of Air," *Science* 97:51 (1943).
(18) Andersen, A.A., "New Sampler for the Collection, Sizing, and Enumeration of Viable Airborne Particles," *J. Bacteriol.* 76:471 (1958).
(19) Dubuy, H.G., A. Hollaender, and M.D. Lackey, "A Comparative Study of Sampling Devices for Airborne Micro-Organisms, "*Public Health Repts. Suppl.* 184 (1945).
(20) Mitchell, R.B., J.D. Fulton, and H.V. Ellingston, "A Soluble Gelatin Foam Sampler for Airborne Micro-Organisms at Surface Levels," *Am. J. Public Health* 44:1334 (1954).
(21) Noller, E., and J.C. Spendlove, "An Appraisal of the Soluble Gelatin Foam Filter as a Sampler for Bacterial Aerosols," *Appl. Microbiol.* 4:300 (1956).
(22) Sehl, F.W., and B.J. Havens, Jr., "A Modified Air Sampler Employing Fiber Glass," *A.M.A. Arch. Ind. Hyg. Occupational Med.* 3:98 (1951).
(23) Thomas, D.J., "Fibrous Filters for Fine Particle Filtration," *J. Inst. Heating Ventilating Engrs.* 20:35 (1952).
(24) Sawyer, K.F., and W.H. Walton, "The Conifuge–A Size Separating Sampling Device for Airborne Particles," *J. Sci. Instr.* 27:272 (1950).
(25) Wells, W.F., "Apparatus for Study of Bacterial Behavior of Air," *Am. J. Public Health* 23:58 (1933).
(26) Kraemer, H.F., and H.F. Johnstone, "Collection of Aerosol Particles in Presence of Electrostatic Fields," *Ind. Eng. Chem.* 47:2426 (1955).
(27) Kethley, T.W., M.T. Gordon, and C. Orr, Jr., "A Thermal Precipitator for Aerobacteriology," *Science* 116:368 (1946).

POLYNUCLEAR AROMATICS

Epidemiological studies indicate a possible relationship between air pollution and lung cancer mortality in the United States and other countries. While the role of cigarette smoking is clearly better defined, the role of air pollution and possibly of organic carcinogens is less certain but is still believed to be present. The existence of urban-rural differentials and migrant-native differentials speak of a possible contribution of air pollution to total lung cancer mortality, as noted by D.A. Olsen and J.L. Haynes (1).

Animal experiments indicate that under certain conditions many of the organic carcinogens found in ambient air can produce tumors. In most of the experiments, the compounds were painted on the skin. However, these investigations have shown that synergistic and antagonistic effects are important. Thus, some compounds, such as phenols, have been found capable of promoting or reinforcing the action of organic carcinogens.

Inhalation experiments indicate that benzo(a)pyrene (BaP) adsorbed on inert particulates (iron oxide) causes a higher incidence of lung tumors in hamsters than just BaP alone.

Moreover, these tumors were of the same types observed in humans. In contrast, many noncarcinogenic polynuclear aromatic hydrocarbons (PAH) can reduce the potency of organic carcinogens, at times to such an extent that no effect is found. No information has been found in the literature concerning the effects of airborne organic carcinogens on plants or materials.

The atmospheric carcinogens fall into three principal categories, namely, PAH, polynuclear heterocyclics and oxygenated compounds, and alkylating agents. There are very little data available on cocarcinogens and anticarcinogens that may be present in the air.

The major sources of PAH are heat generation, refuse burning, industrial processes, and motor vehicles. Of these, heat generation accounts for more than 85% of the PAH emitted, with the other three sources each accounting for about 5% of the total. Similar figures for other organic carcinogens have not yet been estimated.

Sources of aza-heterocyclics and polynuclear carbonyl compounds have not been well established. Combustion products such as coal tar are likely sources. Only limited data are available on the concentration of organic carcinogens in the ambient air. In 1966, the concentrations of BaP in 106 urban and nonurban areas were measured. The average value was 0.00279 μg./m.^{3} for urban areas and 0.00035 μg./m.^{3} for nonurban areas. Other data indicate that BaP constitutes only a small percentage (as low as 5%) of the total PAH present in the atmosphere. Control methods for organic carcinogens are being studied in connection with hydrocarbon and particulate control programs.

Measurement in Air

Several publications have compared methods and provided an overview of the role of analytical chemistry in carcinogenesis studies (2) through (6). The methods of analysis generally consist of the following sequence of steps: (a) sampling, (b) extraction, (c) separation, and (d) analysis.

Sampling Methods: Organic carcinogens are usually associated with particulates and thus are collected as particulates. Sampling is often accomplished with a high-volume air sampler, the particulates (with which the relevant compounds are associated) being collected on a glass-fiber filter. Collection into a liquid (cyclohexane) absorbent has also been employed. The collection process may continue for many days, in order to obtain sufficient quantities of material. A technique which has proved satisfactory for collecting BaP from high-temperature gas streams is the use of a series of water bubblers and condensate traps immersed in an ice-water bath, followed by a high-efficiency filter (7).

Sample collection techniques for polynuclear hydrocarbons have been discussed by R.P. Hangebrauck, D.J. Von Lehmden and J.E. Meeker of the National Air Pollution Control Administration (8).

Extraction Methods: The organic carcinogens are generally removed from the particulates by solid-liquid extraction, although liquid-liquid extraction, sublimation, and distillation (for coal-tar mixtures) have also been used (3).

Particulates are commonly extracted from the atmosphere by solid-liquid extraction with either benzene or cyclohexane (3). Other solvents used include chloroform, acetone, isooctane, methanol, and dimethylformamide, as well as benzene mixtures of aliphatic hydrocarbons or methanol. The solvent-particulate mixtures can be stirred at a certain temperature or a Soxhlet extraction can be used.

Recent studies have been concerned with the efficiency of extraction. Stanley et al. (9) showed that in the analysis of equal weights of air particulates enriched with BaP, benz(c)-acridine, and 7H-benz(de)anthracen-7-one, the percentages of these compounds extracted ranged from 50 to 100, 15 to 100, and 40 to 80, respectively. The solvents were cyclohexane, benzene, methylene chloride, and acetone.

In another study Dubois et al. (10) stated that benzene was of questionable value as an extracting agent for the initial preparation of an air sample. The larger amounts of material extracted by benzene, as compared with cyclohexane, lead to analytical interferences in the subsequent analyses. Aigina and Mints (11) found somewhat similar results in that the extinction of BaP fluorescence called for a controlled content of 1,12-benzoperylene as an internal standard.

Separation: Techniques used for separating the different organic carcinogens include column chromatography, thin-layer chromatography, and gas chromatography (3). Column chromatography is the common method for separation of a complex mixture of PAH.

Column chromatography techniques are sufficiently advanced that predictions can be made with regard to adsorbability of hydrocarbons on alumina. Structural studies (3) have shown that adsorbability is greater for:

(1) A compound with more rings; e.g., chrysene is adsorbed more strongly with phenanthrene;
(2) A compound with more double bonds; e.g., benz(a)anthracene is adsorbed more strongly than pyrene;
(3) An acene (a linearly condensed arene) than for an isomeric phene (an angularly condensed arene); e.g., naphthacene is adsorbed more strongly than benz(a)anthracene;
(4) Fluorenic hydrocarbons than for pericondensed hydrocarbons with the same number of rings and double bonds; 11H-benzo(b)fluorene is adsorbed more strongly than pyrene;
(5) The most nearly coplanar compound of a group; e.g., the decreasing order of adsorbability, 2-phenylanthracene>1-phenylanthracene, follows the increasing angle of twist in these molecules;
(6) A hydrocarbon substituted with an increasing number of sterically unhindered methyl or alkylene groups than for the hydrocarbon itself; e.g., 9-methylanthracene is adsorbed more strongly than anthracene; and
(7) The most extensively conjugated isomer; e.g., 1-(2-naphthyl)cyclopentene is adsorbed more strongly than 3-(2-naphthyl)cyclopentene.

It is postulated that adsorption on alumina involves pi-type complexation where the active sites on the alumina are relatively broad electron-attracting areas to which the electron-donating hydrocarbon substrate is held monomolecularly and preferentially in a planar configuration parallel to the surface, if such arrangement is sterically possible.

In general, the columns and techniques used are compromises between better separations and analytical speed (12) through (17). For example, the cyclohexane elution of an air sample can take two weeks and require 500 to 600 fractions totaling 5,000 ml. To speed up the process it is customary to add increasing amounts of ethyl ether. This speeds the removal of solutes from the column but at the expense of resolution (12). The chromatographic separation can also be speeded by the use of alumina deactivated by the addition of 1.6 to 1.8% water. If the alumina were not deactivated to this extent, the PAH would come off the column too slowly (12).

Hydrocarbons are eluted in the following order (17): aliphatics, olefins, benzene derivatives, naphthalene derivatives, dibenzofuran fraction, anthracene fraction, pyrene fraction, benzofluorene fraction, chrysene fraction, benzopyrene fraction, benzoperylene fraction, and coronene fraction. For example, with alumina containing 13.7% water, the pyrene fraction was found in the beginning of the 3% ether eluent; the chrysene fraction followed in the last part of the 3% ether eluent; the benzopyrene fraction appeared in the beginning of the 6% ether eluent, followed by the benzoperylene fraction in the start of the 9% ether eluent, and the coronene fraction shortly afterward in the same 9% ether eluent.

Although the relative location of the fractions was always the same, unknown variables sometimes caused the fractions to elute sooner or later than expected. The fractions were fairly well separated, although test tubes containing the tail end of one fraction usually contained small amounts of the next fraction. Most fractions were found in three to six tubes. However, the benzoperylene and coronene fractions were each spread over 6 to 10 tubes.

Column chromatography studies have also been reported by Sawicki et al.,(14)(15), in which the separation of polynuclear aza-heterocyclic hydrocarbons and polynuclear aromatic amines was considered. The procedures are essentially the same as for PAH. However, the alumina does cause some decomposition of the amines; e.g., 9-aminoanthracene is recovered in 70% yield as anthraquinone, while 2-aminoanthracene is in only 26% yield.

In his 1964 review Sawicki (3) noted that although thin-layer chromatography had been applied only to a small extent in the separation of PAH, it would find increased use: since that time there have been a sizable number of thin-layer chromatography (TLC) studies reported (14)(15)(18) through (39). The advantages of this technique are that it is simple, rapid, inexpensive, sensitive, requires little working material, and can resolve complex mixtures (3).

Various adsorbents, including silica gel, alumina, cellulose, and cellulose acetate, along with differing types of developing solvents have been evaluated for use in TLC. Several authors (12)(15)(21)(39) state that the complete analysis procedure generally consists of extraction followed by separation using either TLC or paper chromatography, and then spot location by fluorescence or color development (24). One reagent used for color development is 7,7,8,8-tetracyanoquinodimethan, which reacts with polynuclear compounds to form colored pi complexes.

White and Howard (39) give R_F values (the ratio of the migration distance of the substance to the migration distance of the solvent front) for 29 PAH on cellulose and cellulose acetate adsorbents. Such values, however, are quite specific for a given adsorbent and developing solvent. The R_F values alone are inadequate for characterizing polynuclear compounds (36).

The efficiency of separation and recovery by TLC has also been studied (18)(31)(35)(38). Typical results are those reported by Sawicki et al. (31)(35) in which recoveries of 65% for 9-acridanone and 85% for BaP were obtained. Assays for these same compounds were also run using fluorometric procedures after TLC operation. It was possible to detect 9-acridanone in concentrations of 0.0004 μg./m.^{3}. (31). For BaP the identification limit ranged between zero and 0.04 μg. (35).

Previously Sawicki and Johnson (23) had reported the following identification limits: anthracene, 0.01 μg.; phenanthrene, 1.0 μg.; fluoranthrene, 0.01 μg.; chrysene, 0.01 μg.; pyrene, 0.001 μg.; and BaP, 0.001 μg.

The separation and characterization of polynuclear aza-heterocyclic hydrocarbons and polynuclear aromatic amines have also been reported (19)(20)(33)(37). The identification limits are in the same range as for PAH; examples are: carbazole, 0.02 μg.; 11H-benzo(a)-carbazole, 0.02 μg.; 1-azacarbazole, 0.2 μg.; and 1,2-dinaphthylamine, 0.1 μg.

A limited number of gas chromatography studies have been conducted (40) through (45). Although the procedures are still in the process of being worked out, the method promises to be relatively rapid, provided the various chromatographic peaks have been identified. As an example of the capabilities of the method, the following amounts of PAH in soot have been reported (42): acenaphthylene, 4,700 μg./g. soot; phenanthrene, 5,500 μg./g. soot; fluoranthrene, 5,200 μg./g. soot; chrysene, 6,300 μg./g. soot; and BaP, 2,900 μg./g. soot.

Electrophoresis has also been considered a possible means of separating pollutants after collection (46)(47). Reports show that separations of polynuclear phenols (47) and of romolecular material (46) are feasible; however, no quantitative information is given.

Analysis: Analyses of the eluate fractions from chromatography are conducted chiefly via ultraviolet absorption measurements or fluorescence measurements (3). Presently, the National Air Pollution Control Administration (48) analyzes airborne particulates for BaP by a fluorescence method [a modification of the one described by Sawicki et al. (17)] using thin-layer chromatography for separation, see (35) and (38).

The following discussion briefly reviews the various analytical procedures that are applicable to the study of airborne carcinogens. Methods based on ultraviolet-visible absorption spectra and fluorescence spectra play a prominent role in the identification and determination of polynuclear hydrocarbons. The advantages of absorption spectra are that the wavelength maximum of a compound is unaffected by the presence of other compounds, quenching effects are absent, and the absorption curves of many PAH are available in the literature.

The main disadvantage, compared to fluorescence, is the lower sensitivity. For many fluorescent compounds, the excitation and emission spectra are 10 to 10,000 times more sensitive than their absorption spectrum. The disadvantages of fluorescence spectra are the poorer reproducibility, the fewer available quantitative spectra, and quenching that can drastically affect the sensitivity and shape of the spectral curve (3).

Several articles have dealt with spectral methods. Of these, one was a preliminary communication concerned with the detection of impurities in commercial polycyclic hydrocarbon samples (49). The remaining articles are concerned with the methods of ultraviolet-visible spectra (50) through (54) and fluorescence spectra (25)(34). The results of these studies are summarized in Table 42.

In one study of other than spectral methods for analysis Sawicki et al. (54) also investigated the application of the piperonal test to the benzene-soluble fraction of airborne particulates. Piperonal reacts with aromatic compounds to give a colored product which obeys Beer's law. The correlation coefficients between BaP concentrations and the piperonal test were 0.95 for 174 urban samples and 0.89 for 25 nonurban samples.

Two papers deal with thermochromic tests for determining the amounts of polynuclear compounds containing the fluorenic methylene group (55) and polycyclic p-quinones (32).

TABLE 42: SUMMARY OF SOME INVESTIGATIONS IN SPECTRAL METHODS

Method	Compound(s)	Procedure	Normal Range of Sensitivity	Reference
Ultraviolet	Benzo(a)pyrene	Column chromatography	1 to 6 ug/ml	(52)
Ultraviolet	Dibenz [a,j]-anthracene	Hexane extraction	0.5 to 5 μg/ml	(50)
Ultraviolet	Benzo(a)pyrene	Benzene extraction	0 to 45 ug/1,000 /m³ air	(54)
Ultraviolet	Benzo(a)pyrene	Hexane extraction	10^{-2} μg/ml	(51)
Ultraviolet	Benzo(a)pyrene Pyrene Perylene	Modified Piperonal tests	0.003 μ moles/ml 0.01 μ moles/ml 0.001 μ moles/ml	(53)
Fluorescence	Benzo(a)pyrene Perylene	Acetophenone-trifluoroacetic acid	2.5×10^{-6} M 10^{-6} M	(34)
Fluorescence	Perylene Anthanthrene Benzo(a)pyrene	Pentane extraction	0.3 ug/ml 0.3 ug/ml 0.4 ug/ml	(25)

Source: Report PB 188,090, September 1969

In both cases borohydrides were used at reflux temperatures for color development. Identification limits ranged from 3 to 40 μg. A somewhat similar study was reported in which 4-azobenzenediazonium fluoroborate was used to develop color in the determination of aniline, naphthylamine, and anthramine derivatives. The identification limits were in the range of 2 to 20 μg.

A direct method of determining BaP has been devised using its characteristic quasi-line emission spectrum at liquid nitrogen temperatures (56). The method has been used to determine BaP in the air of submarine crew quarters, typically about 1 ppm.

Bioassay techniques have been reported in which the response of a microorganism is correlated with carcinogenic activity (57) through (60). These techniques are based on the phenomenon known as photodynamic action where a combination of light energy and chemical sensitizer (in this case PAH) causes the immobilization and death of *Paramecium caudatum.* Definitive assay for BaP can be performed in the 0.001 to 100 μg./ml. region (59).

Details of PNA determination in atmospheric particulate matter by column chromatography followed by spectrophotometry or fluorimetry have been published (60a) as a product of deliberations by an Intersociety Committee.

Measurement in Water

A report on "The Analyses of Aromatic Compounds in Water Using Fluorescence and Phosphorescence" has been prepared by D.W. Ellis of the Water Resources Research Center of the University of New Hampshire (61).

The purpose of this work was to develop a method for the determination of trace amounts of PNA present in natural water. The analytical scheme as envisioned initially was composed of four major steps.

(1) Continuous liquid-liquid extraction of 5 liters of water with n-pentane followed by concentration of the extract to several milliliters.
(2) Column chromatography of the resultant concentrate on acidic alumina to remove interfering basic components.
(3) Thin-layer chromatography (TLC) of the concentrated eluent to separate the individual PNA.
(4) Quantitative fluorescence analysis of the separated compounds.

Four different polynuclear aromatics were identified in the samples tested. Fluoranthene was the only PNA to be found in all samples tested. Anthracene, methyl anthracene and dimethyl anthracene were identified in one sample. No correlation could be made between rivers and seasons.

The banks or the tributaries of one river tested do not contain towns, cities or industrial complexes above the sampling site. The types of compounds identified from that river were not associated with agricultural fertilizers or pesticides. It must therefore be concluded that the PNA compounds identified in this study originated largely from natural sources.

References

(1) Olsen, D.A. and Haynes, J.L., "Air Pollution Aspects of Organic Carcinogens," *Report PB 188,090,* Springfield, Va., Nat. Tech. Info. Service (Sept. 1969).
(2) Rothwell, K., and J.K. Whitehead, "Complex Formation, Isolation and Carcinogenicity of Polycyclic Aromatic Hydrocarbons," *Nature* 213:797 (1967).
(3) Sawicki, E., "Separation and Analysis of Polynuclear Aromatic Hydrocarbons Present in the Human Environment," *Chemist-Analyst* 53:24, 28, 56, 88 (1964).
(4) Sawicki, E., T.W. Stanley, W.C. Elbert, J. Meeker, and S.P. McPherson, "Comparison of Methods for the Determination of Benzo(a)pyrene in Particulates from Urban and Other Atmospheres," *Atmos. Environ.* 1:131 (1967).
l, W., "The Role of Analytical Chemistry in Carcinogenesis Studies," *Anal. Chem.* 41:22A (1969).

(6) Wettig, K., "Essay on the Problematics Concerning the Determination of Benzpyrene in Atmospheric Air," *Neoplasma* 14:181 (1967).
(7) Stenburg, R.L., D.J. Von Lehmden, and R.P. Hangebrauck, "Sample Collection Techniques for Combustion Sources–Benzopyrene Determination," *Am. Ind. Hyg. Assoc. J.* 22:271 (1961).
(8) Hangebrauck, R.P., Von Lehmden, D.J. and Meeker, J.E., *Publ. No. 999-AP-33,* Durham, N.C., Nat. Air Pollution Control Admin. (1967).
(9) Stanley, T.W., J.E. Meeker, and M.J. Morgan, "Extraction of Organics from Airborne Particulates," *Environ. Sci. Tech.* 1:927 (1967).
(10) Dubois, L., A. Zdrojewski, and J.L. Monkman, "Comparison of Three Methods for Trace Analysis of Polycyclics," *Mikrochim. Acta* 5:903 (1967).
(11) Aigina, E.P., and I.M. Mints, "Determination of Low 3,4-Benzpyrene Concentration by Means of the Sbpol'skii Effect," *Hyg. Sanitation* 31:264 (1966).
(12) Dubois, L., A. Zdrojewski, C. Baker, and J.L. Monkman, "Some Improvements in the Determination of Benzo(a)pyrene in Air Samples," *J. Air Pollution Control Assoc.* 17:818 (1967).
(13) Moore, G.E., R.S. Thomas, and J.L. Monkman, "The Routine Determination of Polycyclic Hydrocarbons in Airborne Pollutants," *J. Chromotog.* 26:456 (1967).
(14) Sawicki, E., H. Johnson, and K. Kosinski, "Chromotographic Separation and Spectral Analysis of Polynuclear Aromatic Amines and Heterocyclic Imines," *Microchem. J.* 10:72 (1966).
(15) Sawicki, E., T.W. Stanley, and W.C. Elbert, "Characterization of Polynuclear Aza Heterocyclic Hydrocarbons Separated by Column and Thin-Layer Chromatography From Air Pollution Source Particulates," *J. Chromatog.* 18:512 (1965).
(16) Sawicki, E., W. Elbert, T.W. Stanley, T.R. Hauser, and F.T. Fox, "The Detection and Determination of Polynuclear Hydrocarbons in Urban Airborne Particulates. I. The Benzopyrene Fraction." *Intern. J. Air Pollution* 2:273 (1960).
(17) Sawicki, E., W. Elbert, T.W. Stanley, T.R. Hauser, and F.T. Fox, "Separation and Characterization of Polynuclear Aromatic Hydrocarbons in Urban Airborne Particulates," *Anal. Chem.* 32:810 (1960).
(18) Bender, D.F., and E. Sawicki, "Differentiation of Benzofluorenes on Thin-Layer Substrates," *Chemist-Analyst* 54:73 (1965).
(19) Bender, D.F., E. Sawicki, and R.M. Wilson, "Characterization of Carbazole and Polynuclear Carbazoles in Urban Air and in Air Polluted by Coal Tar Pitch Fumes by Thin-Layer Chromatography and Spectrophotofluorometry," *Intern. J. Air Water Pollution* 8:633 (1964).
(20) Bender, D.F., E. Sawicki, and R.M. Wilson, "Fluorescent Detection and Spectrophotofluorometric Characterization and Estimation of Carbazole and Polynuclear Carbazoles Separated by Thin-Layer Chromatography," *Anal. Chem.* 36:1011 (1964).
(21) Lindstedt, G., "A Simplified Chromatographic Method for the Routine Determination of 3,4-Benzopyrene in Air Samples," *Atmos. Environ.* 2:1 (1968).
(22) Sawicki, E., and H. Johnson, "Characterization of Aromatic Compounds by Low-Temperature Fluorescence and Phosphorescence: Application to Air Pollution Studies," *Microchem. J.* 8:85 (1964).
(23) Sawicki, E., and H. Johnson, "Thin Layer Chromatographic Characterization Tests for Basic Polynuclear Compounds. Application to Air Pollution," *Mikrochim, Acta* 2:435 (1964).
(24) Sawicki, E., C.R. Engel, and W.C. Elbert, "Chromatographic Location and Colorimetric Determination of Mercaptans, Prolines, and Free Radical Precursors," *Talanta* 14:1169 (1967).
(25) Sawicki, E., T.R. Hauser, and T. W. Stanley, "Ultraviolet Visible and Fluorescence Spectral Analysis of Polynuclear Hydrocarbons," *Intern. J. Air Pollution* 2:253 (1960).
(26) Sawicki, E., J.E. Meeker, and M.J. Morgan,"Polynuclear Aza Compounds in Automotive Exhaust," *Arch. Environ. Health* 11:773 (1965).
(27) Sawicki, E., J.E. Meeker, and M.J. Morgan, "The Quantitative Composition of Air Pollution Source Effluents in Terms of Aza Heterocyclic Compounds and Polynuclear Aromatic Hydrocarbons," *Air Water Pollution* 9:291 (1965).
(28) Sawicki, E., J.L. Noe, and F.T. Fox, "Spot Test Detection and Colorimetric Determination of Aniline, Naphthylamine and Anthramine Derivatives with 4-Azobenzenediazonium Fluoroborate," *Talanta* 8:257 (1961).
(29) Sawicki, E., T.W. Stanley, and W.C. Elbert, "Quenchofluormetric Analysis for Fluoranthenic Hydrocarbons in the Presence of Other Types of Hydrocarbons," *Talanta* 11:1433 (1964).
(30) Sawicki, E., T.W. Stanley, and W.C. Elbert, "Direct Fluorometric Scanning of Thin-Layer Chromatograms and Its Application to Air Pollution Studies," *J. Chromatog.* 20:348 (1965).
(31) Sawicki, E., T.W. Stanley, and W.C. Elbert, "Assay for 9-Acridanone in Urban Atmosphere by Thin-Layer Chromatography Fluorimetric Procedures," *Talanta* 14:431 (1967).
(32) Sawicki, E., T.W. Stanley, and T.R. Hauser, "Ultraviolet, Visible and Fluorescence Spectral Analysis of Polynuclear Hydrocarbons," *Anal. Chem.* 30:2005 (1958).
(33) Sawicki, E., T.W. Stanley, and H. Johnson, "Direct Spectrofluorometric Analysis of Aromatic Compounds on Thin-Layer Chromatograms," *Microchem. J.* 8:257 (1964).
(34) Sawicki, E., T.W. Stanley, and H. Johnson, "Quenchofluorometric Analysis for Polynuclear Compounds," *Mikrochim. Acta* 1:178 (1965).

(35) Sawicki, E., T.W. Stanley, W.C. Elbert, and J.D. Pfaff, "Application of Thin-Layer Chromatography to the Analysis of Atmospheric Pollutions and Determination of Benzo(a)pyrene," *Anal. Chem.* 36:497 (1964).
(36) Sawicki, E., T.W. Stanley, S. McPherson, and M. Morgan. "Use of Gas-Liquid and Thin-Layer Chromatography in Characterizing Air Pollutants by Fluorometry," *Talanta,* 13:619 (1966).
(37) Sawicki, E., T.W. Stanley, J.D. Pfaff, and W.C. Elbert, "Thin-Layer Chromatography Separation and Analysis of Polynuclear Aza Heterocyclic Compounds, "*Anal. Chim. Acta* 31:359 (1964).
(38) Stanley, T.W., M.J. Morgan, and J.E. Meeker, "Thin-Layer Chromatographic Separation and Spectrophotometric Determination of Benzo(a)pyrene in Organic Extracts of Airborne Particulates," *Anal. Chem.* 39:1327 (1967).
(39) White, R.H., and J.W. Howard, "Thin-Layer Chromatography of Polycyclic Aromatic Hydrocarbons," *J. Chromatog.* 29:108 (1967).
(40) Cantuti, V., G.P. Cartoni, A. Liberti, and A.G. Torri, "Improved Evaluation of Polynuclear Hydrocarbons in Atmospheric Dust by Gas Chromatography," *J. Chromatog.* 17:60 (1965).
(41) Carugno, N., and S. Rossi, "Evaluation of Polynuclear Hydrocarbons in Cigaret Smoke by Glass Capillary Columns," *J. Gas Chromatog.* 5:103 (1967).
(42) Chakraborty, B.B., and R. Long, "Gas Chromatographic Analysis of Polycyclic Aromatic Hydrocarbons in Soot Samples," *Environ. Sci. Tech.* 1:828 (1967).
(43) DeMaio, L., and M. Corn, "Gas Chromatographic Analysis of Polynuclear Aromatic Hydrocarbons with Packed Columns: Application to Air Pollution Studies," *Anal. Chem.* 38:131 (1966).
(44) Ettre, L.S., R.D. Condon, F.J. Kabot, and E.W. Cieplinski, "Dual Column-Differential Flame Ionization Detector System and Its Application with Packed and Golay Type Columns," *J. Chromatog.* 13:305 (1964).
(45) Wilmhurst, J.R., "Gas Chromatographic Analysis of Polynuclear Arenes," *J. Chromatog.* 17:50 (1965).
(46) Goppers, V., and H.J. Paulus, "Macromolecular Compounds Isolated from Airborne Particles by Electrophoresis and Paper Chromatography," *Am. Ind. Hyg. Assoc. J.* 23:181 (1962).
(47) Sawicki, E., M. Guyer, R. Schumacher, W.C. Elbert, and C.R. Engel, "Electrophoretic and Chromatographic Separation and Fluorimetric Analysis of Polynuclear Phenols. Application to Air Pollution," *Mikrochim. Acta* 5:1025 (1968).
(48) Clements, J.B., National Air Pollution Control Administration, Cincinnati, Ohio, (Aug. 1968).
(49) Tipson, R.S., A. Cohen, and A.J. Fatiadi, "Air Pollution Studies," *Natl. Bur. Std. Technical Note 427,* p. 11 (October 1967).
(50) Korotkov, P.A., N.N. Serzhantov, Yu P. Tsyaschchenko, and N. Ya. Yanysheva, "Determination of Small Concentrations of 1,2,5,6-Dibenzanthracene," *Hyg. Sanitation* 29:59 (1964).
(51) Korotkov, P.A., N.N. Serzhantova, and V.B. Timofeev, "A Photoelectrometric Method for the Determination of Low 3,4-Benzpyrene Concentrations," *Hyg. Sanitation* 28:47 (1963).
(52) Nenasheva, S.K., "Spectrophotometric Determination of 3,4-Benzpyrene in the Presence of Pyrene," *Hyg. Sanitation* 31:213 (1966).
(53) Sawicki, E., and R. Barry, " New Color Tests for the Larger Polynuclear Aromatic Hydrocarbons," *Talanta* 2:128 (1959).
(54) Sawicki, E., T.W. Stanley, T.R. Hauser, H. Johnson, and W.C. Elbert, "Correlation of Piperonal Test Values for Aromatic Compounds with the Atmospheric Concentration of Benzo(a)pyrene," *Intern. J. Air Water Pollution* 7:57 (1963).
(55) Sawicki, E., and W. Elbert, "Thermochromic Detection of Polynuclear Compounds Containing the Fluorenic Methylene Group," *Chemist-Analyst* 48:68 (1959).
(56) Parker, C.A., and W.T. Rees, "Determination of 3,4-Benzpyrene in the Atmosphere of a Submarine," *AML Report No. A/68(M)* Admirality Materials Laboratory, Holton Heath, Poole.
(57) Epstein, S.S., "Photoactivation of Polynuclear Hydrocarbons," *Arch. Environ. Health* 10:233 (1965).
(58) Epstein, S.S., M. Small, J. Kiplan, N. Mantel, H.L. Falk, and E. Sawicki, "Photodynamic Bioassay of Polycyclic Air Pollutants," *Arch. Environ. Health* 7:531 (1963).
(59) Epstein, S.S., M. Small, E. Sawicki, and H.L. Falk. "Photodynamic Bioassay of Polycyclic Atmospheric Pollutants," *J. Air Pollution Control Assoc.* 15:174 (1965).
(60) Won, W.D., and J.F. Thomas, "Developmental Work on Bioassay Technique for Atmospheric Pollutants," *Natl. Cancer Inst. Monograph No. 9.* p. 59 (1962).
(60a) American Public Health Assoc., *Methods of Air Sampling and Analysis,* Wash., D.C. (1972).
(61) Ellis, D.W., *Report PB 212,168,* Springfield, Va., Nat. Tech. Info. Service (June 1972).

POTASSIUM

Measurement in Water

The monitoring of potassium in water may be accomplished by the instrumental techniques which follow (1): Atomic (flame) emission; ion selective electrode; spark source mass spectrometry and x-ray fluorescence.

Reference

(1) Lawrence Berkeley Laboratory, "Instrumentation for Environmental Monitoring," *Publ. LBL-1,* Vol. 2, Berkeley, Univ. of Calif. (Feb. 1, 1973).

RADIOACTIVE MATERIALS

Radiation has been observed to produce somatic effects such as leukemia; lung, skin, thyroid, and bone cancer; cataracts; and life-span shortening. In addition, it is responsible for significant genetic effects. Although some estimates of the dose-time relationships to these effects have been reported, there is some uncertainty in safe levels of exposure to radiation, as pointed out by Miner (1).

There is at present a generally wide acceptance of the biological concept which holds that there is no level of radiation exposure below which there can be absolute certainty that harmful effects will not occur to at least a few individuals. This concept is based to a large extent on considerations of potentially harmful genetic effects. While many of the acute and long-term biological effects of high doses of radiation are known, there is a lack of information on the biological effect of low doses and low-dose rates of radiation. In general, somatic effects are less likely to occur at low-dose rates. Much more information is required to fill the information gaps in the area of low doses and low-dose rates, which are of primary concern in air pollution.

Animals suffer effects similar to those observed in man, and all of the effects observed in man have been confirmed with experimental animals. Plants are suspected of undergoing genetic mutations. However, the experiments have been carried out at radiation doses far in excess of those encountered in ambient air. No material damage has been observed by the radiation found in ambient air.

On the basis of recommendations from the International Commission on Radiological Protection (ICRP), the National Committee on Radiation Protection (NCRP), and the Federal Radiation Council (FRC), the AEC has established standards of maximum permissible concentrations (MPC) of nuclides that can be released from nuclear plants.

The nuclear industry has expanded rapidly in the past decade and will continue to expand. With this rapid expansion, there has been an increase in potential radioactive pollution of the atmosphere. Experience to date has shown that the radiation dose to the general public from nuclear plant emissions has been insignificant when compared with that from natural radioactivity. The dose to the population from nuclear weapons testing was more significant, amounting to levels about 5 to 10% higher than the levels of natural radioactivity.

Recent investigations have indicated that krypton 85 releases from fuel processing may add significantly to the general public radiation dose rate (50 to 100 mrad./yr.) by the year 2060. Krypton 85 is a radioactive gas with a long half-life and at present time is vented to the atmosphere. Methods must be developed for preventing the release of this noble gas.

The projected growth of the nuclear industry in localized areas may in the future produce higher than desired radiation levels in the local air basin. The total emissions from these concentrated facilities may be excessive, even though the emissions from each new facility alone are well within their discharge limits. This problem will require careful review in the future. Fossil fuels contain natural radionuclides that are released from the fuel by combustion. Therefore, radioactivity is released from fossil-fuel-fired power plants that in some cases can amount to more than that released from a similar-sized nuclear power plant. Accidents have occurred in the nuclear industry, and in some cases resulting in releases of appreciable amounts of radioactivity. In other instances, the result has been temporary atmospheric pollution. Most of these incidents were caused by human error rather than the failure or inadequacy of the air cleaning systems.

Environmental radiation monitoring programs are conducted by State, local, and Federal agencies external to the nuclear facility site perimeter to monitor radioactivity releases. The low levels of radioactivity from all phases of the nuclear industry are accomplished by rigidly controlling the plant emissions.

Control of radioactive pollution is accomplished by a variety of methods. Radioactive particulates are removed by filtration, electrostatic precipitation cyclones, or scrubbers. Gases and vapors are often removed by absorption or chemisorption. Storage is effective in eliminating those radionuclides which have a comparitively short half-life. Most reactors are required to have containment buildings to preclude the possibility of atmospheric contamination from an accident.

Measurement in Air

Radioactive materials are produced and dispersed in a variety of ways. In most cases, the radioactive pollutants occur as solid particles dispersed in air. They rarely occur dispersed in air as liquids. However, some of the products are gaseous such as radon, elemental radioiodine and some of its organic compounds, radiocarbon as carbon dioxide, and radioargon.

The method of sampling and monitoring for radioactive material dispersed in air depends on the physical form of the material. Techniques for measuring radiation have been developed which are sensitive to extremely minute amounts. As a result, the amounts of radioactive material that can be detected and measured quantitatively with a high degree of accuracy are much smaller than almost any other atmospheric pollutant.

The monitoring of radioactive environmental pollutants is based on the special characteristics of radioactive materials as described by Vernon E. Andrews of the Environmental Protection Agency Western Environmental Research Laboratory (2). Because of the variety of radioactive elements all facets of the environment are subject to radioactive pollution and may require various degrees of surveillance. As a result of several decades of development of equipment for detecting and measuring radioactivity, a wide range of systems of high sensitivity is now available.

Radioactive materials provide the source of their own detection and measurement. Radioactive decay normally occurs by the emission of charged particles, alpha and beta particles. The decay of some isotopes is accompanied by gamma rays. The three types of radioactive emissions are physically different, and although most radiation detectors can be made to detect or measure all three, specialized systems and techniques are generally applied to each type. This specialization tends to increase the amount of equipment required for complete environmental surveillance; however, it also serves as a means of differentiating between the types of radioactivity.

Three basic types of radiation detectors, photographic film, gas counters, and scintillation detectors, have been in existence since the early days of radiation experimentation. Two relatively new developments, solid state detectors and thermo-luminescent dosimeters (TLD), have had profound effects on radioactivity measurements.

Photographic emulsions are used for detection and quantitation of all types of radioactivity. Autoradiographs of environmental samples are used to locate and quantitate alpha and beta emitting particles. Of major interest in this discussion, is film dosimetry. Film packets, or film badges, are used to measure personnel or area exposures to gamma, and occasionally beta, radiation. The direct physical action of radiation on the film results in a darkening which provides an easily calibrated, reproducible measure of the amount of radiation to which the film was exposed. Three families of gas detectors are widely used. Ionization chambers measure the ionization of gas molecules in the detector. The chambers are commonly filled with air at atmospheric pressure or argon under a pressure of several atmospheres. Radiation exposure is defined as the amount of ionization produced in air. The ionization chamber instrument, therefore, provides a true measure of gamma exposure.

A second form of gas detector is the proportional counter. Rather than measuring the ionization produced, the proportional counter measures the number of ionizing events occurring within the chamber volume. The signal produced by each event is proportional to the amount of ionization produced by each event, which in turn is proportional to the energy of the event for alpha and beta particles. Proportional counters generally exhibit low sensitivity to gamma radiation. By adjusting gas ionization amplification and discrimination on the signal produced, proportional counters can be made specific for alpha or beta radiation.

Geiger counter systems, employing a Geiger-Mueller tube detector, produce a signal pulse for all ionizing events occurring within the detector volume. Gas amplification is adjusted so that the output signal is the same for all ionizing events. This detector is used most widely as a gamma detector, although it will detect any ionization occurring within the chamber. Certain Geiger counters are used as beta counters.

Scintillating materials are widely employed as radiation measuring devices. These materials have the property of emitting light in amounts proportional to the energy of the radiation particle or ray absorbed within them. In various forms, they are used to detect and quantitate all forms of nuclear radiation.

One of the earliest radiation detectors was a ZnS screen which scintillates when struck by an alpha particle. Since the ZnS phosphor is very sensitive and is specific for alpha radiation it is still widely employed as an alpha counter.

Inorganic crystals of metal-halide salts activated with small amounts of impurities are widely used as extremely sensitive radiation detectors. The two most widely used inorganic crystal scintillators are NaI(T1) and CsI(T1). By varying physical dimensions of the crystals they can be made more or less sensitive to gamma radiation. Very thin detectors are insensitive to gamma radiation, but very sensitive to beta radiation. Scintillating crystals are used as gross radioactivity detectors and to measure the energy of individual radiation events.

Organic crystal scintillators, the most popular of which is anthracene, serve the same general purposes as inorganic scintillating crystals. Their special characteristics such as nonhygroscopicity, physical durability, and relatively low cost make them attractive for many applications in radiation measuring even though they are inherently less sensitive than inorganic scintillators.

Several liquids also have the property of scintillating upon absorbing radiation. These liquids are especially suited for measuring low levels of beta radioactivity in specially prepared samples. Mixing the sample with the scintillator allows for a detection efficiency approaching 100%. In addition, the proportional response to radiation energy permits some discrimination on the signal allowing the simultaneous counting of two or more isotopes which have differing beta energies.

A hybrid scintillation detector is the plastic scintillator. These detectors use essentially the same scintillating compounds used in liquid scintillators. However, the scintillating medium is mixed with a clear plastic which is cast as a solid. Plastic scintillators are less dense and contain a smaller fraction of scintillating material than crystal scintillators, thus have a lower efficiency. This low efficiency is improved somewhat in some plastic scintillators in which lead is incorporated to increase density. Plastic scintillators are insensitive to physical and thermal shock, are nonhygroscopic, and are relatively inexpensive. Their primary advantage lies in their machinability. They can be cast and machined in a variety of shapes for unique and special applications.

The advances in electronics during the past decade have resulted in a family of radiation detectors known as solid-state devices. These detectors, such as lithium-drifted germanium diodes are used for measuring all three types of radiation. Their main advantage is extremely high resolution of radiation energies. This provides an excellent method of spectroscopy, or radiation energy identification, of alpha and gamma radiation.

Another recent development in detectors employs the principle of thermoluminescence. Electrons in the crystal matrix of the TLD are displaced by the energy of radiation absorbed in the TLD. Upon heating the TLD, the electrons return to their lower energy levels, emitting light in the process. The light is proportional to the absorbed radiation. Because of their high sensitivity TLD's are replacing film in many dosimetry applications.

Because of the differing physical characteristics of nuclear radiation, two basic modes of expressing quantities are used in surveillance. The first mode, which corresponds to that used in relation to toxic pollutants, expresses the rate of radioactive decay relative to the unit volume or mass of the medium of interest.

Because of the amounts of radioactivity generally encountered in the environment, concentrations are most commonly expressed as picocuries (2.22 alpha or beta disintegrations per minute) per unit mass or volume. These concentrations are generally determined from laboratory analysis of samples; however, some media can be monitored for the concentration of some radionuclides on a real-time basis. In some cases the alpha and beta particles are counted directly. In others the gamma rays are counted and are related by their known abundance to the alpha or beta decay rate.

In addition to the potential for internal exposure to radiation via inhalation or ingestion, external radiation exposure can occur at a distance from radioactive materials due to the range of emitted beta particles and gamma rays. Radiation exposure rates are defined in terms of the ionization produced in air by gamma or x-radiation. The unit of exposure is the Roentgen. Absorbed doses in tissue are expressed in terms of rads and are used for all types of radioactivity. Absorbed doses, in rads, are essentially equal to exposure in Roentgens for gamma radiation.

Because of its penetrating nature, external gamma radiation results in a whole body exposure, whereas beta radiation delivers a radiation dose only to the skin or surface of the eye. Portable radiation survey devices measure gamma exposure rates in Roentgens per hour or, integrated exposures in Roentgens. Some also have the ability to detect beta radiation, but the indications only provide a measure of relative intensity.

Sampling Methods: Air sampling requires collecting three sample types: particulates, reactive gases, and inert gases. The same sampling techniques used for other airborne particulates are applied to sampling for radioactive particulates. Depending on the analytical techniques to be used, the filter media may be varied. Since alpha and beta counting are routine techniques, a prime consideration is in using a filter medium which exhibits surface collection characteristics as opposed to depth-type filters. Commonly used filters are glass fiber and organic membranes.

Both portable and stationary samplers are available to collect samples over any desired period of time. In addition to collecting samples for laboratory analysis, some samplers provide an analysis of the samples during or immediately after collection. In one type, a moving filter tape collects particulates for a preset period of time after which the collecting surface is moved to an adjacent radiation counter for counting while a new sample is being collected.

In another type a Geiger counter-detector is mounted adjacent to the filter surface and measures the collected particulate radioactivity. A unique sampler for monitoring the airborne concentration of radon daughters uses a washer-shaped thermoluminescent dosimeter (TLD) mounted near a membrane filter. Shielded against outside radiation, the TLD exposure provides a measure of the airborne radon daughter concentration.

Reactive gases are efficiently collected on activated charcoal. These gases, primarily radioiodines, are measured by their emitted gamma rays; therefore, collection in a depth medium does not detract from the analysis. The activated charcoal may be in the form of beds or cartridges inserted in a sampler secondary to a filter, or may be incorporated into the filter medium.

Inert gases are monitored in several ways which are unique to radioactive sampling. A basic technique which has wide application is the collection of compressed samples of air which are returned to the laboratory for direct counting or for separation and counting of the gas of interest. These samples may be either short-term grab samples or long-term integrated samples covering periods of up to a week. A method employed at the Western Environmental Research Laboratory is cryogenic sampling. Air is drawn through a bed of molecular sieve at slightly above liquid nitrogen temperature. Noble gases trapped on the sieve are removed and analyzed in the laboratory.

The types of collecting devices used to sample radioactive particulates in air are filters, impactors, impingers, and settling trays as described by Miner (1). Large particles can be collected on settling trays. However, sampling for radioactive particulates is usually accomplished by pulling the air at a measured flow rate through the collecting device (3) through (6).

Filters — Filtration through paper is the most widely used technique for sampling radioactive particulates (7). The types of filter paper used by some various air sampling networks throughout the world have been listed by Lockhart et al. (8). They have listed cellulose-glass fiber, glass fiber, polystyrene, and membrane filters. Glass filters are probably used more extensively than the other filter types.

However, certain inherent advantages are obtained from using other filter media. For example, the synthetic organic filters and cellulose filters are easily burned and essentially ash-free, where the glass and asbestos filters leave a residual ash when burned. This may be an advantage during analysis because of the presence of a finite amount of material for observation and manipulation. Chemical processes are available to dissolve the ash from the glass or asbestos filters or to dissolve the filter media without ashing.

The membrane-type filters are readily soluble in a wide variety of organic solvents, and they can easily be ashed. Thus, when chemical operations are to be performed on the collected dust, the dust can be easily separated from the filter. In addition, a drop of the proper immersion oil in contact with a filter on a microscope slide makes the filter completely transparent for microscopic examination of collected material (9). Techniques have been developed for transferring collected material from membrane filters to electron microscope grids so that very small particles may be observed (10).

Where direct counting of the filter media is to be used to measure the collected radioactivity, radioactive particle penetration of the filter paper should be minimized; highly compacted filters which are essentially surface collectors should be utilized. Lockhart et al. (8) have made measurements on penetration of various filter media by smoke.

Collection efficiencies of 100% in a sampling system are not necessary provided the efficiencies are at least 90% and are known for the material to be collected. Lockhart et al. (8) have listed measurements made on collection efficiencies of various filter media for natural radioactive aerosols and airborne fission products.

Impactors — In impactors the airstream is speeded up by a jet and then impinged or impacted on a surface coated with a sticky material to catch the dust. The material is collected in a small area immediately in front of the jet and the size range collected is a function of the jet velocity and the system dimensions. Impactors are rarely used for pollution monitoring involving radioactive materials because of the long collection times required under outdoor conditions where natural dust exists. Moving slides and tapes have been used for this purpose but are only satisfactory for short periods of sampling (6).

The Anderson sampler (although an impactor similar to the cascade impactor) collects more material and more fractions. This impactor consists of a series of perforated plates and collecting plates. The air is forced through a perforated plate onto a collecting plate, where the fraction is collected (6).

Impingers — Impingers use impaction under a liquid surface and are rarely used in air pollution studies. They occasionally have been used for sampling stacks emitting hot, wet gases. The impinger may be immersed in ice water for this purpose and the aerosol then trapped in the liquid (6).

Settling Trays — Settling trays are widely used in air pollution work and have been used for radioactive materials. The fallout tray is a standard instrument in radioactive air pollution monitoring. The tray is a metal sheet coated with a sticky material or lined with sheets of gummed paper (6)(11).

After exposure, the metal sheet can be placed in a counter for direct counting of radioactivity, or the material can be removed from the tray and the radioactivity determined. The collected material can be washed off with a solvent and the material wet- or dry-ashed for analysis. After exposure, gummed paper can be stripped off and ashed out. Radiochemical analyses for various elements can then be performed.

Another method of evaluation is by autoradiography of the tray. The sticky surface is covered with a thin plastic sheet, placed in contact with a sheet of x-ray film, and kept in the dark for a fixed period. After development of the film, the dark spots reveal the presence of radioactive particles, which can then be evaluated (12). Instead of sticky trays, a shallow tray filled with water can also be used. The water can then be evaporated or filtered for direct counting.

Radioactive washout by precipitation is evaluated by collecting precipitation in stainless steel trays. The water is then evaporated and the residue is counted for radioactivity.

A device for sampling radioactive aerosols has been developed by C. Lasseur, C. Tonnelier and J.C. Zerbib (13). It is a device for sampling aerosols contained in a pipe. The device comprised means for circulating the fluid through an aerosol filter and a conduit for sampling the fluid upstream of the filter. The first portion of the device has a conical cross-section whose apex opens into a pipe for the admission of fluid, and a second portion having a cylindrical cross-section which forms an extension of the first portion and contains a filter. A fluid discharge chamber disposed within the second portion of the sampling conduit has the shape of a cone whose base is occupied by the filter and whose apex opens into a discharge pipe.

Radial fins disposed within the annular space which is defined by the discharge chamber and the second cylindrical portion of the sampling conduit ensure upstream of the filter a homogeneous distribution of the fluid and a laminar flow.

Quantitative Methods: Analysis of Collected Particulate Samples for Activity — Direct radioactive counting of filter paper and other samples involves considerable electronic equipment. The size of the probe or counting chamber should match that of the collection medium, which usually is filter paper. Special probes that can be used with standard scalers or count-rate meters are built to handle most filter paper sizes (6)(11).

Proportional counters are widely used for activity analysis but can give erratic results with filter papers because the filter paper, being an insulator, distorts the electric field in the counting chamber. Scintillation counters are more widely used at present for counting all types of air samples than proportional counters. For alpha counting, the scintillation surface is placed very close to the filter.

Low-level radioactivity can be counted, using small disks of scintillating material on clear plastic placed in actual contact with the deposited material. The counting device is a photomultiplier which sees the light flashes inside a scintillating medium (14). Gamma activity is usually counted with a crystal as a scintillator although Geiger tubes with end windows have been used. Beta counting can be done with scintillating crystals (or powders) on plastic films or with thin window proportional counters. Multichannel analyzers are used, particularly with gamma emitters, to give qualitative information on the isotopes present. As noted previously, membrane filters are best for collecting alpha emitters. These are then counted with solid-state detectors connected to a multichannel analyzer.

The air usually contains appreciable quantities of naturally occurring radioactive particulates. These particulates are collected on filter paper at the same time that other radioactive contamination is being measured. If the samples are counted immediately after the end of the sampling period, the results are high because of the presence of these short-lived natural radioactive materials. Counting can be delayed for several days to permit the decay of the natural products or several counts can be made and a correction calculated.

Combined sampler-counter units are available that use a scintillation counter probe placed near the filter paper during the sampling period. The counter used is a count-rate instrument and the output is connected to a recorder, which then measures the buildup of activity on the filter paper. These types of instruments are rarely used, however, for monitoring alpha emitters. Instruments also have been built using filter tape, moving intermittently or continuously as a collector so that one sample is counted while another is being collected (15)(16).

Radioactive Particle Size Analysis — The mass concentration of a radioactive contaminant in air usually is so minute, even at concentrations above permissible levels, that it cannot be seen on the collection media. Therefore, it is seldom possible to use optical techniques. The concentration of ordinary dust is always much greater than that of the radioactive dust. In addition, there is seldom any visible characteristic of the radioactive dust by which it can be distinguished under the microscope. Therefore, indirect sizing methods usually are used.

A widely used indirect method for sizing uses the cascade impactor. This instrument draws air through a series of progressively smaller jets. After each jet, the air is allowed to impact on a plate coated with an adhesive or dust-retaining material. Since the jet velocities increase as the jet size decreases, progressively smaller particles are impacted and retained. If the impactor has been properly calibrated, the size ranges deposited on each stage will be known (6)(17); and if the cascade impactor has been properly calibrated using an aerosol similar to the one being sampled, it is fairly accurate.

Particle shape, density, and size affect the stage constants. Other errors may be introduced by leakage of air into various parts of the impactor, by deposition inside the instrument body, and by resuspension of deposited aerosol from heavily loaded slides.

There are several aerosol spectrometers that can be adapted for use with radioactive materials. In Timbrell's aerosol spectrometer, the air passes horizontally in a thin film above a long surface and the particles settle on the surface. Since the larger the particle, the sooner it settles, the distance that the particle is located from the entrance is a measure of the particle size, and the amount settled out at various distances can be measured to give the size distribution (18). This system, satisfactory only for particles larger than 10 microns, is seldom used for air pollution work.

In the Conifuge, centrifugal force is used to speed up the settling. The aerosol-laden air is passed through hollow space between two cones which are rotating rapidly (19). Therefore, the particles are driven to the outside wall, where they are deposited on an adhesive-coated surface. Distance down the wall from the entrance is again a measure of size. This instrument is expensive, difficult to build, and primarily used in laboratories.

Another laboratory instrument, the Goetz aerosol spectrometer (20), is similar to the Conifuge but the air traverses a spiral down the annular space between the cones. The air is not guided into the deposition space in a thin layer and therefore, the distance from the entrances is only a measure of the maximum size particle deposited there. Interpretation of the resulting data is quite complex.

Other methods for sizing radioactive particles depend upon placing the collected sample in contact with film for some time, developing the film, and examining it under a microscope. The particle can be left in place during development or the film can be developed separately and then placed in contact with the particles again. When examined under the microscope, the radioactive particles can be identified by the darkened spots under the particles on the film and can then be sized.

For measuring alpha-emitting particles, the collected aerosol is placed in contact for a period of time with a nuclear track film, which is then developed. When the film is examined under a microscope, tracks can be seen where alpha particles were emitted, and the number of tracks emanating from a single point is a measure of the amount of radioactive material in the particle at that point. From the calculated mass of material, the particle size can be estimated (21).

Analysis of Radioactive Gases — Radioactive gases require special handling for analysis depending on their chemical and physical properties. Iodine is collected on activated charcoal (15), although chemical absorbers also have been used. The samples collected can then be analyzed by placing the absorber directly on a scintillation crystal or in a well counter for gamma counting. By using discriminator circuits in a gated single-channel analyzer, a high degree of sensitivity can be obtained.

When the iodine is completely gaseous and entirely in elemental form, the charcoal absorption method gives reliable results. At ordinary temperatures, however, some iodine may be present as solid particles, or atoms may attach themselves to other solids in the atmosphere. Such materials can penetrate the absorbent. For this reason, filter paper is usually placed in front of or behind the collection cartridge during air sampling. Both should be counted when measuring the iodine concentration.

Some iodine has been found to penetrate various absorbents and filters. There appear to be several compounds of iodine having different diffusion characteristics (22)(23). Some materials such as silver-coated copper mesh have been used as traps for iodine, and their efficiency seems to be dependent on humidity. Silver-coated filter papers and charcoal-loaded filter papers give high efficiencies with iodine formed in the laboratory, but varying efficiencies with iodine produced by reactors or industrial fuel-processing operations (24). Scrubbers containing sodium hydroxide can also be used in sampling air for iodine (25).

Tritium is usually present in the form of gaseous molecular hydrogen or as water vapor. When dispersed in air as molecular hydrogen, it gradually oxidizes to tritium oxide or water as a result of self-activation. Ambient tritium consists mainly of water vapor (HTO) (26).

Low-level counting of tritium can be conveniently and accurately accomplished by liquid scintillation counting systems. Tritium samples are collected from the air by freezing out the water vapor from the air with a cold trap, then melting the collected sample. Water is then mixed with liquid scintillation solution. The mixture is then counted by a liquid scintillation counter. All operations involving the scintillation solution are performed under red light to avoid phosphorescence resulting from excitation of the scintillation solution by white light (27).

The usual method of monitoring for noble gases such as argon 41, krypton 85, xenon 133 and xenon 135 is by means of a simple thin-window Geiger counter in the atmosphere. The Kanne chamber or other ion chamber can also be used. For measurement of very low concentrations of xenon and krypton, a charcoal-freeze-out pump is used for trapping the gases, which can then be released into an ion chamber or a chamber containing a Geiger tube for measurement. Since permissible air concentrations of these gases are relatively high, such techniques are rarely required (6)(23).

Gases such as carbon 14 dioxide and sulfur 35 dioxide may be formed as a result of operations in an isotope laboratory or through incineration of radioactive wastes. These gases are sampled by liquid scrubbers containing sodium hydroxide or barium chloride with an oxidant. The determination of collected radioactive material is easily made by liquid scintillation counting. The precipitated barium carbonate or barium sulfate also can be filtered off and the filter paper counted in a suitable instrument, or a sniffer can be used for determining the gas directly (6).

Oxygen and nitrogen can become radioactive if exposed to intense radiation. The half-lives of these irradiated materials are short; therefore, they are not an air pollution hazard. Unshielded Geiger counters or other detectors can be used for direct measurements of radiation where this is necessary.

When reactor fuel elements are dissolved in highly oxidizing solutions, ruthenium, which is formed by the fission process, may be oxidized to the volatile tetroxide and released. Ruthenium 106 is the most hazardous isotope of this element. Air containing ruthenium can be sampled by passing it through an absorber containing a dry organic material such as polyethylene pellets, and the ruthenium content determined by gamma counting (6)(11).

Air Quality Monitoring — Generally, monitoring for radioactive substances is done in much the same way as for nonradioactive materials. Sampling locations are determined both by meteorological and demographic factors and the specific information to be obtained. Although airplanes, rockets, and high-altitude balloons are all employed in measuring radioactive fallout, the sampling equipment for each uses the same principles. Since such pollutants are widely distributed, exact sampling locations are not critical (6)(11)(28).

Duration and frequency of sampling are also similar to those employed in all air pollution work. In some cases, sampling times must be limited because of the short half-life of the pollutant being measured. The high sensitivity of radioactivity measurements and the ready conversion of the radioactive emissions to electronic pulses make continuous monitoring possible in most cases.

Continuous monitoring of reactor installations is effected by a chain of stations suitably arranged around the site. Many different techniques and types of equipment are utilized at various facilities throughout the country. The Division of Radiological Health has reviewed the monitoring techniques and equipment used with the intention of developing uniform measurement techniques (29). One method presently used at many facilities to measure radioactivity in air is to pass the air through a tape of filter paper that is continuously fed to a discharge or scintillation counter. Ionization chambers and filter detectors give instantaneous information on pollution with radioactive gases and dusts. When used in conjunction with recording equipment, they enable the average pollution at the measurement point to be determined; when fitted with alarm devices, they can give a warning if the maximum permissible concentrations are exceeded.

Probably the widest variety of systems exists in radiation monitors. Most monitors can be defined as in situ monitors as noted by Andrews (2). Monitoring, or measuring of external exposures or exposure rates, is generally conducted near the source of radiation. Monitoring of external exposure rates is conducted with survey meters employing Geiger counter, ionization chamber, and scintillation detectors. The most widely used are Geiger counter survey meters. They usually respond to levels of radioactivity down to the background level and operate to several Roentgens per hour. The probes are shielded to permit measurement of gamma radiation only. A sliding shield can be opened to permit detection of beta radiations also.

Accuracy of the exposure rate measurement depends on the comparability of the average gamma energy being monitored to the average gamma energy of the calibration source. A common calibration source is cesium 137. The 0.66 mev energy gamma approximates the average energy of mixed fission products.

Ionization chamber survey meters in common use are generally applied in situations where the exposure rates vary from several milliroentgens per hour to tens or hundreds of Roentgens per hour. Ionization chamber designs using pressurized argon chambers respond to below-normal background ranges. Large volume ionization chambers, such as the Shonka chamber provide accurate measures of exposure rates at background levels, but are sensitive to environmental changes. They serve as research devices, or to compare to exposure rate measurements made by other sensitive survey methods.

Scintillation survey meters using small NaI (T1) crystal detectors provide extreme sensitivity at background levels. Accuracy is very dependent on gamma energies. Because of the required detector packaging they detect gamma radiation only. Large scintillation detectors are used in aircraft for locating and tracking radioactive plumes. These large detectors are also used in vehicles to survey large areas on the ground while mobile on highways and streets.

Several types of integrating radiation monitors or dosimeters are used for in situ monitoring. Film badges are the most widely used at the present time. Their low cost makes them attractive for continuous large scale monitoring. Film badges are unable to measure environmental background levels of radiation, but are relatively accurate at exposure levels above 30 milliroentgens exposure. TLD's are replacing film badges in many areas because of their ability to measure exposures from one milliroentgen or less to several Roentgens.

Various types of TLD's are available, including powder, chips and enclosed types with the medium baked on a heating coil to facilitate readout. Although the initial cost is high, the TLD's in use provide long service and do not require the processing facilities needed for film processing. Another type of dosimeter is the personal pocket dosimeter. An electrical charge applied to a small galvanometer is dissipated in a radiation field. In self-reading types the accumulated exposure is visible on an internal scale. Others require a reader to observe the exposure. These types of dosimeters are used mainly in industrial application.

Remote radiation monitoring falls into two classes, (1) telemetry from a detector at the source of radiation to a receiver and readout at some remote location, and (2) measurement of gamma radiation at a location some distance from the source of the radioactivity. Although not in wide use, the latter technique has some important applications to environmental surveillance. Specially designed systems, such as the EG&G Airborne Radiation Monitoring System measure ground levels of radioactivity from an aircraft at approximately 500 feet above the surface. This allows rapid monitoring of large areas in a short time.

The detectors are large NaI (T1) crystals with an automatic subtraction of background cosmic radiation and correction for altitude. Portable systems have at times been available to do a similar job, but with much lower sensitivity.

Hard-wired remote systems, employing any form of radiation detector, are used in a number of industrial applications. These systems can be connected to meters, recorders, or alarms. Radio-telemetry of radiation monitoring data is limited in application. A radio-telemetry system using a Geiger counter detector may be dropped by parachute from an aircraft flying above a radioactive cloud. The radiation data is radioed to the aircraft where it is displayed on a strip chart recorder in real time as the parachute is descending. These systems can also be dropped into areas inaccessible to mobile monitors to obtain ground level telemetered radiation exposure rate measurements.

Directional radiation detectors, sometimes referred to as gamma telescopes, are used in both ground and aerial applications of remote monitoring. Sensitive radiation detectors, usually scintillators, are shielded or collimated so that only radiation from a preselected sector can enter the detector. These are used to locate an airborne radioactive cloud relative to the detector. A comprehensive review on instrumentation for radiation monitoring is provided by a loose-leaf volume prepared by the Lawrence Berkeley Laboratory (30) which is constantly updated by periodically-issued inserts.

Details of a number of analytical techniques for radioactive materials have recently been published (31). The methods for atmospheric analysis covered are as follows:

Gross Alpha Radioactivity	Radon 222
Gross Beta Radioactivity	Strontium 89
Iodine 131	Strontium 90
Lead 210	Tritium
Plutonium	

An apparatus developed by P.L. Gant, B.G. Motes and R.S. Brundage (32) provides means for detecting, distinguishing and indicating low-level activity of radioactive gases, such as tritium and/or krypton 85, in a selected gas, even in the presence of electronegative gases. The apparatus consists of a detection chamber of a type which is energy selective through inclusion of plural ionization collector circuits providing two separate detector outputs.

The two detector outputs are applied then to parallel signal processing channels which serve to separate detected energy through pulse amplitude discrimination and/or coincidence selection to determine absolute quantities of radioactive events for selected contaminants in the gas sample.

References

(1) Miner, S., "Air Pollution Aspects of Radioactive Substances," *Report PB 188,092,* Springfield, Va., Nat. Tech. Info. Service (Sept. 1969).
(2) Andrews, V.E., in *Proceedings of the EPA Environmental Quality Sensor Workshop,* Las Vegas, Nevada (Nov. 30–Dec. 2, 1971).
(3) Adams, J.W., et al., "Containment of Radioactive Fission Gases by Dynamic Adsorption," *Ind. Eng. Chem.* 51:1467 (1959).
(4) Dahle, E.W., Jr., "Annual Report of the Bureau of Industrial Hygiene," 1966, Baltimore City Health Department, Bureau of Industrial Hygiene, Baltimore, Md. (1966).
(5) Lippman, M., "Air Sampling Instruments," American Conference of Governmental Industrial Hygienists, Cincinnati, Ohio (1967).
(6) Stern, A.C., *Air Pollution* (New York: Academic Press, 1968).
(7) Merriam, G.R., and E.F. Focht, "A Clinical Study of Radiation Cataracts and Relationship to Dose," *Am. J. Roentgen, Radium Therapy Nucl. Med.* 77:759 (1957).
(8) Lockhart, L.B., Jr., et al., "Characteristics of Air Filter Media Used for Monitoring Airborne Radioactivity," *NRL Report 6045,* U.S. Naval Research Laboratory, Washington, D.C.
(9) Rost, D., "New Method of Particle Size Analysis with Membrane Filters," Translated from German. Report of the Department of Physics and Technology of the Federal Center for the Protection Against Radioactivity. (1968).
(10) Kalmus, E., "Preparation of Aerosols for Electron Microscopy," *J. Appl. Phys.* 25:87 (1954).
(11) Jammet, H.P., "Radioactive Pollution of the Atmosphere," *World Health Organ. Monograph Ser. 46* (Air Pollution) (1961).
(12) Skillern, C.P., "How to Obtain Beta Activity of Fission Particles," *Nucleonics* 54 (1955).
(13) Lasseur, C., Tonnelier, C. and Zerbib, J.C., U.S. Patent 3,528,279; Sept. 15, 1970; assigned to Commissariat a l'Energie Atomique.
(14) Hallden, N.A., and J.H. Harley, "An Improved Alpha-Counting Technique," *Anal. Chem.* 32:1861 (1960).
(15) Foelix, C.F., and J.E. Gemmel, "The Use of Activated Charcoal Iodine Monitors During and Following a Release of Fission Product Iodines," Presented at 8th Atomic Energy Commission Air Cleaning Conf., TID-7677 (1963).
(16) Helgeson, G.L. "Determination of Concentrations of Airborne Radioactivity," *Health Phys.* 9:931 (1963).
(17) Anderson, A.A., "Sampler for Respiratory Health Hazard Assessment," *Am. Ind. Hyg. Assoc. J.* 27:160 (1966).
(18) Timbrell, V., "The Terminal Velocity and Structure of Airborne Aggregates of Microscopic Spheres," *Brit. J. Appl. Phys. Suppl.* 3:886 (1954).
(19) Sawyer, K.F., and W.H. Walton, "The Conifuge, A Size-Separating Sampling Device for Air-borne Particles," *J. Sci. Instr.* 27:272 (1950).
(20) Goetz, A., H.J.R. Steveson, and O. Preining," Design and Performance of the Aerosol Spectrometer," *J. Air Pollution Control Assoc.* 10:378 (1960).
(21) Leary, J.A., "Particle Size Determination of Radioactive Aerosols by Radio Autograph," Los Alamos Sci. Lab. Rept. LAMS-905 (1949).

(22) Browning, W.E., Jr., R.D. Ackley, and M.D. Silverman, "Characterization of Gas-Borne Products," Presented at the 8th Atomic Energy Commission Air Cleaning Conference, TID-7677 (1963).
(23) Terrill, J.G., Jr., et al., "Environmental Surveillance of Nuclear Facilities," *Nucl. Safety* 9:143 (1968).
(24) Ettinger, H.J., "Iodine Sampling with Silver Nitrate Impregnated Filter Paper," *Health Phys.* 12:305 (1966).
(25) Soldat, J.K., "Monitoring for Air-Borne Radioactive Materials at Hanford Atomic Products Operation," *J. Air Pollution Control Assoc.* 10:265 (1960).
(26) Wilkening, N.H., "Natural Radioactivity as a Tracer in Sorting of Aerosols According to Mobility," *Rev. Sci. Instru.* 23:13 (1952).
(27) Moghissi, A.A., et al., "Low Level Counting by Liquid Scintillation. Tritium Measurement in Homogeneous Systems," *Intern. J. Appl. Radiation Isotopes* 20:145 (1969).
(28) Terrill, J.G., Jr., "Public Health Radiation Surveillance," *Health Phys.* 2:917 (1965).
(29) Weaver, C., Director of the Division of Environmental Radiation, Bureau of Radiological Health, Environmental Control Administration of the Consumer Protection and Environmental Health Service, Dept. of Health, Education, and Welfare (July 1969).
(30) Lawrence Berkeley Laboratory, "Instrumentation for Environmental Monitoring-Radiation," *Publ. LBL-1,* Vol. 3, Berkeley, Univ. of California (May, 1972).
(31) American Public Health Assoc., "Methods of Air Sampling and Analysis," Wash., D.C. (1972).
(32) Gant, P.L., Motes, B.G. and Brundage, R.S., U.S. Patent 3,746,861; July 17, 1973; assigned to Continental Oil Co.

SEDIMENT

A wide variety of materials and substances which are the by-products of man's activities are eventually carried into lakes, reservoirs, lagoons, bays, estuaries and the oceans. Such materials generally remain suspended for some period of time, are sometimes acted upon by organisms within the water body and eventually settle to the bottom of the water body. Some of these substances, such as phosphate, nitrate, mercury, lead, DDT, PCB, as well as some bacteria, are known to have a harmful effect on the life or condition of the water body. Also, by accumulating in the tissues of fish or other organisms, such substances may also have a harmful effect on human life.

The amount of potentially harmful substances that is accumulating in the water bodies of the world is creating a serious pollution problem, whose magnitude is so great that many independent scientists and private and governmental institutions have called for regular programs of sampling and monitoring aquatic sediment and pollution to determine the level of pollution and to detect changes in the volume or rate of pollution. See, for example, the chapter on monitoring in the report of the Study of Critical Environmental Problems in Man's Impact on the Global Environment, MIT Press, 1970.

Measurement in Water

Present aquatic sampling and monitoring systems have many problems associated therewith. The most common technique for the periodic sampling of water and suspended matter makes use of plankton nets and sampling bottles. The collected material is returned to the laboratory for analysis by standard chemical and biological methods. However, such direct water sampling has several serious limitations and disadvantages associated therewith.

In the first instance, such techniques require frequent trips to the collecting site which is often in a remote locality or is accessible only after considerable travel by boat. Another disadvantage is that the material obtained during such trips represents the condition of the water body only at the moment it is sampled and such techniques do not provide a continuous record.

A few reported measuring or monitoring studies have placed a collecting box or bottle directly on the bottom of a water body or suspended from a cable or frame. Such collecting device is then recovered after a short period of time, generally after one month.

While this technique permits a continuous record, it requires large containers and frequent visits to the collecting site. In addition, the material collected is generally only a thin film on the bottom of the collecting vessel, which film is easily disturbed by organisms, such as fish.

Attempts have also been made to use funnels to magnify the rate of accumulation of aquatic sediment and pollution. However, these attempts have been generally unsatisfactory because turbulence caused by currents in the water body enters the large opening provided by the funnel and disturbs and distorts the quantity of material entering the collecting vessel. In addition, the accumulated material is susceptible to being acted upon by organisms that scavenge and burrow and mix the material so that the time relationships are destroyed. Finally, this method has the disadvantage of requiring frequent trips to the collecting site.

A device developed by R.Y. Anderson (1), is an aquatic sediment and pollution monitor adapted to be positioned in a body of water. It consists of an elongated, vertically alignable, collecting tube having an open upper end and a closed lower end for collecting, over a long period of time, the natural materials and polluting substances that accumulate in the body of water. A generally funnel-shaped magnifying cone is positioned with the small diameter end thereof extending into the open end of the collecting tube to magnify the amount of sediment and pollution collected.

A baffle is positioned in the magnifying cone adjacent the large diameter end thereof for minimizing turbulence in the collecting tube and for preventing entrance thereinto of large organisms. Means are also provided for automatically marking, at regular intervals, the quantity of sediment and pollution accumulated in the collecting tube during such intervals.

References

(1) Anderson, R.Y., U.S. Patent 3,715,913; February 15, 1973.

SELENIUM

Selenium compounds, particularly the water-soluble compounds, are toxic to humans and animals. In humans, mild inhalation of selenium dusts, fumes, or vapors irritate the membranes of the eyes, nose, throat, and respiratory tract, causing lacrimation, sneezing, nasal congestion, coughing, etc. Prolonged exposure through inhalation can cause marked pallor, coated tongue, gastrointestinal disorders, nervousness, and a garlicky odor of breath and sweat. In animals, subacute selenium poisoning produces pneumonia and degeneration of the liver and kidneys. Furthermore, experiments with rats indicate that selenium may cause cancer of the liver, as noted by Stahl (1).

The biochemical effects of elemental selenium and its compounds on humans is not as yet thoroughly understood. The selenium deficiency diseases found in animal species, as well as some of the frank selenium poisoning, have not been observed in man. Similarly, the carcinogenic hazard of selenium and the antagonistic effect of arsenic for selenium seen in animals are yet to be shown in humans. These are important factors that need clarification to properly evaluate the role of selenium and its compounds in air pollution.

There is no information indicating that atmospheric selenium has any detrimental effect on plants or materials. Some plants contain large amounts of selenium that can be toxic to the plants themselves, as well as to humans and animals who ingest the plants.

Samples of snow, rain, and air taken in Boston, Mass., showed that the selenium content of the air averaged 0.001 μg./m.^{3}. Based on the selenium-to-sulfur ratio in these samples, the atmospheric selenium was probably from terrestrial sources, including the fuels and ores used by industry. A rough estimate as to the magnitude of selenium in the atmosphere might be made from the concentration of sulfur in the atmosphere. This method would be valid if the sources of these two pollutants are sulfide ores, fossil fuels, or igneous and sedimentary rocks, since in these materials the average weight ratio of selenium to sulfur is 1×10^{-4}.

Another source of selenium in air may be the burning of trash containing paper products. Some papers when analyzed contain as much as 6 ppm selenium. Selenium in paper may come from accumulation by the original tree or plant, or possibly from the manufacturing of the paper (from the use of pyrites in the process). Any vegetation which is burned may be a possible source of atmospheric selenium. Another source could be the refining of sulfide ores, particularly copper and lead ores.

As part of a national inventory of sources and emissions, W.E. Davis and Assoc. of Leawood, Kansas have reported on the nature, magnitude and extents of the emissions of silenium in the United States for the year 1969 (2). Emissions of selenium and its compounds can be effectively controlled by use of electrostatic precipitators and water scrubbers.

Essentially no information is available in the water pollution literature, the industrial waste literature, or the technical literature of the industries which use selenium, on levels of selenium in industrial wastewaters, treatment methods for selenium wastes or costs associated with removal of selenium from industrial wastewaters as noted by Patterson and Minear (3). Industries using selenium include paint, pigment and dye producers, electronics, glass manufacturers, and insecticide industries. Available information relating to the general subject of selenium wastewaters and potential method of treatment is discussed below.

Johnson (4) has reported that selenium is present in almost all conceivable types of paper, and it might be concluded from this information that pulp and paper mill wastes could contain selenium. Johnson's primary concern was with the release of selenium as an air pollutant, upon incineration of paper products. However, he did report selenium wastewater levels as follows. For incinerator fly ash quench water, soluble selenium was measured at 5-23 μg./l., while incinerator residue quench water contained 3 μg./l. of soluble selenium. The incinerator handled a solid waste containing approximately 55 to 69% paper.

Hutchinson (5) reports that selenium occurs as an impurity in the form of selenide, Se^{-2}, in metallic sulfide ores. When the sulfide ores are roasted in air (to convert the sulfide and drive off SO_2) the selenide is oxidized to selenium dioxide and is released in the flue gas. If the flue gas is quenched, selenium dioxide reacts with the quench water to form selenious acid, H_2SeO_3. The selenious ion appears to be the most common form of selenium in wastewater except for pigment and dye wastes, which contain the selenide (e.g., yellow cadmium selenide). Other forms decompose to yield the SeO_3^{-2} ion. Hutchinson also reports that selenium dioxide is readily reduced, to precipitate finely divided elementary selenium. It is probable that this same reaction might occur for selenious ion at acidic pH. Selenide may be precipitated as the metallic salt, which is highly insoluble (5).

Secondary municipal sewage treatment plant effluents containing 2 to 9 μg./l. of selenium have been reported (6). A tertiary sequence of treatment, which included lime treatment to pH 11, sedimentation, mixed media filtration, activated carbon adsorption and chlorination yielded selenium removal of 0 to 89%. Best removal was obtained at higher initial selenium concentrations. Raible (7) has noted the presence of selenium in a pharmaceutical wastewater, but did not report concentrations. Linstedt, et al. (8), measured 2.3 μg./l. of selenium in the effluent from a secondary sewage treatment plant.

The permissible selenium concentration in domestic water supplies is 0.01 mg./l. and it is desirable that selenium be absent as noted earlier in Table 3. For continuous water use in irrigation, the selenium tolerance is 0.05 mg./l. and the permissible concentration for short-term use in irrigating fine textured soil is also 0.05 mg./l.

The maximum allowable concentration of selenium in effluents being discharged to storm sewers on streams is 0.01 ppm.

Measurement in Air

Sampling Methods — Air samples have been collected by means of electrostatic precipitators (9), filters (10), and liquid impingers (9)(10)(11). All of these methods can be used for selenium dusts and fumes, while only the last can be used for collection of vapors or gaseous compounds.

The use of filters and liquid impingers are the most commonly used sampling methods today. Some air samples were taken by passing 100 $m.^3$ of air at the rate of 1 $m.^3$/hr. through a 1 micron pore diameter Millipore filter. Because selenium is found in many filter papers, Millipore filters or other selenium-free collection media (e.g., glass fibers) should be used. However, some filter papers can be washed with sodium sulfide and water to remove traces of selenium. The liquid used in the impingers has been water (10)(11), and a solution of 40 to 48% hydrobromic acid with 5 to 10% bromine. The water can be used with nonvolatile water-soluble compounds such as selenium dioxide (9).

The advantage of the hydrobromic acid-bromine solution is that it converts selenium and most of its compounds to the soluble selenium tetrabromide. This solution, after collection, can be analyzed directly after decolorization and neutralization (11), or the selenium may be separated by distillation. It has also been suggested that soda lime or silica gel could be used to collect vapors of selenium compounds (9).

Qualitative Determination Methods — A quick qualitative estimation of selenium can be made by taking advantage of the reaction of aromatic 1,2-diamines with selenious acid [or Se(IV) cation] to form colored piaselenoles. Elemental selenium and selenides readily oxidize in nitric acid or hydrogen peroxide to selenious acid. Selenate salts, or Se(VI) cations, give only weak colors.

The most common diamine employed is 3,3'-diaminobenzidine (12), which forms a yellow dipiaselenole in acid and is sensitive to 10 ppm. The 4-dimethylamino and 4-methylthio derivatives of 1,2-phenylenediamine (13) will also react with selenious acid to give stable bright red and blue-purple colors, respectively. Reaction with 2,3-diaminonaphthalene has been used as a spot test (14). Other ions, particularly copper, iron, tellurium, chromium, nickel, and cobalt, may cause interference in high concentration, but some may be masked by complexing agents such as EDTA. Other methods of detection of selenious acid that have been reported are reaction with (a) iodides (15), (b) thiourea (16), (c) diphenylhydrazine (17), and (d) pyrrole (18), as well as the catalytic effect on the reaction of methylene blue with alkali sulfide (19).

Quantitative Determination Methods – Several quantitative methods of analyzing selenium are based on the reaction of selenious ion with 3,3'-diaminobenzidine. West and Cimerman (20) used this reagent in connection with the ring-oven technique. This colorimetric method is basically free from interferences when masking reagents are used; limit of identification is 0.08 μg., is applicable in the range of 0.1 to 0.5 μg. (20)(21). Kawamura and Matsumoto (11) spectrophotometrically determined the selenium in air samples collected in impingers by measuring the absorbance of the solutions treated with this reagent. The benzidine reagent has also been used in fluorometric determination of selenium in biological materials (22)(23). Detection in these methods was on the order of 10 to 50 ppb selenium.

Walkinson (24), as well as Allaway and Cary (25), found that the product from 2,3-diaminonaphthalene and selenium showed a fluorescence sensitivity in the neighborhood of 0.5 to 5 ppb selenium, a sensitivity greater than that with the benzidine product.

Neutron activation methods have been used to determine selenium from air samples (via filters and aqueous bubbler) and from samples of snow and rain (10). A sensitivity of 0.01 μg. of selenium was obtained when gamma radiation was measured with a scintillation spectrometer. This technique has also been used to determine traces of selenium in biological materials (26) and in metals (27).

More recently West and Ramakrishna (28) have developed a method for determining trace amounts of selenium based on its catalytic effect in the reduction of methylene blue by sodium sulfite. Color comparisons are made in the range of 0.1 to 1.0 μg. of selenium. There is serious interference from copper if present in an excess of 10 μg. Kawashima and Tanaka (29) also developed a catalytic method based on the reduction of 1,4,6,11-tetraazanaphthacene; this method is subject to interference from several ions.

Walkinson (30) has discussed some of the newer methods that are used in the analysis of selenium in biological material, including polarography, x-ray fluorescence, and atomic absorption spectrophotometry.

Measurement in Water

Selenium may be monitored in water by molecular fluorescence (31). As noted earlier in Table 11, selenium may be measured by flame atomic absorption or by neutron activation down to detection limits of 400 to 600 ppb. Flameless atomic absorption or polarography may be used to measure selenium down to detection limits in the 10 to 20 ppb range.

References

(1) Stahl, Q.R., "Air Pollution Aspects of Selenium and Its Compounds," *Report PB 188,087,* Springfield, Va., Nat. Tech. Info. Service (Sept. 1969).
(2) Davis, W.E. and Assoc., *Report PB 210,679,* Springfield, Va., Nat. Tech. Info. Service (May 1972).
(3) Patterson, J.W. and Minear, R.A., "Wastewater Treatment Technology," 2nd ed., *Report PB 216,162,* Springfield, Va., Nat. Tech. Info. Service (Feb. 1973).
(4) H. Johnson, "Determination of Selenium in Solid Waste," *Envir. Sci. Tech.,* 4:850-853, 1970.
(5) E. Hutchinson, *Chemistry: The Elements and Their Reactions,* W.B. Saunders Co., Philadelphia, 1959.
(6) D.G. Argo and G.L. Culp, "Heavy Metals Removal in Wastewater Treatment Processes: Part 2 – Pilot Plant Operation," *Water Sew. Wks.,* 119:28-132, 1972.
(7) J.A. Raible, "Fluorometric Determination of Selenium in Effluent Streams with 2,3-Diaminonaphthalene," *Envir. Sci. Tech.,* 6:621-622, 1972.
(8) K.D. Linstedt, C.P. Houck, and J.T. O'Connor, "Trace Element Removals in Advanced Wastewater Treatment Processes," *Jour. Wat. Poll. Control Fed.,* 43:1507-1513, 1971.
(9) Patty, F.A., (Ed.), *Industrial Hygiene and Toxicology,* vol. II, 2nd ed. (New York: Interscience, p. 887, 1963).
(10) Hashimoto, Y., and J.W. Winchester, "Selenium in the Atmosphere," *Environ, Sci. Technol.* 1:338 (1967).
(11) Kawamura, M., and K. Matsumoto, "Determination of Small Amounts of Hydrogen Selenide in Air," *Japan Analyst* (Tokyo) 14(9):789 (1965).
(12) Hoste, J., "Diaminobenzidene as a Reagent for Vanadium and Selenium," *Anal. Chim. Acta.* 30:1504 (1952).
(13) Sawicki, E., "New Color Test for Selenium," *Anal. Chem.* 29:1376 (1957).
(14) Feigl, F., *Qualitative Analysis by Spot Tests,* 3rd ed. (New York: Elsevier, p. 266, 1946).
(15) Poluektov, N.S., "Detection of Selenium and Tellurium in the Presence of One Another," *Mikrochemie* 15:32 (1934).
(16) Deniges, G., "Detection of Selenium, Application to Natural Waters," *Bull. Soc. Pharm. Bordeaux* 75:197 (1937).
(17) Feigl, F., and V. Demant, "Microchemical Detection of Selenium," *Mikrochim. Acta,* 1:322 (1937).
(18) Suzuki, M., "Colorimetric Method for Determining Selenium in Crude Copper," *J. Chem. Soc. Japan* 56:323 (1953).
(19) Feigl, F., and P.W. West, "Test for Selenium, Based on a Catalytic Effect," *Anal. Chem.* 19:351 (1947).
(20) West, P.W., and C. Cimerman, "Microdetermination of Selenium with 3,3'-Diaminobenzidine by the Ring Oven Technique and Its Application to Air Pollution Studies," *Anal. Chem.,* 36:2013 (1964).
(21) West, P.W., "Chemical Analysis of Inorganic Pollutants," *Air Pollution* II:163 (1968).
(22) Cheng, K.L., "Determination of Traces of Selenium, 3,3'-Diaminobenzidine as Selenium (IV) Organic Reagent," *Anal. Chem.* 28:1738 (1956).
(23) Cummins, L.M., et al., "A Rapid Method for Determination of Selenium in Biological Material," *Anal. Chem.* 36:382 (1964).
(24) Walkinson, J.H., "Fluorometric Determination of Selenium in Biological Material with 2,3-Diaminonaphthalene," *Anal. Chem.* 38:92 (1966).
(25) Allaway, W.H., and E.E. Cary, "Determination of Submicrogram Amounts of Selenium in Biological

Materials," *Anal. Chem.* 36:1359 (1964).
(26) Bowen, H.J.M., and P.A. Cawse, "The Determination of Selenium in Biological Material by Radio-activation," *Analyst* 88:721 (1963).
(27) Conrad, F.J., and B.T. Kenna, "Determination of Selenium by Activation Analysis and Dry Volatilization," *Anal. Chem.* 39:1001 (1967).
(28) West, P.W., and T.V. Ramakrishna, "A Catalytic Method for Determining Traces of Selenium," *Anal. Chem.* 40:966 (1968).
(29) Kawashima, T., and M. Tanaka, "Determination of Submicrogram Amounts of Selenium (IV) by Means of Catalytic Reduction of 1,4,6,11-Tetraazanaphthalene," *Anal. Chim. Acta.* 40:137 (1968).
(30) Walkinson, J.H., "Analytical Methods for Selenium in Biological Material," in *Symposium: Selenium in Biomedicine,* O.H. Muth, Ed. (Westport, Conn.: AVI Pub. Co., 1967).
(31) Lawrence Berkeley Laboratory, " Instrumentation for Environmental Monitoring-Water," *Publ. LBL-1,* Vol. 2, Berkeley, Univ. of Calif. (Feb. 1, 1973).

SILVER

McKee and Wolf (1) have reported that silver, as the solid metal, is used in the jewelry, silverware, metal alloy, and food and beverage processing industries. Little soluble silver waste would be expected to result from use of the solid metal. The only appreciably soluble common silver salt, silver nitrate, is used in the porcelain, photographic, electroplating and ink manufacturing industries and has also found application as an antiseptic (1). The two major sources of soluble silver wastes are the photographic and electroplating industries.

The permissible concentration of silver in domestic water supplies is 0.05 mg./l. as noted earlier in Table 3. The maximum allowable concentration of silver in an effluent being discharged to a storm sewer or stream is 0.05 ppm.

Measurement in Water

The high and rising price of silver makes it increasingly important to recover silver from various waste products, e.g., the waste products from film processing. It is particularly profitable to recover silver from consumed and spent fixing baths. It is therefore very important to be able to provide processes and agents to monitor recovery plants and fixing baths in order to detect the presence of silver therein and semiquantitatively determine the amount of silver in the fixing baths.

The use of a test paper which can be employed for these purposes is known. However, this paper still exhibits a number of drawbacks. For example, yellow test paper containing cadmium sulfide, attains a darker shade only gradually after immersion into a used fixing bath. Moreover, a definite final color does not occur within a practicable period of time. Thus, the inability to obtain a constant color within a short period of time mades it very difficult to determine the silver content of a fixing bath with the aid of a color scale. Besides, this conventional semiquantitative silver determination is conducted preferably in the presence of the light of an incandescent lamp.

Another disadvantage is that when a conventional test paper is immersed into a fixing bath containing a low silver concentration, e.g., about 100 to 200 mg./l. of Ag(I), the presence of the silver ions is indicated by the darkening of the test paper several minutes after immersion therein.

Silver may be measured in water down to a detection limit of 2 ppb by flame atomic absorption or by neutron activation as noted earlier in Table 11. Flameless atomic absorption or polarography may be used to extend the detection limit to 0.1 ppb or below.

A process developed by D. Schmitt, A. Stein and W. Baumer (2), is one in which an indicator for the colorimetric detection of silver ions is formed by impregnating an absorbent carrier with an intimate mixture of cadmium sulfide and selenium.

By employing this process the disadvantages of prior art processes using test papers, as cited above, are overcome. Using this method, it is possible to obtain the maximum color depth based upon the silver ion concentration of a solution, e.g., fixing baths, in a very short period of time. The color produced thereby remains constant for a long period so that a comparison with a color scale is readily possible. Thus, it is possible to conduct semiquantitative tests in order to determine, with ease, the amount of silver ions in a solution. In addition, the test is capable of being used by those not skilled in laboratory techniques. It is also possible to detect with certainty silver ion concentrations as low as 10 to 100 ppm.

A technique developed by T.N. Hendrickson (3), involves monitoring the silver content of the effluent from a silver recovery process. The effluent electrolyte is passed through a voltaic cell wherein one of the electrodes is iron in the form of steel wool. When the recovery vessel has lost its power to efficiently remove silver, silver ions appear in the effluent electrolyte in increasing concentration. Upon the generation of a predetermined value of electrical current in the voltaic monitoring cell representing a predetermined silver concentration, the recovery unit is reactivated. Silver may be monitored in water by x-ray fluorescence (4).

References

(1) J.E. McKee and H.W. Wolf, *Water Quality Criteria,* 2nd ed., California State Water Quality Control Board Publication No. 3A, 1963.

(2) Schmitt, D., Stein, A. and Baumer, W., U.S. Patent 3,661,532; May 9, 1972; assigned to Merck Patent GmbH.

(3) Hendrickson, T.N., U.S. Patent 3,705,716; December 12, 1972; assigned to Eastman Kodak Co.

(4) Lawrence Berkeley Laboratory, " Instrumentation for Environmental Monitoring-Water," *Publ. LBL-1,* vol. 2, Berkeley, Univ. of Calif. (Feb. 1, 1973).

SODIUM

Measurement in Water

Sodium may be monitored in water by the following instrumental methods (1): atomic (flame) emission; ion selective electrode; and spark source mass spectrometry.

References

(1) Lawrence Berkeley Laboratory, " Instrumentation for Environmental Monitoring-Water," *Publ. LBL-1,* vol. 2, Berkeley, Univ. of Calif. (Feb 1, 1973).

STRONTIUM

Measurement in Water

The determination of strontium 90 in water has been discussed by J.O. Johnson and K.W. Edwards of the U.S. Geological Survey (1). A method is described for the determination of strontium 90 in waters free of strontium 89, or for determination of gross radiostrontium activity as strontium 90. The method is suitable for use on natural water samples and other aqueous systems in the absence of high concentrations of organic substances which form stable complexes with strontium. The minimum detection limit at the 95% confidence level is 0.55 picocuries/liter for the recommended 500 ml. sample.

As described by N.R. Anderson and D.N. Hume in a Symposium on Trace Inorganics in

Sea Water (2), strontium and barium may simultaneously be determined. The technique employed was a combination of ion exchange concentration and flame photometry. Strontium may be monitored in water by atomic (flame) emission spectroscopy (3).

References

(1) Johnson, J.O. and Edwards, K.W., *Geol. Survey Water Supply Paper No. 1696-E,* Wash., D.C., U.S. Geological Survey (1967).
(2) Anderson, N.R. and Hume, D.N. in *Advances in Chemistry Series No. 73,* 296-307, Wash., D.C., Amer. Chem. Soc. (1968).
(3) Lawrence Berkeley Laboratory, "Instrumentation for Environmental Monitoring-Water," *Publ. LBL-1,* vol. 2, Berkeley, Univ. of Calif. (Feb. 1, 1973).

SULFATES

Sulfates are significant air pollutants. They may occur in the form of so-called secondary particulates, those formed in the atmosphere by chemical and physical processes. They make up a significant proportion of the aerosol particles found in the atmosphere, perhaps 10% of their weight.

It has been noted by the EPA as of Dec. 1973 that catalytic converters for automobiles, while eliminating hydrocarbons and carbon monoxide, will increase the amounts of sulfates that will come from the tailpipes of cars. Thus, the EPA is accelerating the development of a test procedure for measuring auto sulfate emissions and plans to consider all feasible measures for controlling such emissions from cars including possible regulation of sulfur content in gasoline and other alternatives, according to EPA Administrator Russell Train (1).

Sulfates in domestic water supplies have a permissible concentration of 250 mg./l. as shown earlier in Table 3. The desirable concentration of sulfates is below 50 mg./l., however.

Measurement in Air

As noted by W.E. Ruch (2), sulfates in concentrations above 0.06 ppm can be determined by ultraviolet spectrophotometric analysis with chloranilic acid. The technique, as described by R.J. Bertolacini and J.E. Barney (3), is one in which an air sample containing sulfates is collected in midget impingers containing distilled water. The sample is then passed through a Dowex resin column to remove interfering cations. The solution is then adjusted to a pH of 7.0 and buffered with potassium acid phthalate. Then barium chloranilate is added and the sample is filtered. The absorbance of the filtrate is measured at 322 mμ in a spectrophotometer and the quantity of sulfate read from a calibration curve.

Measurement in Water

Sulfates can be monitored in water by the following instrumental techniques (4): automated chlorimetric and turbidimetric. As noted earlier in Table 12, sulfates may be measured using the Technicon CSM-6 in the nominal range of 0-500 ppm with a detection limit of 5.0 ppm.

A technique developed by C.A. Noll and L.J. Stefanelli (5) permits determining the sulfate ion concentration of an aqueous solution. The mechanism of the method resides in the use of a chelate of a metal which under acidic conditions dissociates to release metal ions which will react with the sulfate ion present to produce a colloidal metal sulfate precipitate.

In order to ensure that the free chelant does not precipitate or react to form a precipitate, there is also added to the solution a quantity of a metal compound, the metal of which will replace the original metal of the chelate to form a water-soluble second chelate.

The turbidity or color density of the aqueous solution is measured and compared to the color density values obtained for solutions containing known quantities of sulfate precipitate of the same metal.

References

(1) Train, R. in *Citizens Bulletin,* Wash., D.C., Environmental Protection Agency Dec. 1973).
(2) Ruch, W.E., *Quantitative Analysis of Gaseous Pollutants,* Ann Arbor, Mich., Ann Arbor Science Publishers (1970).
(3) Bertolacini, R.J. and Barney, J.E., *Anal. Chem.* 30, 202 (1958).
(4) Lawrence Berkeley Laboratory, "Instrumentation for Environmental Monitoring-Water," *Publ. LBL-1,* vol. 2, Berkeley, Univ. of Calif. (Feb. 1, 1973).
(5) Noll, C.A. and Stefanelli, L.J., U.S. Patent 3,622,277; November 23, 1971; assigned to Betz Laboratories, Inc.

SULFIDES

Measurement in Water

Sulfides can be monitored in water by the following instrumental techniques (1): colorimetric and ion selective electrode.

References

(1) Lawrence Berkeley Laboratory, "Instrumentation for Environmental Monitoring-Water," *Publ. LBL-1,* vol. 2, Berkeley, Univ. of Calif. (Feb. 1, 1973).

SULFITES

Measurement in Water

Sulfites can be monitored in water by colorimetric instruments (1).

References

(1) Lawrence Berkeley Laboratory, "Instrumentation for Environmental Monitoring-Water," *Publ. LBL-1,* vol. 2, Berkeley, Univ. of Calif. (Feb. 1, 1973).

SULFUR COMPOUNDS

Measurement in Air

The flame photometric detector has received considerable attention for measurement of sulfur dioxide, hydrogen sulfide, and other sulfur compounds (1) through (3). This technique involves a response to both inorganic and organic sulfur compounds because these substances form S_2 species in the flame zone which are responsible for the emission observed. Several field evaluation studies already completed have demonstrated the effectiveness of this sensor, both as a sulfur analyzer and as a detector in a monitoring gas chromatograph (3). This approach is more attractive than most of the other techniques available for measuring sulfur compounds. The flame photometric sensor has the following characteristics: (a) high specificity for sulfur compounds if emission is measured between

0.39 and 0.40 micron; (b) high sensitivity (0.005 ppm); (c) linear response in ambient air concentration range; (d) fast response characteristics; and (e) a gas flow system (no liquid reagents). The flame photometric sulfur detector is also the only detector sensitive to sulfur compounds that can be used practicably in a gas chromatographic analyzer. When sulfur dioxide is the predominant sulfur species present, the flame photometric analyzer can be used as essentially a sulfur dioxide monitor.

The gas chromatographic analyzer has been used as a specific and sensitive measuring instrument for sulfur dioxide, hydrogen sulfide, and methyl mercaptan (3). The gas chromatographic system has been readily modified to measure a wider variety of organic sulfur compounds where appropriate, such as in the vicinity of kraft paper mills.

A method developed by R.W. Pierce (4) permits the rapid, sensitive, nonelectrolytic detection of specific chemical species in fluid streams, and in particular relates to an apparatus for and a method of detecting and measuring sulfur-containing compounds in a gas stream.

Briefly, the method comprises means for detecting a compound which forms a reversible, poisoning or chemi-absorption or reaction product with a sensing element. The sensing element comprises a material, the surface of which is reversibly poisoned or otherwise effected by the specific chemical species to be detected in a gas stream.

For example, in the detection of sulfur dioxide in an air stream employing a palladium sensing element as part of a balanced bridge circuit, and employing hydrogen as the gas to be added to the air stream, a rapid, sensitive means for measuring and detecting sulfur dioxide in the air stream is provided. In practice, a known partial pressure of hydrogen is added to the air sample stream to be monitored and in which the sulfur dioxide level is to be detected and determined.

Any sulfur-containing material in the air stream such as sulfur dioxide will then poison the surface of the palladium element to some greater or lesser extent, depending in increasing degree upon the ratio of the partial pressure of sulfurous material to that of the hydrogen present and added to the air stream. The accompanying change in resistivity of the sensing element is then used to indicate the sulfur level in the air stream under test. The detecting system may be made specific for a particular sulfur compound by resorting to an absorption train or other devices to remove all but the compound whose particular detection and measurement is desired.

A method developed by C.A. Plantz (5) provides a colorimetric indicator which is responsive to carbon disulfide. The indicator is formed of an inert granular solid such as silica gel carrying a reagent containing a cupric salt, piperazine or an organosubstituted piperazine, and if desired, a copper complexing agent and humectant. A bed of the colorimetric indicator together with a bed of desiccant is contained in an openable transparent tube and atmosphere to be tested in passed sequentially through the desiccant and indicator beds; the length of indicator bed over which a color change occurs is dependent on the concentration of carbon disulfide.

A method developed by E.A. Eads (6) permits accurately sampling ambient air containing sulfur compounds in the low parts per billion range and separating, identifying and quantitatively monitoring each sulfur compound. The gaseous sample is collected in substantially sulfur-free methanol. The methanol solution is then passed through a gas chromatograph column packed with octylphenoxypolyethylene or oxyethanol polytetrafluoroethylene to separate the sulfur compounds. From the gas chromatograph the separated sulfur compounds go through a pyrolysis furnace where they are oxidized to sulfur dioxide, and then into a microcoulometer for titration.

A process developed by J.E. Mitchell (7) is one in which the concentration of inorganic sulfur compounds in a gas is determined by measuring the ion concentration decrease of a flame, produced by the combustion of a mixture comprising a hydrocarbon, hydrogen and oxygen, which results from introducing the inorganic sulfur-containing compound into the flame. Preferably the inorganic sulfur compound is chosen from the group consisting of SO_2, SO_3, H_2S, COS, S, H_2SO_4 and H_2SO_3, and the ion concentration decrease is measured by a decrease in current flowing through an electrical circuit of which the flame is an integral part.

In a particularly preferred method, a gas chromatograph fitted with a flame ionization detector and a recorder is used to measure the concentration of the inorganic sulfur compound by introducing the inorganic compound into a reference mixture, comprising H_2, O_2 and hydrocarbon, and measuring the ion concentration decrease.

References

(1) Stevens, R.K. and O'Keefe, A.E., *Anal. Chem.* 42, 143 A (1970).
(2) Stevens, R.K., O'Keefe, A.E. and Ortman, G.C., *Envir. Sci. Technol.* 3, 652 (1969).
(3) Stevens, R.K., Mulik, J.D., O'Keefe, A.E. and Krost, K.J., *Anal. Chem.* 43, 827 (1971).
(4) Pierce, R.W., U.S. Patent 3,437,446; Apr. 8, 1969; assigned to Abcor, Inc.
(5) Plantz, C.A., U.S. Patent 3,412,038; Nov. 19, 1968; assigned to Mine Safety Appliances Co.
(6) Eads, E.A., U.S. Patent 3,650,696; May 21, 1972; assigned to Lamar State College of Technology.
(7) Mitchell, J.E., U.S. Patent 3,692,481; September 19, 1972; assigned to Esso Research and Engineering Co.

SULFUR DIOXIDE

The maximum allowable concentration of sulfur dioxide in air (threshold limit value, TLV) is generally accepted as 5 ppm. However, sulfur dioxide may be toxic to vegetation in concentrations of about 2 to 3 ppm and is corrosive to metallic construction materials in concentrations of less than 1 ppm. It is extremely irritating to the respiratory system, causing coughing at 8 to 12 ppm and eye irritation at 20 ppm. Although as little as 4 ppm can be detected by odor, the olfactory senses become quickly insensitive to it and the concentrations necessary for detection increase. Because of the aforementioned hazards, coupled with the enactment of specific legislation setting sulfur dioxide tolerance at 0.1 ppm levels and below, it is important to have a continuous, automatic, accurate and legally acceptable method and apparatus for the monitoring of sulfur dioxide in the atmosphere.

Sulfur oxides are common atmospheric pollutants which arise mainly from combustion processes. This section, drawn substantally from (1), will briefly summarize the characteristics, sources, effects, and control techniques of the sulfur oxides. An excellent, well-detailed report on the sulfur oxides to which the reader is referred for further information is the EPA document "Air Quality Criteria for Sulfur Oxides" (2).

The United States alone emitted in 1968 an estimated 33 million tons of sulfur oxides (3). Most of this came from the burning of fossil fuels containing sulfur compounds. Studies have shown that while the sulfur compounds in the atmosphere comprise sulfur dioxide (SO_2), sulfur trioxide (SO_3), hydrogen sulfide (H_2S), and various other sulfur compounds such as the mercaptans (CH_3SH and C_2H_5SH), by far the most predominant form is SO_2.

It is a fact that atmospheric SO_2 is damaging to man; this is perhaps the most compelling reason for its study. Numerous laboratory and correlation studies have demonstrated its effect on man. Other experiments have found SO_2 damaging to animals, plants, building materials, and textiles. Associated problems include decrease of visibility, corrosion of electrical wiring, damage to artworks, and increase in the hardening time of paints. Some of these problems can be translated into an annual cost, such as for agricultural damage to alfalfa or for the need to use corrosion-resistant gold electrical contacts. Many are not as easily assessed, such as for an increased mortality rate. It is likely that any amount of SO_2 that man adds to the atmosphere is an added threat to his health. It is imperative that a maximum degree of control be exerted to limit emissions.

Sulfur dioxide (SO_2) is a nonflammable, colorless gas. It can be tasted by most people at concentrations from 0.3 to 1 part per million by volume (ppm). A vile, pungent odor is experienced at concentrations above 3 ppm (2) as noted above. Sulfur dioxide partially oxidizes in air to form sulfur trioxide (SO_3), which readily combines with water vapor to form sulfuric acid (H_2SO_4). Sulfur dioxide also partially combines with water vapor to form sulfurous acid (H_2SO_3), which oxidizes to form sulfuric acid. The degree of conversion of SO_2 into its derivatives depends upon numerous atmospheric conditions such as humidity, concentration of hydrocarbons, and intensity of sunlight. Several studies describing the reaction kinetics of SO_2 in the atmosphere have been described in the literature, and a good review is the article by Bufalini (4).

Ordinary combustion of fossil fuels forms SO_2 and SO_3 in a ratio of 30 to 1, and in power plants with controlled reaction conditions, it forms SO_2 and SO_3 in a ratio of 60 to 1 (2)(5). A study of ambient air in Cincinnati showed practically no differences between SO_2 and total gaseous sulfur concentrations (6). In Los Angeles, a similar study also found that the total gaseous sulfur in ambient air was at least 90% SO_2 (7). Therefore, measurement of SO_2 is usually considered synonymous with measurement of total gaseous sulfur pollution.

Recorded sulfur dioxide levels may be essentially instantaneous values or averaged values representing such periods as five minutes, an hour, 24 hours, or a year. Because of variations in meteorological conditions and emission rates, instantaneous levels of SO_2 may be correlated with either long-period or short-period average values. Maximum 24 hour averages range from 4 to 7 times the annual mean; hourly averages range from 10 to 20 times the annual mean (2).

The Air Pollution Control Office (APCO), now the Office of Air Programs (OAP) of the EPA, in conducting the Continuous Air Monitoring Project (CAMP), monitored six cities over the period from 1963 to 1967. The project found mean annual concentrations ranging from 0.01 ppm in San Fracisco to 0.18 ppm in Chicago (5).

The APCO's National Air Surveillance Network (NASN) found annual average concentrations ranging from 0.002 ppm in Kansas City, Missouri to 0.17 in New York City; the highest 24 hour average concentration was 0.38 ppm in New York City (2). Short-period average values were sometimes considerably higher than the annual and 24 hour averages.

Table 43 (5) lists the sources and their respective contributions of the estimated 33.2 million tons (3) of man-made sulfur oxides emitted into the atmosphere of the United States during 1968.

TABLE 43: SOURCES OF SULFUR OXIDES EMISSIONS

Process	Approximate Percent of Total Man-made Emissions
Electric power generation:	
Coal	42
Oil	4
Other coal combustion	17
Other petroleum combustion:	
Residual oil	12
Other petroleum	3
Refinery operations	5
Smelting of ores	12
Coke processing	2
Sulfuric acid manufacture	2
Other	1

Source: Lawrence Berkeley Laboratory; Publ. LBL-1, May 1, 1972

Two-thirds of the sulfur oxides are emitted in urban areas, with seven industrial states in the Northeast accounting for nearly half.

Studies of the effect of sulfur oxides are usually conducted in a laboratory-controlled atmosphere or in the field by correlation. An example of the first type may be the effect of 0.02 ppm SO_2 at 50% humidity on the corrosion rate of steel. An example of the latter may be the correlation of SO_2 concentration and the incidence of bronchial congestion in an area of London.

Both types of study have value. The laboratory tests demonstrate isolated effects of sulfur oxides and the effects of sulfur oxides when combined with other pollutants or agents. For example, there is evidence that the combined effect of sulfur dioxide and particulates is much greater than the effect of either alone or both in sequence, this combined action is called synergism. The reactants present and their concentrations can be carefully controlled in the laboratory situation.

The correlation studies are most useful in the field. Effects of sunlight, nitrogen oxides, acid mist, particulates, oxidants, and whatever, all contribute to the possible synergisms with sulfur oxides. Experiments attempting to duplicate the atmosphere have had limited success primarily because of the static nature of the experimental environment as compared to the dynamic nature of the atmosphere and secondarily because of atmospheric constitutional complexity (4).

For certain sensitive individuals, a concentration of 1 to 2 ppm SO_2 will cause a bronchial response. Most individuals in laboratory observations will respond after a 30 minute exposure to 5 ppm or above. Response is usually indicated by an increase in pulmonary flow resistance, which is indicative of bronchoconstriction and its associated increase in airway resistance (2). Substantial increase in pulmonary flow resistance could possibly cause increased cardiac load.

Correlation studies found that a rise in the daily death rate in London occurred when SO_2 climbed sharply to 0.25 ppm in the presence of 750 $\mu g./m.^3$ of particulates. A more distinct rise occurred when SO_2 exceeded 0.35 ppm for one day in the presence of 1,200 micrograms per cubic meter particulates. A daily exposure to 0.52 ppm SO_2 with 2,000 micrograms per cubic meter particulates caused an increase of 20% in the daily death rate. Such episodes have also occurred in New York City. A Rotterdam study indicated increased mortality arising from a few days' exposure to a 24 hour average concentration of 0.19 ppm SO_2 (6).

New York City hospitals reported a rise in respiratory infections and cardiac diseases during a 10 day period when 24 hour averages ranged form 0.07 to 0.86 ppm SO_2; the rise in hospital admissions was evident before 0.25 ppm was reached. A similar occurrence was noted in Rotterdam during the few days of the study when 24 hour averages ranged from 0.11 to 1.19 ppm SO_2 (2).

Effects are more severe on the aged and on those suffering from chronic respiratory disease. One must consider that while pronounced effects are apparent at the stated concentrations, deleterious effects occur at lower concentrations.

As for long-term studies, one in Genoa, Italy, demonstrated a higher incidence of respiratory disease in an area with an annual mean of 0.092 ppm SO_2 versus areas with annual means of 0.037 ppm and 0.028 ppm, respectively. Another study, done in areas near a common smelter, demonstrated again the correlation between SO_2 concentration and resiratory symptoms; after the smelter installed a higher stack for better dispersion, the respiratory symptoms disappeared (2).

Chronic plant injury is associated with an annual concentration of 0.03 ppm SO_2. Because of excessive accumulation of sulfate in the leaves, SO_2 may cause leaf-drop. Acute plant injury signified by leaf discoloration occurs for some trees and shrubs after an 8 hour average greater than 0.3 ppm SO_2. Evidence suggests that ozone and nitrogen dioxide may each act synergistically with SO_2 (2).

Tests on steel panels conducted in Chicago and St. Louis have shown a strong correlation between corrosion rate and SO_2 concentration. While there was some correlation with particulates, the SO_2 correlation was dominant (2). General corrosion of other metals also results from sulfur oxides in air. Electrical hardware in overhead power lines, electrical contacts, and electrical wire are some of the common susceptible materials. The corrosion rate of zinc in Pittsburgh was reduced nearly four-fold with a drop in sulfur dioxide annual concentration from 0.15 to 0.05 ppm (2)(5).

Besides metals, SO_2 also attacks limestone, marble, roofing slate, mortar, and any other carbonate-containing stone in which the carbonate is converted to sulfate. Cotton, rayon, nylon, and leather lose strength upon exposure to sulfur oxides. Discoloration in paper, fading of dyes, increase in the drying time of paint, suspended particulates of sulfuric acid and other sulfates in smog, all arise from sulfur oxides (2)(5).

The principle methods used to limit sulfur oxides emission are (1) use of low-sulfur fuels, (2) removal of sulfur from fuel prior to combustion, (3) removal of sulfur oxides from gases after combustion, and (4) combustion modification processes. Good reviews on sulfur oxide control techniques are the EPA document, "Control Techniques for Sulfur Oxide Air Pollutants" (8), *Sulfur and SO_2 Developments* published by the American Institute of Chemical Engineers (9) and *Abatement of Sulfur Oxide Emissions from Stationary Sources* (10).

The national air quality standards pertaining to sulfur oxides are presented in Table 44 (10). Sulfur oxide emissions as described in "Standards of Performance for New Stationary Sources" (12) applies to two sources: fossil-fuel fired steam generators and sulfuric acid plants.

This regulation specified performance test methods and continuous emission monitoring procedures. In the case of fossil-fuel fired steam generators, fuel monitoring procedures are also outlined. It should also be noted that the use of gaseous fuels negates the requirement for continuous emission monitoring devices. The stationary source regulations are given in the Federal Register (11).

TABLE 44: NATIONAL AMBIENT AIR QUALITY STANDARDS FOR SO_2

Averaging Time	Primary [b] Standards	Secondary [c] Standards
Annual Arithmetic Mean	80 μg/m^3 (0.03 ppm)	60 μg/m^3 (0.02 ppm)
24 hours [a]	365 μg/m^3 (0.14 ppm)	260 μg/m^3 (0.10 ppm)
3 hours [a]	---------	1300 μg/m^3 (0.5 ppm)

a. National standards, other than those based on annual arithmetic or geometric means, are not to be exceeded more than once per year.
b. National Primary Standards: The levels of air quality necessary, with an adequate margin of safety, to protect the public health.
c. National Secondary Standards: The levels of air quality necessary to protect the public welfare from any known or anticipated adverse effects of a pollutant.

Source: Lawrence Berkeley Laboratory; Publ. LBL-1; May 1, 1972

Measurement in Air

The topic of the sampling and analysis of gaseous sulfur dioxide pollutant resulting from the combustion of fossil fuels has been reviewed at length by Walden Research Corp. (13).

Sampling: Sampling of SO_2 for analysis may be conducted with devices of the static, mechanical, or continuous type. A static device is one without moving parts or fluids. A mechanical sampler is one which obtains a number of sequential samples by bubbling air into solution bottles. The statically and the mechanically collected samples are subsequently analyzed by wet chemistry in the laboratory.

Continuous instruments are able to sample and analyze simultaneously, thus giving rapid, continuous data. The continuous instruments allow on-the-spot analysis without need for a laboratory and yield nearly real-data. If provided with a chart recorder or other data storage device, some models can operate unattended for extended periods, depending upon the need for calibration or reagent refills.

Static Samplers — Static samplers based on the lead dioxide (PbO_2) sulfation plate or candle have been in common use for many years; PbO_2 candles were used as early as the 1930's. The technique depends upon the reaction of SO_2 with lead dioxide (PbO_2) to form lead sulfate ($PbSO_4$).

$$SO_2 + PbO_2 \longrightarrow PbSO_4$$

The lead dioxide in paste form is painted in a thin layer on a gauze-wrapped cylinder, a glass plate, or other support, and allowed to dry. After exposure to atmospheric SO_2, the lead dioxide layer is removed, and the sulfate content is quantitatively ascertained by a traditional gravimetric procedure or a newer, more sensitive turbidimetric procedure. Both analysis methods convert the sulfate SO_4^{-2} to insoluble barium sulfate ($BaSO_4$).

The result is reported as a sulfation rate in units such as mg. SO_3/100 cm.^{2}/month or mg. SO_3/100 cm.^{2}/day. Conversion of such a rate to conventional average concentration in ppm SO_2 is highly dependent upon the conditions of exposure and upon the physical form of the lead dioxide, although in practice a single conversion factor is often used for convenience, resulting in an average concentration which is within a factor of three of an instrumental measurement average. For example, Corning Laboratories, Inc., suggest using 0.03 as a conversion factor as shown below; the figure was taken from a 1969 paper presented by Huey (14).

$$\text{ppm } SO_2 = (0.03)\ \text{mg. } SO_3/100\ \text{cm.}^2/\text{day}$$

In some instances the sulfation rate may be a better indicator SO_2 effect than the simple average concentration. The effect of a given concentration of SO_2 on a sulfation plate on a calm day should be different than the same on a windy day. The effect of a given concentration SO_2 on zinc and on steel panels has been shown to be milder on calm days than on windy days. Agitation of the air by wind would seemingly allow replenishment of the SO_2 about the sulfation device and also a continual cleaning of the device surface; but studies have shown an apparent decrease in sulfation rate with greater wind speeds, the sulfation rate being proportional to the fourth root of the wind velocity (15).

In general, gravimetric analysis has been demonstrated to be more precise than turbidimetric analysis. However, in sulfation analysis, variations in atmospheric and lead dioxide conditions tend to lower the precision of the gravimetric procedure to that of the turbidimetric.

Analyses using turbidimetric procedures have obtained standard deviations of 5, 7, and 13% with sulfation candles, and of 4 and 8% with sulfation plates (16). An analysis using a gravimetric procedure has obtained a standard deviation of 7% with sulfation candles (17). The American Society for Testing and Materials (17) has proposed a method using the sulfation candle with subsequent gravimetric analysis. The Intersociety Committee on

Methods for Ambient Air Sampling and Analysis has proposed two methods: sulfation candle with subsequent gravimetric or turbidimetric analysis (18) and sulfation plate with subsequent turbidimetric analysis (19).

Because of its higher sensitivity and easier, fewer manipulations, the turbidimetric method is popular. A study comparing the two methods has indicated that there is little difference in accuracy, in eleven samples the average deviation between the turbidimetric and gravimetric procedures was 2.5% (20).

As opposed to the typical 30 day exposure for gravimetric analysis, the turbidimetric analysis can be used after 1 to 2 days of exposure. The duration is determined by the quantity of sulfate needed for analysis, the SO_2 concentration, and the need to retain a constant reactivity of the lead dioxide layer. It is believed that if the total lead dioxide reduced to lead sulfate is less than 15% (17)(18), the reactivity of the lead dioxide layer is constant, and, therefore, the sulfation rate is proportional to the SO_2 concentration.

The sulfation process for SO_2 measurement is susceptible to two different classes of interferences. Firstly, any species which appreciably reacts with the lead dioxide is a chemical interferent. Hydrogen sulfide (H_2S), sulfur, and sulfate aerosols such as ammonium sulfate (NH_4SO_4) and sulfuric acid mist (H_2SO_4) can contribute to the sulfation, causing a higher rate (18). Methyl mercaptan (CH_3SH) also interferes with the reaction by reducing the PbO_2 to lead oxide (PbO), which results in less available PbO_2 for reduction by SO_2.

Secondly, any species which inhibits or enhances the SO_2-lead dioxide contact is a physical interferent. The lead oxide (PbO) formed by methyl mercaptan will, if formed on the surface, somewhat mask the lead dioxide underneath from SO_2. Dust will also somewhat mask the underlying lead dioxide. The agitation of air, as by wind, apparently decreases the sulfation rate as mentioned earlier.

The shape of the shelter in which candles are placed affects the sulfation rate, round shelters allowing more sulfation than square shelters (18). Temperature increases the rate of reaction by approximately 0.4% per 1°C. rise; moisture on the surface of a sulfation device also increases the reaction rate (15). Affecting the surface area of the lead dioxide are the size of the lead dioxide particles, the density of the layer, and the thickness of the layer; homogeneity within one batch of devices and consistency from batch to batch are difficult to control. Other factors which might affect the sulfation rate are humidity and purity of reagents.

Sulfation plates and candles can be quite readily fabricated and analyzed with minimal laboratory facilities. Procedures for preparation are given in the ASTM Method D2010-65 (17), the Intersociety Method 42410-01-70T (18), and the study by Huey (16). Consistency in procedure and in homogeneity of mixtures will maximize precision. For analysis the turbidimetric procedure is preferred unless one desires to obtain a 1 month average.

Commercially prepared plates and candles are readily available at costs typically around thirty-five dollars per twenty. Holders and shelters are additional initial costs. A protective plywood shelter (18) is recommended to minimize air turbulence and dust. Many manufacturers and distributors of sulfation devices offer analytical services at typically three to five dollars per device or sell analysis kits.

With a standard deviation of 5 to 10% and an indeterminant factor for conversion to concentration, the sulfation devices are recommended for rough, noncritical applications. Despite their limited accuracy, the sulfation devices are sufficiently precise for relative measurements. The pattern of SO_2 concentration around the perimeter of a valley, for example, may be assessed by sulfation plates. At such low cost and without need for attendance, the sulfation devices may be liberally used for one-month average sulfation rates. They are also useful for occasional daily samples to determine relative variations, as the costs associated with a continuous analyzer may not be warranted for infrequent use.

Four other static sampling methods are (1) filter paper impregnated with $KHCO_3$-glycerine-water (21), (2) membrane filter impregnated with $KHCO_3$ (22), (3) Whatman No. 1 filter impregnated with KOH-glycerol (23), and (4) the K_2CO_3 alkaline plate (24)(25).

Mechanical Samplers — Sampling of SO_2 in ambient air can be performed with mechanized bubbler devices, which bubble an air sample through an absorbing solution. The solution stored in a glass or polyethylene bottle is later analyzed in the laboratory, usually by the pararosaniline method. A list of manufacturers of mechanized samplers may be found in the *Environmental Science and Technology Annual Pollution Control Directory* under "Samplers." Gas bubbler samplers are recommended by EPA in their implementation plan regulations.

It is not the object here to discuss these mechanical samplers in detail. A discussion of air sampling in general can be found in a book by Leithe (26).

Continuous Sampling — Correct sampling is often the key to any analytical investigation and this certainly holds true in air monitoring for SO_2 content. A sampling system may be of the type in which air is collected in containers which are carried to the laboratory for subsequent analysis, or the sampling system may be of the type in which the air sample is introduced directly into a continuous analyzer. The present section is limited to those sampling systems interfaced with continuous analyzers.

General guidelines are usually more easily drawn for ambient air sampling systems than for stationary source sampling systems. In the latter case each source must be approached, as a new problem. Some general comments, however, applicable to both ambient and source sampling can be made. Factors for consideration in the design of a sampling system include the following:

(1) Sampling Location — The selection of a sampling site and of the sampling points needed in order to obtain representative samples.
(2) Gas Flow Characteristics — Velocity measurements, temperature, pressure, flow problems, etc.
(3) Removal of Undesirables — The use of selective filters to remove undesirable species (particulates, interfering gases, water vapor, etc.) likely to be encountered.
(4) Sampling Train Design — Consideration of the various components of a system (probes, gas collectors, flow-measurement devices, air pumps, etc.)
(5) Sampling Train Materials — Consideration with respect to sample adsorption, chemical inertness, etc., of the types of material (glass, stainless steel, teflon, etc.) used in the various components of the sampling train.

Detailed discussions of sampling may be found in the literature (26) through (30). A recent review on stationary source sampling is the article by Morrow, et al. (31). The reader is also referred to the ambient air sampling and stack sampling methodology as recommended by EPA (32)(33).

Continuous Analyzers: Continuous analyzers have been in use for many decades. Also known as automatic analyzers, their convenience as compared to laboratory wet-chemical methods is obvious. Ideally, it is a small, portable unit, and all that needs to be done is to turn on the power, allow the sample air to flow, and observe the output. Realistically, the portable unit may weigh up to 150 pounds and require mechanical maintenance, frequent reagent refilling, interference filters for dust and gases other than SO_2, and more.

A lengthy section in *Instrumentation for Environmental Monitoring–Air* (1) is devoted to discussing the merits and drawbacks of each type of continuous analyzer, each type having a different principle of operation. Different principles are here listed.

- Conductivity
- Colorimetry
- Coulometry
- Electrochemical transducers
- Flame photometry and gas chromatography
- Thermal conductivity
- Nondispersive absorption spectroscopy
- Correlation spectroscopy
- Second derivative spectroscopy
- Condensation nuclei formation
- Mass spectrometry
- Microwave spectroscopy
- Chemiluminescence
- Mercury substitution with nucleonic detection
- Laser techniques
- Dispersive absorption spectroscopy

Not all of the above named principles of operation have been found in commercial analyzers. Conductivity, colorimetry, and coulometry are well-studied wet chemical methods. Continuous analyzers using these three principles have been used for many years and have been thoroughly tested under field conditions.

In an effort to reduce the need for automating wet chemistry, electrochemical transducers have been introduced. More field tests of electrochemical-transducer analyzers should appear within the next few years.

The other principles listed are even less dependent on chemical principles and are considered physical principles of detection. Flame photometry, flame photometry with gas chromatography, nondispersive absorption, and thermal conductivity are readily found in numerous models. Some of the other physical principles are found in one or two commercial models. Several of the physical methods have demonstrated great potential for wide range, high sensitivity, good linearity, and excellent selectivity in SO_2 monitoring. Electro-optical methods are capable of long-line average measurement and of remote sensing.

Sulfur dioxide may be monitored in air by the following instrumental techniques (1):

- Colorimetric (reference method)
- Sulfation plates and PbO_2 candles
- Conductimetric
- Coulometric
- Electrochemical transducers
- Flame photometric detection
- Gas chromatography-flame photometric detection
- Infrared absorption
- Ultraviolet absorption

The EPA-designated reference method for SO_2 in air is the pararosaniline method.

Sulfur dioxide monitoring systems can be classified for either ambient air monitoring or stationary source monitoring, and can be of either the manual or automatic type. Each system for ambient air or stationary source monitoring is quite unique and must accordingly be considered individually. Tables 45 and 46 summarize data on ambient air monitors and stationary source monitors respectively.

Manual ambient air monitoring systems may be of the static (e.g., sulfation plates, PbO_2 candles) or mechanical (e.g., gas bubbler devices) type as noted above. These systems involve ambient sampling with subsequent laboratory analysis (usually by a manual method). Automatic ambient air monitoring systems may be of the type in which a sample is extracted from the ambient air and analyzed on-site with a continuous analyzer (e.g., conductimetric instrument) or of the type in which the continuous analyzer requires no sampling system (e.g., correlation spectrometer).

Manual stationary source monitoring systems are of the mechanical type. The laboratory analysis is usually performed away from the sampling site. Automatic stationary source monitoring systems may be (1) the stack monitoring type in which a sample is extracted from the stack and analyzed on-site, usually with a continuous analyzer; (2) the in-situ type in which there is no sampling system, and analysis is by across-the-stack-electro-optical methods; (3) the remote monitoring type in which there is no sampling system, analysis is by single-ended electro-optical methods, and monitoring is to a remote point such as a stack plume; or (4) the long-path monitoring type in which there is no sampling system, analysis is by double-ended electro-optical methods, and monitoring is between two points such as across the envelope of a plume.

TABLE 45: TABLE OF SO_2 INSTRUMENTS-AMBIENT AIR MONITORS

Principle of Operation	Instrument Note	Range* (ppm)	Response Time	Cost	Multi-Parameter Capability	Remarks
			Static Samplers			
Lead Peroxide Candle						
	Precision Scientific 1			$ 90	Sulfation rate	Shelter, stand, candle
	Research Appliances 1			$ 61.50	Sulfation rate	Shelter, 4 candles
	Silver-top 1			$104	Sulfation rate	Shelter, stand, candle
Sulfation Plate						
	Corning 1			$ 30/20 plates	Sulfation rate	
	Harleco 1			$ 40/25 plates	Sulfation rate	
	Research Appliances 2			$ 35.25/20 plates	Sulfation rate	
			Intermittent Analyzers			
Colorimetric						
	Bacharach 2	1-80, 2700	25 sec	$ 76.20	CO, CO_2, etc.	hand-held
	Bendix 3	1-80, 40,000	3 min	$ 75	H_2S, mercaptans	hand-held
	Cenco 1	0-100%		$595	NO_2, O_3, NH_3, H_2S, etc.	
	Edmund 1	.1-400	several minutes	$ 17.50	CO, H_2S, NO_2, etc.	hand-held
	Lovibond 1	1-20		~ $150	several	
Conductimetric						
	Combustion Equipment 1	0-0.5, 5	~ 1 min	$495		hand-held, H_2O_2

* The minimum and maximum ranges

(continued)

TABLE 45: (continued)

Principle of Operation	Instrument Note	Range* (ppm)	Response Time	Cost	Multi-Parameter Capability	Remarks
			Continuous Analyzers			
Conductimetric						
	Bacharach 1	0-5, 20	5 min		H_2S, NH_3, NO_2, HCl	H_2O
	Calibrated Instruments 1	0-0.4, 4.0	1.5 min	$4,900	None	H_2O_2
	Devco Engineering 1	0-1	20 sec		H_2S, NH_3, Cl_2, CO_2	H_2O
	Intertech 4	0-0.35, 10	60-100 sec	$5,450	NO, NO_x, H_2S, etc.	H_2O_2
	Kimoto 1	0.002-0.2, 0.5	3.9 min	$1,490 (Japan)		H_2O_2
	Leeds and Northrup 1	0-0.5, 2.0	2-6 min	~ $2,500	None	H_2O_2
	Scientific Industries 1	0-0.5, 2.0	30 sec	$2,575	None	H_2O_2
	Scientific Industries 2	0-0.1, 2.0	20 sec	$1,500	None	H_2O_2
	Scott Aviation 1	0-1.0, 5.0	13-90 sec	$2,250	None	H_2O
Colorimetric						
	Atlas 2	0-0.25, 10	1 to 6 min	$1,950	Aldehydes	$R'NH_2$ dye
	Houston Atlas 1	0-0.5, 100%	2 min	$4,420	Total sulfur, H_2S	Lead acetate tape
	Litton 1	0-0.5, 2.0	< 10 min	$2,950	NO_2	Pararosaniline
	Monitor Labs 1	0-0.2, 2.5	5 min	$2,000	NO, NO_2	Pararosaniline
	Pollution Monitors 1	0-0.25, 4.0	8 min	$2,090	NO_2	Pararosaniline
	Pollution Monitors 2	0-0.25, 4.0	8 min	$1,895	NO_2	Pararosaniline
	Technicon 1	0-0.1, 1.0	14.5 min	$5,535	NO_2, NO_x, H_2S, etc.	Pararosaniline
	Wilkens-Anderson 1	0-0.5, 8.0	5 min	$2,805	NO, NO_2, NO_x	Pararosaniline

* The minimum and maximum ranges

(continued)

TABLE 45: (continued)

Principle of Operation	Instrument Note	Range* (ppm)	Response Time	Cost	Multi-Parameter Capability
			Continuous Analyzers		
Coulometric					
	Atlas 1	0-0.1, 10	< 2 min	$1,775	O_3
	Barton 1	0-1.0, 1000	~ 5 min	$4,150	H_2S, mercaptans
	Beckman 1	0-0.5, 4.0	4 min	$2,760	None
	Philips 1	0-0.115, 3.83	5 min	$5,290	H_2S, mercaptans
	Process Analyzers 1	0-1.0, 3	2 min	$4,000	H_2S, mercaptans, etc.
Electrochemical Transducer					
	Dynasciences 1	0-1.0, 10	3 min	$2,250	Several
	Envirometrics 1	0.01-100, 20,000	5-10 sec	$1,750	NO_x, NO_2, H_2S, CO, RCHO
	Mast 1	0-0.2, 5.0	2-3 min	$ 950	
	Theta Sensors 3	0-1.0	23 sec	$1,950	
Flame Photometric Detector					
	Bendix 1	0-1.0	30 sec	$3,950	Sulfur compounds
	Meloy 1	0.01-10	28 sec	$3,250	Sulfur & phosphorus compounds
	Meloy 2	0.01-10		$3,995	Sulfur & phosphorus compounds
	Meloy 3	0.01-1.0	12 sec	$4,750	Sulfur & phosphorus compounds
	Meloy 4	0.01-1.0	28 sec	$4,225	Sulfur & phosphorus compounds
	Meloy 5	0.01-10	28 sec	$5,270	Sulfur, hydrocarbon & phosphorus compounds

* The minimum and maximum ranges

(continued)

TABLE 45: (continued)

Principle of Operation	Instrument Note	Range* (ppm)	Response Time	Cost	Multi-Parameter Capability
- - - - - -	- - - - - -	- - - - - -	Continuous Analyzers	- - - - - -	- - - - - -
Gas Chromatography-Flame Photometric Detector					
	Analytic Instrument 1			\$5,950	Sulfur compounds
	Bendix 6	0-1.0	5 min	\$7,200	Sulfur compounds
	Tracor 1	0-2.0	~ 3 min	\$5,275	Sulfur & phosphorus compounds
	Varian 1	0-0.05, 10	~ 3 min	\$6,200	Sulfur compounds
Correlation Spectroscopy					
	Barringer 1	1-1000 ppm-meters	1 to 32 sec	\$21,000 (Canada)	NO_2
Second Derivative Spectroscopy					
	Spectrometrics 1	0-0.33, 10	24 sec	\$12,500	NO, NO_2, O_3, NH_3, etc.
Condensation Nuclei Formation					
	Environment | One 1	0-0.005, 5.0	45 sec	\$5,325	NH_3, Hg, CO, NO_x
Mass Spectrometry					
	Uthe (no note - see text)		< 1 sec	~ \$20,000	Several
	Varian (no note - see text)			~ \$20,000	Several
Microwave Spectroscopy					
	Hewlett-Packard (no note - see text)				Several

*The minimum and maximum ranges

Source: Lawrence Berkeley Laboratory; Publ. LBL-1; May 1, 1972

TABLE 46: TABLE OF SO_2 INSTRUMENTS-STATIONARY SOURCE MONITORS

Principle of Operation	Instrument Note	Range* (ppm)	Response Time	Cost	Multi-Parameter Capability	Remarks
			Intermittent Analyzers			
Colorimetric						
	Bendix 2	200-3000	5 min	$ 325	None	
	Bendix 3	1-80, 40,000	3 min	$ 75	H_2S, mercaptans	hand-held
			Continuous Analyzers			
Conductimetric						
	Calibrated Instruments 2	0-500, 8000	2 min	$5,615	CO_2	H_2O_2
Colorimetric						
	Houston Atlas 1	0-0.5, 100%	2 min	$4,420	Total sulfur	
Coulometric						
	Barton 1	0-1.0, 1000	5 min	$4,150	H_2S, mercaptans	
	Barton 2	0-1000	~ 5 min	$5,425	H_2S, mercaptans	
Electrochemical Transducer						
	Dynasciences 2	0-500, 5000	90 sec-3 min	$2,000	Several	
	Envirometrics 1	0.01-100, 20,000	5-10 sec	$1,750	NO_x, NO_2, H_2S, CO, RCHO	
	Theta Sensors 1	0-500, 5000	20 sec	$3,500	NO_x	
	Theta Sensors 2	0-500, 5000	13-20 sec	$1,150	NO_x	
Flame Photometric Detector						
	Meloy 6	25-2500, 10,000	12 sec	$8,500	Sulfur & phosphorus compounds	
Thermal Conductivity						
	Intertech 2	0-15,000, 100%	$\geq$1 sec	~ $1,660	H_2, NH_3, CO_2	
	Intertech 3	0-15,000, 100%	~ 2 sec	$1,000	H_2, NH_3, CO_2	
Nondispersive Absorption Spectroscopy						
	Beckman 2			~ $2,900	CO, etc.	NDIR
	Bendix 4	0-200, 500,000	3 sec	$3,403	CO, etc.	NDIR
	Bendix 5	0-2000	4 sec	$2,680	CO, etc.	NDIR

(continued)

TABLE 46: (continued)

Principle of Operation	Instrument Note	Range* (ppm)	Response Time	Cost	Multi-Parameter Capability	Remarks
Continuous Analyzers						
	Calibrated Instruments 3	0-500, 500,000	~15 sec	$2,995	CO, etc.	NDIR
	DuPont 1	0-200, 5000	15 sec	$8,175	NO_2, NO_x, NH_3, H_2S, etc.	UV
	Intertech 1	0-200, 100%	0.5 sec	$3,280	CO, etc.	NDIR
	Mine Safety Appliances 1		5 sec	$3,570	CO, etc.	NDIR
	Mine Safety Appliances 2	0-5000, 200,000	5 sec	$1,795	CO, etc.	NDIR
	Peerless	0-50, 50,000	< 5 sec	$2,800-3,800	CO, NO_2, O_3, etc.	NDIR/UV
Dispersive Absorption Spectroscopy						
	Environmental Data 1				NO, CO, CO_2	
	Wilks 1	0-3000	30 sec	$3,950	CO, etc.	IR
Correlation Spectroscopy						
	Combustion Equipment 2	0-200, 5000	20 sec	$7,800	None	Barringer Spectrometer
Second Derivative Spectroscopy						
	Spectrometrics 2	0.1-100%		$9,000	NO, NO_2, O_3, O_2, NH_3, CO	

*The minimum and maximum ranges

Source: Lawrence Berkeley Laboratory; Publ. LBL-1; May 1, 1972

Understanding the pollution problem begins with the ability to accurately monitor its presence. As better monitoring systems are established, one will be able to identify build-up areas for SO_2 and related pollutants, identify flow and dispersement patterns, and eventually learn to predict and avoid serious health episodes.

In designing an SO_2 monitoring system several factors must be considered. These include sampling, calibration, analysis, data acquisition and reduction, and finally data interpretation. At the core of such a system is the analytical instrument or technique, and it is the major concern of this section to discuss those instruments and techniques which are designed for the analysis of SO_2. It must be remembered, however, that the monitoring instrument by itself does not complete the system.

The most promising new instrumental developments for air monitoring are those techniques based upon the application of well-known spectroscopic methods. The techniques discussed here are either in early commercial stages or in an advanced prototype stage. They have, in general multi-parameter capability and are not limited to only SO_2 monitoring.

Mass spectrometers are marketed by many manufacturers and a few have recently developed instruments specifically for air pollution monitoring. The sensitivities of such instruments are on the order of 0.1 ppm and response times are less than 1 second. Two such instruments are the Precision Gas Sampling System, Model Q 30 developed by UTI, Uthe Technology of Sunnyvale, Calif, and the Chemical Vapor Analysis (CVA) System developed by Varian Associates, Analytical Instrument Division, Palo Alto, Calif.

Both instruments use a quadrupole mass analyzer and cost approximately $20,000. Microwave techniques can also be applied to air monitoring (35), and Hewlett-Packard has recently introduced a commercial microwave spectrometer designed for air monitoring: Molecular Rotational Resonance Spectrometer Model 8460A, Hewlett-Packard, 1505 Page Mill Road, Palo Alto, Calif. 94304.

Instruments based on the chemiluminescence technique are already commercially available for air monitoring of NO/NO_x and O_3. The chemiluminescence technique is also feasible for the detection of other gaseous pollutants (36)(37). Several manufacturers have prototype instruments in various development stages which should extend the chemiluminescence technique to the monitoring of such air pollutants as SO_2, CO, H_2S, and NH_3.

A sulfur dioxide monitoring instrument based upon a mercury substitution and nucleonic detection technique has been developed (38). The instrument is based on a stoichiometric substitution of mercury for sulfur dioxide in a reaction cell, the transfer of the mercury into a measurement cell, and the measurement of the mercury by low energy x-radiation absorption.

Laser techniques show good promise for the remote monitoring of air pollutants. Several recent studies have considered the application of laser resonance absorption techniques for air monitoring (39)(40)(41). Resonance scattering and Raman scattering techniques also show potential use for gaseous air pollution monitoring (42)(43). A recent paper by Kildal and Byer (44) thoroughly analyzes these three techniques (resonance absorption, resonance scattering, and Raman scattering). The reader should refer to the above papers for thorough discussions of laser applications to air monitoring.

Several varieties of electrochemical or electroanalytical techniques have been utilized in sulfur oxide monitors as noted by O'Keefe (45). Research and development continues on these approaches. Electroanalytical techniques are rarely specific. Liquid or solid substrates are often used to confer specificity. One type of electrochemical transducer utilizes a selective membrane along with a galvanic cell having an electrode potential selected to reduce interferences. Work has been done on selection of electrode and electrolyte materials and membranes for development of sulfur dioxide and nitrogen oxide analyzers (46). This sulfur dioxide analyzer is a compact, relatively low-cost instrument. Its sensitivity at present is adequate for short-term emission measurements but is marginal for continuous measurement of sulfur dioxide. The sensor utilizes 1 N H_2SO_4 electrolyte, a gold sensing electrode and a lead dioxide counter-electrode with a 1 mil polyethylene membrane. With potentiometric control, interference from nitric oxide is low, but hydrogen sulfide constitutes a significant interference. Response time is less than a minute. This analyzer requires improvement for use in ambient air analysis. Because of its mode of operation, simplicity of construction, compactness, and low cost, it offers possibilities for use in large monitoring networks for measurements averaged over extended time intervals, such as 24 hour periods.

A sulfur dioxide pollution monitor developed at the National Bureau of Standards, U.S. Dept of Commerce (47) is based on the principle of measuring the fluorescence of SO_2 in air. It is rapid, continuous, nearly specific to SO_2 and linear in response up to 1,600 ppm of SO_2.

The most obvious application of the new device is in the monitoring of smokestack gases. A recent California law limits SO_2 concentration in stack gases to 500 ppm, and similar laws are being considered by other states. No present instrumentation measures such concentrations with high reliability.

The lower limit of detection of the monitor in its present form is 20 parts per billion, and a special light source is being developed that should make it possible to reach even lower levels (48).

Ozone, H_2S, NO_2, CO, CO_2 and H_2 do not interfere with SO_2 detection in the monitor, while large concentrations of CS_2 (500 times the SO_2 concentration), NO (500 times), and C_2H_4 (4,000 times) do interfere. Water vapor decreases the SO_2 signal, at room temperature, a relative humidity of 100% reduces the SO_2 signal by 25% and a correction must be applied according to the concentration of water vapor in the air.

The fluorescence of SO_2 is produced in the cell attached to the detector by radiation from either a zinc or cadmium vapor lamp operating on alternating current. Before entering the fluorescence cell the light passes through a 10 cm. chlorine filter that transmits more than 90% of the exciting zinc (213.8 nm) or cadmium (228.8 nm) line but absorbs almost completely source emission lines from 270 to 390 nm. (This is necessary because the fluorescence from SO_2 covers the region from 240 to 420 nm.) The light is then focussed at the center of the fluorescence cell, and the SO_2 fluorescence is detected by a photomultiplier and displayed on a strip chart. A glass filter between the fluorescence cell and photomultiplier rejects those source emission lines passed by the chlorine cell.

The response of the detector up to 10 ppm was calibrated with SO_2 permeation tubes, available from NBS under the Standard Reference Material program (49), using air flow rates from 0.3 to 3 l./min. Above 10 ppm the device was calibrated with static air-SO_2 mixtures.

A method developed by N.A. Lyshkow (50) utilizes a colorimetric reagent for detecting sulfur dioxide in air including concentrated hydrochloric acid as the bleaching agent, p-rosaniline hydrochloride as the dyestuff, formaldehyde and ethylene glycol.

The use of colorimetric reagents for the analysis of sulfur dioxide in air has been proposed heretofore. Thus, the well-known West-Gaeke reagent has been used to detect sulfur dioxide by removing the sulfur dioxide from an atmosphere sample by scrubbing the sample with sodium tetrachloromercurate (II) to form a sample solution, and then adding a reagent of hydrochloric acid-bleached pararosaniline hydrochloride and formaldehyde to the sampling solution to form a characteristic red-violet color. The level of sulfur dioxide in the sample can then be determined by measuring the color intensity of the red-violet color.

Although the West-Gaeke type of reagents have been suitable for certain applications, their relatively high ultimate color intensity has made it difficult to achieve the desired level of the final color intensity in a short response period. Consequently, the conventional West-

Gaeke type reagents have been not satisfactory for use in the extremely rapid-sensing continuous monitoring instruments being developed today. Moreover, the nonlinear nature of the color development characteristic of the West-Gaeke type reagents has made it difficult to accurately calibrate the modern monitoring instruments. Also, the use of the West-Gaeke reagents results in color pockets because the rate of color development does not appreciably lessen after the initial response period.

In accordance with this method, ethylene glycol is added to the conventional West-Gaeke reagent in order to produce an improved reagent which is suitable for use in the most modern rapid sensing continuous monitoring instrument. At least about 0.8 milliliter, preferably 0.8 to 1.2 milliliters, of ethylene glycol are added per liter of conventional reagents. It has been unexpectedly found that the resulting reagent containing only a minute portion of ethylene glycol provides a lower final color intensity that the conventional West-Gaeke reagents, and yet the color intensity of the inventive reagent after about one minute is essentially the same as that of the conventional West-Gaeke reagent. Consequently, a relatively high percentage of the final color intensity can be achieved in an extremely short response period.

Moreover, after the initial response period, the color development characteristic of the reagent is substantially linear so that the instrument in which the reagent is used can be accurately calibrated. The rate of color development after the initial one minute response period is also relatively slow, which has the advantage of minimizing the development of color pockets due to different stages of color development within the optical cells used to measure the color intensity within the instrument. The lower color intensity of this reagent also achieves a beneficial effect in that the reacted reagent will conform more closely to Beer's law, with the attendant advantages being thereby realized.

A device developed by J.J. McGee, T.J. Kelly and J.N. Harman III (51) is an improved scrubber apparatus for removing constitutents in a gas stream which interfere with the analysis of SO_2 in the gas stream by an analytical instrument.

A solid phase chemical scrubber containing either alkylthio mercury chloride, dialkylthio mercury, alkylthio lead chloride or dialkylthio lead, or a mixture thereof, selectively removes ozone from a sample gas stream containing SO_2. The scrubber is intended for use with an SO_2 analyzing instrument which is sensitive to both ozone and SO_2. The scrubber may also contain mercuric chloride or lead chloride to remove hydrogen sulfide and mercaptans from the sample gas.

A technique developed by H. Dreckmann (52) permits the rapid analysis of trace (ppm) concentrations of a gas such as SO_2 in a gas mixture using the measuring burette and leveling bottle of an Orsat type gas analyzer for measuring a chosen volume of a gas sample. Valved conduit means between the burette and a gas sample source provides two passages in parallel, of which one is valved and the other includes a removably mounted transparent container of a material which is color sensitive to the trace gas to be measured.

The well-known Orsat apparatus is widely used for the accurate analysis of carbon dioxide, oxygen, carbon monoxide and other gases in volume percent concentrations in a gas sample. It is frequently desirable to test the same gas sample for trace (ppm) concentrations of a gas in a gas mixture, such as sulfur dioxide, which commonly is a by-product of combustion processes and the like in connection with which measurements of carbon dioxide and carbon monoxide are made.

The analysis of gases to ascertain the presence of trace concentrations has heretofore required the use of apparatus different than Orsat apparatus, for example, apparatus such as disclosed in U.S. Patents 3,131,030, April 28, 1964 and 3,223,487, December 14, 1965. The requirement for use of separate apparatus to measure different components of a gas is inconvenient and time-consuming and entails a duplication of expense for the apparatus required. The primary object of this technique is to adapt the Orsat type gas analyzer, which in its known form is suited for the measurement of volume percent concentrations, to the measurement of traces, i.e., ppm concentrations.

In a process developed by F. Schulze (53) sulfur dioxide in gas mixtures is determined by absorbing the sulfur dioxide in an alkaline solution containing a p-aminophenylazobenzene dye, e.g., p-(p-aminophenylazobenzene) sulfonic acid salt, then reacting with an acidic solution containing formaldehyde to form the more highly colored N-methyl sulfonic acid derivative of the dye and spectrally measuring the dye-derivative solution to determine the amount of sulfur dioxide.

The method is highly suitable for an apparatus for the continuous analysis of air, as in air pollution control, by providing pumping, metering and mixing of the gas and reagent streams in the proper sequence and photometric scanning of the resultant solution.

Figure 45 is a schematic diagram showing the type of apparatus which may be employed for the conduct of this process.

TABLE 45: CONTINUOUS COLORIMETRIC APPARATUS FOR SO_2 DETERMINATION IN AIR

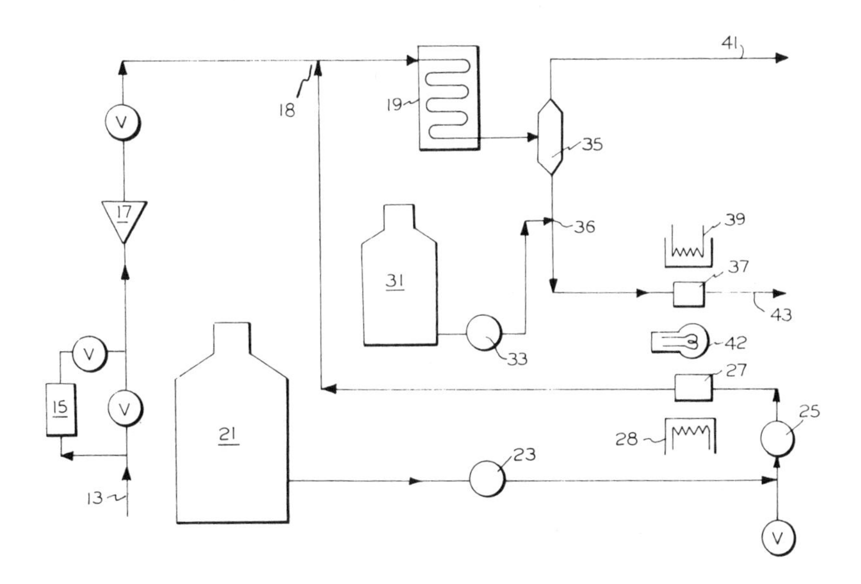

Source: F. Schulze; U.S. Patent 3,567,392; March 2, 1971

The air sample is continuously fed at **13** to the rotameter **17**, either directly or through the filter **15**. Simultaneously, a continuous flow of the first reagent solution containing alkaline dye is fed from reservoir **21** to a metering pump **23**, through measuring bulb **25** and measuring flow cell **27** which is equipped with photometer device (lamp **42** and photocell **28**, and to the junction **18** where it meets the flow of gas sample. The combined gas and first reagent stream pass through the helix **19** for mixing and to the air-liquid separator **35**. The air of the sample gas is eliminated from the system at **41**.

The second reagent solution is fed from reservoir **31** to a metering pump **33**, and joins the deaerated first solution (containing absorbed sulfur dioxide) at the junction **36**. The two solutions are mixed and reacted during their passage to measuring flow cell **37** which is provided with a photometer device (lamp **42** and photocell **39**) and is equipped with a continuous recording device. The measured solution is evacuated from the system at **43**.

The following is one specific example of the conduct of this process. An air sample containing sulfur dioxide, metered at one liter per minute, is contacted with reagent containing 0.02 g./l. of p-(p-aminophenylazobenzene) sulfonic acid sodium salt and 0.2 g./l. of sodium hydroxide and 0.5 g./l. of sulfamic acid sodium salt metered in at 1 ml. per minute, in a glass helix. Air and reagent flow concurrently down through the helix.

At the outlet of the helix air and reagent are separated. Immediately below the separator a second reagent containing 200 ml. of concentrated hydrochloric acid, 200 ml. of 40% formaldehyde, and 600 ml. of water is metered into the main reagent stream at 0.05 ml./min. These proportions are designed to yield 10 ml. each of concentrated hydrochloric acid and formaldehyde per liter of composite reagent.

The color-forming reaction takes place between the junction of the two reagent streams and the flow cell of a photometer. A 6 in. length of ⅛" i.d. tubing has been found sufficient to give adequate lag time for completion of the color-forming reaction. The colored solution is measured in a photometer previously calibrated with solutions containing known amounts of sodium metabisulfite or containing known amounts of absorbed sulfur dioxide.

A process developed by B.Y. Cho & L.B. Anderson (54) permits the quantitative determination of relatively small or trace amounts of gaseous sulfur dioxide and involves a reaction between mercurous chloride and sulfurous acid, the latter formed when sulfur dioxide is dissolved in water. One of the reaction products is a water-soluble bis-sulfitomercurate complex, the mercurate being easily removable from the remainder of the reaction mixture. The conversion of sulfur dioxide to the bis-sulfitomercurate complex is about 95% complete. By nucleonically detecting the amount of the bis-sulfitomercurate complex, which contains a metal having a high Z number (atomic number), using a monoenergetic source of nuclear radiation in the range of 15 to 25 kev., the amount of the complex is determinable and easily compared against a known standard sample either electronically or visually.

This system may be used on a continuously flowing gas such as stack gas, in which event it is unnecessary to scrub the gas, and permitting accurate determination of trace amounts of sulfur dioxide by monitoring or controlling the gas flow rate. Figure 46 shows a suitable form of apparatus for the conduct of the present process.

Referring to the drawing, a reaction chamber **10** is shown with multiple inlet and outlet conduits for introducing reagents and samples and for withdrawing reaction products. Mounted above the reaction chamber is a reactant reservoir **12** interconnected to the chamber by conduit **13** provided with a simple metering device formed by the enlarged ball section **14** and valves **15** and **16**. By opening valve **15** to fill the metering ball **14**, then closing **15** and opening **16**, a predetermined amount of reactant is introduced into the chamber **10**. The valves **15** and **16** may be manually or automatically controlled.

A sample of gas to be analyzed is introduced into the chamber **10** through an inlet conduit **17** controlled by valve **18**, the end of the conduit being below the reaction mixture **20** which is in the chamber and at a level approximately as shown. Thus, incoming gas is bubbled through the reaction mixture **20**. Water or other aqueous solution is introduced into the chamber through conduit **22** controlled by valve **23**, the conduit **22** being connected to a supply of water such as a tank **24**, as shown or a water main. The chamber is also provided with a gas outlet conduit **25** and a valve controlled purging line **26** for emptying the chamber in the event this is needed.

The chamber is also equipped with a side arm **30** through which soluble portions of the reaction mixture may flow, the arm **30** being connected to a measurement cell **35** provided with a trap **36** and emptying into a drop chamber **37** which also runs the effluent from the purge line. Positioned in the side arm is a filter **40** which prevents solid elements of the reaction mixture from passing into the measurement cell.

Nucleonic determination of the material in the measurement cell is conducted by a source **41** and a detector **42** connected to a signal processing apparatus, not shown. In practice a

FIGURE 46: SO_2 MONITOR BASED ON MERCURY SUBSTITUTION AND NUCLEONIC DETECTION

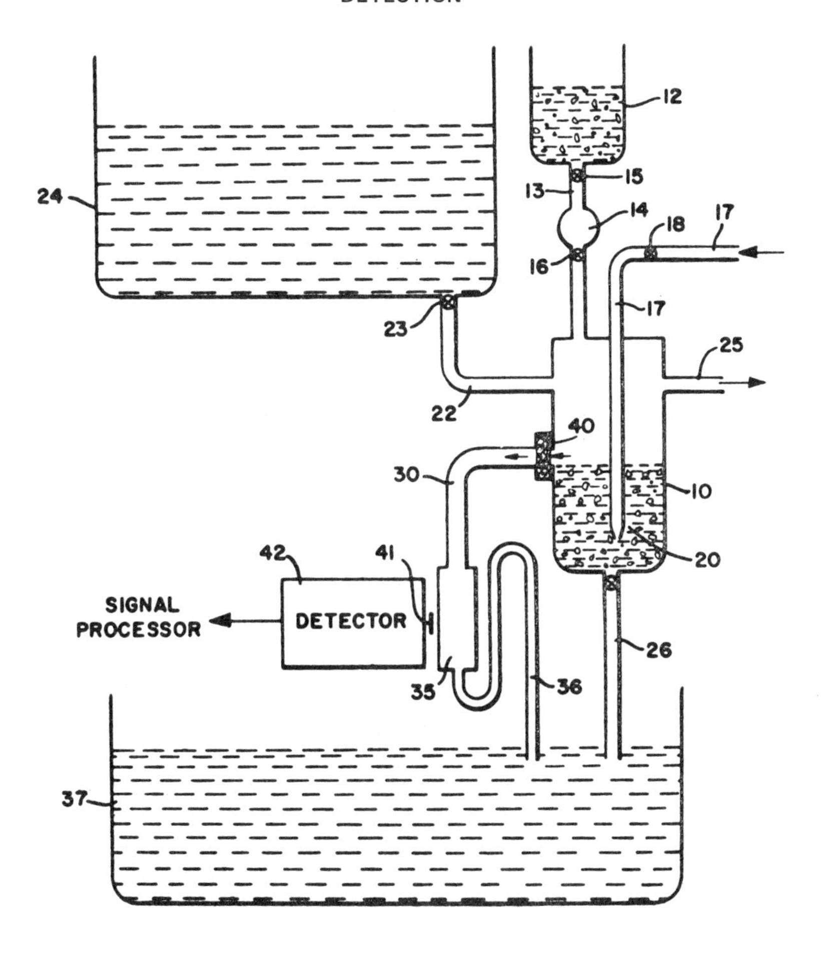

Source: B.Y. Cho and L.B. Anderson; U.S. Patent 3,578,406; May 11, 1971

predetermined amount of mercurous chloride is measured into the reaction chamber. This is done by first opening valve **15** to fill the ball **14** between valves **15** and **16**. After valve **15** is closed, valve **16** is opened to fill the reaction chamber with a fixed quantity of mercurous chloride. Water or aqueous solution is fed into the reaction chamber from the reservoir **24** and the solution flow rate is controlled by a proper adjustment of the valve **23**.

After the reaction chamber is filled with the solution to the level indicated, stack gas is bubbled through the solution using gas inlet and outlet tubings **17** and **25**, respectively. As the gas is bubbled through the solution, the following reaction occurs:

$$Hg_2Cl_2 \text{ (solid)} + 2H_2SO_3 \underset{}{\overset{K_T}{\rightleftharpoons}} Hg(SO_3)_2^{=} + 2Cl^- + Hg \text{ (metal)} + 4H^+$$

Mercurous chloride or calomel is a white solid that is practically insoluble in water (2 to 10 μg. of Hg_2Cl_2 is soluble in 1 cm.^{3} of water). When sulfite ions are introduced into the solution, however, the above reaction will proceed, forming a water-soluble bis-sulfitomercurate complex, $Hg(SO_3)_2^=$.

The solution is then fed into the measurement cell through the filter, separating unreacted mercurous chloride and metallic mercury from the solution of $Hg(SO_3)_2^=$ ions. The level of the solution mixture in the absorption cell is maintained so that the cell is filled with at least 8 cm.^{3}. As mercurous chloride is consumed, the volume of water will gradually increase and reach a maximum. When the original charge of mercurous chloride is completely consumed, some metallic mercury will remain in the cell. This is drained off through valve in line **26** together with any solid contaminants that might have been collected from the stack gas during the operation.

For example, if stack gas is bubbled through the absorption cell at a rate of 0.15 liter per minute, and at an SO_2 concentration of 2,000 ppm, about 0.8 mg. of SO_2 is absorbed per minute. This will consume Hg_2Cl_2 at a rate of 2.8 mg./min. and dissolve 1.2 mg. of mercury each minute. Thus, 30 grams (about 4 cm.^{3}) of Hg_2Cl_2 measured into the cell initially will last over a week, and the filtered solution will have a mercury concentration of 1.2 mg./cm.^{3}.

The detection of the mercury concentration in the measurement cell can be made by counting mercury L-x-ray photons that are excited by a source **41**. The source is preferably a monoenergetic source in the range of 15 to 25 kev., the lower end of this range being just above the L absorption edge of mercury.

Other sources which may be used include an americium source 241 isotope with a silver target to produce x-radiation of an energy of about 22 kev., or Am-241 with a molybdenum target which provides x-radiation of about 17.5 kev. The Cd-109 source also provides silver K-x-rays. The mercury L-x-ray photons are detected by a sodium iodide crystal in the detector, the latter including a photomultiplier tube to provide an electrical signal. A proportional counter may also be used to provide an electric signal. The use of a monoenergetic source has the advantage of reducing noise which exists with a continuous energy source.

The signal from the detector provides an indication of the quantity of heavy metal, mercury, in the sample. Since the conversion of SO_2 to the bis-sulfitomercurate complex is about 95% complete, the signal is representative of the amount of SO_2 in the sample being analyzed. Lesser amounts of conversion may be used but for maximum sensitivity the reaction should be at least 70% complete. Since the present process is concerned with small or trace amounts of SO_2, the temperature dependent solubility of SO_2 is not a factor since the solubility of 22.8 grams of SO_2 per 100 ml. of water at 0°C. is far greater than the trace amounts to be determined.

A portion of the stack gas may be analyzed using a standard sampling technique such as through a sampling tube, and where a continuous system is used, the flow rate of the gas should be controlled or monitored. This is easily accomplished by means well-known, per se, so that knowing the rate of flow of the sample, the amount of gas in the sample may be calculated electronically or by other means.

A technique developed by F.W. Van Luik, Jr. and P.E. Coffey (55) permits the accurate detection and measurement of extremely dilute concentrations of sulfur dioxide (0.005 ppm) in samples of air containing carbon monoxide in concentrations in excess of 1 ppm.

In accordance with this process, a permselective membrane is employed in the detection means to reduce the concentration of the interfering gas to a level which will no longer interfere with the detection and measurement of the selected gas while maintaining the concentration of the selected gas at such a level as to still be measurable.

The concentration of SO_2 in the air sample passed through the membrane is determined by converting the SO_2 to condensation nuclei, which are then measured electrically. Condensation nuclei is a generic term applied to small airborne particles which are characterized by the fact that they serve as the nuclei upon which a fluid such as water, for example, condenses to form droplets. Such nuclei, as this term is understood in the art, include microscopic and submicroscopic particles ranging from 10^{-4} to 10^{-8} cm. radius. This mixture takes place in the presence of a converter unit which might be in the form of a hot platinum wire, a source of ultraviolet light or a spark gap. The gas mixture is then converted to an aerosol forming substance SO_3 by the reaction of the SO_2 with the O_2 contained in the clean air. A humidifier in the line provides water vapor which reacts with the SO_3 thus converted to produce H_2SO_4 in the form of particles or condensation nuclei.

The sample containing the condensation nuclei may then be reunited with the remainder of the original measured quantity of gas to be studied to insure volumetric uniformity of the sample. The gas mixture is then fed into a condensation nuclei counter (CNC). The CNC includes an expansion chamber where the condensation nuclei are subjected to an expansion process which causes water droplets to form having the condensation nuclei as centers. By irradiating these water droplets an electrical indication of the intensity of the scattered irradiation can be obtained. This indication provides an accurate measurement of the concentration of SO_2 in the gas sample being studied.

A method developed by N.A. Lyshkow (56) provides an improved colorimeter for the determination of SO_2 in air. A method for determining SO_2 concentration in air using similar apparatus is described by Lyshkow in "The Continuous Analysis of Sulfur Dioxide in Gaseous Sample," *Journal of the Air Pollution Control Association,* 17, pp. 687-9, (October 1967) in which use is made of a reagent composition formulated to include pararosanaline which is capable of reacting with the sulfite ion for the development of color.

As is pointed out in the aforementioned article, one of the more difficult problems that exist with respect to colorimetric determinations of sulfur dioxide is that sulfur dioxide is oxidized to sulfur trioxide at a fairly rapid rate in the presence of oxygen and under high humidity conditions. For this reason, it is necessary to entrap the SO_2 in the reagent solution as rapidly as possible to prevent or substantially minimize oxidation SO_2 and SO_3. Thus, when use is made of packed column scrubbers or the like for scrubbing SO_2 from air, in which there exists maximum agitation to facilitate intimate contact of the SO_2-containing air sample with the reagent composition, it is necessary to employ sodium tetrachloromercurate (II) as a trapping solution to prevent oxidation of SO_2 in the scrubber. The use of sodium tetrachloromercurate (II) is generally disadvantageous in that it is not completely stable over long periods of time, and it frequently causes staining of the photocells, thereby distorting the results of the analysis.

Thus, in order to avoid the use of sodium tetrachloromercurate (II), it is necessary to provide a scrubber system having minimal agitation and yet having maximum surface area to promote intimate contact of the SO_2 in the air sample with the reactive reagent solution.

One such scrubber is described in the aforementioned article, and is formed of a plurality of rotating discs which promote intimate contact with the SO_2 in the air sample with a minimum of agitation. However, the scrubber as described has a large volume, and consequently cannot be used as part of a compact, portable colorimeter assembly.

The photocell assembly as described in the aforementioned article is similarly subject to many drawbacks. The SO_2 colorimeter photocell described includes a dual beam colorimeter to compare the relative percent transmission of the reacted and unreacted pararosaniline reagent solutions. As can be appreciated by those skilled in the art, the complex prism and filter systems which must be used with a dual beam colorimeter present a number of difficulties from the standpoint of manufacture and use. In addition, since the output of the photocell assembly is measured in terms of the imbalance of the photo-resistors, the photocell assembly is susceptible to the same drift characteristics described above as a result of temperature and light level instabilities.

In this process, a sample scrubber is provided which includes a helical coil into which an air sample is drawn and admixed with a liquid absorbent whereby the liquid flows gravitationally through the coil in the form of wave fronts to provide complete contact between the air sample and the liquid absorbent with minimum agitation. The photocell assembly includes a pair of spaced photoresistors and a light source spaced therebetween, with the light source having a lamp housing enclosing the light source and defining a pair of spaced openings whereby the light source projects optical spots to the photoresistors to illuminate the photoresistors with light, the relative intensity of which is dependent upon the color developed in solutions passed in front of the photoresistors as an indication of the pollutant gas concentration in the air sample.

A device developed by F.J. Salzano, A.M. Davis, H.S. Isaacs and L. Newman (57) is a sulfur oxide activity meter for measuring directly in an electrochemical cell the changes in SO_2 activity in a sample gas being supplied continuously to the cell. The electrolyte consists of molten fused Li_2SO_4, K_2SO_4 and Na_2SO_4. The sample gas forms part of one electrode while a reference gas having a fixed concentration of SO_2 forms part of the other electrode.

A device developed by H. Dahms (58) is a measuring cell for determining the concentration of SO_2 in a fluid. The cell includes an electrode covered with a thin layer of an electrolyte containing silver ions, and a counterelectrode. A voltage is applied across the electrodes and the resulting current is a measure of the concentration of SO_2.

References

(1) Lawrence Berkeley Laboratory, "Instrumentation for Environmental Monitoring–Air," *Publ. LBL-1,* Vol. 1, Berkeley, Univ. of Calif. (May 1, 1972).

(2) "Air Quality Criteria for Sulfur Oxides," (National Air Pollution Control Administration, Washington, D.C., 1969) *EPA Publication AP-50* (APTIC, EPA, Research Triangle Park, N.C. 27711).

(3) M.J. Kerbec, *Your Government and the Environment* (Output Systems Corporation, Arlington, Va., 1971).

(4) M. Bufalini, "Oxidation of Sulfur Dioxide in Polluted Atmospheres, A Review," *Env. Scie. & Tech.,* 5, 685 (1971).

(5) *Air Pollution by Sulfur Oxides* (National Industrial Pollution Control Council, U.S. GPO, Washington, D.C., Feb., 1971).

(6) J.T. Middleton, "Air Pollution Does Threaten Human Health," *National Tuberculosis and Respiratory Disease Association Bulletin* 57, 5 (1971).

(7) R.K. Stevens, L.F. Ballard, and C.E. Decker, "Field Evaluation of Sulfur Dioxide Monitoring Instruments," *Report of the Air Pollution Control Office (EPA),* Raleigh, N.C.

(8) "Control Techniques for Sulfur Oxide Air Pollutants," (National Air Pollution Control Administration, Washington, D.C. 1969) *EPA Publication AP-52* (APTIC, BPA, Research Triangle Park, N.C. 27711).

(9) *Sulfur & SO_2 Developments* (American Institute of Chemical Engineers, New York, N.Y., 1971).

(10) *Abatement of Sulfur Oxide Emissions from Stationary Sources,* report of Committee on Air Quality Management, Ad Hoc Panel on Control of Sulfur Oxides from Stationary Combustion Sources, National Academy of Engineering, National Research Council, Washington, D.C. (1970).

(11) *Federal Register,* "National Primary and Secondary Air Quality Standards," 36, No. 84, 8186, April 30, 1971.

(12) *Federal Register,* "Standards of Performance for New Stationary Sources," 36, No. 247, 24876, December 23, 1971.

(13) Walden Research Corp., *Report PB 209,267;* Springfield, Va., Nat. Tech. Info. Service (June 1971).

(14) R.N. Corning, "SO_2 Conversion Factor," form-letter from Corning Laboratories Inc., Box 625, Cedar Falls, IA 50613 (received 1971). Discussed a paper (No. 69-133) presented by N.A. Huey at the 1969 National Meeting of the Air. Pol. Control Asso., N.Y.

(15) B.H. Wilsdon and F.J. McConnell, *J. Soc. Chem. Ind.* (London) 53, 385 (1934), as referenced in M. Katz, "Analysis of Inorganic Pollutants," Chap. 17 in *Air Pollution,* Vol. II, A.C. Stern (ed.), Academic Press, N.Y., 1968, p. 73.

(16) N.A. Huey, "The Lead Dioxide Estimation of Sulfur Dioxide Pollution," *J.APCA,* 18, 610 (1968).

(17) "D 2010-65, Standard Method for Evaluation of Total Sulfation in Atmosphere by the Lead Peroxide Candle," *1971 Annual Book of ASTM Standards,* Part 23, pp. 514-517, American Society for Testing and Materials, 1916 Race St., Philadelphia, PA 19103 (1971).

(18) Intersociety Committee, "Tentative Method of Analysis for Sulfation Rate of the Atmosphere (Lead Dioxide Cylinder Method) 42410-01-70T," *Health Lab. Sci.,* 7, 164 (1970).

(19) Intersociety Committee, "Tentative Method of Analysis of the Sulfation Rate of the Atmosphere (Lead Dioxide Plate Method-Turbidimetric Analysis) 42410-02-71T," *Health Lab. Sci.,* 8, 243 (1971).

(20) M. Boulerice and W. Brabant, "New PbO_2 Support for the Measurement of Sulfation," *J. APCA,* 19, 432 (1969).

(21) V.M. Buck, "Methods for the Determination of Hydrogen Fluoride and Sulfur Dioxide in the Atmosphere," (1961), as referenced in S. Hochheiser, "Methods of Measuring and Monitoring Atmospheric Sulfur Dioxide," Publ. Hlth. Serv. Publication No. 999-AP-6 (1964).

(22) J.B. Pate et al., "The Use of Impregnated Filters to Collect Traces of Gases in the Atmosphere," *Anal. Chim. Acta.,* 28, 341 (1963).

(23) C. Hugen, "The Sampling of Sulfur Dioxide in Air with Impregnated Filters," *Anal. Chim. Acta.,* 28, 349 (1963).

(24) P.K. Mueller, E.L. Kothny, L.B. Pierce, T. Belsky, M. Imada, and H. Moore, "Air Pollution," *Anal. Chem.,* 43, 1R (1971).

(25) Y. Tokiwa and P.K. Mueller, "Status of Measuring Air Quality," *J. Env. Sci.,* 14, 10 (1971).

(26) W. Leithe, *The Analysis of Air Pollutants,* Ann Arbor-Humphrey Science, London 1970.

(27) A.C. Stern, *Air Pollution,* Vol. II (Academic Press, New York, N.Y., 1968).

(28) M. Katz, *Measurement of Air Pollutants* (World Health Association, Geneva, Switzerland 1969).

(29) *Air Sampling Instruments,* 4th ed., (American Conference of Governmental Industrial Hygienists, Cincinnati, Ohio, 1972).

(30) *Methods for Air Sampling and Analysis* (American Public Health Association, Washington, D.C., 1972).

(31) N.L. Morrow, R.S. Brief and R.R. Bertrand, "Sampling and Analyzing Air Pollution Sources," *Chem. Eng.,* 79, 84 (1972).

(32) *Federal Register,* "Standards of Performance for New Stationary Sources," 36, No. 247, 24876, December 23, 1971.

(33) *Federal Register,* "National Primary and Secondary Air Quality Standards," 36, No. 84, 8186, April 30, 1971.

(34) J.S. Nader, "Stationary Source Measurement Techniques and Instrumentation," presented at the 1972 Pittsburgh Conference in Analytical Chemistry and Applied Spectroscopy, Cleveland, Ohio, March 9, 1972. (J.S. Nader, EPA-DCP, National Environmental Research Center, Research Triangle Park, N.C. 27711).

(35) L.W. Hurbesh, "Microwave Rotational Spectroscopy Applied to Trace-Gas Pollutant Monitoring," presented at the Joint Conference on Sensing of Environmental Pollutants, Palo Alto, CA (Nov. 8-10, 1971). Available as UCRL-73197 report, Lawrence Livermore Laboratory, Livermore, CA 94550.

(36) A.D. Snyder and G.W. Wooten, "Feasibility Study for the Development of a Multifunctional Emission Detector for NO, CO, and SO_2," report of Monsanto Research Laboratory, prepared for the National Air Pollution Control Administration, Contract CPA 22-69-8 (Oct. 1969).

(37) J.A. Hodgeson et al., "Application of a Chemiluminescence Detector for the Measurement of Total Oxides of Nitrogen and Ammonia in the Atmosphere," presented at the Joint Conference on Sensing of Environmental Pollutants, Palo Alto, CA (Nov. 8-10, 1971).

(38) R.J. Pfeifer et al., "Mercury Substitution-Nucleonic Detection Instrumentation for Sulfur Dioxide Measurement," *ISA Trans.,* 9, 9 (1970).

(39) C.B. Ludwig et al., "Study of Air Pollution by Remote Sensors," report of General Dynamics Corporation, prepared for the National Aeronautics and Space Administration, NASA publication CR-1380, July, 1969.

(40) P.L. Hanst, "Infrared Spectroscopy and Infrared Lasers in Air Pollution Research and Monitoring," *Appl. Spect.,* 24, 161 (1970).

(41) L.R. Snowman et al., "Infrared Laser System for Extended Area Monitoring of Air Pollution," presented at the Joint Conference on Sensing of Environmental Pollutants, Palo Alto, CA (Nov. 8-10, 1971).

(42) H. Inaba and T. Kobayasi, "Laser-Raman Radar for Chemical Analysis of Polluted Air," *Nature* 224, 170 (1969).

(43) D.A. Leonard, "Feasibility Study of Remote Monitoring of Gas Pollutant Emissions by Raman Spectroscopy," report of Avco Everett Research Laboratory, Everett, MA, prepared for the National Air Pollution Control Administration, EPA publication APTD-0658, Dec. 1970.

(44) H. Kildal and R.L. Byer, "Comparison of Laser Methods for Remote Detecting Atmospheric Pollutants," *Proc. IEEE,* 59, 1644 (1971).

(45) O'Keefe, A.E., "Modern Methods for Monitoring Atmospheric Sulfur Dioxide," paper presented at 8th Indiana Air Pollution Control Conference, W. Lafayette, Ind. (October 14, 1969).

(46) Chand. R. and Marcote, R.V., "Development of Portable Electrochemical Transducers for the Detection of Sulfur Dioxide and Oxides of Nitrogen," Report on Contract CPA-22-69-118, Durham, N.C., Nat. Air Pollution Control Admin. (Sept. 1970).

(47) Anon., "Sulfur Dioxide Pollution Monitor Developed by NBS," *Industrial Heating* 39, (II), 2052, 2054 (Nov. 1972).

(48) Okabe, H., Splitstone, P., and Ball, J., "Ambient and Source SO_2 Detector Based on a Fluorescence Method," paper 72-14, presented at the Air Pollution Control Association Meeting (June 20, 1972).

(49) Information on SO_2 permeation tubes and on all of the Standard Reference Materials available from the Office of Standard Reference Materials, National Bureau of Standards, Washington, D.C. 20234.
(50) Lyshkow, N.A., U.S. Patent 3,433,597; Mar. 18, 1969; assigned to Precision Scientific Co.
(51) McGee, J.J., Kelley, T.J. and Harman, J.N. III, U.S. Patent 3,495,944; Feb. 17, 1970; assigned to Beckman Instruments, Inc.
(52) Dreckman, H., U.S. Patent 3,539,302; Nov. 10, 1970; assigned to The Hays Corp.
(53) Schulze, F., U.S. Patent 3,567,392; Mar. 2, 1971.
(54) Cho, B.Y. and Anderson, L.B., U.S. Patent 3,578,406; May 11, 1971; assigned to U.S. Atomic Energy Commission.
(55) Van Luik, F.W., Jr. and Coffey, P.E., U.S. Patent 3,674,435; July 4, 1972; assigned to Environment/One Corp.
(56) Lyshkow, N.A., U.S. Patent 3,708,265; Jan. 2, 1973; assigned to Pollution Monitors, Inc.
(57) Salzano, F.J., Davis, A.M., Isaacs, H.S. and Newman, L., U.S. Patent 3,718,546; Feb. 27, 1973; assigned to U.S. Atomic Energy Commission.
(58) Dahms, H., U.S. Patent 3,756,923; Sept. 4, 1973.

SULFUR TRIOXIDE

Measurement in Air

Sulfur trioxide may be monitored in air by the use of a paper tape impinger (1). A technique developed by J. Sieth and H.G. Heitmann (2) relates to the continuous analysis of smoke gases which, after mechanical separation of fly ash and other solids, essentially consist of a mixture of sulfur dioxide and sulfur trioxide, aside from carbon dioxide.

The technique is predicated upon the recognition that when such gas mixtures pass through condensing layers of water vapor, only sulfur trioxide is eliminated, whereas the other components of the gas mixture, particularly sulfur dioxide and carbon dioxide pass through such layers without being absorbed.

In combustion plants, particularly steam-boiler plants, the flue gases contain more or less considerable quantities of sulfur dioxide as well as traces of sulfur trioxide, stemming from the combustion of sulfurous fuel such as coal and oil. When the temperature of the flue gases drops below the dew point, the gases condense and may cause serious damage by corrosion in the boiler. The dew point is influenced substantially by the proportion of sulfur trioxide in the waste gases. The quantity of sulfur trioxide produced depends to a considerable extent upon the nature of the combustion.

For better control of the combustion processes and satisfactorily determining when the temperature drops below the dew point, it is of great importance to measure and record the SO_2 and the SO_3 content of the flue gases.

Heretofore it has been possible to continuously measure only SO_2 separately from SO_3. Employed for this purpose have been devices for measuring electric conductance, or devices based upon the principle of colorimetric or potential measurements. These are not suitable for continuously ascertaining the concentration of SO_3 gases, particularly in mixture with other gases.

However, it is desirable to provide means, based upon a suitable separation method, for measuring the SO_3 concentration in a gas mixture, particularly for such purposes as the detection and quantitative indication of harmful emission from a waste gas source into the atmospheric air, and it is one of the objects of the development to devise such apparatus and to reliably afford the desired, continuous and selective analysis with a high degree of accuracy.

According to this technique, a quantitative and continuous concentration measurement with respect to a selected component of a gas mixture is carried out in apparatus wherein the incoming gas mixture is brought in contact with condensing vapors of an auxiliary medium, preferably water, whereby one or more gas components are separated from the gas mixture and become dissolved in the auxiliary medium; and the concentration of this medium, in continuous flow, is measured preferably by electric conductance measurement, as being indicative of the concentration of the separated component in the original gas mixture. This process can be carried out in stages so that the concentrations of different components of the original gas mixture are continuously analyzed in the manner just described.

Accordingly, the concentration of sulfur trioxide and sulfur dioxide in a flow of smoke gas, can be measured continuously by performing the following steps:

(a) Treating the flowing gas mixture continuously with condensing water vapor to selectively absorb sulfur trioxide from the mixture;
(b) Continuously measuring the concentration of the sulfuric acid solution resulting from the reaction of the condensing water and the sulfur trioxide, this concentration being indicative of the sulfur trioxide concentration in the gas mixture;
(c) Continuously treating the residual flow of gas, now free of sulfur trioxide, with water to absorb sulfur dioxide; and
(d) Measuring the concentration of the sulfurous acid solution resulting from the reaction of water and sulfur dioxide, as indicative of the sulfur dioxide concentration in the gas mixture.

The concentration of the sulfuric acid solution and/or the sulfurous acid solution is advantageously measured by electric conductivity measurement.

The joint measurement of sulfur dioxide and sulfur trioxide may be important in pollution studies related to sulfuric acid manufacturing plants as discussed by the Environmental Health Service (3).

The Shell Development Method was used by a joint MCA-PHS field test team for sample collection and analysis of both sulfur dioxide and sulfur trioxide (4). Sample gas was first drawn through a glass-wool filter, then passed through a heated glass probe into a system of three sintered glass plate absorbers. The first absorber was immersed in an ice bath. The first two absorbers contained an isopropyl alcohol-water solution for absorption of sulfur trioxide. The third absorber contained dilute hydrogen peroxide in water for absorption of sulfur dioxide. Purified air was passed through the absorbers at the end of the run to remove any dissolved sulfur dioxide in the first two absorbers. Any sulfur dioxide removed was absorbed by hydrogen peroxide in the third absorber. Analysis was conducted by titrating each solution with standard barium chloride using thorin indicator.

The Chemical Construction technique was used to determine sulfur dioxide and sulfur trioxide concentrations. (5). Sample gas was drawn through a glass probe and tray into a system consisting of two glass-fiber filters held by a fritted glass disk. Two bubblers with coarse-fritted glass gas distributors and a flow meter completed the sampling train. Each bubbler contained a standardized solution of hydrogen peroxide in water. At the conclusion of sampling, the absorbing solution in the two impingers was transferred to a volumetric flask and diluted to the mark. Half of this solution was titrated with standard potassium permanganate for unused hydrogen peroxide.

The difference in titration between the standard hydrogen peroxide solution and the sample yielded the sulfur dioxide concentration. The other half of the sample solution was titrated with standard caustic for total concentration of sulfur dioxide and sulfur trioxide. The sulfur trioxide cooncentration was determined by difference.

References

(1) Lawrence Berkeley Laboratory, "Instrumentation for Environmental Monitoring-Air," *Publ. LBL-1,*

Vol. 1, Berkeley, Univ. of Calif. (May 1,1972).
(2) Sieth, J. and Heitmann, H.G., U.S. Patent 3,367,747; February 6, 1968; assigned to Siemens-Schuckert-werke AG, Germany.
(3) Environmental Health Service, *Publ. No. 999-AP-13,* Durham, N.C., Nat. Air Pollution Control Admin. (1965).
(4) Shell Development Co., "Determination of Sulfur Dioxide and Sulfur Trioxide in Stack Gases," *Emeryville Method Ser. 4S16/59a* (1959).
(5) Chemical Construction Corp., "Gas Analysis of Sulfuric Acid Plants," Tech. Method, R. and D. Lab., New York, N.Y. (Aug. 1961).

SULFURIC ACID

Measurement in Air

The air pollution aspects of sulfuric acid manufacturing have been reviewed in a bibliography with abstracts prepared by the Office of Air Programs of the EPA (1). As noted by the Environmental Health Service (2) a variety of stack-sampling and analytical procedures have been used by various sulfuric acid manufacturers, by air pollution control districts, by the joint Manufacturing Chemists' Association and Public Health Service field-test team in obtaining emission data.

Many of the contact acid plants use the Monsanto Company Method for collection and analysis of sulfuric acid mist (3). This technique, with several modifications, was also used by the joint MCA-PHS field test team. Effluent gas samples were collected in the duct or exit stack just above the acid absorber. Pitot tube traverses were made to determine the velocity profile of the gases in the duct. Sampling was performed isokinetically at a number of traverse points.

The stack gases were drawn through a glass sampling train consisting of a probe, a cyclone collector, and a glass-fiber mist collector. Both the probe and enclosed sample train were heated to preclude condensation in the sample gas stream. The cyclone collected acid-mist particles larger than 3 microns. The particles smaller than 3 microns were collected on the fine glass-fiber filter. Analysis for sulfuric acid mist in both the cyclone and the glass-fiber filter tube was performed by titrating with dilute caustic to a phenolphthalein end point.

A few of the plants for which data are reported employed medium-porosity fritted glass disks, or millipore or Whatman filters for collection of sulfuric acid mist. In each case analysis was performed by titrating with dilute caustic to a phenolphthalein end point.

References

(1) Environmental Protection Agency, *Publ. AP-95,* Research Triangle Park, N.C. (May 1971).
(2) Environmental Health Service, *Publ. No. 999-AP-13,* Durham, N.C., Nat. Air Pollution Control Admin. (1965).
(3) Patton, W.F., and J.A. Brink, Jr., "New Equipment and Techniques for Sampling Chemical Process Gases." *JAPCA,* 13:162-66 (Apr., 1963).

SURFACTANTS

Measurement in Water

A process by J.K. Kerver (1) develops a quantitative technique for analyzing for small concentrations of surface active materials. The solution of surface active material, an adsorbent composed of particles on which the surface active material is adsorbed, two immiscible fluid phases such as oil and water, and a colorant are mixed in a glass vial. The surface active material which adsorbs on the adsorbent alters the surface wettability of the adsorbent, and portions of the adsorbent are preferentially attracted by either the water or oil phase. The relative amount of adsorbent attracted by each phase is a measure of the concentration of surface active material. Standard vials are prepared from known concentrations of the surface active material to facilitate quantatively determining the concentration of an unknown sample by comparison.

Reference

(1) Kerver, J.K., U.S. Patent 3,393,051; July 16, 1968.

TIN

Feeding 5 ppm tin in drinking water to mice produced no increase or decrease in tumor formation (1). Tin is not known to be toxic to mice or to affect their growth rate (2). The 24 hour LC_{50} of stannic chloride is 78 ppm. At 46 ppm none of the fish died in a test on tin exposure. Stannous sulfate failed to kill fish in 24 hours at a concentration of 553 ppm. This comparatively low toxicity in hard water is attributed to much of the material being in suspension (3). The maximum allowable concentration of tin in effluents being discharged to a storm sewer or stream is 2.0 ppm according to L.E. Lancy (4).

Measurement in Air

Tin may be monitored in air by emission spectroscopy (5).

Measurement in Water

Tin may be measured in water by: flame absorption spectroscopy, flameless absorption spectroscopy, neutron activation and polarography. The detection limits for these various methods were shown in Table 11.

References

(1) Schroeder, H.A. and Balassa, J.J., *J. Nutrition* 92, 245 (1967).
(2) Konisawa, M. and Schroeder, H.A., *Cancer Research* 27, 1192 (1967).
(3) Her Majesty's Stationery Office, *Water Pollution Research,* London (1969)
(4) Lancy, L.E. in *Industrial Pollution Control Handbook,* H.F. Lund, Ed., New York, McGraw-Hill Book Co. (1971).
(5) Lawrence Berkeley Laboratory, "Instrumentation for Environmental Monitoring-Air", *Publ. LBL-1,* Vol. 1, Berkeley, University of Calif. (May 1, 1972).

TITANIUM

Tetravalent titanium when added to the drinking water in a concentration of 5 ppm increased the growth of mice but did not significantly affect mortality (1). Christie et al. (2) found no evidence of pulmonary effects when rats inhaled titanium dioxide. Titanium accumulates with age in man, but there is no evidence that it is an essential metal (1). Thus it appears that in ppm and smaller amounts, titanium is not a serious health hazard.

Measurement in Water

Titanium may be monitored in water by the following instrumental techniques (3): spark source mass spectrometry and x-ray fluorescence. Titanium may be measured in water by atomic absorption (flame) spectrometry down to a detection limit of 60 ppb as noted earlier in Table 11.

References

(1) Schroeder, H.A., Virton, W.H. and Balassa, J.J., *J. Nutrition* 80, 39 (1963).
(2) Christie, H., MacKay, R.J. and Fisher, A.M., *Am. Ind. Hygiene Assoc. J.* 24, 47 (1963).
(3) Lawrence Berkeley Laboratory, "Instrumentation for Environmental Monitoring-Water," *Publ. LBL-1,* Vol. 2, Berkeley, Univ. of Calif. (Feb. 1, 1973).

URANIUM

The permissible concentration of uranyl ion in domestic water supplies is 5 mg./l. as noted earlier in Table 3. It is desirable, however, that the uranyl ion be absent.

Measurement in Water

A process developed by E. Jungreis and L. Ben-Dor (1) is one in which uranium is detected by a reagent which is prepared by reacting p-dimethylaminoaniline hydrochloride and salicylaldehyde which is then reacted with ammonia, the product being yellow. Uranyl ions added to the yellow product give a color change from yellow to red.

Reference

(1) Jungreis, E. and Ben-Dor, L., U.S. Patent 3,403,004; September 24, 1968; assigned to Yissum Research Development Co., Israel.

VANADIUM

Vanadium is toxic to humans and animals, especially its pentavalent compounds. Exposure of humans through inhalation of relatively low concentrations (less than 1,000 μg./m.^{3}) has been found to result in inhibition of the synthesis of cholesterol and other lipids, cysteine, and other amino acids, and hemoglobin. Low concentrations also act as strong catalysts on serotonin and adrenaline, as discussed by Athanassiadis (1).

Chronic exposure to environmental air concentrations of vanadium has been statistically associated with the incidence of cardiovascular diseases and certain cancers. Human exposure to high concentrations of vanadium (greater than 1,000 μg./m.^{3}) results in a variety of clinically observable adverse effects whose severity increases with increasing concentrations. These effects include irritation of the gastrointestinal and respiratory tracts, anorexia, coughing (from slight to paroxysmal), hemoptysis, destruction of epithelium in the lungs and kidneys, pneumonia, bronchitis and bronchopneumonia, tuberculosis, and effects on the nervous system ranging from melancholia to hysteria.

No information has been found on adverse effects of atmospheric vanadium concentrations on vegetation or on commercial or domestic animals. What is known about the effects of vanadium on materials relates mostly to the corrosive action of vanadium, acting (together with sulfur dioxide) on oil- and coal-fired boilers, especially those using vanadium-rich residual oils and coals.

The major sources of vanadium emissions are the metallurgical processes producing vanadium metal and concentrates; the alloy industry; the chemical industry; power plants and utilities using vanadium-rich residual oils and; to a lesser extent, the coal and oil refining industries.

Vanadium production is concentrated in the states of Colorado, Utah, Idaho, and New Mexico, while the highest concentration of industries producing vanadium chemicals is found in New Jersey and New York.

In communities in the United States in which vanadium concentrations were measured, the average values (quarterly composites) ranged from below detection (0.003 μg./m.^{3}) to 0.30 (1964), 0.39 (1966), and 0.90 (1967) μg./m.^{3}.

Little information is available on the economic losses due to vanadium air pollution or on the costs of abatement. One report indicated that measures taken to reduce the loss of vanadium to the atmosphere from an oil-fired steam generator resulted in recovery of commercially valuable vanadium pentoxide, thereby producing a profit from air pollution abatement. No other information was noted in the literature on control procedures specifically intended to reduce air pollution by vanadium. However, customary methods used to contol particulate emissions in general are considered suitable to the industrial processes using vanadium or vanadium-containing fuels.

There is no known reason why vanadium is essential to man (2). It is not a toxic trace metal for man but in large doses (of about 160 ppm) it will kill a rat. No other trace metal has so long had so many supposed biological activities without having been proven to be essential.

Exposure to vanadium pentoxide dust produces a clinical syndrome characterized by irritation of the eyes, nose and throat followed by rales throughout the lungs and acute bronchospasm similar to bronchial asthma (3). Vanadium pentoxide inhaled at the level of 100 ppm lowered the free cholesterol and phospholipid content of the liver in rabbits (4). It also has been found that vanadium enhanced the activity of monoamine oxidase, an enzyme which has been shown to be acutely antihypertensive in rats and dogs and which inactivates angiotensin (5).

Vanadium has been found to inhibit the synthesis of cholesterol in rat liver cells (6)(7). However, most of the ingested vanadium is excreted within 24 hours (8). Congestion and fine droplets of fat in the liver have been produced by the administration of vanadium compounds to rats. Reduction in fat and lipid content of adrenal cortex occured at the same time (4). Additional microgram quantities of vanadium acetate or metavanadate to minced liver slices caused increased oxygen uptake.

In man, intramuscular injection of sodium tetravanadate resulted in increased catabolism as indicated by increased output of all nitrogen, sulfur and phosphorous constituents determined in the urine (4). Ingestion of vanadium pentoxide at dietary levels beginning at 100 ppm vanadium caused lessening of cystine content of rat hair (4). Thus it appears that the presence of vanadium at the ppm level is not a health hazard to man.

The 24 hour LC_{50} (the amount that kills 50% of the fish within 24 hours) of orthovanadate was 14 ppm. For metavanadate it was 50 ppm. Concentrations lower than those tested might have been lethal to fish after prolonged exposure (9). No toxicity was demonstrated for tetravalent vanadium by rats. Vanadium pentoxide in drinking water was highly toxic to rats at 49 ppb.

The allowable concentration of vanadium in irrigation water is 10.0 mg./l. on a continuous use basis and is also 10.0 mg./l. on a short-term use basis in fine textured soil as noted earlier in Table 5.

Measurement in Air

Sampling Methods: At low concentrations of vanadium in air, high volume samplers are used which operate during the 24 hour period sampling 50 cubic feet of air per minute, or 2,200 cubic meters of air. Preweighed glass-fiber filters (about 10 inches in diameter) are used. The filters should be equilibrated at a standard temperature (75°F. or less) with relative humidity of 50% or less, and then weighed to determine the concentration of particulate matter. Afterwards, an aliquot of the sample is ashed (at 100°C.) and then extracted with nitric and hydrochloric acids. For nonurban samples, extracts are made that are up to five times more concentrated than these for urban samples (10).

Kuz'micheva (11) described the following sampling method for the colorimetric determination of aerosols of vanadium and its compounds in metallurgical plants. Air samples were passed through filters of polyvinyl chloride fabric and then placed in porcelain dishes, treated with 2 ml. of a 50% HNO_3 solution, and evaporated to dryness. Ashing was done in a muffle furnace at 500°C., and the residue was treated with 2 ml. of a 10% NaOH solution which dissolved the vanadium, leaving iron in the residue.

Membrane ultrafilters (having a pore width of 0.6 to 0.9μ) were used in a study by von Jerman (12), who used a polarographic method to determine vanadium concentrations in the air of alloy and chemical plants.

Quantitative Methods: Colorimetric Methods — A very simple, inexpensive, and specific method for the determination of vanadium in air is the ring-oven technique. The relative error is said to be within the range from 5 to 10%, which at the microgram level compares well with other more sophisticated methods. The limit of identification is 0.01 μg., and its range 0.01 to 3.0 μg. The wavelength used is 3184A, and the concentration giving 1% absorption is 1.5 μg. of vanadium per ml. (12).

Two other colorimetric techniques have been recently described. The first is based on the oxidation of vanadium and its compounds to vanadium pentoxide and its further reaction with hydrogen peroxide in acid medium. This method has a sensitivity of 17.8 μg. of vanadium pentoxide or 10 μg. of vanadium. The second method is based on the development of a greenish-yellow color when pentavalent vanadium reacts with sodium tungstate in neutral medium. The method is not specific and there is interference from alkalies and mineral acids. Kuz'micheva (11) used a method based on the formation of yellow phosphotungstovanadic acid when vanadium or vanadium compounds react with phosphoric and sodium tungstate.

The determination was made on a 5 ml. aliquot placed in colorimetric test tubes. The sensitivity was 5 μg. of vanadium pentoxide in 5 ml. No interference was observed by aluminum, calcium, silicon dioxide, or iron. Colored chromium compounds were found to interfere when present in amounts greater than 40 μg. In general, colorimetric methods are being replaced by more sophisticated and sensitive methods.

Atomic Absorption Spectroscopy — Vanadium in the range of 500 to 1,000 μg. per liter can be determined by atomic absorption spectroscopy in an oxyacetylene or nitrous oxide-acetylene flame. For use with oxyacetylene flames, vanadium is extracted as vanadium cupferrate into a mixture of ketone and acid, and the resultant product is aspirated by the flame. For use with the nitrous oxide-acetylene flame, an aqueous solution of vanadium is aspirated directly (13).

Polarography — Only one paper described a polarographic method for the determination of vanadium in air. The method described was designed to be used for determining vanadium in the air of the working environment in the alloy industry, where vanadium is used as input, and in chemical manufacture, where vanadium is used as a catalyst. Dusts in such environments contain iron, aluminum, and magnesium, but these elements are not expected to interfere with the test.

Following mineralization of the sample in 45% nitric acid, the polarographic levels of vanadium were recorded from a conductive solution of borax, ammonia, and chelaton III. The method is said to be sensitive to 1.5 μg./ml. of vanadium pentoxide (12).

Emission Spectrography — This method is used by the National Air Sampling Network (14) for the determination of vanadium concentration in aliquots obtained from 24 hour samples after ashing and extraction. In 1966, improvements in sensitivity made this method accurate enough for the determination of vanadium in many nonurban air samples. The minimum detectable concentration of vanadium by this method is 0.003 μg./m.^{3} for urban samples and 0.0005 μg./m.^{3} for nonurban samples.

Other Methods — Other analytical methods, used mostly for the determination of vanadium in biomaterial, are paper chromatography (15)(16), neutron radioactivation (17)(18), electrophoresis (19), low-energy x-ray mass absorption (20), and autoradiography (21). Determinations of vanadium content have been made in erythrocytes (22), bones (21), organ tissues (23), urine (24)(25), and biomaterial in general (17)(18)(20)(26).

Vanadium may be monitored in air by the following instrumental techniques (27): colorimetric, emission spectroscopy and neutron activation.

Measurement in Water

Vanadium may be monitored in water by the following instrumental techniques (28): neutron activation and spark source mass spectrometry. Vanadium may be measured in water by flame atomic absorption down to a detection limit of 2 ppb as noted earlier in Table 11.

A process developed by D.G. Biechler and D.E. Jordan (29) is one in which the vanadium-containing sample is dissolved in a suitable acid or mixture of acids and the vanadium is then oxidized to the pentavalent state by the use of suitable oxidizing agents. Either prior to or after the oxidation of the vanadium to the pentavalent state, phosphoric acid and a suitable tungstate salt are added to the solution to form the phosphotungstate complex of vanadium.

The complex thus formed is extracted from the acid solution using aliphatic alcohols containing from 6 to 12 atoms. The vanadium in the alcoholic extract is then determined colorimetrically by use of a photoelectric colorimeter or spectrophotometer.

Several analytical methods have been utilized for determining trace or microgram amounts of vanadium in the presence of other metallic ions. Many of these procedures employ the step of forming a pentavalent vanadium complex compound which can be at least partially extracted with a suitable solvent, or can be separated by precipitation from the solution in which the complex is formed. The previous separatory and analytical methods each have disadvantages which distract from the universality of their usefulness.

For example, many of the solvents which have been proposed for extraction of pentavalent vanadium compounds do not completely or consistently extract the vanadium from the solution in which it is contained. Moreover, the form in which the vanadium is extracted frequently does not permit the extract to be subjected to simple and rapid analysis, or, in some instances, extraneous undesirable materials are extracted with the vanadium which interfere with subsequent analysis or render the analysis inaccurate.

In some of the analytical procedures proposed, the pentavalent vanadium is complexed with other materials, and the vanadium determined by colorimetric methods while the vanadium is still present in the original solution. One of the most successful of these methods, insofar as the determination of trace amounts of vanadium is concerned, is based upon the formation of a phosphotungstovanadate complex in an acid solution by the addition to an acidic solution of pentavalent vanadium, of phosphoric acid and a suitable tungstate salt.

The complex, the precise chemical nature of which is obsure, is a stable yellow color and the vanadium content of the solution from which it is derived is generally determined by spectrophotometric procedures while the complex remains in the original sample solution.

The latter aspect of the analytical procedure does not, of course, lend itself to the separation or removal of the vanadium from the other materials in the sample, and while the analysis is recognized to be characterized by excellent accuracy and repeatability, other or additional procedures must be adopted if the vanadium is to be recovered from the sample. Thus, in summary, in many instances, the analytical technique adapted does not ultimately result in the separation of the vanadium from the material in which it occurs, and conversely, the separatory technique does not place the vanadium in a state suitable for analysis, or else concurrently with the extraction of the vanadium, extracts other materials which interfere with the vanadium analysis. Moreover, the analytical techniques heretofore proposed generally are not both rapid and accurate.

References

(1) Athanassiadis, Y.C., "Air Pollution Aspects of Vanadium and Its Compounds," *Report PB 188,093,* Springfield, Va., Nat. Tech. Info. Service (Sept. 1969).
(2) Schroeder, H.A., Balassa, J.J. and Tipton, I.H., *J. Chron. Dis.* 16, 1047 (1963).
(3) Zen, Z.C. and Berg, B.A., *Arch. Environ. Health* 14, 709 (1957).
(4) Mountain, J.T., Stockell, F.R., Jr. and Stokinger, H.E., *Proc. Soc. Exp. Biol. Med.* 92, 582 (1955).
(5) Perry, H.M., Jr., Teitelbaum, S. and Schwartz, P.L., *Fed. Proc.* 14, 113 (1955).
(6) Snyder, F. and Cornatzer, W.E., *Nature* 182, 462 (1958).
(7) Aiyar, A.S. and Srinavasan, A., *Proc. Soc. Biol. Med.* 107, 914 (1961).
(8) Schroeder, H.A., Virton, W.H. and Balassa, J.J., *J. Nutrition* 80, 39 (1963).
(9) Her Majesty's Stationery Office, "Water Pollution Research," London (1969).
(10) Stern, A.C. (Ed.), *Air Pollution,* Vol. II (New York:Academic Press, 1968).
(11) Kuz'micheva, M.N., "The Determination of Vanadium in Air," text in Russian, *Gigiena i Sanit.* 31:229 (1966).
(12) von Jerman, L., et al., "Polarographische Bestimmung von Vanadin in der Luft von Arbeitsraumen," *Zeit. Hyg.* (Berlin) 14(1):12 (1968).
(13) Sachdev, S.L., et al., "Determination of Vanadium by Atomic Absorption Spectrophotometry," *Anal. Chim. Acta* 37:12 (1967).
(14) *Air Quality Data, 1966,* National Air Sampling Network, U.S. Dept. of Health, Education, and Welfare, Public Health Service, Cincinnati, Ohio (1968).
(15) Miketukova, V., "Detection of Metals on Paper Chromatograms with Rhodamine B," *J. Chromatog.* 24:302 (1966).
(16) Yamakawa, K., et al., "Organometallic Compounds. II. Gas Chromatography of Metal Acetylacetonates," *Chem. Pharm. Bull.* (Tokyo) 11:1405 (1963).
(17) Comar, D., et al., "Concentration of Vanadium in the Rat and Its Influence on Cholesterol Synthesis. Studies by the Technic of Neutron Radioactivation and the Method of Isotopic Equilibrium," text in French, *Bull. Soc. Chim. Biol.* 49:1357 (1967).
(18) Livingston, H.D., et al., "Estimation of Vanadium in Biological Material by Neutron Activation Analysis," *Anal. Chem.* 37:1285 (1965).
(19) Kaser, M.M., et al., "The Separation and Identification of Vanilmandelic Acid and Related Compounds by Electrophoresis on Cellulose Acetate," *J. Chromatog.* 29:378 (1967).
(20) Carter, R.W., et al., "Low Energy X-Ray Mass Absorption Coefficients from 1.49 to 15.77 kev for Scandium, Titanium, Vanadium, Iron, Cobalt, Nickel, and Zinc," *Health Phys.* 13:593 (1967).
(21) Soremark, R., et al.,"Autoradiographic Localization of V-48-Labeled Vanadium Pentoxide (V_2O_5) in Developing Teeth and Bones in Rats," *Acta Odontol. Scand.* 20:225 (1962).
(22) Valberg, L.S., et al., "Detection of Vanadium in Normal Human Erythrocytes," *Life Sci.* 3:1263 (1964).
(23) Stephan, J., et al., "Spectrographic Demonstration of Vanadium in Some Human and Animal Organs and Alteration of Its Presence," text in German, *Med. Exptl.* 4:397 (1961).
(24) Rockhold, L.W., et al., "Vanadium Concentration of Urine; Rapid Colorimetric Method for Its Estimation," *Clin. Chem.* 2(3):188 (1956).
(25) Sartosova, Z., "Determination of Vanadium in Urine," text in Czech, *Pracovni Lekar* 11:518 (1959).
(26) Kuz'micheva, M.N., "Determination of Vanadium in Biologic Materials," text in Russian, *Gigiena i Sanit.* 31:70 (1966).
(27) Lawrence Berkeley Laboratory, "Instrumentation for Environmental Monitoring-Air," *Publ. LBL-1,* Vol. 1, Berkeley, Univ. of Calif. (May 1, 1972).
(28) Lawrence Berkeley Laboratory, "Instrumentation for Environmental Monitoring-Water," *Publ. LBL-1,* Vol. 2, Berkeley, Univ. of Calif. (Feb. 1, 1973).
(29) Biechler, D.G. and Jordan, D.E., U.S. Patent 3,345,126; Oct. 3, 1967; assigned to Continental Oil Co.

ZINC

It is not possible to assess fully the role of zinc and its compounds as air pollutants. Despite the fact that specific effects attributed to certain compounds of zinc have been noted, the common association of zinc with other metals, and the frequent presence of toxic contaminants (such as cadmium) in zinc materials, raise questions which have yet to be answered concerning the synergistic effects of these metals.

The most common effects of zinc poisoning in humans are nonfatal metal-fume fever, caused by inhalation of zinc oxide fumes, and illnesses arising from the ingestion of acidic foods prepared in zinc-galvanized containers. Zinc chloride fumes, though only moderately toxic, have produced fatalities in one instance of highly concentrated inhalation. Zinc stearate has been mentioned as a possible cause of pneumonitis. Zinc salts, particularly zinc chloride, produce dermatitis upon contact with the skin, as noted by Athanassiadis (1).

Accidental poisoning of cattle and horses has occurred from inhalation of a combination of lead- and zinc-contaminated air. Zinc oxide concentrations of 400 to 600 $\mu g./m.^3$ are toxic to rats, producing damage to lung and liver, with death resulting in approximately 10% of the cases. Although dogs and cats were found to tolerate high concentrations (up to 1,000,000 $\mu g.$/day) of zinc oxide for long periods, evidence of glycosuria and damage to the pancreas became apparent. Concentrations of 40,000 to 50,000 $\mu g./m.^3$ of zinc ammonium sulfate produced no appreciable effects on cats.

Some evidence exists of damage to plants from high concentrations of zinc in association with other metals. No information was found on damage to materials from zinc or its compounds in the atmosphere.

The primary sources of zinc compounds in the atmosphere are the zinc-, lead-, and copper-smelting industries, secondary processing operations which recover zinc from scrap, brass-alloy manufacturing and reclaiming, and galvanizing processing. Average annual production and consumption of zinc in the United States have increased steadily during this century, and it is predicted that this trend will continue.

As the emission of zinc into the atmosphere in most of these operations represents an economic loss of the zinc material, control procedures are normally employed to prevent emission to the atmosphere. In those industries where zinc is a by-product, control procedures for zinc are not as effective, and greater quantities of zinc therefore escape into the environment.

Measurements of the 24 hour average atmospheric concentrations of zinc in primarily urban areas of the United States reveal an average annual value of 0.67 $\mu g./m.^3$ for the period 1960 to 1964; the highest value recorded during that period was 58.00 $\mu g./m.^3$, measured in 1963 at East St. Louis, Ill.

Extensive air pollution abatement methods are in general use by the zinc industry. Control devices include precipitators, scrubbers, baghouses, and collectors. The efficiency of the various control methods varies widely. However, in many instances air pollution control devices are not used in the general metals industries. Thus it is very likely that relatively large quantities of zinc or zinc compounds are emitted into the atmosphere by industrial plants processing zinc or other compounds containing zinc. No information has been found on the economic costs of zinc air pollution or on the costs of its abatement.

As part of a national inventory of sources and emissions, W.E. Davis and Assoc. of Leawood, Kansas have provided information regarding the nature, magnitude and extent of the atmospheric emissions of zinc in the United States for the year 1969 (2).

Industries discharging wastewater streams which carry significant quantities of zinc include: steel works with galvanizing lines, zinc and brass metal works, zinc and brass plating, silver and stainless steel tableware manufacturing, viscose rayon yarn and fiber production, groundwood pulp production and newsprint paper production. In addition, recirculating cooling water systems employing "Cathanodic Treatment" contain zinc, which is discharged during blowdown (3).

Rock (3) estimates that 10,000 tons of waste zinc per year are discharged from viscose rayon production, and that 5 tons of zinc may be used per day in a large groundwood pulp mill. Similar quantities were estimated for a typical brass company (4).

The primary source of zinc in wastewaters from plating and metal processing industries is the solution adhering to the metal product after removal from pickling or plating baths. The metal is washed free of this solution, referred to as dragout, and the contaminants are thus transferred to the rinse water. The pickling process consists of immersing the metal (zinc or brass) in a strong acid bath to remove oxides from the metal surface. Finished metals are brightened by submergence in a bright dip bath containing strong chromate concentrations in addition to acid.

Plating solutions typically contain 5,000 to 34,000 mg./l. of zinc. These concentrated solutions may be discharged periodically, due to contamination. The concentration of zinc in the rinse water will be a function of the bath zinc concentration, drainage time over the bath, and the volume of rinse water used. Zinc and brass plating and rinse solutions generally also contain cyanide. Waste concentrations of zinc range from less than 1 to more than 1,000 mg./l. in various waste streams described in the literature. Average values, however, seem to be between 10 and 100 mg./l. Table 47 summarizes values reported for various zinc bearing wastewaters.

TABLE 47: CONCENTRATIONS OF ZINC IN PROCESS WASTEWATERS

Industrial Process	Zinc Concentration, mg/l		Reference
	Range	Average	
Metal Processing			
Bright dip wastes	0.2-37.0		(5)
Bright mill wastes	40-1,463		(4)
Brass mill wastes	8-10		(4)
Pickle bath	4.3-41.4		(5)
Pickle bath	0.5-37		(4)
Pickle bath	20-35		(6)
Aqua fortis and CN dip	10-15		(6)
Wire mill pickle	36-374		(7)
Plating			
General	2.4-13.8	8.2	(8)
General	55-120		(9)
General	15-20	15	(10)
General	5-10		(6)
Zinc	20-30		(6)
Zinc	70-150		(5)
Zinc	70-350		(11)
Brass	11-55		(5)
Brass	10-60		(11)
General	7.0-215	46.3	(5)
Plating on zinc castings	3-8		(6)
Galvanizing of cold rolled sttel	2-88		(12)
Silver Plating			
Silver bearing wastes	0-25	9	(5)
Acid waste	5-220	65	(5)
Alkaline	0.5-5.1	2.2	(5)

(continued)

TABLE 47: (continued)

Industrial Process	Zinc Concentration, mg/l Range	Average	Reference
Rayon Wastes			
General	250-1000		(13)
General	20*		(14)
General	20-120		(15)
Other			
Vulcanized fiber	100-300		(16)
Cooling tower blowdown	6		(17)

*After process recovery of zinc by ion exchange.

Source: PB 216,162

It has long been known that zinc deficiency had an adverse effect on human health (18). For example, it leads to subnormal growth and impairment of intestinal absorption. Protein and RNA are markedly affected by zinc deprivation. The normal intake of zinc in the diet is 90 mg. of zinc for one kg. of the animal. When this intake is reduced to twelve milligrams per kilogram, serious effects are exhibited by protein and RNA (19).

However, excess amounts of zinc (in the 0.5 to 1.0% range of zinc oxide) also affect growth rate and decrease both the weight and fat content of the liver (this was observed on a high fat-low protein diet). It also interferes with the development and mineralization of bones. Zinc at these same levels affects metabolism to a considerable extent. It causes increased excretion of nitrogen suggesting a general wastage of tissue proteins and a decreased excretion of phosphorous and sulfur (18). This latter phenomenon is suggested as the mechanism for poor bone development and decreased liver weight. This intake of zinc oxide brings about a significant decrease in the phosphate activity of the intestine while increasing it in the liver and kidneys. Phosphate assimilation is known to be related directly to fat production.

Some information is available concerning the effects of atmospheric zinc on humans. Inhaled zinc in the form of zinc ammonium sulfate or zinc sulfate causes increased pulmonary flow resistance (20).

Adding a supplement of zinc (0.25 to 0.75%) for a few weeks to the diet produced a decrease in the normal deposition of calcium and phosphorous in the bones of young rats which was alleviated with calcium and phosphorous supplements. However, the calcium and phosphorous supplements prevented the accumulation of zinc in the bone. These levels of zinc have an adverse effect on both the absorption and utilization of magnesium (21).

Excess dietary zinc causes a hypochromic, microcytic anemia in rats (22). This anemia is wholly or partially alleviated by feeding additional copper. This suggests that excess zinc causes decreased copper resulting in anemia by affecting hemoglobin formation (at levels of zinc in the 1.0 to 1.5% range) (23). The 1.0% zinc reduced hemoglobin concentration by 40 and 0.03% copper restored the hemoglobin level. Excess dietary zinc also causes a lowering of the enzyme activity of cytochrome oxidase. Adding copper increased this enzyme activity of cytochrome oxidase. Adding copper increased this enzyme activity. As little as 0.3 ppm of zinc is toxic to some aquatic insects and 0.3 to 0.7 ppm is the toxicity range for fish.

Reproduction in the fathead minnow was almost totally inhibited at zinc concentrations that had no effect on survival, growth or maturation of these fish. At 0.18 ppm no effect on the survival or growth of eggs or fry was noted but 83% less eggs were produced than at a concentration of 0.03 ppm (24). At sublethal concentrations of zinc goldfish behaved aberantly, failed to reproduce, had poor growth rates or functioned poorly in other respects (25). Thus it can be concluded that zinc is an essential element, but in fairly low concentrations it can cause serious effects not only in fish and insects but also in mammals.

The permissible concentration of zinc in domestic water supplies is 5 mg./l. as noted earlier in Table 3. The desirable situation is that zinc be virtually absent. The tolerance limit for zinc in irrigation water is 5.0 mg./l. on a continuous use basis and is 10.0 mg./l. on a short-term use basis in fine textured soil as noted earlier in Table 5. The maximum allowable concentration of zinc in effluent being discharged to a storm sewer or stream is 1.0 ppm.

Measurement in Air

Sampling Methods: Dusts and fumes of zinc compounds may be collected by any method suitable for collection of other dusts and fumes; the impinger, electrostatic precipitator, and filters are commonly used. The National Air Sampling Network uses a high volume filtration sampler.

Quantitative Methods: The ring-oven technique of analysis has been adapted to the determination of zinc in the range 0.05 to 1 μg./m.^{3}. The reagent, o-mercaptothenalaniline is used as coloring agent. The method is specific for zinc and may be used in air pollution studies. The limit of detection is 0.04 μg, although previous studies (26) had suggested 0.1 μg. as the lower limit.

A spectrophotometric method has been developed for the determination of zinc; concentrations down to 0.1 μg. can be measured using 1-(2-thiazolylazo)-2-naphthol (27). An instrument capable of monitoring and recording air concentration of zinc sulfate pigment (a phosphorescing tracer) on a real-time basis has been developed. The instrument has a limit of detection of about 0.25 μg./m.^{3} (28).

Emission spectroscopy has been used by the National Air Pollution Control Administration for zinc analysis of samples from the National Air Sampling Network (29). The samples are ashed and extracted to eliminate interfering elements. The minimum detectable zinc concentration by emission spectroscopy is 0.24 μg./m.^{3} for urban samples and 0.08 μg./m.^{3} for nonurban samples. The difference in sensitivities results from the different extraction procedures required for urban samples (30).

Thompson et al. (30) have reported that the National Air Pollution Control Administration uses atomic absorption to supplement analyses obtained by emission spectroscopy. The method has a minimum detectable limit of 0.0002 μg./m.^{3} based on a 2,000 m.^{3} air sample.

Zinc may be monitored in air by the following instrumental techniques (31): emission spectroscopy, atomic absorption and x-ray fluorescence.

Measurement in Water

Zinc may be monitored in water by the following instrumental techniques (32): atomic fluorescence, electrochemical, spark source mass spectrometry, and x-ray fluorescence. As noted earlier in Table 11, zinc may be measured down to a detection limit of 2 to 5 ppb by atomic (flame) absorption or by polarography. Neutron activation is effective only to a detection limit of 400 ppb but flameless atomic absorption permits the detection of zinc down to about 0.02 ppb.

A technique developed by C.J. Overbeck and J.J. Hickey (33) is one whereby analysis for small amounts of zinc in waters which also contain high hardness levels may be made by using as a titrant diethylenetriaminopentaacetic acid and a dithizone indicator.

A method developed by C.A. Noll (34) involves determining the zinc concentration in an aqueous medium. More specifically, the method is directed for use to aqueous mediums which contain contaminated and perhaps interfering metal ions, such as the cations of aluminum, iron and copper.

According to the method, the sample of the aqueous medium is acidified, treated with a buffered complexing agent to complex any aluminum and iron ions present, treated with an organo-sulfur compound which will complex with any copper ions present and discriminate against any zinc ions present. An indicator compound which will react with the zinc to produce a color has been added to the aqueous medium and the color intensity of the resulting solution is ascertained. The color intensity is then compared with the color intensity of known quantities of zinc and the indicator compound, and the concentration of the zinc is ascertained accordingly.

References

(1) Athanassiadis, Y.C., "Air Pollution Aspects of Zinc and Its Compounds," *Report PB 188,072,* Springfield, Va., Nat. Tech. Info. Service (Sept, 1969).
(2) Davis, W.E. & Assoc., *Report PB 210,680,* Springfield, Va., Nat. Tech. Info. Service (May 1972).
(3) D.M. Rock, "Hydroxide Precipitation and Recovery of Certain Metallic Ions from Waste Waters," presented at Annual Meeting, American Institute for Chemical Engineers, Chicago, Illinois, Nov.-Dec., 1970.
(4) F.X. McGarvey, R.E. Tenhoor, and R.P. Nevers, "Brass and Copper Industry: Cation Exchangers for Metals Concentration form Pickle Rinse Waters," *Ind. Engr. Chem.,* 44:534-541, 1952.
(5) N.L. Nemerow, *Theories and Practices of Industrial Waste Treatment,* Addison-Wesley Publishing Co., Reading, Mass., 1963.
(6) W. Lowe, "The Origin and Characteristics of Toxic Wastes with Particular Reference to the Metal Industries, " *Wat. Poll. Control (*London), 1970:270-280.
(7) Volco Brass and Copper Co., *Brass Wire Mill Process Changes and Waste Abatement, Recovery and Reuse,* Water Pollution Control Research Series, No. 12010 DFP, U.S. Environmental Protection Agency, Washington, D.C., No., 1971.
(8) G.E. Barnes, "Disposal and Recovery of Electroplating Wastes," *Jour. Wat. Poll. Control Fed.,* 40:1459-1470, 1968.
(9) O.W. Nyquist and H.R. Carroll, "Design and Treatment of Metal Processing Wastewaters," *Sew. Ind. Wastes,* 31:941-948, 1959.
(10) H.L. Pinkerton, "Waste Disposal," in *Electroplating Engineering Handbook,* 2nd ed., A. Kenneth Graham, ed. in chief, Reinhold Publishing Co., New York, 1962.
(11) Anonymous, *State of the Art: Review on Product Recovery,* Water Pollution Control Research Series, U.S. Dept. Interior, Washington, D.C., Nov., 1970.
(12) E.J. Donovan, Jr., "Treatment of Wastewater for Steel Cold Finishing Mills," *Wat. Wastes Eng.,* 7:F22-F25, 1970.
(13) F.X. McGarvey, *The Application of Ion Exchange Resins to Metallurgical Waste Problems,* Proc. 17th Purdue Industrial Waste Conf., pp 289-304, 1952.
(14) C.P. Sharda and K. Namwannan, "Viscose Rayon Factory Wastes and Their Treatment," *Technology, Sindri,* 3:58-60, 1966; Wat. Poll. Abstr. 41: No. 1698, 1968.
(15) American Enka Co., *Zinc Precipitation and Recovery from Viscose Rayon Wastewater,* Water Pollution Control Research Series, No. 12090 ESG, U.S. Environmental Protection Agency, Washington, D.C., Jan., 1971.
(16) Anonymous, "Reclaiming Zinc from an Industrial Waste Stream," *Env. Sci. Technol.,* 6:880-881, 1972.
(17) J.M. Culotta and W.F. Swanton, "Recovery of Plating Wastes: Selection of Lowest Cost Evaporator," *Plating,* 57:1121-1123, 1970.
(18) Sadasivan, V., *Biochem. J.* 48,527 (1951); 52, 452 (1952).
(19) Oberleas, D. and Prasad, A.S., Unpublished Report, Detroit, Mich., Dept. of Medicine, Wayne State University (1969).
(20) Freeman, H.E., *J. Am. Med. Assoc.* 119, 1016 (1942).
(21) Stewart, A.K. and Macgee, A.C., *J. Nutrition* 82, 287 (1964).
(22) Gray, L.F. and Ellis, G.H., *J. Natr.* 40, 441 (1950).
(23) Duncan, G.D., Gray, L.F. and Daniel, L.S., *Proc. Soc. Expl. Biol. Med.* 83, 625 (1953).
(24) Brungs, W.A., *Trans. Am. Fisheries Soc.* 98, 2 (1969).
(25) Cairn, J. and Shirer, H., "Project Completion Report, Kansas Resources Research Institute" (June 1968).
(26) West, P.W., et al., "Microdetermination of Zinc by Means of Reagent Crayons and the Ring-Oven Technique," *Anal. Chim. Acta* 37:246 (1967).

(27) Kawase, A., "2-(2-thiazolylazo)phenol Derivatives as Analytical Reagents. VII. Spectrophotometric Determination of Zinc with 1-(2-thiazolylazo)naphthol," *Talanta* 12:195 (1965).
(28) Nickola, P.W., "A Sampler for Recording the Concentration of Airborne Zinc Sulfide on a Real Time Scale," Pacific Northwest Laboratory, Richland, Washington (1965).
(29) "Air Pollution Measurements of the National Air Sampling Network-Analyses of Suspended Particulate, 1957-1961," U.S. Dept. of Health, Education, and Welfare, *Public Health Service Publication No. 978,* U.S. Government Printing Office, Washington, D.C. (1962).
(30) "Air Quality Data from the National Air Sampling Networks and Contributing State and Local Networks," 1966 ed., U.S. Dept. of Health, Education, and Welfare, National Air Pollution Control Administration *Publication No. APTD 68-9,* U.S. Government Printing Office, Washington, D.C. (1968).
(31) Lawrence Berkeley Laboratory, "Instrumentation for Environmental Monitoring-Air," *Publ. LBL-1,* Vol. 1, Berkeley, Univ. of Calif. (May 1,1972).
(32) Lawrence Berkeley Laboratory, "Instrumentation for Environmental Monitoring-Water," *Publ. LBL-1,* Vol. 2, Berkeley, Univ. of Calif. (Feb. 1, 1973).
(33) Overbeck, C.J. and Hickey, J.J., U.S. Patent 3,479,152; Nov. 18, 1969; assigned to Nalco Chemical Co.
(34) Noll, C.S., U.S. Patent 3,706,532; Dec. 19, 1972; assigned to Betz Laboratories, Inc.

ZIRCONIUM

Compounds of zirconium have a low order of toxicity to animals. The only recorded effect of toxicity of zirconium to human beings involves cutaneous exposure from deodorant sticks and poison-oak lotions (1). When mice were fed 5 ppm in their drinking water (2)(3) it was found that no carcinogenic or tumorigenic effects were produced. The mice did show a slightly shortened life span (about one month in 30 months).

Animals do absorb some zirconium if it is broken down enough to pass through membrane walls. There is no good analytical evidence that zirconium is consistently absorbed in rats. Also no evidence exists that zirconium as fed to rats through their drinking water has any biological activity except possibly to effect weight of older animals (4). Schroeder and Balassa (1) suggest that zirconium is an abnormal (nonessential) element in man.

Measurement in Water

Zirconium may be measured in water by x-ray fluorescence as described by C.R. Cothern (5).

References

(1) Schroeder, H.A. and Balassa, J.J., *J. Chron. Dis.,* 19, 573 (1966).
(2) Schroeder, H.A., Mitchener, M., Balassa, J.J., Kanisawa, M. and Nason, A.P., *J. Nutrition* 95,95 (1968).
(3) Kanisawa, M. and Schroeder, H.A., *Cancer Res.* 29, 892 (1969).
(4) Schroeder, H.A., Mitchener, M. and Nason, A.P., *J. Nutrition* 100, 59 (1970).
(5) Cothern, C.R., "Determination of Trace Metal Pollutants in Water Resources and Stream Sediments," *Report PB 213,369,* Springfield, Va., Nat. Tech. Info. Service (1972).

FUTURE TRENDS

As water is reused to a greater and greater degree in the future, more sophisticated water resource management programs will be necessary. To function, they must have the capability to acquire timely information on water quality over vast areas on a day-to-day basis. This capability does not exist today. To achieve this capability, further research and development are needed on:

(1) Low cost automated (portable and fixed) water quality sensors that can measure a wide variety of indices over long periods with minimal maintenance.
(2) Aerial remote sensing techniques for broad-scale evaluation of water quality conditions over vast geographical areas.
(3) New, more inclusive water quality indices that can better lead themselves to automated and remote sensing techniques. Instead of describing water quality changes using indices such as dissolved oxygen, BOD, etc., perhaps consideration should be given to the use of indices indicative of changes in emission or reflectance of certain energy spectra from a given water body. As long as the changes in emission or reflectance were a result of changes in water quality conditions and were generally proportional to the overall water quality changes, this would be a useful approach. In fact, this may be all the monitoring needed after 1985 to enforce water quality standards on navigable waters.

In summary, much of the water quality monitoring as carried out today is not drastically different from the ways in which it was conducted at the beginning of this century. Most of the improvements in methodology made over the past 50 years have been in laboratory as opposed to field techniques. The development of automated electronic sensors for the measurement of a few parameters has, nevertheless, proven to be a significant step toward meeting the field needs of tomorrow. Recent developments in remote sensing also offer much promise.

One point is certain; if we want to begin thinking of ourselves as true water quality managers we must act to prevent water quality problems, not merely react to water quality problems that have been permitted to occur. Prevention of water quality problems requires early detection of emerging adversities, and early detection is not possible without adequate monitoring tools and a fully implemented monitoring program. Efforts are presently under way to provide the necessary tools and an adequate monitoring program. These efforts must be intensified, however, if we are going to make the transition from the reaction to the action phase by the time the present backlog of pollution problems has been corrected.

If we allow ourselves to dream, it is exciting to picture an effective satellite-based remote monitoring system for a wide range of materials such as pesticide residues. This same system could also be easily adapted to identify other chemical contaminants and possibly heavy metal residues. The results from the satellite could be telemetered back to a central processing point. This would be a virtual real-time monitoring system. If properly managed, it would provide an instantaneous picture of pollutants in the environment.

NOTICE

Nothing contained in this Review shall be construed to constitute a permission or recommendation to practice any invention covered by any patent without a license from the patent owners. Further, neither the author nor the publisher assumes any liability with respect to the use of, or for damages resulting from the use of, any information, apparatus, method or process described in this Review.

POLLUTANT REMOVAL HANDBOOK 1973

by Marshall Sittig

The purpose of this handbook is to provide a one volume practical reference book showing specifically how to remove pollutants, particularly those emanating from industrial processes. This book contains substantial technical information.

This volume is designed to save the concerned reader time and money in the search for pertinent information relating to the control of specific pollutants. Through citations from numerous reports and other sources, hundreds of references to books and periodicals are given.

In this manner this book constitutes a ready reference manual to the entire spectrum of pollutant removal technology. While much of this material is presumably available and in the public domain, the locating thereof is a tedious, time-consuming, and expensive process.

The book is addressed to the industrialist, to local air and water pollution control officers, to legislators who are contemplating new and more stringent control measures, to naturalists and conservationists who are interested in exactly what can be done about the effluents of local factories, to concerned citizens, and also to those eager students who can foresee new and brilliant careers in the fields of antipollution engineering and pollution abatement.

During the past few years, the words "pollution", "environment" and "ecology" have come into more and more frequent usage and the cleanliness of the world we live in has become the concern of all people. Pollution, for example, is no longer just a local problem involving litter in the streets or the condition of a nearby beach. Areas of the oceans, far-reaching rivers and the largest lakes are now classified as polluted or subject to polluting conditions. In addition, very surprisingly, lakes and streams remote from industry and population centers have been found to be contaminated.

This handbook therefore gives pertinent and concise information on such widely divergent topics as the removal of oil slicks in oceans to the containing of odors and particulates from paper mills.

Aside from the practical considerations, including teaching you where to look further and what books and journals to consult for additional information, this book is also helpful in explaining the new lingo of pollution abatement, which is developing new concepts and a new terminology all of its own, for instance: "particulates, microns, polyelectrolytes, flocculation, recycling, activated sludge, gas incineration, catalytic conversion, industrial ecology, etc."

In order to have a safe and healthful environment we must all continue to learn and discover more about the new technology of pollution abatement. Every effort has been made in this manual to give specific instructions and to provide helpful information pointing in the right direction on the arduous and costly antipollution road that industry is now forced to take under ecologic and sociologic pressures. The world over, technological and manpower resources are being directed on an increasing scale toward the control and solution of contamination and pollution problems.

In the United States of America we are fortunate in receiving direct help from the numerous surveys together with active research and development programs that are being supported by the Federal Government to help industry and municipalities control their wastes and harmful emissions.

A partial and condensed table of contents is given here. The book contains a total of 128 subject entries arranged in an alphabetical and encyclopedic fashion. The subject name refers to the polluting substance and the text underneath each entry tells how to combat pollution by said substance:

CARBON MONOXIDE
CARBONYL SULFIDE
CEMENT KILN DUSTS
CHLORIDES
CHLORINATED HYDROCARBONS
CHLORINE
CHROMIUM
CLAY
COKE OVEN EFFLUENTS
COLOR PHOTOGRAPHY EFFLUENTS
COPPER
CRACKING CATALYSTS
CYANIDES
CYCLOHEXANE OXIDATION WASTES
DETERGENTS
DYESTUFFS
FATS
FERTILIZER PLANT EFFLUENTS
FLOUR
FLUORINE COMPOUNDS
FLY ASH
FORMALDEHYDE
FOUNDRY EFFLUENTS
FRUIT PROCESSING INDUSTRY EFFLUENTS
GLYCOLS
GREASE
HYDRAZINE
HYDROCARBONS
HYDROGEN CHLORIDE
HYDROGEN CYANIDE
HYDROGEN FLUORIDE
HYDROGEN SULFIDE
IODINE
IRON
IRON OXIDES
LAUNDRY WASTES
LEAD
LEAD TETRAALKYLS
MAGNESIUM CHLORIDE
MANGANESE
MEAT PACKING FUMES
MERCAPTANS
MERCURY
METAL CARBONYLS
MINE DRAINAGE WATERS
NAPHTHOQUINONES
NICKEL
NITRATES
NITRITES
NITROANILINES
NITROGEN OXIDES
OIL
OIL (INDUSTRIAL WASTE)
OIL (PETROCHEMICAL WASTE)
OIL (PRODUCTION WASTE)
OIL (REFINERY WASTE)
OIL (TRANSPORT SPILLS)
OIL (VEGETABLE)
ORGANIC VAPORS
OXYDEHYDROGENATION PROCESS EFFLUENTS
PAINT AND PAINTING EFFLUENTS
PAPER MILL EFFLUENTS
PARTICULATES
PESTICIDES
PHENOLS
PHOSGENE
PHOSPHATES
PHOSPHORIC ACID
PHOSPHORUS
PICKLING CHEMICALS
PLASTIC WASTES
PLATING CHEMICALS
PLATINUM
PROTEINS
RADIOACTIVE MATERIAL
RARE EARTH
ROLLING MILL DUST & FUMES
ROOFING FACTORY WASTES
RUBBER
SELENIUM
SILVER
SODA ASH
SODIUM MONOXIDE
SOLVENTS
STARCH
STEEL MILL CONVERTER EMISSIONS
STRONTIUM
SULFIDES
SULFUR
SULFUR DIOXIDE
SULFURIC ACID
TANTALUM
TELLURIUM HEXAFLUORIDE
TETRABROMOETHANE
TEXTILE INDUSTRY EFFLUENTS
THIOSULFATES
TIN
TITANIUM
TRIARYLPHOSPHATES
URANIUM
VANADIUM
VEGETABLE PROCESSING INDUSTRY EFFLUENTS
VIRUSES
ZINC

ISBN 0-8155-0489-6

528 pages

WASTEWATER CLEANUP EQUIPMENT 1973

Second Edition

Water pollution is becoming more of a problem with every passing year. Plants engaged in all types of manufacture are being more and more carefully watched by federal, state, and municipal governments to prevent them from pouring their untreated effluents into the nation's waterways, as they used to do. The sewage treatment plants of many municipalities are becoming too small for the burgeoning population, and many communities once served by individual septic tanks are having to build sewers and treatment plants.

Water pollution will be solved primarily by application of techniques, processes, and devices already known or in existence today, supplemented by modifications of these known methods based on advanced technology. This book gives you basic technical information and specifications pertaining to commercial equipment currently available from equipment manufacturers. Altogether the products of 94 companies are represented.

This second edition of "Wastewater Cleanup Equipment" supplies technical data, diagrams, pictures, specifications and other information on commercial equipment useful in water pollution control and sewage treatment. The data appearing in this book were selected by the publisher from each manufacturer's literature at no cost to, nor influence from, the manufacturers of the equipment.

It is expected that vast sums will be spent in the United States during the remaining portion of this decade for control and abatement of water pollution. Much of the expenditure will be for the type of equipment described in this book.

Today's environmental control is taken to mean a specialized technology employing specialized equipment designed to process the discarded and excreted wastes of human metabolism and human activity of any sort.

Next to air, water is the most abundant and utilized commodity necessary for the maintenance of human life. The average consumption of water per person in residential communities in the United States is between 40 and 100 gallons in one day. In highly industrialized communities the average consumption pro head can be as high as 250 gallons per day.

The reuse of wastewater after cleanup is not only becoming a cogent necessity, but it is also becoming more attractive economically. The degree of purity required for industrial water use is in many cases greater or vastly different from that acceptable for potable water.

Special equipment for cleanup of wastewater is therefore an absolute must, and this book is offered with the intention of providing real help in the selection of the proper equipment.

The descriptions and illustrations given by the original equipment manufacturer include one or more of the following:

1. **Diagrams of commercial equipment with descriptions of components.**
2. **A technical description of the apparatus and the processes involved in its use.**
3. **Specifications of the apparatus, including dimensions, capacities, etc.**
4. **Examples of practical applications.**
5. **Graphs relating to the various parameters involved.**

Arrangement is alphabetically by manufacturer. A detailed subject index by type of equipment is included, as well as a company name cross reference index.

ISBN 0-8155-0487-X **372 pages**

POLLUTION ANALYZING AND MONITORING INSTRUMENTS 1972

Before air, water and other pollution problems can be solved, the pollutants must be identified correctly and then measured accurately. Only after qualitative detections and quantitative determinations have been carried out, can a full scale attack be planned. Failure to observe these rules of war against pollution will lose the war, since the pollutant enemies fight dirty and insidiously.

This book gives basic technical information of what may be called step one in a systems approach to total war on pollution. Here are the descriptions and specifications of what is avaiable in ready made, on-the-line commercial equipment for sampling, measuring and continuously analyzing the multitudinous types of pollutants found in the air, in the water, in the soil and food or feedstuffs, or making themselves evident as noise and radiations.

The book is based directly on information supplied by makers of chemical and physical laboratory instruments or by manufacturers of special sampling, analyzing and monitoring equipment especially designed as adjuvants for the abatement of pollution of every type.

The equipment and instrumentation described in this book comprises the following facets of step one in the systems approach to pollution control:

1. **Sampling of air, stack emissions, exhaust emissions, water, sewage and factory effluents.**

2. **Analysis of pollutants, both qualitative and quantitative, by methods which vary from simple colorimetric tests and pH measurements to the use of the most sophisticated chromatographs, spectrometers, and electronic counters of particulates.**

3. **Monitoring of various pollutants, including hydrogen sulfide, sulfur dioxide, chlorine, ozone, nitrogen oxides, dusts and particulates, including the more insidious poisons such as lead and carcinogens.**

4. **Direct measurement of air, water, noise, heat, and radiation pollution.**

The data appearing in this volume were selected by the publisher from each manufacturer's literature at no cost to, nor influence from, the manufacturers of the materials, instruments, and other equipment.

All together 157 companies are represented. About 350 instruments or other analytical equipment pieces are described. Many other materials, including small test kits and sampling devices are also listed.

The major listings include one or more of the following:

(1) **Diagrams of the apparatus with a description of its components and accessory equipment.**

(2) **Technical discussion of the analytical reactions involved.**

(3) **Specifications of the instrumentation.**

(4) **Brief statement about the specific and all-around uses of the instrument or apparatus.**

Governments on all levels have been setting standards for acceptable pollution levels and their control. This constitutes step two in the systems approach which cannot be taken without step one—in other words:

> No company can intelligently cope with its pollution problems without first setting up an accurate sampling, measuring and analyzing program to find the amount and character of the pollutants it discharges.

This book supplies detailed technical data on the types of measurement and analyses which can be made and the companies which provide such instrumentation.

354 pages